## THIS BOOK IS DEDICATED TO:

Baijnath & Muknee Dhani, Durban, South Africa

Mary Jordan, Valencia, Spain

First edition 2018 published by Durban Botanic Gardens Trust

ISBN: 978-0-9947221-2-6

Copyright © in published edition: Durban Botanic Gardens Trust 2018

Copyright © in text: Himansu Baijnath and Patricia A. McCracken

Copyright © in illustrations and photographs: Institutions, individual artists and photographers. Plant specimens from British Museum (Natural History) reproduced under Creative Commons Attribution Licence CC-BY

Design and production management: Clinton Friedman

Printed and bound by TWP Sdn Bhd, Malaysia

All rights reserved. The Durban Botanic Gardens Trust, authors and contributors have asserted their moral right in accordance with the Copyright Act 98 of 1978 to be identified respectively as owners of this work in published form, its text and illustrations and photographs. Any unauthorised reproduction of this work will constitute copyright infringement and render the doer liable under both civil and criminal law.

While every effort has been made to ensure that the information published in this work is accurate, the authors, publishers and printers take no responsibility for any loss or damage suffered by any person as a result of relying upon information herein.

Front cover: Basia Hitchcock Swiel, detail from *Strelitzia juncea*
Back cover: Angela Beaumont: detail from *Strelitzia nicolai*
Endpapers: Angela Beaumont: *Strelitzia reginae* leaves
Title & closing pages: Pierre Sonnerat, detail from *Ravenala madagascarensis*
Backgrounds to front & back matter: Clinton Friedman, clintonfriedman.com

# STRELITZIAS
## OF THE WORLD
A HISTORICAL & CONTEMPORARY EXPLORATION

by Himansu Baijnath & Patricia A. McCracken

Clinton Friedman *Strelitzia nicolai,* 120 film photograph, 2005

| | |
|---|---|
| 4 | CHAIRMAN'S FOREWORD |
| 5 | AUTHORS' PREFACE |
| | CHAPTERS |
| 8 | 1. A flower fit for a queen: *Strelitzia reginae* (1788) |
| 28 | 2. A rarer find: *Strelitzia alba* (1792) |
| 44 | 3. A king's gift to an empress: *Strelitzia nicolai* (1858) |
| 64 | 4. Art, artifice and the strelitzia family: *Strelitzia depicta* (1818) to **citrine strelitzia** (1887) |
| 84 | 5. Secrets of the heavenly tree: *Ravenala madagascariensis* (1782) |
| 102 | 6. Myth, taboo and reality: *Ravenala madagascariensis* (1782) |
| 122 | 7. The enigma of the Amazon: *Phenakospermum guyannense* (1845) |
| 140 | 8. Phenakospermum, Amazonia's tree of life: **The big palulu** |
| 158 | 9. Splitting the difference: *Strelitzia caudata* (1946) and *Strelitzia juncea* (1974) |
| 176 | 10. The flower of the angels: **America's bird of paradise coup** (1956) |
| 190 | 11. Home-grown hopes: **South Africa's strelitzia growers** |
| 206 | 12. Finding flower paradise: **Strelitzias around the world** |
| 222 | 13. New strelitzia tributes: **'Mandela's Gold'** (1996) and **'Centenary Gold'** (2013) |
| 236 | 14. Botanical exploration in the 21st century: **The Mzimvubu strelitzia** (2007) |
| 248 | 15. Conserving paradise plants: **Saving human lives** |
| 266 | REFERENCES |
| 284 | SELECT BIBLIOGRAPHY |
| 294 | INDEX |

# CHAIRMAN'S FOREWORD

It is a delight and a pleasure to introduce our long-awaited book, *Strelitzias of the world*. With its fascinating, wide-ranging text and beautiful illustrations, it exemplifies the striving for botanical excellence and the fostering of relationships between people and plants that are a key mission for all botanic gardens across the world.

Durban Botanic Gardens was founded by and for the people of the city nearly 170 years ago in 1849. It has become a green haven within eThekwini's urban context, an iconic venue giving treasured memories to millions of visitors over the years. It is both the city's oldest public institution and the continent of Africa's oldest surviving botanic gardens.

The Durban Botanic Gardens Trust, founded in 1993, has supported the gardens' botanical, horticultural and heritage endeavours. It has raised several million rand to contribute to many plant-related and physical improvements in the Durban Botanic Gardens, including related publications celebrating and recording its heritage.

As an important part of the Trust's mandate to research and publication, the Trust supported the research and preparation of the *Strelitzias of the world* over the past 20 years. A publishing grant from the Stanley Smith Horticultural Trust (UK) has assisted in this addition to the Trust's umKhuhlu publication series, a worthy companion to the opening volume, *The Durban forest*.

*Strelitzias of the world* highlights the role of botanic gardens globally in bringing together people and plants by displaying and enabling the study of the planet's vast range of plantlife. We are delighted that this publication is also an opportunity to showcase some of the rich heritage of botanical art, as well as a selection of illustrations recently created by 12 contemporary botanical artists from Southern Africa.

Both *Strelitzias of the world* and *The Durban forest* are proud inheritors of the historic tradition of publishing at our botanic gardens and globally. Like many other botanic gardens internationally, Durban Botanic Gardens published its annual reports in the 19th century, which can still be found safeguarded at major botanical research institutions around the world. Its landmark publication of the period was *Natal plants*, seven volumes of which were compiled by curator John Medley Wood. Six of these appeared from 1898 to 1912. More recently, the Durban Botanic Gardens Trust published a history of the gardens, as well as books on the trees of the gardens and of the city.

It is interesting to see how often these early publications exemplify the key threads of the Durban Botanic Gardens Trust's current mission: biodiversity, botanical and horticultural excellence, community relevance, education, heritage, research, environmental sustainability and green innovation. *Strelitzias of the world* picks up on many of these threads, from tracing the adventures of the early explorers of our world's biodiversity and their accounts of the difference that these well-loved plants make to the daily lives of people and animals to the recent discovery of the plants' potentially life-changing medical benefits.

Whether your interest in plants is botanical, historical, horticultural, environmental or biochemical, I wish you every pleasure in engaging with this much-anticipated book.

**IVOR DANIEL**

Chair: Durban Botanic Gardens Trust

# AUTHORS' PREFACE

The distinctive and dramatic-looking strelitzia plants delighted each of the authors long before the concept of this book was born. When Himansu Baijnath began this project two decades ago, he was prodded by fond boyhood memories. He can still easily recall the thrill of being sent with the men of the family to gather wild strelitzia leaves among the sand dunes in southern Durban. Then there was the excitement of seeing the leaves used to decorate tables, knowing that this signalled a big cultural celebration was about to take place. Patricia McCracken immediately agreed to join the project in 2012 as the leaves of treelike strelitzias bowing and swaying in the wind, the flowers of the smaller strelitzia species blazing bright in the hot, shimmering light, contrast for her with the restrained size, shape and colours of indigenous plants in the British Isles where she grew up.

We combine our backgrounds in science and the arts to draw together as much as possible of the history of adventures and botanical detective work that have made these plants so prominent and well known around the world today, one of the great plants of the modern world. This is not a standard field guide book, but rather a plant narrative. The fact that the strelitzia family, Strelitziaceae to botanists, has very few members compared to the hundreds of heather-type ericas (*Ericaceae*), for example, meant we had the unusual opportunity to reconstruct each plant's biography in terms of when and why it became known to the wider world, the people who investigated it and how it made its mark globally. In their own right, the main southern African branch of the strelitzia family makes up what botanists call a genus. This has only six members:

- *Strelitzia reginae*, with its two subspecies *reginae* and *mzimvubuensis*
- *Strelitzia alba*
- *Strelitzia caudata*
- *Strelitzia nicolai*

The more distant relatives of these are different enough for each to form a genus of their own within the strelitzia family. This farflung family came about after the breakup of the Gondwana supercontinent more than 100 million years ago prompted the first major strelitzia diaspora. These species are the traveller's tree (*Ravenala madagascariensis*) of Madagascar and the big palulu (*Phenakospermum guyannense*) found in the Amazon jungles.

Strelitzias are very special plants to South Africa in particular, where they are seen every day on the country's 50-cent coins. The bird of paradise flower is the emblem of both the province of KwaZulu-Natal and also of the South African National Biodiversity Institute, chosen because it 'would reflect the unique beauty of South Africa's botanical treasures', according to then CEO Professor Brian Huntley.

The strelitzia story is a saga of people and plants. Recording it included a detailed search of the history of how and when the different plants were discovered and described for science; how they were portrayed and the excitement their discovery engendered. It has also been a quest to find the uses that people had made of these special plants and their status in the modern world of global floriculture.

The plant we now know as *Strelitzia reginae* made its international debut nearly 250 years at the Royal Botanic Gardens, Kew. Legendary botanist Joseph Banks named it in honour of Queen Charlotte of Mecklenburg-Strelitz and ensured that it was the prize specimen in a gift of plants sent to the Empress Catherine the Great of Russia. Decades later, another strelitzia was named after a Russian aristocrat, part of the human drama and debate that has surrounded identifying and naming the strelitzias.

Over the centuries, plants have often been used by people not just to provide food or tools but also as vehicles for the display of wealth and power. There was the urge to possess and even flaunt the newest and most expensive species. In the 18th and 19th centuries, members of the strelitzia family such as *Strelitzia reginae*, the bird of paradise flower, and *Ravenala madagascarensis*, the traveller's tree, were 'wow factor' purchase of their time. They were prized for their sleek, architectural looks and coveted by the elite who could afford to pay their inflated prices in much the same way that other objects from cars to mobile phones have captured the imagination today.

Developing garden technology with revolutionary new hot-house and conservatory techniques, such as new iron-smelting methods and the invention of curvilinear glass, made it possible for those in the wealthy north to grow these tropical and subtropical plants. Faster ship transport and the invention of the Wardian case to transport plants added to this revolution. The desirability of exotic novelties gave strelitzias design icon status, with illustrations in botanical and popular (often nursery-sponsored) gardening magazines across Britain, Europe and North America, attracting more fascinated readers and more strelitzia enthusiasts. Since the family was first named, its members have been portrayed by history's important artists, from Pierre-Joseph Redouté, botanical artist to Napoleon's empress Josephine, to Georgia O'Keeffe and Vladimir Tretchikoff in the modern era.

Strelitzias were harnessed as easily to the tropical boom of the 1950s and 1960s as they were to the clean lines of minimalist architecture and design from the 1970s onwards. Strelitzias now have international status and are seen as plants of the world. They are commercially cultivated in North America, South America, Asia, Australia and New Zealand, sold on international wholesale flower markets and have been used to create tribute cultivars such as the worldwide hit, 'Mandela's Gold'.

The strelitzia family makes an unusually flamboyant example of botanical history. Botanical history tracks the trends that shape cultures and countries through plants, with plant hunters adding a bravado flourish through their intrepid and sometimes tragic escapades. Botanical history also tracks how the plants of the world have been discovered, preserved and brought back to herbaria, laboratories and gardens over generations by both amateur and professional collectors. In such locations they were and still are studied to help us understand the complexity of the plant world and the possible economic and medicinal uses of plants. Researching the history of how strelitzias became so globally admired and widely cultivated round the world revealed a dramatic pattern of adventure including piracy and warfare which stretches from the Caribbean and the Amazon to the African veld and the Indian Ocean islands. The strelitzia family's rich history features kings and princes, profiteers, pirates and the tragic deaths of promising young botanists among distant jungles and swamps, who joined the pantheon of 'martyrs of botany'.

In 1992, the discovery of a fossil of a possible strelitzia ancestor in the Arctic region showed how this plant family's relatives stretch back at least 65 million years. Finding near the tip of the northern hemisphere the forebear of a plant now known to occur naturally only in the southern hemisphere is an extraordinary reminder of the sweeping changes that have occurred in the earth's climate and ecology. As well as covering the contribution of the strelitzia to the future of the planet, we have also included relevant material illustrating their recently discovered pharmaceutical uses that can potentially improve human health. Today, from the traveller's tree to the bird of paradise, strelitzias have become instantly recognisable flagship species of environmental conservation and even emblems of the struggle to preserve our planet's biodiversity, especially as a new subspecies was discovered in a remote river gorge of South Africa's Eastern Cape as recently as 2002.

The wealth of scientific and cultural connotations prompted us to write a book that seeks to combine garden history and botany, to chart the place of strelitzias over the centuries in politics, science, the arts, commerce and floriculture as well. This interdisciplinary survey aims to enlighten and satisfy strelitzias' many admirers, from gardeners, environmentalists and artists to historians and botanists.

During our historical-botanical quest, we have aimed to answer questions such as:

- When and where did plant hunters first find the different strelitzia species in the wild?
- What did they learn from indigenous peoples about the everyday and cultural uses of strelitzias?
- How were strelitzias first introduced to cultivation and why were certain species preferred to others?
- What is the future of strelitzias on a planet where wild-growing plants are fast disappearing in the face of agriculture and urbanisation?

In this strelitzia project, we delved into manuscript and herbarium material on three continents but focused our attention particularly on: the Pretoria Herbarium of the South African National Biodiversity Institute; and the herbaria and archives of the British Museum at the Natural History Museum, London, and at the Royal Botanic Gardens, Kew, both collection centres and clearing houses for plant specimens for some centuries. In the process, we discovered how even the labels on herbarium specimens could open up the stories of where plant hunters found members of the strelitzia family, the many hardships they faced and the risks they took.

Like freesias and gerberas, strelitzias are frequently treated as flowers of the world. Many international strelitzia admirers are unaware the flower's home primary habitat is in South Africa. Visitors can even be surprised to see it popping up as an emblematic flower in South African national displays such as the award-winning stands designed by Kirstenbosch National Botanic Gardens for the annual Chelsea Flower Show in London.

This survey is an account of our adventures in search of strelitzias. As American botanist John Kress has written of the Zingiberales, the overarching order of plants within which the strelitzias fall, 'The history of the classification of the Zingiberales, like many tropical plant groups, is one of continual refinement and division.' That new chapters to the saga will be added by those who follow we have no doubt, for the botanists' and historians' quest is never done. Our aim has always been to present a narrative which we hope will serve as a foundation to those who follow. We also hope that this publication will further globally promote the popularity of these extraordinary plants.

## ACKNOWLEDGEMENTS

We should particularly like to thank Ivor Daniel, chair of the Durban Botanic Gardens Trust and the other trust members for supporting this project. The Trust adopted this project in 1996 and since then had made various financial contributions towards the research and publication endeavour. Among the trustees, we especially thank Professor Donal McCracken for his untiring support and direction during the research and writing of this book and the late Professor Patricia Berjak for her guidance.

The boundaries of this project were flung wide by the global move to make available digitised source material. We sincerely thank the library and herbarium digitisation projects mentioned in the references and in particular: Biodiversity Heritage Library, Bibliothèque Nationale, France, Missouri Botanic Gardens and US Archives. We would also like to thank the following institutions for giving us access to their herbarium, manuscript and library collections. These were in South Africa: KwaZulu-Natal Herbarium, Durban; National Archives of South Africa; South African National Botanical Institute, Pretoria; University of KwaZulu-Natal; Ward Herbarium, Durban; in London: British Library; National Archives; Natural History Museum of the British Museum; Royal Botanic Gardens, Kew, Herbarium and Archive; in Dublin: National Botanic Gardens of Ireland, Glasnevin, and Trinity College Dublin Herbarium.

We have been fortunate to find so many fellow botanical and historical enthusiasts to help us find our way through the maze of taxonomy and mountains of sources. For help in answering specific queries, assisting us to consult various sources and obtain images, we should particularly like to thank: Alida Alberts; Susyn Andrews; Dr Sue Boinski; Clayton Burne; Gillian Condy; Dr Gillian Feeley-Harnik; Anne-Lise Fourie; Dr Annette Hladik; Dr Claude Marcel Hladik; Dr Hiltje Maas; Dr Paul Maas; Alice Notten; Dr Gustavo Politis; Adam Riley; Yutaro Suzuki; Dr Vikash Tatayah; Pieter Teunissen; and Siela Teunissen.

In addition, we thank: Jameel Adams; Dr Angela Beaumont; Eileen Bass; Suzy Bell; Sibonelo Chiliza; the late Sue Cochrane; Dr Nik Cole; Delyse Comins; Dr Richard Criley; Dr Neil Crouch; Maxime Dechelle; Pearl de Chalain; Gail de Smidt; Graham Duncan; William Forgrave; Peter Frost; Dr Coert Geldenhuys; Dr Irina Filatova; Elizma Fouche; Dr Hugh Glen; Teddy Govender; the late Mary Gregory; Dr Dennis Hansen; Wanda Hennig; Alan Heslop; Bruce Hopwood; Brian Huntley; Dr Johan Ingels; Farhat Iqbal; Mary Jones; Mary Jordan; Jacek Kropinski; Gigi Laidler; Craig & Gail Lewis; Linda Loffler; Dr David Lorence; Dr David Mabberley; Dr Christopher Martius; Morag McGuire; Dr Mike Mellano; Dr Katherine Milton; Dr Eugene Moll; Dr Pedro Moraes; Atsushi Nakamoto; Dr Sershen Naidoo; Mkhipheni Alfred Ngwenya; Geoff Nichols; Dr Maurizio Paoletti; Dr John Parnell; Cecilia Pienaar; Dr Cary Pirone; Dr Gustavo Politis; Dr Serban Proches; Dr David Pryce; Asok Rajh; Dr Tendro Tondrasoa Ramaharitra;Dr Paulo Sampaio; Steve Sayers; Dr Brian Schrire; Jean Senogles and colleagues; Liberty Shuro; Dr Yashica Singh; Hannelie Snyman; Dr Pablo Stevenson; Dr Douglas Stone; Basia Hitchcock Swiel; Sally Townshend; Dr Albertus van de Venter; Ernst van Jaarsveld; the late Roddy Ward; Julie Wilson; the late John Winter; and Phakamani Xaba.

On the final stretch, we were delighted by Clinton Friedman's beautiful design and all his work in facilitating printing and also thank: Kerry Phillips of the Durban Botanic Gardens Trust for logistical assistance; Helen Holyoake of Helco for publicity; and Martin Clement, curator of the Durban Botanic Gardens, and to the Durban Parks Department for establishing Africa's only strelitzia garden in Africa's oldest surviving botanic gardens.

## NOTE ON TERMINOLOGY AND REFERENCING

We appreciate that strelitzias appeal as much to general readers as to academics and professionals so this overview seeks to offer clear summaries of scholarly research and key popular sources, making both more accessible and available. Similarly, we have used accessible language and simplified some academic conventions. A species name is used in full as *Strelitzia reginae*, for example, rather than abbreviated as *S. reginae*, unless this usage occurs in direct quotation. We have adapted a basic form of referencing to ensure clarity both for general readers and for those from widely different disciplines which meet in this study. We have tried wherever we could to give readers the flavour of the original texts in the original language. Quotations reflect the spelling, use of capitals, abbreviations, punctuation and so on of the original, accompanied by translation into English where necessary. The text has undergone a dual external scholarly peer-review process.

## ABOUT THE AUTHORS

Himansu Baijnath, a Fellow of the University of KwaZulu Natal since 1996, was born into a plant-focused family in Merebank, Durban. He began studying botany formally at the University of Durban-Westville (now University of KwaZulu-Natal). After being awarded a Ph.D at the University of Reading, researched while working at the Jodrell Laboratory, Royal Botanic Gardens, Kew, he joined the staff of the University of Durban-Westville, serving as curator of its Ward Herbarium for 20 years. He has authored more than 80 publications and is particularly interested in the biosystematics of higher plants. He was awarded the accolade of eThekwini Living Legend in 2012. As an Honorary Research Professor and Senior Research Associate in the School of Life Sciences, University of KwaZulu Natal, he is still an active researcher and supervisor of postgraduate students.

Patricia A. McCracken began her first collection of local plants when she was nine years old. She retained an interest in natural history while studying Modern and Medieval Languages at Cambridge University and later while working in investment banking and then journalism. She has won more than 40 journalism awards, primarily in the fields of finance health, consumer affairs and environmental reporting. She is now a freelance writer, specialising in scientific and business fields. Her publications include *Natal, The garden colony, Disease profile for Vhembe district, Limpopo* and work on the history of the media and of the Irish diaspora in South Africa. She is a member of the Credit Ombud Council of South Africa and has been a judge of the Vodacom Journalist of the Year Awards since 2013.

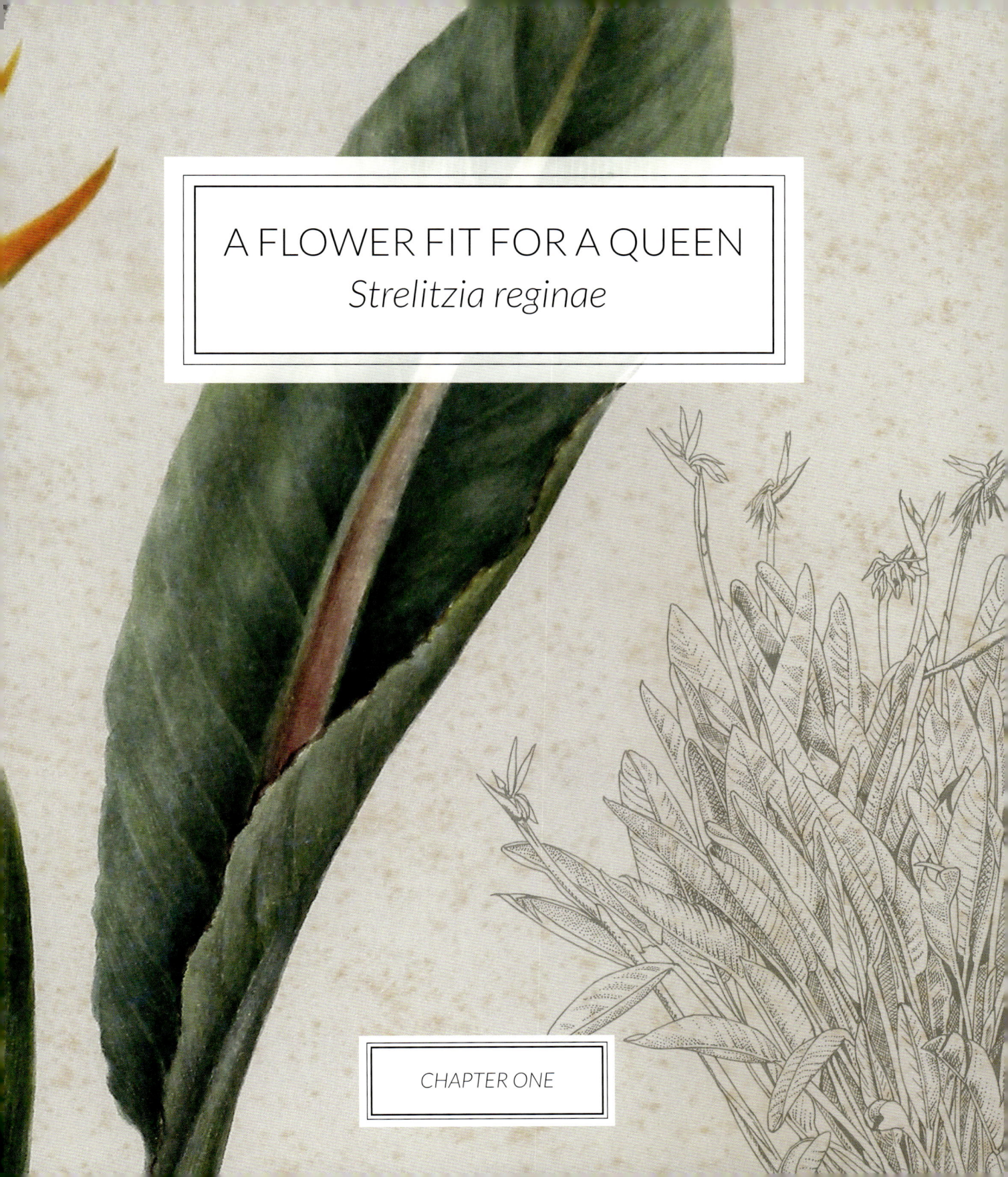

# A FLOWER FIT FOR A QUEEN
*Strelitzia reginae*

CHAPTER ONE

# A FLOWER FIT FOR A QUEEN: *Strelitzia reginae*

The glowing, jewel-like colours of its crested flower prompted garden enthusiasts to call it the bird of paradise flower. Its sharp outlines and splashes of colour make florists rush to use it either singly or as the focus of an arrangement. Landscape designers plant it in lushly dramatic sweeps. It has been chosen as the floral emblem of cities such as Los Angeles on the other side of the globe from its original habitat. It has become one of the most chic and iconic flowers of our time. Its biochemical makeup has amazed scientists. Its primeval origins in the ancient super-continent of Gondwana fascinate ecologists and conservationists. Across the world *Strelitzia reginae* is the best-known member of the strelitzia family, known botanically as the Strelitziaceae.

Yet even on home ground, the *Strelitzia reginae* and its African strelitzia cousins can also be surprisingly elusive. Looking from a distance, the loftier family members happily blend into a background of other treelike plants, including cycads and date palms (*Phoenix reclinata*). Even *Strelitzia reginae*'s vivid flowers, topping a clump of lush, curving leaves often about a metre high, can be almost discreet. Studded among its rocky grassland home or even clinging to clifftops, it is easily outshone from a distance, for instance, by a blazing candelabra of flowering aloes.[1]

The moment when *Strelitzia reginae* began its journey from its heartland in the Eastern Cape of South Africa is similarly elusive. Despite or perhaps because *Strelitzia reginae* is the best-known member of the family, it is surrounded by far more assumptions surrounding its arrival in western Europe from south-eastern Africa than its relatives. Many accepted histories and gardening handbooks pinpoint the 'King's Collector' at the Cape, Francis Masson (1741-1805), as first sending the plant to Kew in the 1770s while Kew's champion at the royal court, Joseph Banks (1743-1820), took the credit for its introduction and naming.[2]

In fact, the plant went without a name of its own for years in Europe. It was the late 18th century before the queen flower, as *Strelitzia reginae* became known to European garden enthusiasts. Its naming and introduction to botany and cultivation were a complex process of imperial adventuring, plant politics and innovative science.

## WEALTH AND FASHION

Joseph Banks is central to the *Strelitzia reginae* saga. From his youth as the son of a Lincolnshire squire and member of parliament, Banks became probably the greatest British patron of botany in his time. He eventually devoted much of his substantial personal fortune to making his passion for botany and natural history felt in the world, to furthering these developing sciences and, increasingly from 1771 onwards, to assisting his king, George III (1738-1820), with transforming the grounds of his palace at Kew into a botanic gardens.

In 1789, *Hortus kewensis*, a catalogue of all the 5 600 or so plants growing at Kew, recorded Banks as having introduced *Strelitzia reginae* in 1773, a year or so after he had begun his role as unofficial, yet all-powerful, director of the royal gardens at Kew. Members of the strelitzia family often take years to mature or resettle sufficiently to produce flowers so when the *Strelitzia reginae*'s 'magnificent' bloom suddenly appeared from its cradle of lush green leaves, Banks was delighted. The type specimen used by Aiton and the team for their description in *Hortus kewensis* is kept at the herbarium of London's Natural History Museum.[3]

This extraordinary new bloom became one of Banks's favourite flowers. He eventually decided in 1789, as discussed later, that this was a flower fit for his queen, Charlotte of Mecklenburg-Strelitz, and ensured it was formally named in her honour. It was both a diplomatic and a courtly gesture as it acknowledged Queen Charlotte's love of botany and gardening. She had enjoyed the well-to-do pastime of botanical drawing and painting since she was a child and diligently pursued botanical tutorials from Banks himself and from wealthy amateur botanist Sir James Edward Smith (1759-1828), the founder of London's Linnean Society. This was 'a just tribute of respect to the botanical zeal and knowledge' of the queen, commented Sir James in Abraham Rees' *Cyclopaedia* in 1819, the year after Queen Charlotte's death: 'Few personages of so elevated a rank have ever loved the study of nature more or cultivated it so deeply'. Naming this remarkable flower *Strelitzia reginae* was also a warm gesture of respect to the royal couple in a year when they had endured the king's 'first really serious attack' of what is now believed to have been porphyria-related dementia.[4]

## FLOWER CRAZES

The *Strelitzia reginae* with its extraordinarily shaped flowers burst onto the European stage from south-eastern Africa when expanding trade routes and commercial and military settlements made unknown plants from around the world a focus of excitement and adventure, profit and rivalry. These exotic plants were curiosities, coveted by those wealthy enough to enjoy being part of the burgeoning interest in science and medicine in general and both economic botany and decorative cultivation in particular. Tulip mania swept the Netherlands in the early 17th century after the introduction of the bulbs from Turkey but this was just one symptom of the flower craze of the period. Several other flowers, including hyacinths, had also become astonishingly highly valued. The first double Spanish broom (*Spartium junceum*) to reach England in 1746 cost 'a golden ducat' and was imported via the German town of Nuremberg.[5]

Over the previous three centuries, greater political and economic stability in Europe had at last allowed ornamental gardening to become a status symbol as it had already been for centuries in countries such as China and Persia. Ornamental gardening had become popular among the aristocratic and wealthy of Europe from the late 15th or early 16th centuries because it offered a means of extending the display of wealth beyond great European houses and royal palaces. Kings and queens, lords and bishops created memorable collections of beautifully arranged exotic trees and plants. The young Banks was already an enthusiastic student of natural history as well as aspiring to become part of this aristocratic and fashionable world.[6]

Frederick, Prince of Wales, and his wife, Princess Augusta (1719-1772), roped their visitors into joining them in gardening and 'each of the royal children had his or her own small plot' to tend in the gardens at Kew. Unfortunately, Prince Frederick's death was popularly said to have resulted from pneumonia that developed after standing in the winter cold watching his trees being planted. Even so, Princess Augusta continued to develop the royal gardens at Kew as a contemporary prototype for garden design and plant fashion in the 1760s. John Stuart, Lord Bute (1713-1792), who was prime minister from 1762 to 1763, had his own botanic gardens at Luton Hoo in Bedfordshire and Highcliffe in Hampshire. He became aristocratic adviser on botanical and horticultural matters to Princess Augusta and the royal gardens, creating a foundation and involvement expanded by Banks.[7]

## CHARLOTTE, THE STRELITZIA QUEEN

The sharper-tongued members of London high society may have relished Sir Joseph Banks's decision to name Kew's extraordinary, new flower from faraway South Africa after their queen, Charlotte of Mecklenburg-Strelitz (1744-1818). When chosen at the age of 17 as the bride of England's King George III in 1761, Charlotte was not considered especially pretty. In an age which conventionally valued the blond, thinner, longer faces of the Habsburgs, she had dark hair framing a wide nose and mouth. It was a look that came to be valued in France as *jolie-laide* but in her own time found little favour, as she well knew. George III's negotiator, Colonel David Graeme reported that Charlotte was 'not a beauty' although 'amiable'. Charlotte herself, though, recalled a more negative general reaction to their betrothal: "The English did not like me much because I was not pretty." Charlotte was so conscious of her looks that when she broke her nose in a carriage accident, she joked self-deprecatingly, "I think I was not so ugly after that."[73]

She gave birth to 15 children – 13 of whom survived into adulthood. During the middle years of her life, the marriage was seen to be happy and the king, nicknamed Farmer George for his interests in agriculture, horticulture and botany, was popular in contrast to his father, Frederick, Prince of Wales, who had been perceived as a wastrel. The couple were also

King George III's swan boat on the lake at Kew

well-respected patrons of the arts, favouring the likes of landscape designer Lancelot 'Capability' Brown, German painter Johann Zoffany and composers such as Johann Christian Bach, who tutored the queen, Wolfgang Amadeus Mozart and Joseph Haydn. As an extension of her botanical enthusiasm and perhaps supporting her husband's agricultural interests, Queen Charlotte was so interested in apple cultivation that she was linked to 'the introduction into England of an Old German variety known as Borsdorfer' and she was mentioned in contemporary pomology writing. It is said that the popular pudding Apple Charlotte, which featured in a 1796 English rhyme, was named after the queen for the royal couple's support of apple farmers.[74]

After 1788, however, George III suffered from lengthy bouts of temporary insanity. The cause was then unknown but is now believed to be the genetic metabolic disorder porphyria. Tension within England generally and for Queen Charlotte personally developed from the strain of managing this illness, together with the French Revolution in 1789 which caused the execution of Queen Marie Antoinette, Queen Charlotte's close friend. Biographers note how she aged quickly during this period in portraits and her chamberlain, Colonel Disbrowe, cuttingly commented: "I do think that the *bloom* of her ugliness is going off".[75] She died, at Kew Palace in 1818, 14 months before her husband. It is ironic that the perhaps carefully selected *jolie-laide Strelitzia reginae* is now considered one of the most beautiful flowers of the botanical world.

## STATUS SYMBOLS

If they were to survive in northern Europe, rare plants often needed to be cossetted in greenhouses. This new infrastructure created another layer of status symbols in the high-society world of early ornamental gardening. Forms of greenhouses were developed from the mid-17th century, beginning with winter-houses to protect bulbs and orange trees that would be moved outside in spring. But plants which normally grew in the East or West Indies (Asia and the Caribbean) prompted further horticultural innovators to experiment with stovehouses, usually heated by an external stove that piped in hot air or by stoves below the building, allowing hot air to rise into it. In the early 18th century, the Dutch introduced bark pits that allowed them to provide 'a steady, unchanging heat' in the soil of potted seeds or cuttings placed in it. This improved method allowed Philip Miller (1691-1771) at the Chelsea Physic Garden 'to germinate the seeds of certain tropical plants which he had hitherto failed to do.'[8]

Such technological developments made it possible for the striking *Strelitzia reginae* with its surprisingly flamboyant flower to become a fashion statement in the stovehouses of the late 18th century. We know it was prized by the likes of Bamber Gascoyne (1725-1791), a lord of the admiralty. In 1777, the plant flowered in the stovehouse of his garden in Barking, Essex, because a dried specimen still survives from his garden in the herbarium at London's Natural History Museum, although no date is given for the flowering. Gascoyne's specimen is grouped on the same sheet as a Cape specimen from Masson and the specimen used for the illustration in *Hortus kewensis*, probably from a plant that flowered in Banks's own garden in 1787.[9]

After Gascoyne gave one of these covetable plants as a prestigious gift to Charles Watson-Wentworth (1730-1782), marquis of Rockingham and twice prime minister of Britain, it flowered at Rockingham's home in Hillingdon in the county of Middlesex in 1778. Banks would have been aware of this as he was monitoring rarities for the *Hortus kewensis* project through Jonas Dryander (1748-1810), a former pupil of the revered Swedish botanist and systematist Carl von Linné, or Carolus Linnaeus (1707-1778). Dryander had come to London in 1777 and begun working with Banks at his Soho Square home, library and herbarium.[10] Banks also had Dryander gathering information on rarities within striking distance of the capital:

> From 18 August 1777, when Banks first introduced him to the Royal Botanic Garden at Kew, Dryander was his field worker among all the important gardens of the Home counties as well as the commercial nurseries, observing, collecting and classifying, especially in the months of flowering.[11]

© The Trustees of the Natural History Museum, London

18th-century *Strelitzia reginae* specimens from plant hunter Francis Masson and grown at Kew and by London collector Bamber Gascoyne

Banks may well have viewed the Gascoyne and Rockingham blooms – and felt envious, even thwarted, that such an unusual plant should not have flowered first at Kew to add to his own prestige as well as delighting the king and queen. The plant did eventually flower at Kew a year later in 1779 and Banks ensured that it was featured by John Miller, or Johann Sebastian Müller (1715-c.1790), in his *Icones novae* of five new plants. Miller, however, based the illustration for one of these plates on Gascoyne's plant. Interestingly, this constituted a rarity twice over – the plant itself and the fact that it was shown with an unusual double spathe. As will be discussed shortly, at that stage, Banks apparently did not recognise the error of publishing this illustration under the name *Heliconia bihai*.[12]

Banks also gave Rockingham another live specimen of the plant. When Rockingham died shortly afterwards in 1782, though, Banks asked Rockingham's widow, Mary, to return it to Kew. The implication was that this would ensure the plant was viewed by a greater number of people and also perhaps that it would receive the best possible care. But the dowager marchioness of Rockingham was herself a botanical and gardening enthusiast and refused Banks's pleas on the grounds that 'the gift of it was the last mark of Sir Joseph's attention that her incomparable Lord could receive' which made her prize it and the collection '10,000 times more'. She breezily concluded that she hoped to be able to show it to Banks and others the following summer.[13]

## A PLANT WITH NO NAME

During these years, this high-status plant effectively had no name – or at best was masquerading under an alias. When Banks realised this, it effectively became the prize in a high-stakes, botanical-naming race.

> In the gardens, hot-houses and conservatories of Spring Grove [Banks's home at Isleworth in west London] all kinds of plants indigenous and exotic were carefully cultivated. Sir Joseph was always ready to engage in the cultivation of new plants[14]

… and in 1787, one of his successes was the first flowering of *Strelitzia reginae*. He was able to admire and examine this rare African plant as much as he wished, which seems to have spurred him to commission another artist to depict it. James Sowerby (1757-1822) was both a versatile and highly skilled artist as well as a scientist. Sowerby's painting of *Strelitzia reginae* was used two years later as one of the endpiece plates of the first volume of *Hortus kewensis*. The process of compiling this work also highlighted the fact that this Cape plant needed a revised scientific name, allowing Banks to exploit the situation to enhance his own prestige with his royal tribute to Queen Charlotte. The need for this must have become more apparent to Banks, William Aiton (1731-1793), who had begun managing Princess Augusta's botanic garden at Kew in 1759, and other members of the team preparing *Hortus kewensis*, such as Dryander.[15]

The format used for *Hortus kewensis* required the description of each plant and a derivation of its name. Linnaeus had introduced the name *Heliconia bihai* in his 1771 *Mantissa plantarum*, including a warning note about the species: 'Species *hujus generis forte plures distinguendae ab autoptis* [Species whose genus is distinguished from others on personal examination].' Checking Linnaeus's description would have shown the *Hortus kewensis* team that it referred to a plant that was similar to their mystery flower but having instead 'Spathae *membranaceae, rubellae.* Corollae *croceae* [With *red* spathe and membranes. With *yellow* corolla].' This combination of red and yellow flowers is recognisable to garden enthusiasts to this day as *Heliconia bihai*, also known as the red palulu.[16]

These were not the flowers with orange or gold sepals and blue petals that Gascoyne, Rockingham and Banks observed when their plants bloomed. By the time *Hortus kewensis* was published in 1789, it was also recognised that the *Heliconia bihai* title for one of Miller's seven plates in his 1780 *Icones novae* had been, in fact, the plant that Banks wanted to call *Strelitzia reginae* and it was cross-referenced as such. Intriguingly, the 1779 prospectus for Miller's *Icones novae* had promised instead to portray a plant referred to as *Heliconia caffra*. As *caffra* is an epithet often used to signal plants from southern Africa and especially from Kaffraria (now South Africa's Eastern Cape), this suggests both that botanists had already realised that its home was not the same as other, more familiar heliconia species which had been identified from South America and the plant's name was being reassessed. It is not yet known, however, why Miller's final published illustration ultimately used a name mistakenly picked up from Linnaeus. As a result, the name 'Heliconia caffra' appears never to have been validly published.[17]

In 1785, the dowager marchioness of Rockingham still referred to her own plant as 'Heliconia'. By then, Linnaeus's son Carl (1743-1783) had updated his father's work. In his 1781 edition, his father's problem plant had become a group of four: *Heliconia bihai*, *Heliconia alba*, *Heliconia psittacorum* and *Heliconia hirsuta*. The description of *Heliconia bihai* had been extended – but introduced a major confusion on which Banks and *Hortus kewensis* team must have pounced gleefully.

Linnaeus jnr had firmly rooted *Heliconia bihai* in South America, giving the local name 'Slaapers' from Surinam. He redefined the plant as '*octopedalis*' (eight feet tall; 2.4 metres) instead of '*tripedalis*' (three feet tall; 0.9 metres) but then seems not to have noticed that he was conflating two plants. Instead of the red and yellow flowers noted by his father, Linnaeus jnr referred to '*Flos* croceus, nectario caeruleo [*Flower* yellow, blue nectary]', a recognisable strelitzia colouration, although for the shrublike *Strelitzia reginae* rather than treelike relatives. Linnaeus jnr had received the first (and, as it turned out, only) fascicle of Miller's *Icones novae*, for example commending his depiction of New Zealand flax (*Phormium tenax*) as '*figura optima*'. Miller had depicted the plant we have come to know as *Strelitzia reginae* as *Heliconia bihai* and Linnaeus jnr seems to have taken this shift from his father's description at face value and lost the opportunity to capitalise on creating a new species himself.[18]

Banks and the *Hortus kewensis* team had seen the plant bloom at Spring Grove in 1787 and again at Kew in 1790 so they spotted the anomaly and realised they could fill a naming gap by disentangling two species. They did not hesitate to underline the original misapprehension by Linnaeus jnr, including an observation explaining the rationale for the splitting off of the new species under the *Heliconia bihai* entry:

> *Linnaeus filius in* Supplem. plant. 157. *sub nomine Hiliconae Bihai tres plantas diversas confundit, hanc Americanam et binas Africanas, Strelitziam nimirum Reginae, et aliam eijusdem generis speciem floribus albis et foliis reticulatis.*[19]

> [Linnaeus jnr in *Supplem. plant. 157.* confused three different plants under the name of Heliconia Bihai, this American one [i.e. *Heliconia bihai* itself] and two African ones, Strelitzia surely Reginae, and another one of this genus, a species with white flowers and reticulate leaves.]

The process was a double win for them, garnering professional prestige for astutely pinpointing the overlap and, Banks saw, gaining greater favour with the royal family through a name honouring Queen Charlotte. For a while, their 'sensation' was effectively a plant with no name as Banks tried to seize the opportunity to control its image and publicity. Miller's *Icones novae* had already reached some key European botanists, such as Linnaeus jnr in Uppsala, Sweden, and Johan Anders Murray (1740-1791) in Göttingen, Germany. This may be why, while Aiton and the team continued working on compiling *Hortus kewensis*, Banks could not resist the temptation to pre-empt its publication by sharing his tribute to the queen with his close circle. The plate for *Strelitzia reginae*, as Banks had named the plant, was co-opted to feature in a short series of coloured engravings that he commissioned as a private publication and which he 'presented to his particular friends' during 1788 and 1789. Banks was so anxious to impress one regular correspondent, renowned French botanist Antoine de Jussieu (1748-1836), that he sent copies of the *Strelitzia reginae* and *Phaius [Limodorum] tankervilleae* plates through a trusted messenger to Paris in June 1788. Despite fierce debate over the years that attributing the plant's new name to Banks usurped Aiton's claim to have originated it in the new *Hortus kewensis*, Banks's privately published folio of engravings has more recently been held to consolidate Banks as the originator of the name *Strelitzia reginae*.[20]

*Strelitzia reginae* was becoming renowned as one of Banks's favourite plants, appearing also more publicly as the second of two engravings in the first volume of *Hortus kewensis* in 1789. Just four years later, in 1793, the marchioness of Rockingham was already displaying her up-to-date botanical knowledge by mentioning in a letter to botanist Sir James Edward Smith: 'My *Strelitzia* is advancing into flower'. Banks may have given another plant to member of parliament and social philanthropist Thomas Johnes (1748-1816), 'the most ambitious and enterprising of all Welsh landscape gardeners' but whose extragavant passion for improving his estate bankrupted him. Johnes certainly wrote to Banks in 1796 telling him 'the *Strelitzia* is in high beauty & full flower'. It was nurtured by the gardener employed by Johnes who had previously been in charge of the Edinburgh botanic gardens. That same year controversial botanist Richard Salisbury recorded *Strelitzia reginae* and a Jamaican banana (incorrectly named as *Heliconia augusta*) growing in his garden at Chapel Allerton, west of Leeds, in Yorkshire.[21]

## THE ENVY OF EUROPE

By the time *Strelitzia reginae* was featured in volume four of the exciting new *Curtis's Botanical Magazine* in 1790, there had been at least four significant flowerings in England: in Gascoyne's stovehouse in 1777 (later named as *Strelitzia ovata* in the 1811 *Hortus kewensis*); at Rockingham's home in 1778 (later named *Strelitzia angustifolia* in the 1811 *Hortus kewensis*); at Kew in 1779 – as Miller noted in the text accompanying his illustration in *Icones novae*: 'Flowered in the royal garden at Kew, An. 1779'; at Banks's home in 1787; and in the Chelsea Physic Garden stovehouse in spring 1790. An accident of timing meant that ironically for Banks, the plant depicted in *Curtis's* was not the 1790 flowering from Kew noted in *Curtis's* but the earlier bloom from the Chelsea Physic Garden.[22] Considerable mystique still seemed to surround both the source of *Strelitzia reginae* and the challenges of cultivating and reproducing it. The magazine's publisher William Curtis (1746-1799) noted that *Strelitizia reginae*:

> has not, that we know of, as yet ripened its seeds in this country; till it does, or good seeds of it shall be imported, it must remain a very scarce and dear plant, as it is found to increase very slowly by its roots: plants are said to be sold at the Cape for Three Guineas each.[23]

Five years later, *Strelitzia reginae* was considered so rare that it was selling for £40 in Holland, a year's salary for many people – and the lower end of the estimated value that Banks put on a whole herbarium assembled by wealthy Dutch businessman and collector George Clifford (1685-1760).[24] By 1804, when botanical publisher Henry Andrews (fl.1790s-1830s) featured *Strelitzia reginae* in his magazine *Botanists' Repository*, it was a drawcard at leading nurseries such as Colvill's in the King's Road and fascinated the high-end gardening market, as Andrews observed:

> At the request of several of our subscribers, who compliment us in saying this work contains nearly all the more showy plants now in cultivation; and who moreover are desirous that it should not long want any of them: we here beg leave to present them with that queen of hot-house plants, the superb Strelitzia: although strictly against our rules and plan; a coloured quarto print of it having already been given in the Botanical Magazine.[25]

With *Strelitzia reginae*, Banks had announced a revelation that had some aristocratic and wealthy garden enthusiasts grumbling as they checked their own stovehouses for signs of this precious plant. The Dutch were rueful that the quick-witted Banks had snatched from them the prospect of naming such a prestigious plant. It had, after all, come from the hinterland of the revictualling station set up by their own Dutch East India Company nearly a century and a half earlier:

> Men heest hier thans andermaal het genoegen in onzen Academie-Thuin in bloei te zien, die zeldzaame en overschoone Plant, welke wy, voor drie jaaren by ons bloeiende, verklaard hebben gene HELICONIA BIHAI te zyn, waar voor den Heer J. J. Richard *en andere haar gehouden hadden, en welke wy te dier tyd meenden, dat nooit in Europa gebloeid hadt, dog zedert ontdekt hebben, dat dezelve in Engeland in's Konings Tuin te Kew haare bloem had voorgebragt, en daar na te recht, door den thands reeds betreurd wordenden* Aiton, *in zyn* Hortus Kewensis, Tom. I, p.285, *STRELITZIA REGINAE genoemd, en zeer schoon afgebeeld is.* Aiton verklaart, dat zy door den beroemden Joseph Banks, *in het Jaar 1793, eerst by hem in Europa is overgebragt,* en houdt hare natuurlyke Geboorteplaats, de Caap de Goede Hoop te zyn: waar by wy alleen voegen, dat onze STRELITZIA REGINAE, de voorige keer dertig volle dagen besteed heest om al de pracht van haaren bloei aan den dag te leggen, en zig deze keer thands niet minder schoon begint voor te doen.²⁶

[People here can once again have the pleasure of seeing in bloom in our University Garden the very same beautiful plant that has been thriving here for three years, the former HELICONIA BIHAI, as it was known to Mr *J. J. Richard* and others, but in all the time that it did not ever flower in Europe, we now discover that the very same plant had flowered in England in the King's Garden at Kew and we now regret the opportunity as Aiton in *Hortus Kewensis, Vol.* I, *p.285,* has named it STRELITZIA REGINAE, and depicted it very clearly. Aiton states that it was first brought to Europe by the celebrated *Joseph Banks* in 1773, and gives the Cape of Good Hope as its birthplace: to which we can only add that our STRELITZIA REGINAE lasted for thirty full days in all the splendour of their blooms.]

The Dutch writer seemed to have misunderstood the *Hortus kewensis* entry and believed that Banks had introduced *Strelitzia reginae* to the whole of Europe. However, the 1789 *Hortus kewensis* stated only that *Strelitzia reginae* is '*Nat*. of the Cape of Good Hope' and '*Introd*. 1773, by Sir Joseph Banks, Bart' and did not claim direct cause and effect – that Banks introduced the plant from the Cape himself instead of simply to the royal gardens at Kew. In fact, debate lingered over how the plant reached Kew.

## THE KEW MAESTRO

Banks was clearly a champion of *Strelitzia reginae*. It was both a 'magnificent' revelation and could also be used to suit the goals of his botanical politicking. His claim to have introduced the plant, though, was later overtaken by the Masson myth, a consensus honouring Kew plant hunter Francis Masson with having introduced it. The grounds for this assertion seem to be that Masson was in the right place at the right time – that is, at the Cape of Good Hope from late 1772.²⁷ So is it correct to credit Banks with introducing *Strelitzia reginae*?

Either provenance could be a legend designed to flatter and obscure the reality that the facts of the matter were never recorded or that any records were lost. Record-keeping at the Royal Botanic Gardens in the 1770s was not yet as organised as it would become in later years under the closer direction of Banks and then especially the Hooker dynasty. The earliest surviving Kew record book begins only in 1793, when Aiton senior's son, William Townsend Aiton (1766-1849), took over as superintendent. In 1891, Kew director Sir William Thiselton-Dyer (1843-1928) observed that 'scarcely any authentic records exist of the period prior to 1840', leaving a large gap in the early history of Kew. Even so, the cherished Masson legend does not seem to be upheld either by other early records or by logistics of time and place. It is true that the quantity and quality of the new plants Masson sent home on his first assignment as King's Collector at the Cape of Good Hope made his name. Masson is credited with having introduced more than 400 living plants from the Cape to Kew during this first, shorter stay alone and a new stovehouse was built at Kew in 1778 to accommodate and display his finds. Masson then consolidated this reputation during his second stay at the Cape from 1786 to 1795.²⁸

But Banks loomed large over everything that Masson did. In fact, it was Banks who gave him his place in perpetuity as a plant hunter, lobbying on his behalf for the Royal Society to fund

*Strelitzia reginae* in *Curtis's Botanical Magazine*, 1790

Masson's original post. It is even possible that Banks created Masson's role in the *Strelitzia reginae* legend to boost the benefits of sending out plant collectors as part of his own plan to crest the wave of innovation that he sensed in the natural sciences. As a boy, Banks had explored the countryside around the family's south Lincolnshire estate, Revesby Abbey. In 1761, aged 21, Banks could draw on a fortune inherited from his father, who had died in 1761. At Oxford University, the young Banks concentrated on natural sciences instead of the classics and had arranged and paid for Israel Lyons jnr (1739-1775) as a visiting lecturer from Cambridge to give a series of botany classes.[29]

After graduating from Oxford in 1763, Banks began using his wealth to start pursuing his passion for natural sciences, making the most of his personal charm and enthusiasm to generate a fanfare for his favourite specimens. He was elected to the prestigious Royal Society at the age of just 23 and could comfortably pay his own way to join expeditions that were pioneering understanding of geography and the natural sciences. He went first to Newfoundland and Labrador in 1766. Two years later Banks financed his place on the Royal Society's 'geophysical' voyage around the globe in partnership with the Royal Navy, which provided a ship, the *Endeavour*, and crew under Captain James Cook (1728-1779). Banks would shift the voyage's original focus considerably towards natural history from astronomy and geographical exploration. He and fellow botanist, Dr Daniel Solander (1733-1782), a Swedish pupil of Linnaeus, were assisted by Sydney Parkinson (c.1745-1771), a woollen draper turned draughtsman who was a protégé of Banks; a second artist, Alexander Buchan (d.1769), who specialised in landscapes; and Herman Spöring (1733-1771), a Finnish botanist known to Solander, as a secretary and artist. They had sailed from Plymouth in England in August 1768, heading west past Cape Horn at the foot of South America, and on to Tahiti, the New Zealand coast and Australia. Despite the exciting new sights which Banks had witnessed on the Endeavour's extraordinary journey of nearly three years, he was still enthusiastic about the exciting potential of the Cape flora, which he had singled out in a letter shortly before leaving London as one of the 'places much worth the attention of a Naturalist'.[30]

## BULBS AND BUTTER

For nearly a century and a half, the Cape had been an acknowledged stopover on exploration and trade routes for Dutch, English, French and Scandinavian ships.[31] Banks was well aware of this strategically placed port's intriguing reputation built on oddities and essentials – bulbs and butter.

Banks and his fellow members of the Royal Society knew that Arab, Chinese and Indian navigators had been aware of the approximate outline of southern Africa for some centuries before Bartolomeu Dias (1450-1500) in 1488 took the knowledge of its existence back to Europe in ambivalent form under the less appealing name of the Cape of Storms. European nations were initially unenthusiastic about the region's potential for creating a strategic bridgehead, partly because of the notorious Cape sea route and also because of tales of hostile local inhabitants. In 1510, a Portuguese landing party had been frightened away by a group of Khoi with trained battle oxen.[32]

The French may have rounded the Cape in 1503 and certainly did in 1529. Later the English commander, Sir Francis Drake (c.1540-1596) sighted the Cape in 1579 en route to India, although the first English landing there was not until 1591. About a decade later, in 1602, the Dutch established their Vereenigde Oost-Indische Compagnie (VOC, or United Dutch East India Company), following the English founding their Honourable East India Company two years earlier. But the Dutch scooped the Cape ahead of the English after the proposal by an English sea captain that England should annex the region fell on deaf ears. The English government felt that, like other European powers, its focus remained firmly further east.[33]

Jan van Riebeeck's settlement at the Cape of Good Hope for the Dutch republic's VOC displaced St Helena as their primary revictualling station en route to their prime trading station of Batavia in the Dutch East Indies (now Indonesia). This Cape enclave became a source of staples such as meat and milk but another valuable resource for visitors was the fresh greens and herbs – welcomed both as a relief after months of monotonous shipboard diets at the time and to keep the crew healthy, even though how to prevent scurvy was not discovered in the west until 1747. Despite the demands of wind and tides, the opportunity to explore within sight of shore would have been hard to resist for many, especially given the majestic sight of Table Mountain, though early records suggest visitors did not have to go far to find plants which they realised would be rewarded by wealthy florophiles at home.[34] Later, for strategic and competitive reasons, the VOC restricted access beyond the settlement, especially for the non-Dutch.

These factors restricted early plant hunting geographically and even seasonally, according to the ebb and flow of shipping schedules to and from the Far East. Even so more than 40 VOC representatives, free burghers and visitors appear in Cape plant-collecting records before *Strelitzia reginae* was introduced at Kew in 1773, and more probably went unrecorded. For much of this period, plants from the East Indies and the Americas dominated European curiosity, to the point when in 1597 the first South African plant known to have reached Europe, *Protea neriifolia*, was casually misidentified as having come from Madagascar because it had arrived on a ship which had called earlier at the Indian Ocean island. A woodcut of its dried flowerhead later appeared in the 1605 publication *Exoticorum libri decem*, written by Flemish botanist Charles de l'Ecluse (1526-1609), also known as Carolus Clusius. He was prominent in the new generation of botanists, interested in learning more about plants for their own sake and not simply as sources of pharmaceuticals, which made botany a subsidiary to medicine. Clusius had been a plant hunter for Portugal and Spain before being appointed professor of botany at the University of Leiden in the Netherlands in 1593.³⁵

It has been suggested that this protea was selected for collection – or it might have survived – precisely because of its robust appearance. Following this, several Cape plants, most of which could be transported in their sturdy bulb state, were featured in the lavishly illustrated botanical books becoming more common in the 17th century as printing techniques developed. Such plants tended to be those that were conspicuous in the dry days at the end of the Cape summer when ships made their scheduled calls in March on their voyage back to Europe.³⁶

## HIGHLY PRIZED

While some collected and showed off live specimens of new and exotic plants, others collected dried specimens gathered from the wild. Milestones such as Thunberg's pioneering *Strelitzia reginae*, probably collected at Plettenberg Bay on his 1772 expedition, and Gascoyne's 1777 *Strelitzia reginae* flowering have special historical value. Novelty alone, though, could make early plant specimens from distant lands prize items for interested collectors back in Europe. Some of the earlier specimens at Kew show this process.

A specimen collected in 1825 became part of the Bentham Herbarium and then Bishop Goodenough's herbarium, which was presented to Kew by the Carlisle Corporation in June 1880. The collection assembled by Dr William Gill (1792-1863), an English-born medical doctor, naturalist and pioneer merino farmer who settled in Somerset East in the Eastern Cape, is believed to have been sold to English botanist William Hooker (1785-1865) while he was still in Glasgow and then taken by him to Kew. One specimen carries a note from Dr Gill: 'I think a species of Strelitzia grows in abundance on the Banks of Keiskamma'.⁷⁶

*Strelitzia reginae* remained a trophy for plant hunters, even though creating a good pressed specimen for a herbarium of this large, fleshy plant was difficult and still remains a challenge. This challenge may be what made any specimen valuable even if, as was common at the time, no date or location of collection were recorded – seen, for example, in a specimen collected by Thomas Cooper and bought for the Cape of Good Hope collection in March 1873. Fortunately this does not apply to all early specimens, even if details are sketchy. Botanist and naturalist William Burchell (see chapter 7) collected 'between Blaauw Kraantz and Kowie Poort', nearly 400 kilometres east of Plettenberg Bay, probably during his visit in 1813. Within a decade, James Bowie (c.1789-1869) similarly collected the plant in the Albany division and near the Kowie River, as well as near Uitenhage, just over 200 kilometres east of Plettenberg Bay. German horticulturist and botanical collector Franz Drège (1791-1867) collected the plant also in the Bathurst area, as well as near the Fish River and then further north-east between the Umtata River and the St John's River, more than 800 kilometres north-east of Plettenberg Bay. In the 20th century, the plant was collected near East London on the banks of the Buffalo River (about 500 kilometres away from Plettenberg Bay) in July 1919 by Margaret Heatley (1885-1953), a British-born, American-educated botanist, then lecturing at the Huguenot University College in Wellington in the Western Cape. Her collecting partner was Charles Moss (1870-1930), inaugural professor of botany at the South African School of Mines and Technology (later the University of the Witwatersrand), whom Heatley married in 1922, having joined the staff at the University of the Witwatersrand in 1921. Several decades later, Margaret Gillett (1878-1962) – an enthusiastic English amateur botanist who had joined General Jan Smuts (1870-1950), Kew botanist John Hutchinson (1884-1972) and her botanist son Jan (1911-1995) on their 1930 expedition to Lake Tanganyika – collected *Strelitzia reginae* on 22 March 1937 during a trip through southern Africa in the Port Alfred area.⁷⁷

Bergius Herbarium

*Strelitzia reginae* specimen sent back to Sweden by Carl Thunberg about 1774

## THE CAPE PLANT CRAZE

For a couple of generations preceding Banks, these pioneering books on the flora beyond Europe would be key to igniting professional, academic and collectors' interest in Cape plants across Europe. German physician Paul Hermann (1646-1695) made the first known herbarium collection of plants from the Cape of Good Hope when he called there in 1672 on the way to his new post as medical officer in Ceylon. Hermann set the early pattern of collegial sharing information between fellow enthusiasts rather than clutching the honour and glory to himself, sending some seeds and specimens through fellow surgeon Hieremias Stolle to Thomas Bartholinus (1616-1680) in Copenhagen, for example. Bartholinus later wrote and published what is believed to be the first botanical account devoted only to Cape plants in 1675. Hermann also sent from Ceylon packages of seeds and specimens that might have contained Cape plants. Among the recipients was Jan Commelin (or Commelijn, 1629-1692), an Amsterdam merchant who was one of the founders of the city's Hortus Medicus (medical garden) in 1682 and given the title of the garden's botanist. Hermann also sent seeds and specimens to rare plant enthusiast Jacob Breyne (1637-1697) in Danzig (now Gdansk, Poland). Breyne in turn sent specimens on to London apothecary James Petiver (c.1663-1718), who became a demonstrator at the Chelsea Physic Garden in 1709.[37]

When Hermann returned from Ceylon to Holland in 1680, he became professor of botany and director of Leiden University Botanic Gardens, where he was renowned for building up its collection of rarities. He continued building his international botanical network and established exchanges, including with the Chelsea Physic Garden. Hermann's 1687 catalogue of the Leiden gardens contained 34 Cape plants, several described and illustrated for the first time. When Hermann died in 1695, he was working on an illustrated volume of rare plants from the Cape and the Indies that was published three years later. Over the following century, Hermann's herbarium ended up in London at the British Museum through the efforts of Irish-born doctor, natural-history enthusiast and collector Sir Hans Sloane (1660-1753) – who had also endowed the land for the Chelsea Physic Garden – James Petiver and Joseph Banks.[38]

It was mid-March 1771 when Banks landed in Cape Town, where the *Endeavour* and her crew restocked and recuperated for a month before the final leg of the voyage back to England. The Cape plant craze was catching alight and spreading among the cognoscenti of Europe so it must have been galling for Banks that he was prevented from making as much as he might have hoped from his short time at the Cape. In late 1770, the *Endeavour* had begun the return to England on the Cape of Good Hope route. Nursing their badly damaged ship to Batavia, they stopped for emergency ship repairs. For the next 10 'agonized' weeks, most of the crew and passengers battled tertian malaria (which brought fevers every other day), dysentery and possibly typhoid fever. Deaths had begun in Tahiti and continued in Batavia but another 22 died between there and Cape Town, which they reached on 14 March. Ultimately, more than half of the crew and passengers died. In port at Cape Town, Dr Solander was still 'bedridden for two weeks with a violent and emaciating combination of the prevailing infections, to the acute alarm of Banks more than once during its course'.[39] Banks himself was also convalescing from his own bout of typhoid.

As a result, the two naturalists could not attempt a full scientific expedition during their month in Cape Town, although they did still collect a few hundred specimens around the settlement, where Banks found the vegetation stunted, and visited the Dutch East India Company's garden. Apart from the clipped oak dividing hedges, Banks noted approvingly the shade offered by the oak trees in the garden's central walk, reporting that most of the garden was devoted to growing cabbages, carrots and other vegetables to fulfil the gardens resupply role as a key point in the Cape revictualling station. Captain Cook himself considered the gardens a 'most ravishing spot', probably because he appreciated the fruit it produced, including 'peaches, pomegranates, pineapples [and] bananas'. Banks was more concerned that only 'two small squares' were at that time devoted to live botanical specimens and, given the time of year and the fact that most were annual bulbs, they were generally dormant underground. Though Banks found this small area well and neatly kept, he believed specimen numbers had probably dropped by half since the Cape-based German physician, botanist and land-surveyor Heinrich Oldenland compiled his 'Kruid Boek' (book of dried plants) and wrote his catalogue in the 1690s and that the amount of land devoted to botanical specimens had dropped correspondingly, too.[40]

Departing after a month in the colony, Banks seems to have been tantalised by the potential he guessed its flora might have. He had probably seen little more than he would previously have learned from collections and catalogues already in Europe but he had seen the early plant-collecting infrastructure was in place and had even gathered valuable local logistical knowledge. It would be a pivotal visit for international botany and horticulture because shortly after his return to England, Banks was presented to King George III. Banks began building on their mutual interests in natural science and innovative agriculture. Ultimately, Banks was so successful in implementing his ideas and impressing the king that he was created a baronet in 1781 and became Sir Joseph Banks.

When Princess Augusta died in 1772, Bute resigned his advisory position at Kew and Banks began nearly half a century of work on the gardens. His new status also enabled Banks to put his initiative, understanding and enthusiasm into ensuring that someone else should start exploring the Cape's botanical potential on behalf of England and the king. Banks was not planning to explore the region's flora himself as he intended to join Captain Cook's next circumnavigation. But Banks was thwarted in this plan. The new expedition was to be led by the Royal Navy rather than the Royal Society, leaving him less able to use his influence to insist on the facilities he felt necessary for a full scientific expedition so he ultimately withdrew. Just 18 months after Banks had sailed from the Cape, his protégé Francis Masson landed there from the *Resolution*. Banks had hoped that Masson, a Kew gardener born in Aberdeen, Scotland, would take the place Banks had himself initially intended occupying on the *Resolution's* new voyage of discovery. Instead, Masson was given the more limited brief of King's Collector in the Cape, and sponsored by the Royal Society of which Banks was already a prominent member. Masson arrived at the Cape in October 1772. The following year, Banks was said to have introduced the *Strelitzia reginae* at Kew for the first time. How that living plant material reached him and flourished at the royal gardens remains unclear.[41]

## THE MASSON MYSTERY

In these circumstances, it is perhaps hardly surprising that the assumption developed at Kew and elsewhere that the *Strelitzia reginae*, recorded as having been introduced in 1773, must have been another Masson find. Masson was renowned as 'modest', though delighted when Linnaeus named a genus *Massonia* after him. Surviving correspondence and publications from Masson do not suggest that he claimed *Strelitzia reginae* as one of his finds. None of the available records support this and logistically it would have been difficult for him to achieve in 1773. Masson was not mentioned in the 1790 article on *Strelitzia reginae* in *Curtis's Botanical Magazine* even though this could have been used to reflect on Banks's perspicacity in sending him out to collect among the Cape floral treasurehouse. When Aiton's son, William Townsend Aiton updated and expanded *Hortus kewensis* for a second edition in 1811, numerous new plants were credited to Masson but Banks's claim to have introduced *Strelitzia reginae* remained unchallenged.[42]

It is also curious that if Masson did introduce or even collect *Strelitzia reginae*, recollection of this eye-catching plant did not come to mind as interesting or unusual enough to record in the account of the three plant-collecting expeditions he undertook during his first stay, written between November 1775 and February 1776 for Sir John Pringle, president of the Royal Society.[43] By contrast, naturalist William Burchell, travelling about 40 years later between 1811 and 1815 into the interior as far as the Orange River (Gariep), highlighted the appeal of colourful showy plants, even to the local volunteers who assisted him:

> one of the women of our party, having heard that I was an admirer of flowers, good naturedly brought me a large quantity of a very pretty plant which grows here in abundance. It was quite new, and its deep-orange colored flowers induced her to think it was worth gathering; for Hottentots, like the generality of Europeans, see no beauty in a plant unless it have a showy flower.[44]

Masson had landed at the Cape on 30 October 1772 and quickly started preparing his first plant-collecting expedition, which he undertook from December 1772 to January 1773. In terms of timing alone, this might seem to be a possible source for *Strelitzia reginae* but on this trip Masson travelled no further east than Swellendam, about 220 kilometres east of Cape Town. Currently, suitable habitat for *Strelitzia reginae* is considered to begin around the Humansdorp area, about 450 km further east still. Far from mentioning a plant that might be *Strelitzia reginae*, Masson concludes his account with his lasting impression of this trip: 'It was on this journey that I collected the seed of the many beautiful species of *ericae* [heaths] which, I find, have succeeded so well in the Royal Garden at Kew.'[45]

Masson was invited to join an expedition that Swedish collector Carl Thunberg (1743-1828) was planning into the Eastern Cape region. Masson, his servant, Thunberg and 'four Hottentots' set off in September 1773. Thunberg was a pugnacious observer who often took the opportunity in the

© The Trustees of the Natural History Museum, London

Franz Bauer, *Strelitzia reginae* from *Strelitzia depicta*, 1818 prefaces to his books to belittle potential rivals. In one preface, for example, Thunberg seemed to attack fellow Swede Anders Sparrman (1748-1820), who travelled in the Cape just after Thunberg but managed to publish his account a decade earlier. This made Sparrman the first trained naturalist to publish a personal account of travel in regions beyond the Table Bay and Stellenbosch areas known to astronomer and later VOC secretary Peter Kolb (1675-1726), who lived at the Cape from 1705 to 1713 and produced the landmark *Caput Bonae Spei hodiernum* in 1719. In a second preface, Thunberg perhaps condescendingly described his fellow plant hunter, Masson, as 'a skilful English gardener'.[46] Thunberg also did not hesitate to promote the value of his own two-volume *Travels* over what was, comparatively, a pamphlet produced by Masson. If Masson was to achieve the daunting task set him by Banks, Thunberg seemed to imply, he needed careful mentoring by a professional botanist such as Thunberg. This might even have resulted in the marked difference between Masson's crisper and more detailed account of their journey together between September 1773 and January 1774 compared to the more general tone and content of the account of Masson's solo first trip from December 1772 to January 1773.

During this second journey, some plants seemed so striking to Thunberg and Masson that they apparently named them in the field. These included the amaryllis which they called *Amaryllis disticha* for its fanshaped leaves, although it was commonly named *vergiftboll* for its poisonous qualities, and the 'species of heath remarkable for having its branches and leaves all covered with a fine hoary down or nap,' which they named *Erica tomentosa*. Masson spent nearly a page of his account discussing 'a new palm, of the pith of which the Dutchman told us the Hottentots make bread.' The two plant hunters had encountered cycads that we now know as *Encephalartos longifolius* and *Encephalartos caffer*.[47]

The memoir of the second trip ends quite baldly: '29th [January 1774], We arrived at the Cape Town, after a journey of four months and fourteen days.' There is no reference in Masson's account to the specimens which might have been brought back, though some authors have claimed, 'Masson returned . . . with *Strelitzia reginae* in his oxwagon.'[48] *Strelitzia reginae* generally flowers from May to December so it is possible that Masson might have collected *Strelitzia reginae* seed without having seen the plant's flower. But even if he did bring this back, Masson returned to Cape Town at the end of January 1774 – the beginning of the year after the plant is recorded as having been introduced at Kew by Banks.

## A 'MOST BEAUTIFUL' FIND

The assumption that Masson and Thunberg together brought back *Strelitzia reginae* seems to have come about through conflating the dates of Masson's first, shorter trip in 1772 with this later, longer trip and also with Thunberg's account of his own first of three journeys. Both Thunberg and Masson partnered only in their second and third journeys as Masson did not arrive at the Cape until October 1772, more than a month after Thunberg left Cape Town on his first journey, returning at the beginning of January 1773.[49]

Thunberg was accompanied on this first journey by Johann Auge (1711-c.1805), superintendent of the Dutch East India Company garden, Cape resident D.F. Immelman and a soldier, C.H. Leonardi. In November 1772, Thunberg's companions were resting up in camp and recovering from the shock of a frighteningly close encounter with buffalo when Thunberg and his Khoi tracker found the plant that would become known as *Strelitzia reginae* at the foot of the Robberg (Seal Mountain), near today's Plettenberg Bay in the south-eastern Cape.[50]

That made Thunberg the first botanist to record finding the 'most beautiful' *Strelitzia reginae* in the field. But the discovery did not seem to stand out for him and more than 15 years would pass before he published his account of the event. Although Thunberg

SANBI

Auriol Batten, *Strelitzia reginae*, about 1974

has been hailed as 'the father of South African botany',[51] for him the Cape was a stepping stone to his original goal, exploring the flora of Japan. He had been groomed for this by Linnaeus and was one of the talented students retained in Linnaeus's 'inner circle'.

Linnaeus had introduced Thunberg to Johannes Burman (1707-1779), professor of botany in Amsterdam, and his son Nicolaas (1733-1793). The Burmans negotiated a post for Thunberg as a ship's surgeon in the Dutch East India Company so that he could collect plants in Japan where at the time the Dutch were the only foreigners admitted. Beyond plant collecting, Thunberg's main purpose at the Cape was to learn Dutch well enough to pass himself off as Dutch in Japan.[52] This focus on the key Japanese goal of his overall expedition also meant that Thunberg prioritised publication of his Japanese collection. He did not publish his account of his Cape travels until more than 15 years later, first in Swedish in 1788. In a single, short paragraph, Thunberg describes how he was captivated by the plant he already refers to as 'strelitsia':

> Strelitsia med gul bloma och blåt nectarium växte här I nejden, och var an af de aldratåckaste blomor, hvaraf lokar til ofversånde åt Europa håmtades. Det sades, at Hottentotterne åta dess frugt.[53]

[The strelitsia, with its yellow flowers and blue nectarium, grew near this spot, and was one of the most beautiful plants, of which the bulbs were procured to send to Europe. The Hottentots were said to eat the fruit of it.]

It is likely that the 'bulbs' and dried material from this plant, which reached Cape Town in Thunberg's wagon on 2 January 1773, were among the first specimens of *Strelitzia reginae* to reach Europe – although wind, weather and rats often wreaked severe damage on the long passage back from the Cape.[54] Aiton's *Hortus kewensis* was not published until the year after Thunberg's *Travels*, yet Thunberg still used the name 'strelitsia'. This suggests that Banks had already publicised his intention to give the name *Strelitzia reginae* to the plant which had recently flowered in his own garden and had perhaps even presented a set of the engravings, which included the plant, that he commissioned and distributed privately in 1787 and 1788 to the team carrying on the work of the Linnaeus dynasty in Uppsala.

Thunberg sent duplicates of specimens he collected in the field to the Burmans and to Linnaeus, as well as possibly to other influential collectors in Amsterdam and London. At least three of these *Strelitzia reginae* specimens are known to survive, none of them dated or giving location. Two are in the herbarium of Uppsala. The third is at the Bergius Herbarium in Uppsala, Sweden, where Linnaeus lived. This herbarium is part of the legacy of Swedish Bergius brothers, Bengt (1723-1784) and Peter (1730-1790). Their summer residence eventually grew into a botanic gardens and herbarium, run under the Royal Swedish Academy of Sciences as the Bergius Foundation.[55]

The only way that Masson could be linked to this collecting of *Strelitzia reginae* was if it had been among some duplicates that Thunberg shared with him for passing on to Kew via Banks as a gesture of good faith before their joint journeys further into the Cape. Even so, that does not solve the mystery of the attribution to Banks when so many other plants were directly attributed to Masson.

## THE PLANT TRADERS

Banks had been in Cape Town just the year before Thunberg found *Strelitzia reginae* and two years before it was introduced to Kew. Delay in sorting through the vast collection brought back on the *Endeavour* might in itself account for the introduction date two years after his return. Banks did not, though, record having seen the plant there nor having brought it or a rhizome back to England. Given his later enthusiasm for *Strelitzia reginae*, however, it seems unlikely that Banks would have forgotten the plant if he had seen it in flower in a garden in Cape Town. Yet this in itself is not likely as *Strelitzia reginae* usually flowers between May and December in South Africa.[56] How then could Banks be credited with introducing the plant to Kew – and, by extension in the eyes of some writers, to Europe?

It remains possible that Banks had simply brought the plant home himself from the Cape, either as a plant or plants that were not flowering or in a mixed lot of 'bulbs'. Banks could have bought these or received them as a gift from Auge at the Company garden and, given the short section of his journal that Banks devoted to his Cape visit, might have made no reference to this as he was at the time unaware of what exactly he had received. Auge was certainly known to parcel up bulbs and other curiosities of natural history for gifts or for sale to overseas visitors and collectors, as

*Strelitzia reginae* Banks subsp. *reginae* ●

Distribution of *Strelitzia reginae*

well as responding enthusiastically to VOC Governor Rijk Tulbagh's encouragement to develop the gardens beyond a market garden to supply the garrison and passing ships.⁵⁷ One of Auge's curiosities was the book of dried plant specimens bought by Swedish East India Company director Michael Grubb at the Cape in 1764 and given in Stockholm to Professor Peter Bergius. On this basis, Bergius gratefully named a genus *Grubbia* in his benefactor's honour, unleashing a controversy of attribution to a facilitator rather than the collector that re-echoes to this day. Thunberg later remarked, for example:

> . . . I visited M[onsieur]. Auge, the gardener, who has made many, and those very long, excursions into the interior part of the country, and has collected all the plants and insects which the late Governor Tulbagh sent to Europe to Linnaeus, and to the Professors Burman and Van Royen. And as he still continued his journies yearly into the country, he sold to strangers, as well herbals as birds and insects. It was of him, that M[onsieur]. Grubb, the director of the bank in Sweden, purchased that fine collection of plants . . .⁵⁸

Or perhaps *Strelitzia reginae* had been in Europe or even in Kew for some, possibly many years through the collections sent back by Hermann and others and which became part of the international botanical network of exchange. Banks would have been familiar with Hermann's 1687 catalogue of the Leiden gardens and his later illustrated volume of rare plants from the Cape and the Indies.

In its early days, the VOC settlement at the Cape was relatively inward looking, focused on establishing itself, surviving and fulfilling its task of providing vegetables for the fleets. Pressure from farming pushed the frontiers of the Dutch settlement across the Cape peninsula to the Hottentots Holland Mountains. By then, nearly 40 years after Hermann, Heinrich

*Strelitzia reginae* featured on a gatefold in *Curtis's Botanical Magazine*, 1790

Oldenland (1663-1697), the second trained botanist to collect at the Cape according to Linnaeus, was able to venture significantly further from the settlement than Hermann. Oldenland was a member of the 1689 expedition led by Ensign Isaq Schriver (fl.1689-1705/1706),[59] perhaps because he may have been employed at the VOC's Cape gardens at the time. This expedition allowed him to explore far into territory then largely unknown to Europeans. He reached as far east as Oudtshoorn and the Camdeboo areas, ranging around Uniondale, Willowmore and close to Aberdeen. This is neither far enough south nor east to be strelitzia country so it is unlikely to have yielded the plant, though it would have entrenched excitement among the cognoscenti about Cape plants.

It is believed that the Cape plants illustrated by the Commelin brothers in Amsterdam were grown from seed sent by Oldenland and Jan Hartog (c.1663-1722), both of whom worked at the VOC gardens and accompanied expeditions into the interior of the Cape and the Karoo, though not reaching strelitzia country. At the time, these seeds were attributed out of courtesy to successive VOC governors, Simon (1639-1712) and Willem Adriaan van der Stel (1664-1733). Significantly, Oldenland specimens were also sent to or bought directly from him, through travellers at the Cape or through the international botanical network by Oxford University Herbarium and by Petiver. After Oldenland's death in 1697, most of Oldenland's collection eventually went to the Geneva Herbarium in Switzerland but his 'Kruid Boek' ended up in Holland with Professor Johannes Burman. Burman's son Nicolaas took a substantial number of Cape plants, including some Oldenland specimens, to Linnaeus in the 1750s. Linnaeus was one of the key European botanists with whom Banks corresponded and in 1784, despite protests by the Swedes, Banks persuaded Sir James Edward Smith to buy Linnaeus's herbarium from his widow and ultimately make it the core of the Linnean Society collection.[60]

The same protocol for attributing the source of plants continued to prevail under the Cape governorship of one of van der Stel's successors, Rijk Tulbagh (1699-1771). However, Tulbagh was renowned for his personal interest in natural history and for encouraging exploration beyond the confines of the settlement 'ter opsoeking en versameling van Planten, gewassen, kruijden en Insecten [to search for and collect plants, herbs, fruits and insects]'. Auge had certainly explored the south-eastern Cape by 1772 and had made 'eighteen journies of different lengths into the country', according to Thunberg, which is why he selected Auge as a guide.[61] Tulbagh seems to have included bulbs and seeds collected on this expedition in dispatches to the botanic gardens in Amsterdam and Leiden. In Amsterdam these would have caught the attention of the influential Burmans.

French mathematician and astronomer Abbé Nicolas de la Caille (1713-1762) may have benefited from Auge's plant distribution encouraged by Tulbagh, sending plants, seeds and bulbs back to France during his two-year stay at the Cape from April 1751. De la Caille's main goal, though, was to observe the stars and determine the longitude of the Cape settlement from his observatory in an attempt to settle the contemporary debate as to whether the ball shape of the earth flattened at the poles. Pehr Bladh (1746-1816), who became a fellow of the Swedish Academy of Science in 1779, supplied Bergius and is known to have sent Swedish naturalist Anders Retzius (1742-1821) a 'type of Teucrium trifidum which does not grow within easy distance from Cape Town and would not be found by a traveller calling in during a voyage to or from the East'. Another likely beneficiary was Swedish navigator Carl Gustaf Ekeberg (1716-1784), who became well acquainted with Governor Tulbagh and collected personally around False Bay. Ekeberg was one of several other collectors who supplied Bergius with Cape material, including Sparrman, fellow Swede Pehr Osbeck (1723-1805), the French administrator and naturalist Pierre Sonnerat (1745-1814) and the Latvian Johann König (1728-1785). Linnaeus also obtained specimens from the Cape using contacts such as these, as well as direct contact with Tulbagh. In fact, in just one letter to Linnaeus, Tulbagh speaks of sending 'a keg of 36 kinds of flower bulbs, and a small basket of flower seeds', while a letter from Linnaeus to Tulbagh acknowledges 'a second most and valuable present of more than 200 specimens of plants, numerous bulbs and 50 kinds of seeds'.[62]

## FALSE FRIENDS?

Although it has become common belief that Francis Masson, the King's Collector at the Cape of Good Hope from 1772 to 1775 and from 1786 to 1795, originally collected the plant of *Strelitzia reginae* introduced by Joseph Banks, this chapter discusses reasons why that might not have been the case. Beyond the fact that Masson was almost in the right place at almost the right time, there is no evidence for this longstanding assumption – and the currently known specimen record leaves the question hanging. The only specimen of *Strelitzia reginae* attributed to Masson is in London's Natural History Museum and is undated. It is not even clear whether the specimen was collected in the wild or from a garden, whether Masson's garden, the Company garden or the garden of another wealthy Cape Town horticultural enthusiast resident, nor whether it was collected during Masson's first visit to the Cape or his second.[78]

Equally tantalising is another Natural History Museum specimen – unusually with a complete *Strelitzia reginae* flowerhead – that is simply marked, 'Prom b. spei Nelson'. Two different collectors named Nelson were active for a while in the Cape at very different times. Erring on the side of caution suggests that this might have been Sheffield-born nurseryman William Nelson (1852-1922), who arrived at the Cape of Good Hope in 1876. He took a job as an overseer on the diamond fields of the Orange Free State but during his overland trip by mule wagon, he did not collect plants and did

not pass through strelitzia country. In 1878, William Nelson returned to the coast to sail to England for a visit but chose to travel from the Transvaal (now Gauteng) south east to Durban. Although *Strelitzia reginae* does not occur naturally in Pietermaritzburg or Durban, he might have come across the plant in a garden – although these were relatively few in a colony that was only just over two decades old.[79]

The other candidate is David Nelson, who was cast adrift in an open boat with Captain Bligh after the *Bounty* mutiny and died shortly afterwards of fever in Timor. Before this ill-fated assignment, David Nelson was nominated by London nurseryman James Lee (1715-1795) to collect for Joseph Banks on Captain Cook's third circumnavigation. During this voyage, Nelson spent a few days collecting on the Cape peninsula in November 1776 when Cook's ships were docked at Simonstown. He also called at the Cape again on the *Bounty* voyage in May 1788. As David Nelson did not reach strelitzia country on either visit, if he sent the specimen, he would have had to have collected it from a garden, such as the Company garden, or have been given it or bought it while at the Cape. The notation 'Prom b. spei', short for *Promontorium bonae spei* or Cape of Good Hope, suggests an early date as later specimens were marked 'Cape' or 'CGH' and William Nelson would also probably have used 'Colony of Natal' for any specimens collected there. This circumstantial pointer cannot, however, be considered conclusive.[80]

Most tantalising of all at first glance, though, is a *Strelitzia reginae* specimen in the Kew Herbarium labelled 'C.G.H. 1773'.[81] It is tempting to hope that this might be one of the earliest specimens reaching Europe from Thunberg's expedition. But, unfortunately, there is no other evidence to support such a theory – nor the faint possibility that the specimen was one collected by Masson during his expedition in the last quarter of 1773.

The specimen reached Kew in 1895 after it bought the herbarium of William Gower (1835-1894). Gower was a foreman at Kew, who worked for commercial nursery firms from 1865 and put together a personal herbarium of garden plants.[82] It is more likely that Gower added the label to indicate the date and place of introduction to Kew than that he had been able to acquire one of the original specimens.

## REACHING OUT TO STRELITZIA COUNTRY

It is feasible that any one or more of these traders and travellers could have been a channel through whom *Strelitzia reginae* was introduced into Europe – if the plant was available in the Cape at that time. In just a decade between 1735 and 1744, 'there were 102 recorded hunting trips to the East . . . [and] probably even more unrecorded trips.' The Dutch East India Company held a monopoly on cattle trade, making it against the settlement's law to trade cattle with local peoples. Even so, 'Hunters and traders concealed their journeys from officialdom,

Basia Hitchcock Swiel, *Strelitzia reginae*, watercolour, 2015

smuggling tusks into Cape Town, where they secretly sold them to the captains of Dutch and other ships.' Ivory-hunters particularly were known to range far beyond the settlement's recognised boundaries, sometimes at their peril. Hermanus Hubner (c.1694.1736) led a hunting expedition of 13 wagons up into Pondoland in 1736, reaching as far as the current area of Butterworth, about 9 600 kilometres east north-east of Cape Town in the Transkei area of the Eastern Cape. Hubert never returned, having been killed with six others. It is not known whether the survivors who returned had the opportunity or inclination to bring rootstocks or plants of *Strelitzia reginae* with them, though they would probably have been aware of the premium placed on interesting plants by Governor Tulbagh and by sophisticated callers at the Cape.[63]

The first formal VOC expedition into these areas was led by Ensign August Frederik Beutler. A train of 11 wagons accompanied by more than 70 people set off for what was then called Kaffraria in February 1752. Among them were soldiers, a surveyor and, to act as plant-collector, Hendrik Beenke (d.1745), a German from Celle. Along their route, they met wreck survivors from a French expedition which had left Mauritius to investigate potential areas for French settlement on the south-eastern Cape coast. This party of French sailors had been sent ashore from one of the vessels, *Le Necessaire*, to collect fresh water but their boat had been wrecked in Algoa Bay and they were expected to make their way west across country to meet up with the expeditionary fleet in Cape Town. This incursion into territory that the Dutch considered theirs by right shook Tulbagh, who sent orders post haste that Beutler should pre-empt any French attempt to take possession by erecting VOC marker beacons in the bays along his route.[64]

Among the French fleet, on board the *Glorieus*, was Captain Jean-Baptiste d'Après de Mannevillette (1707-1780), a pioneering French hydrographer whose series of charts of Indian Ocean waters, published together as *Le Neptune Oriental*, were prized internationally, not least because he had used the new navigational technology, the octant. De Mannevillette had spent two years under orders from the French East India Company reconnoitring the coast north easterly from the Cape of Good Hope and including what are now the port areas of Durban and Maputo where he made his most thorough observations. But the fleet was forced to return to the Cape when food and water provisions ran short. De Mannevillette did not apparently make observations or measurements on land but crew expeditions ashore might have brought back some plant specimens. De Mannevillette moved in the circles of France's enthusiastic experimental scientists and explorers, had published on the use of the octant as early as 1739 and on his outward journey had delivered the abbé de la Caille to the Cape in 1751. De Mannevillette would have been aware of the interest and knowledge value to be gained from plants unknown in Europe. As de Mannetteville's expedition focused on coastal areas where *Strelitzia reginae* can be found – as well as other areas where the considerably larger *Strelitzia alba* and *Strelitzia nicolai* occur – so it

could be another possible vector for *Strelitzia reginae* at least reaching Europe in the mid-18th century. As for Beutler and his expedition, they also reached country where they might have seen and collected *Strelitzia reginae* such as St Francis Bay, the Keerom River, Humansdorp and the Great Fish River during May and June 1752, eventually turning back on 9 July near the present Butterworth.[65]

## THE DUTCH CONNECTION

These factors, the apparently garbled identification by both Miller and Linnaeus jnr and the growing volume of specimens and plants being taken back to Europe from the Cape all make it likely that *Strelitzia reginae* was already circulating in Europe, although not yet recognised as a new species, before Thunberg's visit to the Cape. In this period when more and more large private houses were paying close attention to their gardens, plant nurseries had begun flourishing to supply the new craze: 'there were no doubt many . . . undocumented introductions, often through relatives and friends in service overseas.'[66] A legitimate web of correspondence and exchange was also woven between enthusiasts such as Banks and professionals such as Linnaeus in several European countries.

In his attempt to 'raise the Company's Tuin above its original cabbage growing into something like a Botanic Garden', it was said that 'Auge used his utmost diligence to bring into cultivation every rare and curious African species'. So there is also a slender possibility that *Strelitzia reginae* was indeed growing in the Company garden and was supplied by Auge to Masson between his arrival in Cape Town on 30 October 1772 and his departure on his first expedition on 10 December 1772. In this case, Masson could have sent it on to Kew along with an early field collection from the Cape Town area. David van Royen (1727-1799), professor of botany and director of Leiden botanic gardens, as well as nephew of collector and businessman George Clifford, boasted of the Leiden gardens' excellent Cape collection, which included Eastern Cape specials such as *Nerine undulata* and *Dais cotinifolia*. This has been used to emphasise that Eastern Cape plants were reaching Europe much earlier than previously believed, although their safe arrival was complicated further by wars between major European powers.[67] Politics and shipping embargos at the Cape sometimes meant that Masson, for example, had to ship his plants back to Banks via Holland. As well as collectors shipping their plants and specimens home via convoluted routes, collections might wait months or even years to be shipped back to Europe or, worst of all, plants could be seized as spoils of war by privateers.

Even if Masson had been lucky enough to have picked up a plant which had been brought west by a trader, he would still have had to defy the odds of safely delivering a live specimen to Kew. When plants were safely loaded on board ship outside time of war, still only a small proportion of live plants survived the journey back to Europe. Live plants were usually potted up in large tubs or wooden boxes and exposed to the elements on deck, particularly being washed with salty waves during a storm. To be transported below deck usually meant plants were condemned to death from lack of sunlight. There were also all kinds of prowling enemies – people who broke off leaves, flowers and branches; rats and cockroaches; dogs and cats; goats, poultry and even parrots. Serious collectors such as Masson took to modifying their plant-boxes by adding a strong glass lid. Nothing, though, could make the minimum 60-day journey from the Cape go any faster – and it took longer if the ship was becalmed in the doldrums, with plants suffering further from intense heat and humidity.[68]

George Clifford III, a VOC director and heir to a banking fortune, used part of his wealth to build up the plant collection in his garden and conservatory at his home, De Hartekamp, in Heemstede in the north of the Netherlands. Clifford was a central point in the plant-exchange nexus, just as the garden remains a touchstone in the period's botanical and horticultural history, memorialised in *Hortus Cliffortianus*, published in 1738. As the grandson of a settler from Lincolnshire in England, Clifford probably still maintained his links with the extended family who had remained there. These connections also point to how apolitical and deep-rooted plant linkages were maintained in various ways. From East Anglia, prosperous Norwich merchants traded regularly with Holland in the 18th century. Future renowned landscape designer Humphry Repton, a tax-collector's son, was sent at the age of 12 to live in Rotterdam to learn the language for commercial purposes – in the process, he also developed an interest in horticulture.[69]

After becoming bishop of London in 1675, Henry Compton (1632-1713), had created an excellent collection of exotic trees and plants at his Fulham garden, including tender exotics cosseted in stove houses. In 1689 Bishop Compton had crowned England's new Dutch king and queen, William III and Mary. When Compton attended the Amsterdam Congress in 1691, he was presented with paintings of 100 Cape plants, later known as the *Codex Comptoniana*. 'The gardens of Holland were at that time the richest in Europe'[70] with other Williamite supporters, such as William Bentinck (1649-1709) taking great pride in their gardens so it is likely that Bishop Compton also received plant specimens during this visit.

During 1773, when Banks introduced *Strelitzia reginae* to Kew according to Aiton, Banks spent more than a month touring Holland, from 13 February to 20 March. Coming from Lincolnshire, he would have been well aware of such networks and he even immediately commented how similar the landscape was to the fenlands of eastern England, enthusiastically asking for advice on drainage issues. Botany and drainage dominated the trip, with the occasional addition of art and music. In Leiden, he met Swiss naturalist Jean Nicolas Allamand (1713-1787), professor of philosophy and natural history, and Professor David van Royen, director of the Leiden botanic gardens, as well as nephew of collector and businessman George Clifford. Banks also saw Leiden's herbarium, botanical gardens, greenhouse and a plant nursery. In Amsterdam, he met Nicolaas Burman, by then professor of botany, and saw Paul Hermann's herbarium and Oldenland's Cape plants. Banks also visited the botanic gardens in Utrecht and The Hague. He was a guest of honour at a meeting of the Rotterdam Society of Literature.[71] As *Strelitzia reginae* is comparatively inconspicuous when it is not flowering among a mixture of other plants so at any of Banks's later engagements during his Dutch tour, he might have been presented with or bought such a plant or assortment of plants for his or the king's collection.

Even without a clear record of the source of the Kew introduction of *Strelitzia reginae*, Banks's flair for publicity and botanical empire-building meant that assumptions about him and assertions by him would probably have gone unchallenged. This was the force of the Banks phenomenon, which came to transform and rule a significant proportion of the public perception of botany and horticulture in Britain and beyond for nearly 50 years. The belief that Masson introduced *Strelitzia reginae* seems to be a fond homage to a favoured son of Kew that skews circumstantial evidence. It was also probably conflated with Banks putting *Strelitzia reginae* on his wishlist of collections for Masson's second stay at the Cape from 1786 to 1795 so that stock of the plant at Kew could be increased. The plant was certainly flowering at Kew in the first decade of the 19th century and may have been a reference for William Townsend Aiton's 1811 second edition of *Hortus kewensis*. William McNab (1780-1848), a gardener and later foreman at Kew, where he worked from 1801 to 1810, collected a specimen for his personal herbarium in 1808.[72]

As things stand, the source of Banks's 'magnificent' revelation, *Strelitzia reginae*, remains opaque. What is clear is that of all the botanists in Europe, Joseph Banks not only spotted its potential but also shrewdly channelled this to the greater good of botanical endeavour. Indeed, far from simply stepping in to take the credit, Joseph Banks was central to giving it the prestige of being named after the reigning queen to start it on a path to international glory.

## *Strelitzia reginae* and its synonyms, c.1750-1850

| PUBLICATION DATE | SUPERSEDED SYNONYM | MEANING OF NAME/SYNONYM | ORIGINATED |
|---|---|---|---|
| 1788 | *Heliconia bihai* | | London: Banks, *Icones plantarum* |
| 1791 | *Heliconia strelitzia* | | Leipzig: J.F. Gmelin, *Systema naturae* (13th ed.) |
| 1811 | *Strelitzia farinosa* | Mealy-stalked strelitzia | London: W.T. Aiton, *Hortus kewensis* |
| 1811 | *Strelitzia ovata* | Oval-leaved strelitzia | London: W.T. Aiton, *Hortus kewensis* |
| 1821 | *Strelitzia humilis* | Dwarf strelitzia | Berlin: H.F. Link, *Enumeratio plantarum Horti regii botanici berolinensis altera* |
| 1831 | *Strelitzia glauca* | Greyish strelitzia | Bonn: A. Richard, *Nova acta physico-medica academiae caesareae leopoldino-carolinae naturae curiosorum* |
| 1846 | *Strelitzia rutilans* | Ruddy strelitzia | Ghent: C. Morren, *Société Royale de l'Agriculture de Gand* |

## FAMILY FIGHTS

Kew's World Checklist of Selected Plant Families tracks existing names, new names and names which have been introduced and accepted, superseded or rejected, sometimes in vituperative wars of words between botanists. Despite the small number of Strelitziaceae species, by 2018 this amounted to 59 different names. Despite nearly 60 bids to join the strelitzias family, it is currently considered to contain seven species and two subspecies, indicating the scientific jostling as botanists and gardeners searched out small differences between plants to proclaim new species – or had their claims quashed by taxonomic rules. Nearly 20 of the records were for *Strelitzia reginae* alone. The table lists names proposed before 1850, all superseded except Banks's 1788 *Strelitzia reginae*.[83]

Most of these synonyms were reduced to varieties or forms of *Strelitzia reginae*. At the end of the 19th century, for instance, *Strelitzia humilis* was renamed as both a form and a variety of *Strelitzia reginae* within just a three-year period between 1893 and 1895. Between the 1840s and the 1960s, the naming debate created a further nine names now considered invalid.

Angela Beaumont, details from *Strelitzia reginae*

Francis Masson, *Strelitzia alba*, about 1790; and detail from Franz Bauer, *Strelitzia depicta*, 18

# A RARE FIND
*Strelitzia alba*

CHAPTER TWO

## 2

# A RARE FIND: *Strelitzia alba*

The excitement over the revelation of *Strelitzia reginae* was infectious. Within a decade of the plant being scientifically recognised, it became a desirable, rare treasure for the conservatories of royal, aristocratic or simply very rich collectors. As plant hunters and other travellers sent more specimens back to Europe and travelled further east from the Cape of Good Hope to find them, it became clear that *Strelitzia reginae* was not alone. Even before Thunberg had found his *Strelitzia reginae* specimen in the south-eastern Cape near the Robberg in November 1772 (see chapter 1), his party had crossed the nearby Piesang River, which flows into the sea at Beacon Island. As German naturalist and medical doctor Hinrich Lichtenstein (1780-1857) noted when travelling there in 1803, its name was taken from the Cape Dutch name for the abundant *Strelitzia alba*, Cape wild banana, growing in the area. Thunberg's casual reference and the very much larger specimens beginning to reach taxonomists eventually led them to recognise that a much rarer, treelike relative of *Strelitzia reginae* grows in some south-eastern Cape forests. It would later be named botanically as *Strelitzia alba*.[1]

In the meantime, another botanical landmark steadily established itself at the heart of the Company garden in Cape Town. This huge mass of *Strelitzia alba* with its palmy fronds on lofty trunks towers over visitors, who take nearly 60 paces to stride around the whole clump. It is so huge, old and well established that it is considered to be the descendant of the original specimen planted by the 18th-century superintendent Johannes Auge. By 1811, it was already an object of admiration for the likes of British naturalist William Burchell when he visited Cape Town.[2] Auge is an important figure in the strelitzia drama, though the nature of surviving historical records makes his role somewhat shadowy. Lichtenstein paid tribute to Auge:

> He exerted the utmost diligence to store the garden with every sort of rare African plant, so as to convert it into a true botanic garden . . . With equal diligence did he collect wild plants for his herbarium.[3]

Auge had moved to the Cape in 1747, possibly prompted by encountering the significant collection of Cape plants while working at the prestigious Leiden botanic gardens after his training both in Germany and in Holland. Good references from Dutch patrons helped persuade the governor at the time, Hendrik Swellengrebel (1734-1803), to appoint Auge as the assistant superintendent of the Company garden. After succeeding Swellengrebel as governor in 1751, Rijk Tulbagh promoted Auge to superintendent and, perhaps because he shared his natural-history interests, ensured that Auge had opportunities to visit farflung areas of the colony. Auge had made a total of 18 journeys of varying duration before 1772, when Thunberg made use of his experience as a guide.[4]

Despite all the plants that Auge had encountered in those many miles of journeying, the lofty white treelike strelitzia was one that made the most lasting impact. By 1804, Auge was 93 years old and living far from Cape Town on the farm Rotterdam, about an hour away from Swellendam. Governor Jan Willem Janssens (1762-1838) gave Lichtenstein permission to join a group of Dutch officers travelling to that area so he could visit Auge to inform him that Thunberg had commemorated his collaboration with Auge by naming the Karoo plant *Augea capensis* after him. This honour from Thunberg was intensified because he had chosen to name not just a species but a higher-ranking genus after the botanist and plant hunter who had assisted in his field work in the Cape, as well as helping other plant hunters such as Kew's Francis Masson. Lichtenstein noted that Thunberg had hoped the name would mean 'that future botanists might have a lasting memorial of [Auge's] services.'[5]

Lichtenstein was moved to find this already legendary botanist and curator was living mostly on the charity of the Swellendam landdrost, A.A. Fauré (1758-1824) – 'one of the most able and respectable men in the Colony,' Lichtenstein was told. Auge had lost his pension about a decade earlier after the British took the Cape from the Dutch in 1795. By the 1770s, Auge's eyesight was already failing badly and in 1778, just a very few years after working with Thunberg and Masson, Auge had retired at the age of 67 to a friend's farm on the Gamtoos river. After the farm was looted and burned by a raiding party, Auge and his friend fled. Auge, by then completely blind, was taken in by Fauré and installed on his farm beside the Buffeljagts river.[6]

Lichtenstein recalled how he found Auge outside his cottage door:

> He was tall, still tolerably upright, and his hair, as white as snow, hung about his shoulders. The sight of a blind person always excites compassion; here respect was inspired by the sight of dignified old age, to which with me was added veneration for one of the most skilful of botanists; and notwithstanding the homeliness of his dress, he stood in the midst of us as the object of our highest respect and interest.[7]

Auge was excited about this visit from 'the first botanist he had seen for many years, and [said] if he could have the same pleasure once every year, his life would be less tedious to him'. Auge surprised Lichtenstein with his keen questioning:

> . . . it was very evident that he still retained all his love for his favourite science, and I was not a little astonished to find how well he remembered the names of the various African productions. He was rejoiced to hear that I too was a collector of plants, and told me many things relating to those which were to be found in the *Duiwelsbosch* (Devil's-bush), but he called them by the old names given by Burmann . . .

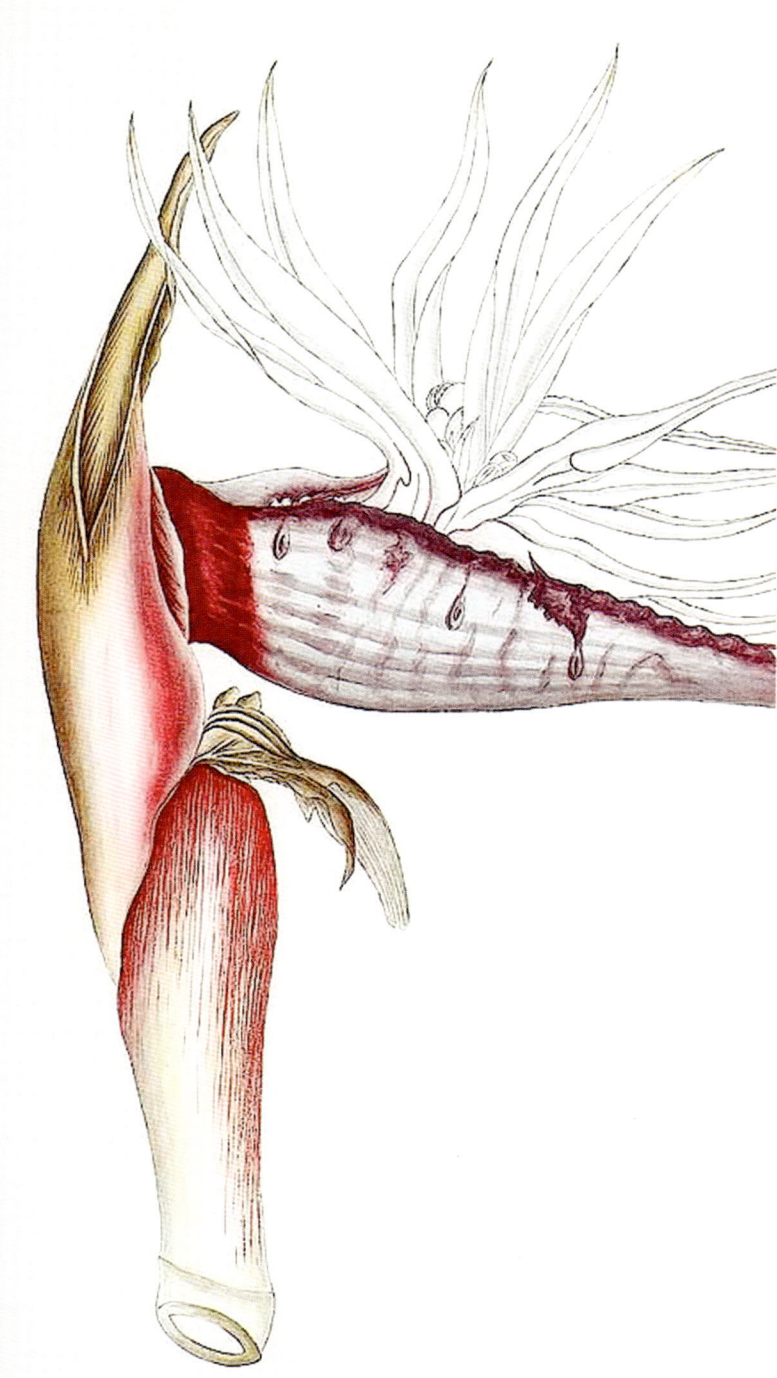

He was almost angry that I could not tell him immediately from recollection to what class [*Augea capensis*] belonged, for he would gladly have known that, and whether it was a plant with which he was acquainted.[8]

## AUGE'S BOTANICAL LANDMARK

Auge particularly asked Lichtenstein about the fate of fondly remembered plants in the Company garden: 'Is my *Heliconia alba* alive? – is my *Corolladendron* as fine a tree as ever?' Auge was referring to the trees we now know as *Strelitzia alba* and *Erythrina* or coral tree. He was eager to hear as much detail as possible from Lichtenstein: 'He begged me to describe them to him, how tall and how thick they were, and he said he should die happier, if he could but feel them once again.'[9]

But the discussion also had its frustrations for both botanists:

> He enquired about several others [plants], concerning which I could not give him any information, either because they were no longer in existence, or that I did not understand the names by which he distinguished them.[10]

Perhaps a wish to make amends for this mutual incomprehension prompted Lichtenstein to make enquiries about Auge's pension:

> I learnt by the way that the worthy Fauré had never applied to get the pension renewed, out of delicacy to the old man, lest it should appear as if he grudged any part of the money spent upon him.[11]

Despite such sensibilities being involved, Lichtenstein appealed to Governor Janssens to restore Auge's pension which

> would at least have found him in cloathing . . . A mere relation of what passed at this interview was sufficient to obtain from the Governor the restoration of the pension with a small monthly addition to it; and if the English, as I am inclined to hope, have continued the benefaction . . . it has contributed towards rendering the evening of so venerable an old man's life somewhat more easy and serene.[12]

Auge would have very little time to benefit from Lichtenstein's well-meant intervention as he died about a year or so after Lichtenstein's visit. By the time Lichtenstein published this account, Auge had been dead for nearly a decade. Auge also left behind him an enduring botanical mystery.

Franz Bauer, *Strelitzia alba* from *Strelitzia depicta*, 1818

## A MYSTERY SOURCE

Reminiscing with Lichtenstein, Auge told him that he had collected the white heliconia, as he knew it, in 'Namaland'. This is not the habitat of *Strelitzia alba*, which has a limited range in the south-eastern Cape between George and Humansdorp at altitudes below 150 metres, roughly following what is now known as the Garden Route.[13] It is possible, though, that Auge picked up the specimen on a south-eastern Cape leg at the beginning or end of the trip.

While superintendent of the Company garden, Auge had been Governor Rijk Tulbagh's choice as plant-collector to accompany Captain Hendrik Hop's expedition, north from the VOC settlement, crossing 'Namaland' (Namaqualand) and what was named the Orange River (now Gariep), reaching an area north of the present Bela Bela (formerly Warmbaths). A party of about 90 set out on 16 July 1761, returning to the VOC castle on 27 April 1762. Among the 18 or so expeditions that Auge undertook, this would have lingered in his memory and, indeed, became celebrated for the new ground it covered and for its fascinating haul of scientific material, including a giraffe skin donated to the Leiden Museum. This may have prompted him in old age to confuse his travels so that it was the first trip that came back to mind although others took him into the area where the plant could have been found. It was Auge's familiarity with this area that prompted Thunberg to put him in charge of leading his 1772 expedition shortly before they reached the Goukamma River and the Knysna area.[14]

However, a composite sheet marked *Strelitzia augusta* in the Natural History Museum in London contains a specimen named *Heliconia albiflora*, dated 1772 and attributed to Frans Oldenburg (1740-1774), a soldier in the Dutch East India Company at the Cape from March 1771. Thunberg claimed to have trained him in botany when they collected together on the Cape peninsula in 1772. Oldenburg accompanied Masson as a Dutch interpreter on his first Cape journey, although this reached only Caledon and Swellendam, not into *Strelitzia alba* country. This was the furthest inland that Oldenburg went and in 1774, he was sent on a mission to Madagascar, where he died of fever. Oldenburg was a correspondent of Solander and Bergius, so he could have simply passed on a specimen he had received from another source. Or a clerk in Europe might have been more familiar with Oldenburg's specimens and entered that name. To complicate matters further, modern understanding of the strelitzia family suggests that the other two later specimens on the sheet are likely to be *Strelitzia caudata* and *Strelitzia nicolai* as they were noted as collected at Lydenburg in the then Transvaal Republic and in the Colony of Natal (see chapters 3 and 9).[15]

Thunberg's decision to commemorate Auge's achievements and his gratitude to this botanical pioneer emphasise how Auge was renowned in his day as a plant collector, collecting many plants that were new to science. So it is possible that Auge originally supplied Oldenburg's *Heliconia albiflora* specimen. Sir Joseph Banks bought Oldenburg's Cape plant collection and some of his plant drawings, all of which later passed to the British Museum in London. Auge's 'large herbarium . . . ultimately fell into the hands of Burmann of Amsterdam', noted Peter MacOwan (1830-1909), director of the Cape Town Botanic Garden and curator of the Cape Government Herbarium. But records are scanty for this founding period of scientific botany so there is no more than circumstantial evidence that Auge personally collected and planted his 'Heliconia alba'. As well as making shorter trips south east of Cape Town, Auge was hired as a botanical guide by Thunberg on his first journey from September 1772 to January 1773. During this trip, Auge and Immelman crossed the Piesang River. Thunberg's party stayed on Jacobus Botha's farm, also known as Piesang Rivier, two kilometres west of Beacon Island at today's Plettenberg Bay. Thunberg himself later sent a specimen of *Strelitzia alba* back to his sponsor Bergius in Sweden, which is still held by the Bergius Herbarium. It is possible that Thunberg collected it on this trip, along with the related *Strelitzia reginae* which he recorded nearby in the vicinity of the Robberg just a few days later.[16]

Lush growth of *Strelitzia alba* near what is now Plettenberg Bay became quite a landmark for travellers exploring the region. John Barrow (1764-1848), the British auditor-general at the Cape who later founded the Royal Geographical Society, came upon these trees in 1797 and commented:

> Not far from Plettenberg's bay, along the banks of a small rivulet [Groot River, 19 km east northeast of Plettenberg Bay] I met with a whole forest of the Strelitzia Alba, whose tall and tapering stems, like those of the Areca nut, or Mountain cabbage [*Cussonia paniculata*], were regular and well-proportioned . . . Many of them ran to the height of five

*Strelitzia alba* (L.f.) Skeels ●

Distribution of *Strelitzia alba*

and twenty or thirty feet [about 7.5m to 10m], without a leaf. It is . . . remarkable that the white species should grow so very abundantly along the side of one stream of water, and not a single plant be found near any of the rest in the same neighbourhood. From the great resemblance of this plant to the Banana tree, the peasantry call it the wild plantain.[17]

Neither of Thunberg's Bergius specimens has a date or location so it is not clear whether Thunberg collected it at this point on his first trip or during a later one. As for Auge, it is possible that he collected a living plant on this trip with Thunberg to take back to the Company garden in Cape Town. Even assisted by Thunberg and Immelman, digging out the treelike plant would have been a major undertaking. This makes it more likely that Auge took a small sucker which he carefully grew to maturity. Collecting the tree that he called 'my *Heliconia alba*' personally would add to the poignancy and possessiveness in the enquiry to Lichtenstein from this blind old man so close to his death.

## FATAL DELAYS

During the three years that Thunberg spent at the Cape before heading on to Japan in 1775, he achieved a strelitzia-collecting double, sending back to Europe specimens of both *Strelitzia reginae* and what is now known as the *Strelitzia alba*.[18] But neither plant contributed to ensuring his immortality as a botanist because he fatally delayed writing up his field notes. Thunberg was not the only one to see some of his botanical ambition wither in the face of a fatal oversight, as Joseph Banks was also soon to discover.

A decade passed after Thunberg's return to Europe before Banks began engineering the naming of the Eastern Cape revelation, *Strelitzia reginae*, updating and clarifying Linnaeus's name of *Heliconia bihai* in honour of Britain's Queen Charlotte of Mecklenburg-Strelitz (see chapter 1). This was first published in the limited edition of *Icones plantarum* distributed by Banks in 1788 and 1789 (see chapter 1). However, the plant's description in *Hortus kewensis* in 1789 contained a fatal error in terms of its other known relative, then named *Heliconia alba* by Linnaeus jnr. This may have been an oversight by Linnaeus's pupil Dr Daniel Solander, considered to have drafted much of the text. It may have been a loosely worded addition by the overall author William Aiton. Even Banks, such a shrewd strategist of botanical politics, either approved or overlooked the significance of the wording used. As a result, only observations, 'Obs', mention the plant's treelike, white-flowering cousin. The observation at the end of the *Heliconia bihai* description, refers to *Strelitzia reginae* 'et aliam ejusdem generis speciem floribus et foliis reticulatis [and another one of this genus, a species with white flowers and reticulate leaves].' At the end of the next entry, that for *Strelitzia reginae*, the observation reads:

*Differentia specifica Heliconiae albae in* Linn. Suppl. 157. *Huius est plantae, sed nomen triviale ad aliam pertinent speciem Africanam, in hortis Europeaeis nondum obviam.*[19]

[Specifically different from Heliconia alba in *Linn. Suppl.* 157. This is a related plant, but the species name belongs to another African species, not found in European gardens.]

These observations remain simply a hint that the name of the plant should be updated to *Strelitzia alba* along with the African *Heliconia* which Banks had renamed *Strelitzia reginae*. They do not clearly state that the *Hortus kewensis* team had also renamed '*Heliconia alba*' – possibly because the fact that they did not know of its existence in European gardens meant they had heard of it but never seen it. No mention was made of this strelitzia relative in the 1790 splash in *Curtis's Botanical Magazine* that drew the attention of the botanical and horticultural world to the fascinating new *Strelitzia reginae*.[20]

This tentative approach might be due to the fact that neither Banks nor Aiton had seen a live specimen of *Strelitzia alba*, nor possibly even a dried specimen. Masson appears to confirm this when he wrote to Banks in December 1789:

I have a fine plant of *Strelitzia alba* which I wish much to send home as I think I have not heard that there is any of that Sp. yet in England.[21]

It is not surprising that Masson was fascinated by this relation of *Strelitzia reginae*, given that it grew about four times taller on a single fibrous stem, instead of in a lower-growing clump. Everything is big about *Strelitzia alba*. Its flowers sit on a 30-centimetre-long spathe, compared to about 20 centimetres long in *Strelitzia reginae*. This spathe is about eight centimetres high and about 4.5 centimetres broad. Each one bears an average of five white flowers with tepals up to 18 centimetres long and 3.5 centimetres broad, filaments that are three centimetres long and anthers up to 5.5 centimetres long.[22] When Banks, like other rich collectors, provided his wishlist of plants for Masson, neither realised just how rare the plant would prove to be.

In 1811, Aiton's son, William Townsend Aiton (1766-1849), was the overall author of the second edition of *Hortus kewensis* and diplomatically split the credit, acknowledging that Masson collected the plant then known as *Strelitzia augusta* in the Cape and that Banks introduced it to Kew in 1791. A specimen in London's Natural History Museum is marked as coming from the garden at Kew in 1797 and a second as coming from Masson's own garden at the Cape.[23] It is not clear whether this is the plant that Masson attempted to send back in December 1789, which might not have survived the voyage or perhaps not thrived at Kew. But in March 1791, Masson had sent Banks another live specimen, notifying Banks:

I had the honor to writ you by Mr. Riou who took charge of a collection of Bulbs and seeds packed in two boxes

containing 386 sorts. A large growing plant of Strelitzia alba and drawing of a compleat plant and fructification of it . . . I have inclosed one seed of Strelitzia alba which I received this day. a seed of singular beauty.[24]

This arrangement underlines how highly plant specimens were prioritised and valued at the time. Captain Edward Riou (1758-1801) had been one of the survivors of the wreck of *HMS Guardian*, his first command. In late December 1789, the ship struck an iceberg about 2 000 kilometres beyond Cape Town en route to New South Wales. Half the crew was evacuated in the ship's boats. As only one of those boats was rescued, profound relief and astonishment greeted Riou's feat of seamanship in bringing the crippled frigate into Table Bay about two months after the disaster. Banks was well aware of this epic journey as he had designed the greenhouse that was installed on the *Guardian's* quarter-deck to transport useful plants, including some from the Cape, to the new British colony of New South Wales. Riou instead spent the year after the wreck in Cape Town waiting for orders. When he was homeward bound at last, he still took care of the seeds and bulbs from Masson, and possibly others, which were to be delivered to Banks's home. Banks made a point of thanking Riou for this in a personal letter and seems to have been as anxious as Masson that his strelitzia should reach Kew safely: he assured Riou specifically that he would 'instruct Aiton' to arrange for the '*Strelitzia alba*' to be collected from Portsmouth.[25]

Masson was so taken with the *Strelitzia alba*, as he called the plant, that he painted it. This is one of more than 100 other South African plant paintings by Masson that were collected by his friend and client, the Hammersmith nurseryman James Lee (1715-1795). Masson became better known as a botanical artist for his stapelia portraits, published in four parts as *Stapeliae novae* (*New Stapelias*) in 1796 and 1797.[26]

Francis Masson, *Strelitzia alba*, about 1790

## WRANGLING BOTANICAL NAMES

By 1789, Masson had been back in the Cape for three years yet his letters show that he was keeping well up to date with developments in the world of botany. The first edition of *Hortus kewensis* had been published in August or September 1789, only three months or so before Masson's earlier, 21 December 1789 letter to Banks on the subject.[27] Masson seems to have acted on the implication of *Hortus kewensis*, discarding the Linnean name *Heliconia alba* in favour of extending Banks's new identification of the strelitzia family. Masson was using a new name, *Strelitzia alba*, perhaps because he anticipated that it was only a matter of time before this was formalised. Sir John Barrow (or his editor) had made the same assumption when his description of the *Strelitzia alba* he had seen in 1797 was published in 1806. However, as noted earlier, the earlier *Hortus kewensis* observation was not an explicit statement so ultimately did not constitute valid publication that would update of the name. Similarly, Masson had used the name *Strelitzia alba* only in a private letter to his patron. Barrow's observations had been published but this did not constitute valid publication of a new name, either, according to the later rules of botanical naming because it was not accompanied by a Latin description of the plant.

Eventually Kew's oversight would ultimately mean that credit for formally updating the name would go more than a century later to a young botanist across the Atlantic from Kew. In the meantime, Masson had not been the only one to conclude that the plant should be called *Strelitzia alba*. In 1806, Henry Andrews (fl.1790s-1830s) published a plate of *Strelitzia reginae* in his magazine, the *Botanical Register*. While discussing the 'majestic genus' to which the queen's flower belonged, he listed the species known at the time, beginning with:

> Strelitzia alba (Hortulanorum), which, towering far above the rest, and quite erect, attains the kingly height of twenty feet or upwards: it ought to have been called S. Regis.[28]

Although Andrews' real point is that he would like the plant to become the king's strelitzia to accompany *Strelitzia reginae*, his use of the term 'Strelitzia alba (Hortulanorum) [of the gardeners]' suggests that it was already known in gardening circles as *Strelitzia alba*.

Banks and Aiton might have been circumspect about their observation because they had neither a live nor a dried specimen of this second, desirable member of their new strelitzia genus at the time of publishing *Hortus kewensis* in 1789. A dried specimen from Masson exists in the collection of the Natural History Museum, London. Although it does not show a date, it was probably not brought back from his Eastern Cape trip with Thunberg when he returned to Cape Town in 1774 because it is marked 'Hortus. Mr Masson'. This suggests various possible sources: it is a specimen taken by Masson from the plant established by Auge at the Company garden; it is a specimen sent

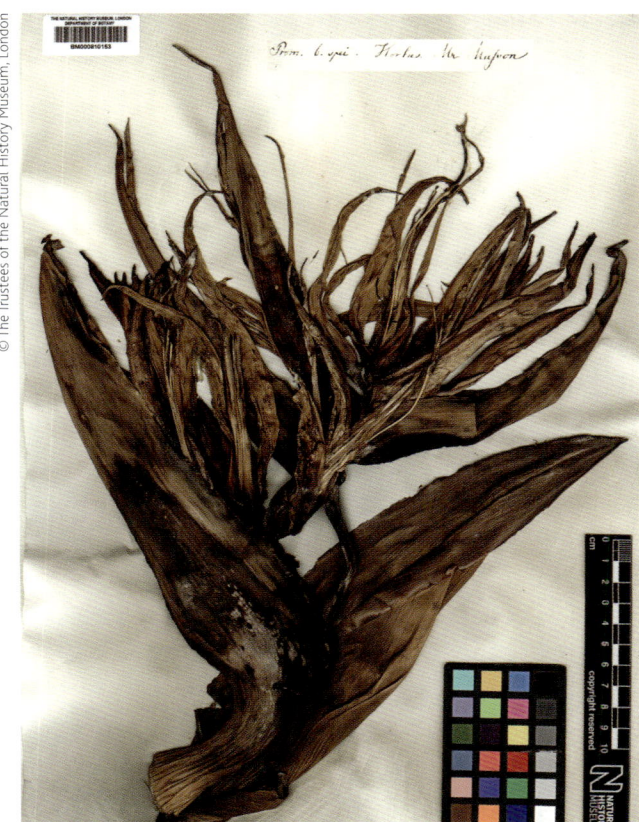

by Masson from a plant in the garden he established during his second period at the Cape, from 1786 to 1795, to keep plants in good health until he was able to ship them back to Banks; or it is a specimen taken at Kew from the live plant sent back by Masson.[29] As well as pointing to Banks and Aiton relying on Masson to provide the plant, the fact that they were not definitive about this species in the 1789 *Hortus kewensis* could also suggest they calculated that they might gain more strategic mileage by timing a dedication of it to another powerful and influential botanical enthusiast carefully. If so, their own fatal delay meant they lost that first race to Thunberg and one of his research students.

The correspondence between Masson and Banks suggests that this race was quite close, with Banks ultimately beating Thunberg by only three years. For Banks to make a splashy announcement as he had with *Strelitzia reginae*, he needed preferably both a live specimen and a well-executed botanical illustration to display. Meanwhile, Thunberg had delayed publishing details of his Cape collection due to the hectic work schedule that he faced after returning to Uppsala in 1779 from his travels in the Cape and Japan. In Thunberg's absence, Bergius published in 1767 a list of new Cape plants that he had received, *Descriptiones plantarum ex Capitae Bonae Spei*, but this did not even include any mention of *Heliconia bihai*, as *Strelitzia reginae*

*Strelitzia alba* specimen from a plant grown by Francis Masson at the Cape of Good Hope

was still then known.[30] Thunberg was working for Linnaeus's son, who had taken over the chair of botany in Uppsala in 1776, when his father retired. Linnaeus senior died in 1778, followed by his son five years later. Their deaths caused upheaval in Uppsala particularly and the botanical world generally as well as affecting their renowned, returned former pupil, Thunberg.

## A BOTANICAL RACE

Thunberg's delay in publication affected his impact on naming *Strelitzia reginae* and ultimately thwarted acceptance of the name, *Strelitzia augusta*, that he wanted to give its new relation. This fatal delay was probably due to his calculated decision to concentrate on his Japanese material to please his patrons. His original commission from Johannes and Nicolaas Burman had been the botanical exploration of Japan, with his stay at the Cape simply envisaged as a means to an end to improving Thunberg's fluency in Dutch. Thanks to his service with the Dutch East India Company and to his effective diplomacy and enthusiasm, Thunberg became one of the few Europeans to carry out any botanical exploration in Japan, probably another factor in his prioritising publication of his substantial *Flora japonica*. This was published in 1784, the year after Linnaeus jnr's death. Thunberg's account of his travels, *Rese uti Europa, Africa, Asia* followed, first in Swedish between 1788 and 1793, and then in English from 1793. Similarly, nearly two decades passed after Thunberg's visit to the Cape before the time he attempted to catch up further on his botanical publishing, with part 1 of *Prodromus plantarum capensis* in 1794. This was essentially a catalogue of Thunberg's Cape collection as the subtitle explained:

> *Prodromus plantarum capensium quas in promontorio Bonae Spei Africes annis 1772-1775 collegit Carol. Pet. Thunberg*
>
> [Precursor of Cape plants collected in the Cape of Good Hope, Africa, between 1772 and 1775 by Carl Peter Thunberg]

Part 2 did not follow until 1800, then the first part of Thunberg's *Flora capensis* in 1807, followed by the remaining parts of the work until 1820. Not rushing into print professionally was characteristically meticulous of Thunberg. Yet though he was trying to use the means at his disposal to change the destiny of *Strelitzia alba*, he ultimately failed.[31]

Thunberg had become adept at using the scientific academic system of the time to help him process the huge collection of nearly 25 000 plant specimens that he had gathered during his travels, 3 100 species from the Cape alone, of which more than a third were new to botany. Thunberg himself had been promoted to doctor of medicine and then to botanic demonstrator during this time. His duties included supervising students and, although publication records for the period seem incomplete, over half a century as professor, he appears to have been *praesus* (promoter) of at least 294 student dissertations. Foreshadowing the research teams that are today built up around prominent and pioneering scientists, Thunberg appears to have given graduate students whom he supervised the task of working with a portion of the specimens he had collected. They described their assigned plants under the most up-to-date names from botanical literature of the time as the basis of the thesis that each would defend publicly. Many of these, perhaps all, were published as pamphlets of about 10 pages or so. According to the protocol of the time, Thunberg's students are each named as second author and any newly named plants in them were attributed to Thunberg as lead researcher and supervisor – a similar protocol still often prevails to this day in some sciences, although much less frequently in taxonomy. As a result, these dissertation publications contributed enormously to Thunberg's tally of publications – estimated at about 400 – as well as to the tally of plant names that he is considered to have authored.[32]

The pamphlets seem to have been published separately as well as bound together in groups. A volume entitled *Nova genera plantarum* ('New plant genera') now at the Natural History Museum in London contains dissertations from 16 different students that were defended over 20 years between November 1781 and December 1801. In the strelitzia section of a listing prepared for a thesis defended on 2 June 1792, Eric Carl Travenfeldt reflected the changing landscape of strelitzia naming at the time – and also what Thunberg considered a further naming enhancement.[33]

## ANOTHER ROYAL COMPLIMENT

Travenfeldt seems to have provided the first, fuller scientific description of the plant which had been referred to as *Heliconia alba* by Linnaeus jnr in 1782 and effectively left as such by the hint, rather than explicit statement, in the 1789 *Hortus kewensis*. Travenfeldt followed the 1789 *Hortus kewensis* updating of the genus name from *Heliconia* to *Strelitzia* but did not simply update the Cape species to give *Strelitzia alba*, as Masson and others seem to have expected.[34] Instead, in a move probably guided by his supervisor Thunberg, Travenfeldt changed the species epithet and called the plant *Strelitzia augusta*. Neither Travenfeldt nor, later, Thunberg in his own *Prodromus plantarum capensis* and *Flora capensis* mention any reason for choosing to use *augusta* as the epithet but its dual connotations of stately looks and a nod towards Banks's choice for *Strelitzia reginae* to honour royalty probably appealed. Thunberg seems to have mirrored a botanical dynasty with a royal dynasty, honouring Princess Augusta, the patroness of Kew and mother-in-law of Queen Charlotte with a diplomatic scientific compliment, rather than honouring the Swedish King Gustav III, to whom Thunberg had devoted a fulsome eight-page dedication in the original Swedish edition of his account of his travels, *Resa uti Europa, Africa, Asia, forråttad åren 1770-1779*.

Perhaps ominously for Thunberg's coining of *Strelitzia augusta*, the name *Strelitzia alba* lingered on. Even in 1806, nearly 15 years after Travenfeldt's thesis, it was used in both the *Botanical Register* and in Sir John Barrow's account of his own southern African travels. By the time the plant was listed as *Strelitzia augusta* in the 1811 second edition of the updated *Hortus kewensis*, Banks at least, though, must have conceded that he had lost the naming race in this instance and accepted Thunberg's preferred name. In addition, the name *Strelitzia augusta* had already been supported by German taxonomist Carl von Willdenow (1765-1812) when he updated the fifth edition of Linnaeus's *Supplementum plantarum* at the end of the 18th century. The plant was later featured as *Strelitzia augusta* in *Curtis's Botanical Magazine* in 1845.[35] Mending an international split between botanists in the latter part of the 20th century had to be mended before the competing claims of prior publication and usage were resolved and the name *Strelitzia alba* was widely accepted.

Some of this name-trading process is reflected by the *Strelitzia alba* specimen at the Bergius Herbarium. The back of the herbarium sheet is annotated: 'Swartz scripsit [wrote] *Strelitzia alba*; Thunberg scripsit *Heliconia alba*'. This suggests that the specimen was both named and renamed by Thunberg and Swartz before Travenfeldt's work. Olof Peter Swartz (1760-1818), the first Professor Bergianus appointed under the brothers' will and ex officio director of the Bergius Foundation, had worked on his doctorate under Linnaeus jnr, graduating in 1781. Swartz travelled to Lapland, North America and the West Indies between 1783 and 1787, amassing a collection of more than 6 000 specimens, and worked in London. Swartz became Professor Bergianus in 1791, the year before Travenfeldt's dissertation was published. This suggests that Swartz updated the name to *Strelitzia alba* during that year but did not return to it after Travenfeldt's renaming.[37]

Swartz's successor was Johan Emanuel Wikström (1789-1856), a doctor who had already been spending a considerable amount of time helping organise the Bergius collections and who had begun pursuing his passion for botany full time.[38] Wikström was initially appointed director of the horticultural school and of the Academy of Sciences collections, now at the Swedish Museum of Natural History. From 1821, as the academy's head of botany, he was required to produce an annual survey reporting recent 'botanical works and discoveries'. In 1823, five years after being appointed director, he also received the title of professor. Wikström is known for having added valuable annotations to specimens by Thunberg, Sparrman and Osbeck in his 'easily read hand' to collections at both the Bergius Herbarium and the Swedish Royal Academy of Sciences. At some point, Wikström also annotated Thunberg's *Strelitzia alba* specimen, adding Travenfeldt's updated name *Strelitzia augusta*, which the specimen still carries in the Bergius Herbarium records.[39]

### *Strelitzia alba* synonyms[36]

| DATE | NAME | ORIGINATED |
| --- | --- | --- |
| 1781 | *Heliconia alba* | Braunschweig: Carl Linnaeus jnr, *Supplementum plantarum* |
| 1792 | *Strelitzia augusta* | Uppsala: C.P. Thunberg & Travenfeldt, *Nova genera plantarum*, part 7 |
| 1839 | *Strelitzia angusta* [considered a spelling variant] | Weimar: D. Dietrich, *Synopsis Plantarum seu enumeratio systematica plantarum*, vol.1 |
| 1959 (published 1960) | *Strelitzia alba* alba subsp. *augusta* (Thunberg) | Paris: R. Maire & M. Weiller, *Flore de l'Afrique du Nord*, no. 6 |

**Richard Salisbury included a 'Heliconia augusta' in his 1796 list of the plants in his Yorkshire garden but the text makes clear that he confused this with a banana plant from Jamaica.**

Franz Bauer, *Strelitzia alba* from *Strelitzia depicta*, 1818

Travenfeldt, the author of the name *Strelitzia augusta*, is not mentioned on the Bergius specimen sheet, however. When *Strelitzia augusta* was eventually featured in *Curtis's Botanical Magazine* in 1845, the work that Travenfeldt and Thunberg had done for *Genera nova plantarum* was overlooked and only Thunberg's sole authored publications, the *Prodromus* and the *Flora capensis*, were cited. But Travenfeldt's work was eventually picked up in the early 20th century on the other side of the world by a American taxonomist, Homer Collar Skeels (1873-1934).

## THE BROTHER NATURALISTS

Whether known by Thunberg's name of *Strelitzia augusta* or Skeels's *Strelitzia alba*, specimens of this plant rarely crossed botanists' benches in herbaria. Oldenburg's 1772 specimen at the Natural History Museum in London is intriguing but, generally, early specimens are scarce, both in Kew Herbarium and in South Africa, for instance. No flood of specimens followed the early ones collected by Thunberg and Masson. This is only partly because specimens from this lofty, treelike strelitzia are more difficult to collect, their fibrous bulk making them difficult to process in the field and at base camp to produce a good herbarium specimen. In addition, several specimens assumed to be *Strelitzia augusta* have been reidentified since the mid-20th century as the related species *Strelitzia nicolai*, which occurs further east (see chapter 3), or occasionally *Strelitzia caudata*, which occurs further north (see chapter 9). The realisation dawned in the 1940s that *Strelitzia alba* has a limited natural range along the south-eastern Cape coast between George and Humansdorp. This emphasises how fortunate early plant hunters had been to find *Strelitzia alba* and makes specimens in the Kew Herbarium from the early 19th century even more interesting.

Kew's earliest dated specimens are both attributed to the German Drège brothers. Carl Friedrich Drège (1791-1867), an apothecary who came to work at the Cape in 1821, became so excited by the region's flora and fauna that within six months of arriving he began making an income by sending bulbs, seeds, birds and skins back to Europe. He lured his brothers, horticulturist Johann Frantz Drège (1794-1881) and watchmaker Wilhelm Eduard Drège, out to the Cape in 1826. Frantz spent about seven years collecting plant specimens while Carl collected zoological and ethnological specimens.[40]

Although it was believed that Carl Drège did 'not appear to have collected herbarium specimens', two sets of specimens at Kew are marked 'Drège 1840', well after Frantz Drège returned to Europe in 1834. Further datestamps show that one set of these specimens reached Kew through the Bentham Herbarium in 1854 and the other through the Hooker Herbarium in 1867. The annotations are slightly different on the various specimens. One of the Hooker Herbarium specimens shows a flower and a leaf, the other a flowerhead and two separate flowers. The label on the first of these reads, '*S. augusta* Thunb. (plus annotation *Strel. alba* Ait a.), Drège, Cape, *Flora capensis*'. The other is similar except that the location is given as 'S. Africa'. The two Bentham Herbarium specimens consist of a leaf and a flowerhead, two flowers and six flower parts. Both sheets are marked 'Drège, 1840', and '*S. augusta*, Thunb.' in the same handwriting as the Hooker specimens. They have an additional annotation (in another hand): '*Strelitzia alba, Afr. austr* [southern Africa]'. Kew also has a specimen of the flower in its carpological collection, marked 'Drège 1834'. This could feasibly be from Frantz Drège but it also raises the question of whether some of these early herbarium specimens might show the date they were acquired rather than the date they were collected. Another curiosity in Kew's carpological collection is a flowerhead, seed pod and seed from a *Strelitzia alba* (although still classified as *Strelitzia augusta*) from a plant flowering in the public gardens in Cairo, Egypt, in October 1914 and supplied by British botanist and agronomist G. St. Clair Feilden.[41] Originally, the plant had probably reached Egypt either from Kew or from the Jardin des Plantes in Paris.

The arrival of the Drège specimens at Kew might have stimulated new interest in the plants as *Strelitzia augusta*, as it was then generally known, was not featured in *Curtis's Botanical Magazine* until 1845 – more than half a century after Thunberg and Travenfeldt had renamed it and two decades after Sir William Hooker had taken over *Curtis's Botanical Magazine*. The plant might not have been a priority because it remained a rarity in Europe, as Hooker pointed out in the text accompanying the two plates. It arrived in France shortly before 1814, Baron F. du Mont de Courset (d.1836) noted in his journal, *Le botaniste cultivateur*. According to Pierre-Aimé Lair (1772-1839), secretary of the Société Royal d'Agriculture et de Commerce de Caen, the baron had a renowned garden at Courset, outside Boulogne in Normandy. It included greenhouses containing nearly 7 000 pots and a botanic garden arranged in accordance with Jussieu's system of botanical naming.[42] It is likely, however, that Du Mont de Courset was simply aware of *Strelitzia augusta* rather than having one: if describing his own plant, it would have been natural to mention this in his reference in *Le botaniste cultivateur*.

There was also a plant growing in Liverpool, as shown by a cultivated specimen from 1838, now in the Kew Herbarium. This is filed together with a specimen from Kew Gardens itself taken in 1861.[43] By 1855, *Strelitzia augusta* was also noted as occurring at the Breslau University botanical gardens in a report in *Flora* by Professor Heinrich Göppert, (1800-1884), the university's professor of botany and director of the botanic gardens. Few could meet the challenge of caring for the plant, Hooker acknowledged: 'Not indeed that it is difficult of increase, for it sends out offsets frequently, but it requires the heat of a stove, and more space than cultivators can generally afford to give it.'[44]

The plant's trunk is typically about 18 feet (about 5,5 metres) high, Hooker's article noted, and measuring to the tip of its leaves, the specimen at Kew stood 23 feet (about seven

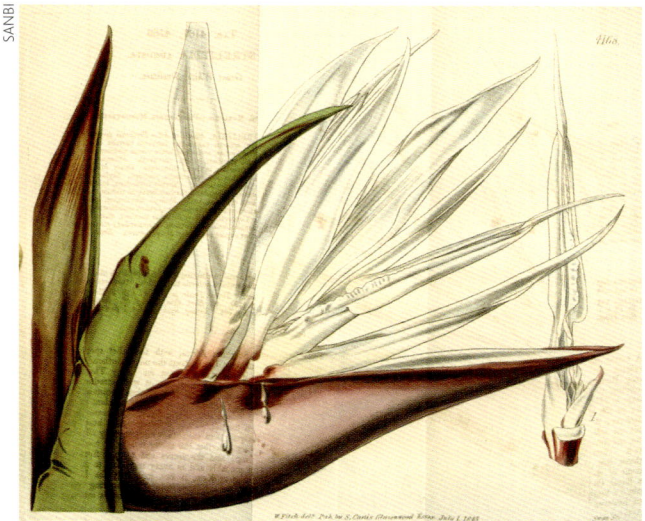

metres) high. Interestingly, a fairly young plant with a much shorter trunk was chosen as the specimen for the illustration, probably because this would be easier for the artist to examine and was generally easier to manoeuvre into a studio. The artist for these plates was the botanical art sensation Walter Hood Fitch (1817-1892), whom Hooker had discovered in 1834 and who moved with Hooker to Kew in 1841. Fitch's output was huge: he created 200 botanical illustrations in the same year as his *Strelitzia augusta* plates and published more than 10 000 illustrations over his career. The following year, Belgian nurseryman Louis van Houtte (1810-1876) launched his magazine, *Flore des Serres et des Jardins de l'Europe*, aimed at interesting wealthy European garden enthusiasts in ordering the exotic plants featured for their gardens and greenhouses and promising '*Descriptions et figures des plantes les plus rares et les plus méritantes* [Descriptions and illustrations of the rarest and most worthy plants]'.[45] Although the magazine also became a showcase for plants brought back by van Houtte's own collectors, he also followed botanical and gardening trends and so featured *Strelitzia augusta* the year after Hooker had included it in *Curtis's Botanical Magazine*, reproducing the illustration and quoting part of Hooker's text. Charles Antoine Lemaire (1800-1871), a former classics professor at the Université de France, enthusiastic botanist and experienced horticultural writer, prepared some of the accompanying text for *Flores des Serres* and commented how different *Strelitzia augusta* appears from other members of its family:

> *par ses proportions grandioses et majestueuses qui dépassent celles des plus hauts bananiers, plantes dont elle rappelle assez bien le port. Les feuilles . . . sont amples et plus robustes que celles des Bananiers. Ce simple aperçu peut donner une idée des nobles dimensions de la plante en question.*[46]

[its grandiose and majestic proportions outstrip the tallest banana trees, though its shape is similar. The leaves . . . are full and more robust than those of banana trees. This simple glimpse gives an idea of the noble dimensions of the plant in question.]

Van Houtte added his own admiring note:

> *Les espèces de* Strelitzia *sont un des principaux ornements de nos serres chaudes, et même de nos serres tempérées, où elles se plaisent assez volontiers. . . . Rien n'égale leur pittoresque aspect, au bord d'un ruisseau factice, dans un jardin d'hiver, quand elles sont mêlées aux* Canna, *aux* Thalia, *aux* Caladium, *aux* Crinum, *etc, toutes plantes qui ne sont jamais plus belles et plus vigoureuses que dans une telle situation.*
>
> *. . . C'est une de ces plantes dont l'acquisition est imposée à tout collecteur un peu sérieux.*[47]

[The *Strelitzia* species are one of the principal ornaments of our hothouses, and even our temperate houses, where they settle fairly happily . . . Nothing equals their picturesque looks, beside an artificial river, in a winter garden, when they are mixed with *Canna, Thalia, Caladium, Crinum,* etc, all plants which are never more beautiful and more vigorous than in such a situation.

. . . It is one of those plants whose acquisition is a must for any even slightly serious collector.]

## REVIVING *STRELITZIA ALBA*

It would be more than half a century before the American taxonomist Skeels revised the name of the plant that van Houtte knew as *Strelitzia augusta* by rigorously applying the newer rule of giving priority to older, valid plant names. Skeels was a member of the 15-strong department of foreign seed and plant introduction at the US Bureau of Plant Industry, part of the US Department of Agriculture. The bureau published a wide range of regular bulletins between 1901 and 1913, including quarterly listings of seeds and plants imported into the country with introductory information on species that could be used in botanical, horticultural and agricultural development. The director at the time was David Fairchild (1869-1954), who was also a plant hunter for the Department of Agriculture, introducing the Japanese flowering cherry tree to Washington, and after whom the Fairchild Tropical Botanic Gardens was named.[48]

Travenfeldt's research work had been unacknowledged apart from his dissertation and his name was not mentioned on the Bergius specimen sheet. However, Skeels still picked it up. Enthusiasm and sympathy for the young Travenfeldt, particularly given Skeels's own more junior position as one of the bureau's two scientific assistants, may have prompted him to attribute to Travenfeldt the name *Strelitzia augusta* rather than, more conventionally, to Thunberg. In 1911, Skeels's naming research

Walter Hood Fitch, *Strelitzia augusta* in *Curtis's Botanical Magazine*, 1845

was limited to cross-referencing available literature and he probably had no way of accessing the Bergius herbarium sheet which Swartz had considered to be *Strelitzia alba* about a century earlier. In addition, the emerging rules of plant naming that Skeels was so keen to apply insisted that for a name to be validly published, it must be in a form that would be widely available to key herbaria – a note on a herbarium sheet can be considered botanical history but does not count as publication. As a result, Oldenburg's 1772 *Heliconia albiflora* also remains a fascinating footnote unless published corroboration comes to light.

Skeels acknowledged that he was dealing with 'a well-known South African plant' but was clearly keen to follow the American preference at the time for the oldest legitimate and valid name taking precedence over any established, if later name. Accordingly, he adjusted the name to *Strelitzia alba*, combining the newer, more accurate name for the genus with the descriptive species epithet applied by Linnaeus jnr, though deliberately varied by Travenfeldt and Thunberg. Skeels effectively swept away the royal connotations of *Strelitzia augusta*, replacing the name with the more prosaic white strelitzia, *Strelitzia alba*, and commenting:

> This beautiful white-flowered plant belonging to the family Musaceae was firs[t] named in 1781 by Linnaeus's son as a species of Heliconia, with the specific name *alba*. Travenfeldt, in 1792, transferred the species to the genus Strelitzia, where it is generally considered to belong, but gave it a new specific name, *augusta*. The binomial *Strelitzia alba*, although the proper name of this plant, according to recognized nomenclatural practice, appears never to have been used heretofore.[49]

Aiton had pointed the way to these deductions by grouping *Heliconia bihai* and its white-flowered relative together in the 1789 *Hortus kewensis*. Skeels does not reference this so it is not clear whether he consulted this source. If he did, he gave it no standing in his detailed interpretation of contemporary botanical naming rules, citing only Linnaeus jnr and Travenfeldt. Until Skeels, nobody had explicitly regularised the plant's name in line with this precedent in a context that would qualify as acceptable publication. It is, of course, unlikely that Travenfeldt had access to sufficient sources to be aware that he was formally adopting the name that had been anticipated by Masson, Andrews and Swartz. This meant that in terms of botanical naming, Skeels took the honour of being recognised as the formal author of the new plant name, *Strelitzia alba*. In the bureau's publication and in perpetuity, his own surname was attached to the plant alongside Linnaeus jnr: *Strelitzia alba* (L.f.) Skeels. The split in botanical taxonomy between the supporters of the rule of precedent and the supporters of the rule of usage meant that it was well into the 20th century before Skeels's renaming was widely accepted.

Skeels's eye seems to have focused on naming rather than broader botanical detail, given that he wrote that *Strelitzia alba* occurs 'from Durban to the Cape of Good Hope'. In fact, *Strelitzia*

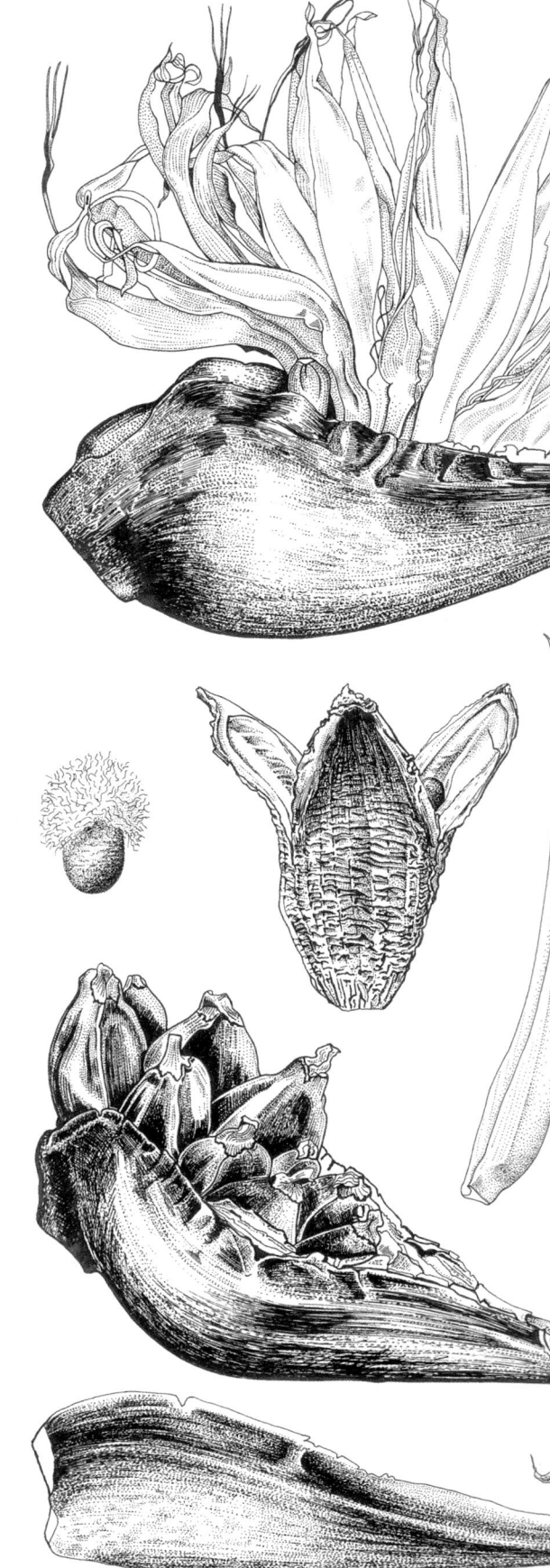

Durban Botanic Gardens Trust

*alba* has one of the most restricted ranges of all the plants in the strelitzia family. As a result, Skeels still perpetuated the confusion with *Strelitzia nicolai* that a German botanist working in St Petersburg in Russia had hoped to resolve about 65 years earlier (see chapter 3).

## THE ELUSIVE FOREST PRINCESS

It was a shame that naturalists in the homeland of *Strelitzia augusta* had not given more attention to the plant, wrote Charles Lemaire in *Flore des Serres et des Jardins de L'Europe* in 1845.[50] Today, though, it is clear that this was probably less an oversight than a lack of opportunity.

Looking through the *Strelitzia alba* files at the SANBI National Herbarium in Pretoria takes researchers into the attempts to document the ecology of the south-eastern Cape from the early 20th century onwards. The files also ultimately emphasise that despite igniting occasional spates of international taxonomic debate, *Strelitzia alba* was and remains elusive in the wild – so much so that in 1926, a young South African assistant research officer based at Diepwalle in the Knysna forests, John Phillips, was proud to announce to his superior, the conservator of forests for the Cape midlands, that several *Strelitzia alba* had been found 14 kilometres west of Plettenberg Bay in the direction of Knysna:

> I know you will be interested to learn that Forester Stevens of Harkerville has located several isolated specimens of *Strelitzia augusta* ("Wild Banana", "Wilde Pisang") as far West as Section 1926-27, Harkerville . . .
>
> Between Harkerville and Kaffirkop, along the "Bosrivier" there are a number of *Strelitzia augusta*, while they also occur at Kaffirdrift, Kaffirkop.[51]

Phillips had personally 'inspected' one of the Harkerville specimens and found it 'an excellently grown plant, its topmost foliage about 30 feet [10 metres] above the ground.' He particularly noted how strelitzia fitted into the forest succession, 'follow[ing] *Hygrophilous Macchia*, forming finally more or less pure consocies, which later are ousted by such spp. [species] as Myrsine melanophleos, Ilex capensis, Sparmannia and Brachylaena neriifolia.'[52]

Also in 1926, German naturalist Herbert Lang (d.1957) took photographs believed to be of *Strelitzia alba* near Willowmore (almost due north of Plettenberg Bay) or Steytlerville, 87 kilometres north-north-west inland from Humansdorp. These photographs appear to have been acquired by the Transvaal Museum in 1957 and passed to the National Herbarium in Pretoria. The next item in the same file at the SANBI Herbarium today is a botanical treasure map. This hand-drawn sketch shows the way to find prize specimens of *Strelitzia alba* and the colourful bulb *Cyrtanthus montanus* in the Baviaanskloof Forest Reserve.

Angela Beaumont, *Strelitzia alba*, pen and ink, 2010

From the 1940s onwards, SANBI's files point to the emphasis of work turning to questions of plant identity. Reassessing the treelike strelitzias was a key interest of Robert Dyer (1900-1987), director at the Botanical Research Institute (BRI). Originally from Pietermaritzburg in then Natal, Dyer was based at the Albany Museum in Grahamstown for eight years from 1925. He was compiling a detailed survey of the vegetation of the Albany and Bathurst districts but was obviously intrigued by the pattern of the distribution of *Strelitzia alba* emerging from reports from field officers and collectors. More photographs, illustrations and specimens were sent to the Botanical Research Institute in Pretoria in 1943 by J.H. Keet, district forest officer at Storms River, west of Port Elizabeth. Keet had found *Strelitzia alba* in 'the Groot River Nature Reserve, which is a portion of the Salt River Demarcated Forest in the Division of Knysna'. Keet also posted Dyer specimens of three flowerheads of *Strelitzia alba* and particularly highlighted in his covering letter the difficulties of obtaining good, comprehensive strelitzia specimens: 'I did not have any implements with which to collect leaves and suckers but I will do so when I collect the fruits, probably next month.'[53]

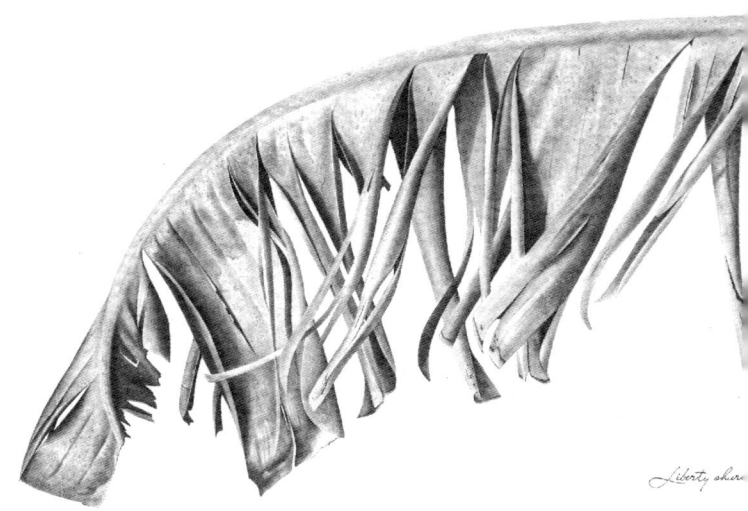

In the meantime, Keet included a precise description of the strelitzia's leaves:

> The blade of the mature leaf is about 6 feet [1.8 metres] long by 18" [inches] to 2 feet [about 45 to 60 centimetres] broad, oblong [in] shape. The juvenile leaves are smaller and lanceolate. Exposed leaves are ribboned by the wind.[54]

Keet did collect specimens later that month and these were used by Edith K. Burges, the BRI's botanical artist at the time, to prepare the illustration to accompany an account of the species in *Flowering Plants of Africa*. Some of these specimens also found their way to the Kew Herbarium. These two sets show the trouble that Keet took to prepare and display a good range of plant parts from what would be large and unwieldy specimens. The first set of three sheets shows a section of a leaf and a flowerhead; two flowerheads and a cross-section of the trunk; and two more flowerheads. The second set of five sheets shows a single flowerhead; a section of leaf by itself; a section of a leaf with two flowers; another single flowerhead; and a section of a leaf with a cross-section of trunk.[55]

Keet collected the specimens just off the main road from Knysna to Humansdorp that ran along the southern slopes of the kloof, at an altitude of about 400 feet (about 120 metres), but his description suggests the plants beyond were relatively inaccessible:

> The *Strelitzia* grows in a deep kloof [gorge] containing a tributary of the Groot River . . . These slopes have a northern aspect and are hot and dry. One or two *Strelitzia* are to be found adjacent to the road within the fringe of indigenous forest which clothes the bottom of the kloof. More are found in the high forest at the bottom of the kloof and the lower portions of the northern side. The northern side of the kloof is very steep, almost vertical, with earthy ledges. On these ledges the biggest *Strelitzia* are to be found but still associated with such forest trees as *Virgilia*, *Podocarpus*, *Sparmannia*, *Pterocelastrus*, etc. Higher up, the cliff slopes away to the north and above the forest, especially in a small side gully, the *Strelitzia* occurs in pure communities in the typical fynbos of this area.[56]

There was a clear difference in height between plants in the two habitats, Keet noted:

> Within the forest the *Strelitzia* is palmlike bearing a tuft of leaves at the top of a clean trunk up to 30 feet high and 3 inches to 4 inches in diameter. The stem has numerous leaf scars. Outside the forest the plants are up to 10 feet in height and the leaf stems extend right down to the ground.[57]

This part of the Knysna forests was the only site where Keet knew of strelitzias growing: 'I have not seen *Strelitzia* anywhere else in this area. Dr Fou[r]cade may know of other localities.' In his retirement, former land surveyor and forest officer Henri Fourcade (1865-1948) had spent several years from 1920 carrying out the Botanical Survey of South Africa for the George, Knysna, Humansdorp and Uniondale districts, with his checklist of flora for the area eventually published in 1941.[58] In 1944, just four years before he died aged 79, Fourcade was still active and sent more photographs, illustrations and specimens to join the body of evidence that supported the new insights into the treelike strelitzias that Dyer would publish two years later in 1946. It was the beginning of Dyer's pivotal reassessment of the treelike strelitzias of southern Africa.

Liberty Shuro, Strelitzia leaf, pencil, 2017

## INTERNATIONAL NAMING BATTLES

Naming a plant or any other natural species is not a one-time, definitive event. It is a process that reflects developing scientific understanding. As a result, a plant's name should now be changed only to reflect:

- Discovery of more of its family members and align their naming
- Fresh understanding of the plant or its relationships thanks to new techniques of analysis, such as DNA testing and cell photomicroscopy
- Discovery of an old, validly published name that has been overlooked

Now and in the past, though, naming and classifying a plant can spark energetic and even vituperative debates. When Carl Linnaeus snr (1707-78) published his contribution to the species classification debate *Systema naturae* in 1735, he was one of several scientists trying at the time to develop more practical and workable ways of naming and classifying natural organisms. Not all scientists, or even all botanists, agreed that Linnaeus's classification system was either correctly based or straightforward to use.

Linnaeus advocated a shorter, double-barrelled naming system in Latin, a language at that time more broadly comprehensible to scientists. But despite his best intentions, it has been drily observed, 'Stability was not a hallmark of the early implementation of the binomial naming system.' The straightforward internal logic of Linnaeus's system was intended to be easy to use, offer an international approach that all scientists could understand and be flexible enough to be adapted with deepening scientific understanding of relationships at all levels between different plants and different plant groupings.[59]

Linnaeus himself found he needed to update names which he had already coined according to his new system. Within a few decades, Linnaeus's original system was already creaking under the volume of scientific naming, the expectations of the process and the huge quantities of plants that collectors such as Thunberg and Bergius were introducing to science.

By 1798, just six years after Travenfeldt included the name *Strelitzia augusta* in his thesis for Thunberg and about half a century after Linnaeus had introduced his new scientific naming system, French zoologist Jean Baptiste de Lamarck (1744-1829) was lamenting not the system itself but the lack of rules for implementing it, leading to confusion over which name to use. Lamarck even 'railed against Linnaeus's own name changes'. He was not alone among scientists in feeling they needed to call a halt, or at least slow down, what seemed to be massively proliferating numbers of species. Those who believed the system is overcomplicated have become known as 'lumpers'. They prefer species and even group definitions based on a wider range of characteristics. 'Splitters', on the other hand, argue that greater clarity is achieved by creating greater numbers of species and groups based on small differences.[60]

By the time Thunberg was publishing volumes of *Flora capensis*, the renaming debate had grown so loud that Swiss botanist Augustin Pyramus de Candolle (1778-1841) proposed prioritising the oldest name over later ones to resolve the issue of which name should take precedence. But from the time that de Candolle's son Alphonse (1806-1893) codified this and other proposals to regulate naming ahead of the 1867 International Botanical Conference in Paris, such an apparently simple solution caused a widening split among the dominant botanical world of the northern hemisphere, generally along east-west lines.

Broadly, most Europeans 'were more pragmatic and advocated stability of usage over strict applications of the rules'. They felt that the American preference for rigidly applying the priority principle, overruling well-known and generally accepted more recent names to reinstate an older name that might suddenly be found in the literature – as Skeels did to coin *Strelitzia alba* – would create unnecessary confusion not only in the botanical world but in related areas such as horticulture.[61]

Opposition between the two approaches was eventually resolved by setting up the International Plant Naming Index to oversee plant naming, approving new names and adjudicating disagreement. It compiles the work of various other institutions, particularly the Royal Botanic Gardens, Kew, Harvard University Herbarium and the Australian National Herbarium, to create a database of scientifically accepted names. By the 21st century, this showed just five accepted strelitzia species – *Strelitzia alba*, *Strelitzia caudata*, *Strelitzia juncea*, *Strelitzia nicolai* and *Strelitzia reginae*. Also accepted are the two subspecies of *Strelitzia reginae*: *mzimvubuensis* and *reginae*.[62]

The strelitzia family is quite small compared to all the hundreds of ericas, for example, but hand-written amendments or new labels on dried plants preserved in herbaria around the world show that strelitzia names have not been immune from botany's long-running debate as professionals again and again try to clarify the answer to the question, 'When is a species not a species?' When discussing their research on strelitzia evolution in 2012, scientists found it necessary to clarify which phylogenetic species concept they were using and to quote the definition two decades earlier by ornithologist Joel Cracraft from the American Museum of Natural History that a species is 'the smallest diagnosable cluster of individuals within which there is a parental pattern of ancestry and descent.'[63]

# A KING'S GIFT TO AN EMPRESS
*Strelitzia nicolai*

CHAPTER THREE

# 3

# A KING'S GIFT TO AN EMPRESS: *Strelitzia nicolai*

In August 1795, a floating garden of plant treasures reached St Petersburg in Russia. This gift from King George III's own precious Kew garden was a diplomatic offering to a fellow monarch, Catherine the Great, an empress with a passion for plants and who had told French philosopher Voltaire in 1772: *'J'aime à la folie présentement les jardins à l'anglaise . . . en un mot l'anglomanie domine dans ma plantomanie* [I currently love English-style gardens madly . . . in a word anglomania dominates my plantmania]'.[1] At the heart of the network of people involved in ensuring that this selection of Kew's most prestigious plants arrived safely was Sir Joseph Banks, often imperious, sometimes peevish as he tried to direct and dictate the enterprise from Kew, across the Baltic via Elsinore in Denmark to St Petersburg and even inside the imperial greenhouses that would be the plants' final home.

The list of plants supplied stretches over five pages in the first Kew record book.[2] To catch the eye placed at the top of the list of the more than 200 species of plants and to underline the royal provenance of the gift was Banks's pride, his tribute to Queen Charlotte, *Strelitzia reginae*. But concealed within the collection there may have been an unexpected gem which had not yet been recognised by science. Much later, the familiar wild banana of the eastern coast of southern Africa was formally named *Strelitzia nicolai* after it was pulled from obscurity by a sharp-eyed German botanist overseeing the St Petersburg gardens founded by Catherine II.

King George III's gift of plants to the Grand Duchess of All Russias, as Catherine II was then generally known, was neither a surprise nor unsolicited. Rare and exotic plants did not long remain the exclusive preserve of botanists and taxonomists. They had become part of the dance of diplomacy, prized as symbols of wealth, power and territorial reach across the world. Documents that tell the story of this particular plant gift had been scattered between libraries across the world and were finally pieced together by Banks's biographer, Humphrey B. Carter, nearly 200 years after the original gift was made.[3]

English merchants had been trading in Russia since at least the 16th century. In 1697, Peter the Great (1672-1721) had travelled mostly incognito in Europe to observe the latest European sociopolitical trends to help him reform and modernise Russia. He saw landscaping of parks contributing as significantly to beautifying his capital, St Petersburg, as artists, craftsmen and furniture designers. To make his Summer Gardens outshine Versailles, 'he ordered peonies and citrus trees from Persia, ornamental fish from the Middle East, even singing birds from India'. In 1714, he founded the apothecary's garden which evolved into the St Petersburg Botanic Gardens. Just a year after the Grand Duchess Catherine took power as Her Imperial Majesty The Empress and Autocrat of All Russias in July 1762, she sent an architect and some gardeners to Kew, looking for inspiration for renovations at Peter the Great's Petershof (Petrodvorets) palace and gardens.[4]

A generation later, in late 1793, Catherine II personally requested plants and seeds from Kew for gardens she was making at her palace of Tsarkoe Selo, south of St Petersburg, and her son's nearby country estate Pavlovsk. She channelled the request through Charles Whitworth (1754-1825), who would be created a baron in 1800 for his diplomatic work, particularly in Russia.[5] In November that year, when the British foreign secretary Lord William Grenville (1759-1834) wrote to King George III in Windsor, he included what appears to be a reminder of the request:

> Lord Grenville begs leave also humbly to request to be honoured with your Majesty's commands respecting the letter from Sir Charles Whitworth on the subject of the seeds and plants desired by the Grand Duchess from your Majesty's garden at Kew.[6]

Both protocol and the king's genuine interest in his garden at Kew dictated that the request from Whitworth, his chief diplomat in Russia, should not be channelled direct from the Foreign Office to Banks. The king's reply underlined his personal interest:

> I shall order the seeds wished for the Russian Empress's garden to be collected at Kew and such plants as in the present early state of cultivation can be spared to be sent at the proper season to Petersburgh. Lord Grenville will notify this to Sir Charles Whitworth.[7]

British plant diplomacy was partly trying to heal the breach in the 1750s when Russia had first sided with the coalition of Austria and France against Prussia and Britain in the Seven Years War (1756-1763), although this had changed with the accession of pro-Prussian Peter III in Russia. Grenville would have been aware that Peter III's successor, Catherine II, ran Russian foreign policy herself but the *forces majeures* of international politics intervened in his attempt to build a warmer relationship. Europe's shock at France's 1789 revolution led Prussia and Austria to declare war on France in spring 1792. By early 1793, Britain, the Dutch republic and Spain were also ranged against France and by September 1794, Britain had allied formally with Russia and Austria against France. This probably prompted the British government to revive the idea of the gift of 'as Compleate a Collection of Exotic Plants as can possibly be Spard' to seal this new alliance with Russia. The king had also commanded – or been prompted by Banks to command – his clerk of works to prepare

Overleaf: George French Angas, *Umvoti Mouth*, 1849

'Plans & Elevations of the Principal Hothouses at Kew' that would also be presented to Catherine II as a guideline. Catherine II's declaration of devotion to English-style gardening had been no throwaway line. Over the decades, a series of British gardeners were imported to Russia to reproduce her preferred gardening fashion.[8]

## THE STRELITZIA'S ROLE IN INTERNATIONAL DIPLOMACY

The overlap of international politics and botanical aims was nothing new for Banks so the request from British diplomats, channelled through the king, was not unusual. It did, however, present particular challenges. Safety during times of international war, for instance, meant that any ship transporting the strelitzia and other plants would have to be part of the Baltic convoy. This usually left only once a fortnight and was coordinated between the Russia Company and the admiralty. In addition, the winter of 1794/5 was 'one of the most severe ever experienced in Britain since records of any kind have been kept or weather conditions in any way noted systematically' so plants were endangered, whether under cover in hothouses or not. As well as coping with such weather and a recurrence of his gout, Banks was working on projects involving China, New South Wales, West Africa and Jamaica as well as running his Lincolnshire estate, Revesby.[9]

When the winter finally broke, Banks met King George III at Kew on 4 April 1795. It is believed they discussed details of the gift to Catherine II. Banks was then put in direct contact with Grenville's deputy, Sir James Bland Burges (1752-1824), undersecretary for foreign affairs. They met on 4 May and Burges promptly found a ship ready to transport the plants by the middle of that month.[10] Banks, though, would not be rushed and was adamant that this would be a mistake in terms of the seasonal climate:

> I by no means advise their being sent to Sea till midsummer at the soonest
>
> if you recollect that we [in] England Shall not for a month to Come venture to expose our most Hardy green house plants to the open air, you will probably agree with me that hot house Plants which are the principal Object of the Grand duchesses request, cannot safely be ventured on the Baltic at this Season of the year unless perhaps the Captain Consents to fit up his Cabbin with Flues for the purpose of Protecting them . . .[11]

Even after further thought, the earliest Banks could advise for the shipment was between 24 June and 24 July: 'before midsummer we are liable to Frost in the nights in this Climate & after the end of August winter approaches Fast in the high Latitude of St Petersburgh'.[12] Plants from the tropics and subtropics, such as the *Strelitzia reginae*, would be desirable additions to Catherine II's collections only as long as they could be helped to survive the northern frosts.

That same day, Burges replied acknowledging that 'I know nothing personally' of the skills of the ship's captain earmarked for the task and 'on the contrary suppose that he is entirely unskilled in the management of Plants'. In fact, he recommended that Banks should appoint an 'experienced' gardener to accompany the plants and ensure they arrived with Catherine II in the best condition possible.[13] Banks's response suggests that he wished Burges to know that this was already taken care of:

> His Majesty . . . has ordered one of his own Gardiners to Proceed to St Petersburgh in the Ship which will Carry them [the plants], who will be instructed to take charge of them during the voyage & to give such information Respecting the English mode of Culture as the Grand Duchesses Gardiners may wish to Receive from him.[14]

*Strelitzia nicolai* in *Flore des Serres et des Jardins de l'Europe* magazine, 1858

## BANKS'S ESTIMATE OF COSTS[16]

| ITEM | AMOUNT (£ s d) |
|---|---|
| A Gardiner who should be taken from the Kings Establishment at Kew next week to Carry messages & Prepare himself &c at one guinea a week Say for 6 months | 27 16 0 |
| For his Extra Maintenance while at Petersburgh where he must appear like a Gentleman Say 6 weeks at a guinea a week | 6 6 0 |
| Gratuity for his cloth[e]s | 30 0 0 |
| For his Passage out & home & his Maintenance while on board | 21 0 0 |
| Gratuity to a man or boy to assist in watering Plants Removing Pots &c on board | 2 2 0 |
| For freight of Plants which out to occupy the whole of the great Cabbin | 52 10 0 |
| Cost of a Stage to place them upon in the Great Cabbin to advantage | 20 0 0 |
| Cost of Carrying the Plants from Kew to the Ship with a Mate to the Gardiner at Kew | 5 0 0 |
| Garden Tools for his use on board & when he assists at St Petersburgh | 3 3 0 |
| Gratuity to Mr Brown for making Plans & Elevations of the Hothouses at Kew | 10 10 0 |
| TOTAL | 188 12 0 |

## THE COST OF COSSETTED PLANTS

Expenses were mounting – plants, plans and ultimately the substantial refit of the ship selected for this special delivery. Of most immediate interest for Banks, though, was the opportunity this gave him to implement one of his favourite items of botanical technology to try to ensure the strelitzia and other plants arrived looking as impressive as when they left Kew.

Banks recommended to Burges that rather than stowing the plants in the hold, they should be cossetted in a bespoke plant cabin to protect the plants from storms, frost and the salty sea. In 1789, Banks had already created 'a special plant "coach"' on the *Guardian's* quarterdeck to carry useful food or medicinal plants out to Australia and key plants new to European botany back to Banks and the Kew staff in London. Others had followed for HMS *Discovery* in 1791 and *HMS Reliance* in 1794. Such experiences might have encouraged Burges to approve Banks's costing for the exercise, of which the cost of freighting the plants was by far the highest amount at £52 10 shillings, or about £360 000 today. This ate up more than a quarter of the total estimate of £188 12 shillings, about £1.3 million today.[15]

## FUNDING BANKS'S FLOATING GARDEN

Banks may have suggested that he would lean on the Russia Company to provide sponsorship in kind for the venture, confident in the response of Edward Forster jnr (1765-1840), a keen botanist and in 1800 became one of the first members of London's newly formed Linnean Society. However, Forster was a shrewd businessman, ultimately serving a 23-year term as governor of the Russia Company.[17] He also made clear that he was not expressing simply his own view:

The Russia Company have as a Company no ships nor strictly speaking any fund, they lay a small duty on Importations for defraying necessary expences, & the surplus is the only sum they have the command of, which can only be disposed of by a resolution of the Council of Assistants where the proposition hinted at by S[i]r Joseph Banks might probably meet with a variety of opinions, & might become the subject of a discussion which perhaps it might be better not to hazard.[18]

Banks did not take the refusal well, grumbling to his diary that the Russia Company should understand why they should share the responsibility and reflected glory of the gift.

Banks tried an appeal that proved fruitless to another member of the Russia Company, Thomas Raikes (1741-1813).[19] But Raikes, a London merchant who would become a director of the Bank of England two years later, was able at least to give Banks some valuable advice on planning the logistics:

He said that when he visited Russia he did not Sail till the 10th of June & that he Experienced some very Cold nights in the north sea but that as soon as he passd the Sound the heat was excessive & uninterrupted he informed me that the Ship in which the Plants are embarkd will not proceed father than Cronstadt [Kronstadt, west of St Petersburg] & that the Plants must be then put on board a galliot & sent to Petersburgh about 20 miles a Galliott he says is a Roomy vessel with wide Hatches & plenty of hold but no Cabbin Conveniences

The Grand Duchesses Garden is at Pavelffski 26 verst [27.75 kilometres] from St Petersburg he believes wholly by Land[20]

Unable to persuade the Russia Company to help bear the costs of 'The Present', Banks quibbled that the ship they earmarked for the voyage, the *Venus*, was too small.

It was, he said, really not more than 70 tons when rated 100 tons and with a cabin just '8 Feet 4 by 6 Feet 3' and just five or six feet high, with '2 Stern Lights . . . & one Small Skylight'. Banks also tried asking the admiralty for use of one of the English East India Company ships which had been refitted for naval use. He particularly preferred the *Calcutta*, commanded by Captain William Bligh (1754-1817):[21]

His Experience in the Care of Plants at Sea will be of infinite use to the undertaking, & his name will add not a little to the Compliment, indeed I think it will be Compleat if the Foreman of the Royal Botanic Garden attends the Plants, as is intended, & Capt. Bligh carries them out.[22]

The negotiations that followed Banks and the admiralty were both lengthy and, indeed, ill-tempered. Banks was generally used to prevailing but this time he did not get his way. Eventually, he was forced to accept the *Venus*.[23]

## SELECTING THE SHIPMENT'S STRELITZIA CENTREPIECE

Banks was so confident of the plant stock at Kew that he did not go there to oversee the selection of plants until Monday, 22 June, a week after his plan to visit there had been diverted by the heated debate over the suitability of the *Venus* and 10 weeks after he had first discussed the gift with George III. The whole undertaking was a momentous task for William Townsend Aiton, who was still not yet 30 and had succeeded his father as head gardener at Kew just two years earlier.[24]

Fortunately, all was proceeding well under Aiton jnr. Banks added only the instruction that the plants should be double potted. On Wednesday, 24 June, Banks returned to check on the weight of the pots for shipping. With 103 larger, 12-inch pots weighing about 30 pounds and 184 smaller nine-and-a-half-inch pots weighing about 20 pounds, there was a total weight so far of 6 770 pounds that needed to be loaded for shipment downriver, then offloaded and reloaded onto the *Venus*.[25]

There was little time to play with plant selection as the carpenter was due to work between Thursday and midday Friday, 26 June. He would install platforms three feet wide (0.9 metres) along each long side, flanking a central platform that would be six feet wide (1.8 metres). He was overseen by Banks, Aiton jnr and George Noe (fl.1790s), the Kew foreman who would accompany the gift to St Petersburg and settle the plants into Catherine II's hothouses. Noe enthusiastically approved the platform. By the time the ship was due to sail, Banks was fired with enthusiasm and pride: 'visited the Plants which were partly Stowd the place in which they are appears to answer admirably & the Carpenter has fitted it up with much Economy & neatness'.[26]

Banks planned that the plants would be brought downriver from Kew on the Friday afternoon of 26 June and loaded on board on the Saturday morning so the *Venus* could catch the evening tide to join the Baltic convoy. He irascibly tried to micromanage the process[27] but it was 2 July, nearly a week after his planned loading date, when Aiton jnr sent Banks for approval a copy of the list of plants to be shipped to Catherine II. Aiton tentatively commented:

I believe it is correct . . . I would have sent a neater Copy but shipping the plants & other directions I have been necessitated to attend to today would not allow of it. I have added a few remarks in pencil as they struck me. The Stove plants do not run Alphabetically because I could not refrain to start off with *Strelitzia*[28]

Together with Banks, Aiton jnr had gathered together 226 different species, with about 30 of them coming from Masson's recent collections at the Cape. There were nearly 30 ericas, which would soon be the height of fashion across Europe, and a group of Australian plants was also included.[29] The collection displayed the British king's territorial power as much as representing a selection of the world's rarest and most desirable plants. Aiton jnr noted which plants had been included in the 1789 *Hortus kewensis* or featured in the recently launched *Curtis's Botanical Magazine*. About half a dozen unnamed plants were included for interest value even though they had not been seen in flower and their genus was undetermined. Banks would probably have regretted this oversight had he suspected what would transpire.

## THE MAN FOR THE JOB

Banks's claim that he would second the Kew foreman himself to oversee the delivery of King George III's gift to Empress Catherine II was no idle boast. Banks was backed in his selection of George Noe by Kew's director, Aiton jnr. As a German working in England, Noe had some experience of moving between countries and working in different cultures. He even came from the same part of Prussia as the empress – Württemberg, now in south-west Germany – so they would speak the same kind of German, remarked Banks.[30]

The assignment must have been an improvement that Noe would not have anticipated in the dark days of his early training at Kew. He had been sent by Duke Charles Eugene of Württemberg to improve his horticultural skills under Aiton snr but was left 'destitute' there after the duke's death, when nobody would take responsibility for maintaining him at Kew or repatriating him after 'the rapid succession of Sovereigns which that house has unfortunately in the last two years sustained'.[31] Fortunately, Noe impressed Aiton snr and jnr enough to be kept on and later given the post of Kew foreman. Now he would be responsible for ensuring the strelitzia and other rare plants survived the difficult sea journey not only alive but also in good condition.

By 2 June, well before Banks began finalising the selection of plants, Noe's appointment had been approved by the king, with his pay for the task dating from 31 May. Banks budgeted a £30 gratuity for Noe to buy suitable clothing before he left, then a guinea a week until he arrived in St Petersburg, where his stipend would be doubled to two guineas a week. Noe was involved in the preparations and when the *Venus* was about to sail received a long letter of instructions from Banks, putting precise boundaries on his responsibility. Noe was to take charge of the plants from Aiton as they left the gardens, would report in St Petersburg to the British ambassador, Charles Whitworth, and should retain charge of the plants at least until they had been delivered to Catherine II's agent. Banks included detailed instructions about care of the plants on the voyage. Salt spray was one of the major risks to plants being shipped by sea so Banks expected Noe to lick the plants' leaves to test them for salty deposits.[32] Clean water supplies were another concern:

> When the Ship Comes to an anchor at Elsinor or Elsewhere you are to Remind the Captain to Fill all your Empty water Casks, which he has agreed to do, with the best Fresh water

Durban Botanic Gardens Trust

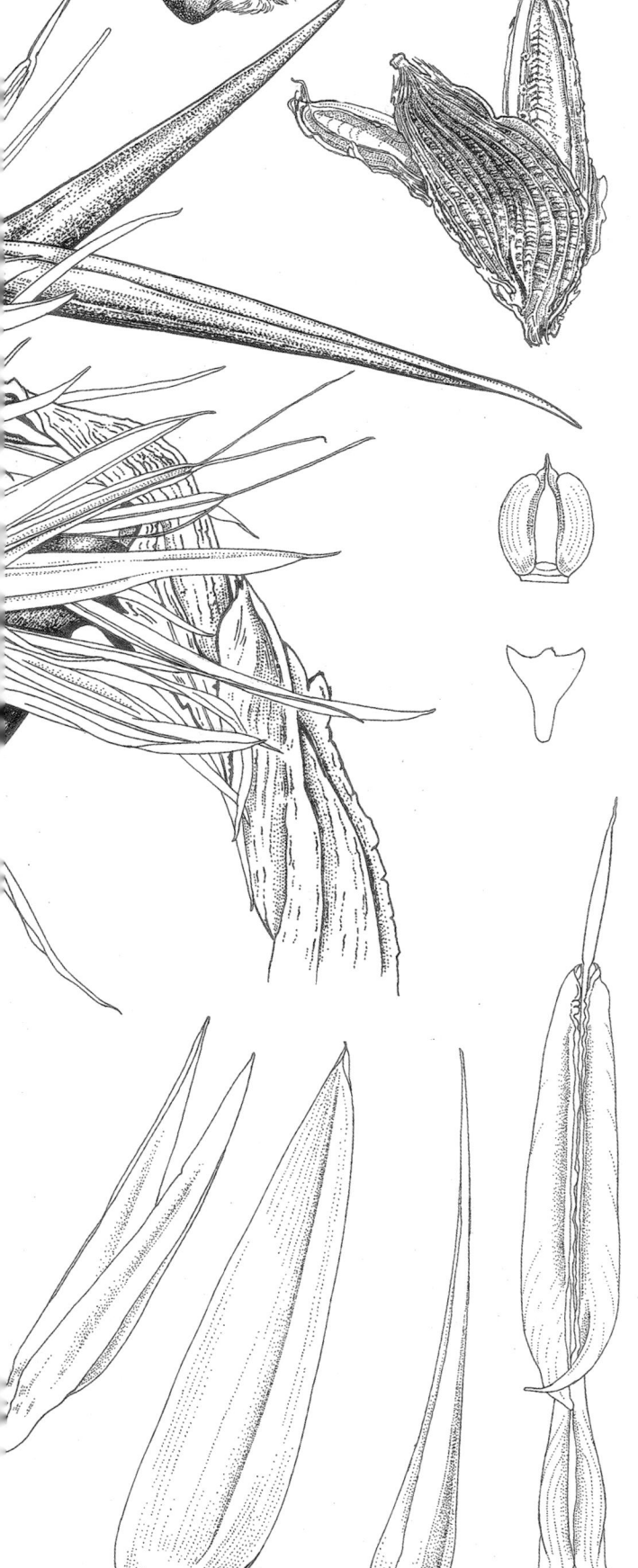

that can be procurd & you are afterwards to use the new water in preference for watering & the Thames water for Sprinkling the Plants[33]

Noe was told to stay on board the *Venus* from the moment the plants were loaded '& by no means to Leave her till her arrival' at St Petersburg. Despite Banks's meticulous attention to horticultural detail, two things he could not control were the Baltic convoy itself and the winds. Nearly 10 days after Banks's instruction letter, Noe wrote to let him know the state of the plants and that they now hoped to sail at last on 14 July:

We arrived at the Nore on the 8th but are obliged to wait for a Convoy till 14th of July. and perhaps not then if the wind does not turn out more favourable. The name of the Convoy is Daedalus of 32 Guns, and an nice little frigate . . .[34]

Noe sent Banks another update from the Danish port of Elsinore two weeks later, reassuring him about his watering routine for the plants, with only three 'a little Sick'.[35]

When they arrived at St Petersburg on 30 July, the quay was so crowded that Noe needed extra help to offload his precious cargo. Whitworth instructed him to store the plants at the Imperial Gardens until instructions came through from Catherine II. On 7 August, Noe was ordered to deliver the plants to the country residence of the empress's son Grand Duke Paul (1754-1801) and his wife Grand Duchess Maria Fyodorovna (1759-1828) at Pavlovsk, an estate presented to them to mark Grand Duchess Maria conceiving Catherine II's first grandchild. Pavlovsk is nearly 35 kilometres outside the city and only about four kilometres away from Tsarkoe Selo, Catherine II's estate. The plants were transported to Pavlovsk in a convoy of 15 coaches, each pulled by four horses. Arriving there at midnight, the gardens were lit by lamps to make the offloading as smooth as possible. The air was already frosty, slightly damaging the *Protea argentea* (now *Leucadendron argenteum*). It lost about three centimetres in height, 'it being too tall to be Covrd Properly'. Noe had lost just three plants – probably the ones that had been 'a little Sick' in Elsinore – with just four more 'rather Sick'. The remainder, though, 'look as well as if the[y] had been at Kew'. Noe also reported to Banks that the hothouses at Pavlovsk were all 'very bad' except for the most modern one, just a year old and '350 feet Long with Ten pits in two Divisions' where the new plants were installed. The Grand Duchess had requested that Noe mark each pot from George III's gift with the king's initials, GR (George Rex) to distinguish them from her own existing plants.[36]

Eager to see the new arrivals, the empress was in the hothouse at six the next morning, before 'Noe could arrange them'.[37] This was one of those occasions that Banks had predicted when it was particularly useful that the empress, Noe and indeed her daughter-in-law, Grand Duchess Maria had all grown up in Württemberg so they could understand each other clearly when Noe sensibly begged for a few hours to marshal the collection that

Angela Beaumont, *Strelitzia nicolai*, pen and ink, 2010

had been 'left in Confusion the night before'. During this time, the empress's excitement and pride prompted her to invite Noe to the drawing room present the engravings of the collection's star specimens, including 'the Botany Bay flax' (*Phormium tenax*, a New Zealand plant cultivated as an economic in the early Australian and other colonies). By 2pm, the empress was back in the hothouse once more, with a huge entourage led by the Grand Duke Paul and 70 courtiers. Only three of the plants presented by George III previously existed in her hothouses and the empress followed up her interest in the new specimens by visiting 'every day for an hour to Learn the names of the Plants'. Ten plants flowered while Noe was in Russia. Even after the empress had left Pavlovsk for Gatchina Palace, about 25 kilometres away, which Count Grigori Orlov (1734-1783) had recently landscaped in the English tradition, Noe had to take each one to the empress so that she could see it and draw 'Every one with her own hand'.[38]

Noe spent a month in Russia training local gardeners to care for the Kew plants and instructing garden staff at the various palaces in Kew horticultural methods, including Kew-style plant labelling. His efforts were so well received that he was offered a permanent post in Russia but he turned this down, despite Banks having gained the king's permission in case this offer were made. But Noe had been recalled to Württemberg, although he told Banks ruefully that he was 'sure he never shall receive so much pay as was offerd him by the grand duchess.' Noe was rewarded in Russia by Catherine II with a gold watch and 100 ducats but had to wait six weeks after arriving back in London on 27 December for his outstanding wages and expenses from Burges. Banks had brought the enterprise in under budget, at £162 against his estimate of £188, but he did not recover the costs owing to him for three years.[40]

Banks had anticipated Noe bringing home Siberian perennials and trees and mused that, 'They will travel more safely in a box packd with sphagnum than in a growing State'. In the event, Noe was himself disappointed in bringing home only 'about 25 plants and 180 different seeds from the Grand Duchess but I

| EN ROUTE TO ST PETERSBURG | NOE'S RUSSIAN EXPENSES[39] | | IN RUSSIA |
|---|---|---|---|
| ITEM | AMOUNT (£ s. d.) | ITEM | RUBLES |
| Trowel | 1 6 | To translator at Kronstadt | 3 |
| 2 watering pots | 8 | Customs duty for clothing | 4½ |
| Truss of hay, London | 4 | To 2 extraordinary customs officers | 2 |
| Pump, Gravesend | 3 2 | Barge, moving plants to Imperial Gardens, St Petersburg | 7 |
| 2 trusses of hay and forage, Sheerness | 1 3 | Carriage to & from Sir Charles Whitworth | 4 |
| Hay and forage, Elsinore | 10 6 | Coach hire to Pavlovsk | 4 |
| Letters from Elsinore & St Petersburg | 6 | Newspaper advert to leave Russia | 1 |
| Pass | 2 | Passport, interpreter & other fees | 10 |
| TOTAL | 28 2 | Coach to Kronstadt | 8 |
| | | 4 weeks' stay at Elsinore | 7 |
| | | (8 rubles = £1 sterling: £9 3 shillings) TOTAL | £11 11s 2d |

believe most of them are already in England, excepting the Rhododendron chrysanthemum'. Catherine II sustained her botanical enthusiasm and Scottish plant hunter and nurseryman John Fraser (1750-1811) had a rare piece of business fortune when he managed to sell every specimen he had with him to her during a visit to Russia the following year. As in Britain, rare plants were probably also presented to high-ranking contacts and filtered into horticultural nurseries. More than half a century later, in 1860, German horticulturist and botanist Eduard von Regel (1815-1892), director of the Imperial Botanic Gardens at St Petersburg, compiled a catalogue of the plants featured in the gardens of Russian soldier and administrator Nikolai Aksakov (1797-1882) at his estate outside Gorodisch. This was in the province of Penska, just over 600 kilometres south-east of Moscow on the River Sura. Both the large library and the greenhouse, stocked with exotic plants by Aksakov, were renowned and attracted many famous personalities of the time to view them. Among the many rare plants listed by von Regel were four strelitzia relations: *Strelitzia augusta*, *Strelitzia humilis*, *Strelitzia ovata* and *Strelitzia reginae*. Today, however, *Strelitzia humilis* and *Strelitzia ovata* are considered synonyms of *Strelitzia reginae*.[41]

Catherine II died in November 1796, just over 12 months after George III's gift of plants had arrived. Its real benefit was enjoyed by her daughter-in-law Grand Duchess Maria Fyodorovna, who had also grown up in Württemberg like Queen Charlotte and Empress Catherine herself. Like them, Grand Duchess Maria had also developed an early enthusiasm for botany, botanical art and gardens. Catherine II's son and successor, Tsar Paul I, died in 1800 but a tenuous link was maintained between Queen Charlotte and Dowager Empress Maria. After the Napoleonic wars ended, peaceful gifts of plants again became possible and in July 1814, Charlotte sent Maria a second smaller gift of plants from Kew, mainly the latest novelties from Australia. The impact of the imperial interest in gardens and exotic plants lingered even longer. Gardens of imperial palaces were renowned right up to the late Tsarist period and there was enough broader interest for a Russian Horticultural Exhibition to be planned in St Petersburg in May 1899, coordinated by the director of the city's imperial botanic gardens, Aleksandr Fischer von Waldheim (1839-1920), with American and other horticulturists invited to send floral exhibits.[42] But in international botanical terms, there had already been a key sequel to the sending of royal plants from Kew.

## THE HIDDEN GIFT

New plants flooding into the herbaria and botanic gardens of Europe from across the world needed identification but also reminded botanists to look more carefully for similarities and differences among the exotic plants they already had to hand. Banks had already proved that by effectively calling the late, great Linnaeus to task over *Strelitzia reginae* in 1790. Soon the tables would be turned, though, as von Regel uncovered a classic case of presumed but mistaken identity that had slipped past Banks and his successors at Kew. Von Regel had worked in botanical gardens at Bonn, Berlin and Zurich, where he was the curator until he moved to St Petersburg in 1855 to become director of the Imperial Botanic Gardens. The gardens was well supplied with interesting plants from around the globe and von Regel was naturally keen to take stock of what was there. The collection's range also gave him the opportunity to publish his descriptions of new species in the magazine that he had founded in 1852, *Gartenflora*. It was in this journal in August 1858 that von Regel published a triumphant and sometimes playful article on the banana-like tree that we would come to know as *Strelitzia nicolai*.[43]

Von Regel noted that Thunberg had described the treelike *Strelitzia augusta* in his *Prodromus florae capensis* and that the same plant had been 'illustrated by Hooker' in *Curtis's Botanical Magazine* in 1845[45] and then reproduced in *Flore des Serres*.[46] Von Regel then did the equivalent of a classic comedy double take:

*Sie is eine in den Gärten allgemein vertbreitete Art und als ausgezeichnet schöne Blattpflanze in den Warmhäusern geachtet. Wer hätte nun aber geglaubt, dass unter den als Str. Augusta in den Gärten verbreiteten Pflanzen, zwei ganz verscheidene Arten enthalten seien?*[47]

[It is a generally distributed species in gardens, and considered to be an excellently beautiful plant in hothouses. Who would have believed, that among the plants distributed in gardens as Str[elitzia] augusta, two completely different species are contained?]

Yet it was only when the two plants flowered side by side that the exciting truth was revealed — St Petersburg botanic gardens had clearly been hosting two different species unawares:

*Die beiden grössten Exemplare kamen in diesem Frühling zur Blüthe, und erweisen sich nicht nur als eine von Strelitzia augusta Thbrg. durchaus verschiedene, sondern auch als eine noch weitaus schönere Art.*[48]

[Both the largest specimens came into flower this spring, and showed themselves not only as a species thoroughly different to St[relitzia] augusta, but far more beautiful.]

It is not clear whether von Regel himself first noticed this subtle but important difference. The fact that he named Friedrich Körnicke (1828-1908) as co-author of the new description suggests that this young German botanist might have made the initial discovery. Körnicke had come to the St Petersburg Imperial Botanic Gardens in 1856, after graduating with a PhD from Berlin's Humboldt University. He stayed until 1858, the year of von Regel's announcement of *Strelitzia nicolai*, then moved back to Germany as a lecturer, becoming an authority on cereal crops and a professor of botany.[49]

The leaves of the new St Petersburg plant were very close to Thunberg's description of *Strelitzia augusta*, although it seemed to have leaves '*mit mehr herzförmingen Grunde*' [with more heart-shaped bases]. But it was the flowerhead that really distinguished the two, von Regel emphasised. While Thunberg described both inner and outer tepals as white, three inner tepals in the Russian plant were blue, '*petalis coeruleis*', and also of a slightly different shape. The flowers on the Russian plant were generally larger and had four spathes of slightly larger flowers arranged one above the other in a single inflorescence. Von Regel conceded that he had not been able to check the herbarium specimens used for Thunberg's description but felt justified in separating the species because he had been able to corroborate the 1845 illustration in *Curtis's Botanical Magazine* against a St Petersburg herbarium specimen of 'S. alba' collected in the Cape by Drège.[50]

## THE SHORT, SWEET LIFE OF A *STRELITZIA NICOLAI* FLOWER

The average lifespan of *Strelitzia nicolai* flowers is five days, researchers in Durban, South Africa, found:[44]

*Day 1*: A 'white lump' of furled sepals appears and large amounts of mucus are secreted to ease the flower's blooming out of the 'tightly packed' spathe.

*Day 2*: The flower is already completely open. With the added 'high rate of nectar secretion', this is the day of peak pollination.

*Day 3*: The stigma is already starting to dry and less nectar is being secreted but pollination is still possible. The anther sheath inside the flower has risen to an angle of 40° to 50°, compared to 20° to 30° on Day 2, which makes pollination more difficult.

*Day 4*: The flower has browned considerably and the anther sheath has risen further out of reach to an angle of 70° to 80°. The flower is also secreting much less nectar and mucus.

*Day 5*: In its final day, the flower is already very dry and turning dark brown to black, with its anther sheath at about 90°. There are signs of the next flower beginning to emerge from the spathe.

The discovery allowed Regel to name this '*herrliche Pflanze*' [splendid plant] *Strelitzia nicolai* in honour of the Grand Duke Nikolai:

*Als einen der wenigen Pflanzen, welche mit dem hohen palmenähnlichen Wüchse und prächtigen grossen Blättern auch noch grosse schöne Blumen vereinen, die sich in reicher Fülle zwei Monate lang ununterbrochen aus den grossen Scheiden entwickeln, achteten vir diese Pflanze würdig, den Namen Sr. Kaiserlichen Hoheit, des Grossfürsten Nicolai Nicolajewitch, des hohen Protectors des Russischen Gartenbau-Vereins in St. Petersburg, zu tragen, dem wir dieselbe hiemit in tiefster Ehrfurcht widmen.*[51]

[As one of the few plants which unites with its tall palm-like habit and attractive large leaves, [its] large beautiful flowers, which develop in rich plenty for two months uninterruptedly from its large spathes, we regard this plant as worthy to bear the name of His Imperial Highness the Grand Duke Nikolai Nikolaievich, the high Protector of the Russian Horticultural Society in St. Petersburg, to whom we dedicate the same herewith in deepest reverence.]

Himansu Baijnath

Von Regel was a co-founder and vice-president of the Russian Horticultural Society so naming the plant to honour its patron was pointedly diplomatic. Sir Joseph Hooker (1817-1911), though, seems not to have checked on this when writing the description for *Curtis's Botanical Magazine* in 1889, four years after he retired as director of the Royal Botanic Gardens, Kew. Hooker apparently mistakenly assumed that a similar tribute was being paid as when Banks had honoured Queen Charlotte and that it 'was named by Regel & Körne after the Emperor Nicholas.'[52]

Matilda Smith, *Strelitzia nicolai* in *Curtis's Botanical Magazine*, 1889

It is possible that Banks in 1795 did not unwittingly authorise the gift of an unknown species to Catherine II and unlikely that it was concealed in the 1814 consignment of plants from Kew. Von Regel implies that the plant might have come via Madeira, where Masson collected between 1776 and 1782 and where he is believed to have introduced members of the strelitzia family. However, Masson never reached further east than the Great Fish River, about 100 kilometres south of East London, considered the southernmost limit of *Strelitzia nicolai*.[53] It is possible, though, that other travellers had brought *Strelitzia nicolai* plants to the Company garden in Cape Town and that Masson had acquired one to send back to Banks.

In his article, von Regel observes:

*Im herbarium von Fischer, des früheren Directors des hiesigen Gartens, finden sich cultivierte Exemplare im Jahre 1849 aus Madeira von Dr. Sebastian Fischer, den verstorbenen Leibarzte Sr. Kaiserlichen Hoheit des Herzogs von Leuchtenberg, als Strelitzia augusta überbracht, die mit unserer Pflanze übereinstimmen.*[54]

[In the herbarium of Fischer, the former Director of the local garden, are found cultivated specimens brought from Madeira in the year 1849 by Dr Sebastian Fischer, the late court surgeon to His Imperial Highness the Duke of Leuchtenberg, as *Strelitzia augusta*, which match our plant.]

Fischer (1806-1871) was doctor to Duke Maximilian of Leuchtenberg, who married Czar Nicholas I's eldest daughter. Dr Fischer was also a keen amateur botanist who had collected in Egypt while working there, mainly at a military hospital, for 10 years from 1830 and would collect in Russia, the Mediterranean and Middle East while travelling with Leuchtenberg. It is not clear whether Dr Sebastian Fischer was related to Dr Friedrich Fischer (1789-1854). After graduating in Germany in 1804, Dr Friedrich Fischer came to Gorenki, outside Moscow, to be the second superintendent of 'a magnificent garden with hothouses' created for Count Alexei Kirillovich Razumovski. Dr Friedrich Fischer even produced a catalogue of this showpiece's 'interesting Siberian plants' and 'many rare exotic plants' in 1808, which ran to a second edition in 1812. After the count's death in 1822, Fischer took the post of director of the Imperial Botanic Gardens at St Petersburg, where he remained until 1850. In 1824 Fischer travelled to England, France and Germany to buy exotic plants to restock the gardens. He 'returned with a magnificent collection of 2320 species', which probably included some from Kew as Fischer began a correspondence with Sir William Hooker. This would have contributed to the interesting collection that von Regel found when he took over in 1855 although no detail is known as Fischer did not keep a register of accessions. So it remains unclear whether the *Strelitzia nicolai* which caused such excitement for von Regel and the botanical world had arrived in the consignment for Catherine the Great, from Dr Friedrich Fischer's 1824 buying trip or among live Madeiran plants from Dr Sebastian Fischer in 1849.[55]

## THE TENDER TRAP

The beautiful flowerheads produced by members of the strelitzia family reward visits from a variety of creatures and so assist in perpetuating the plants. Nearly 40 years after his then controversial *On the origin of species* was published in 1859, 19th-century British naturalist Charles Darwin (1809-1882) published a study of plant fertilisation, *The effects of cross and self fertilisation in the vegetable kingdom*, where he noted that he had been informed that strelitzias were an example of flowers from the Cape that were pollinated by birds. Two years earlier, Naples-born Italian botanist Federico Delpino (1833-1905) had published an article on the relationship between insects and plants that exude nectar and Darwin and Delpino became regular correspondents, exchanging and discussing their ideas, with Delpino publishing a second, related article in 1886.[56]

The idea of birds pollinating plants, rather than the involvement of Delpino's ant pollinators, intrigued British botanist George Scott-Elliot (1862-1934). He spent most of 1888 collecting in the Cape, Eastern Province and then Transvaal, then working in the Cape Government Herbarium. Scott-Elliot had such an interest in bird-pollinated plants that he wrote an article on South African examples that was published in 1890.[57] It was accompanied by another, shorter article specifically looking at the fertilisation of the banana relatives, *Musa*, *Strelitzia reginae* and *Ravenala madagascariensis*.

The 'specialised' *Strelitzia reginae* flower with its 'petaline sheath' was described in detail by Scott-Elliot as 'somewhat like that of an arrowhead with the flanges slightly

Juvenile golden-tailed woodpecker

turned up ... If the flanges are pressed down, the tubular cavity opens and exposes the six anthers.'[58] Though frustrated in his attempts to observe visits by birds to the plant, he recounted information given him by Professor Peter MacOwan, curator of the Cape Government Herbarium at the time of Scott-Elliot's visit:[59]

> the bird walks along the flanges (probably developed for this purpose) and in bending down beneath the dome-shaped odd petal probably causes the petaline sheath to open so as to dust its breast with pollen. The stigma being in front of the petals will of course be touched first.[60]

German botanists saw this mechanism as prompting what they called explosive pollen release in both the southern African strelitzias and their Madagascan relative, ravenala. This general belief that form follows function was supported in the 1914 *Flora of South Africa* by the Cape-based eminent German botanist Dr Rudolf Marloth (1855-1931), considered in his day 'the most outstanding botanist that South Africa has seen'. Marloth quoted the conclusion of Austrian botanist Professor Otto Porsch (1875-1959):

> No-one could imagine that the evolution of the Strelitzia flower and its present perfect adaptation to bird visitors would have been brought about without the long continued co-operation of the bird.[61]

By the 1960s, though, bird observers questioned such 'circumstantial evidence of the flower's shape':[62]

> birds do not seem to do what our deductions would have them do. They do *not* perch on the purple arrow-shaped pollen-vessel, open it under pressure of body weight, small as that is, and let their bellies be dusted in consequence.
>
> The truth is that the birds, or all those I have watched, go out of their way to avoid perching there. They sit readily on the hard, pointed projection of the green sheath from which the flowers emerge one by one, but not on the vital purple part.
>
> ... They perch on the side of the flower. Their feet grip the rim of the green sheath at the point where the orange petals and purple column emerge. They then insert their bills between the doubly-folded purple membranes and find the nectar. Neither their heads nor their bodies are ever closer than half an inch [1.25 centimetres] from the wings of the spearhead, usually they are much further. A friend even reports seeing a Sunbird hovering while probing.[63]

Sunbirds seem to choose this perch to avoid picking up sticky mucus on their feathers — even paper wasps and ants that come into contact with this sticky substance are agitated and clean themselves of it as soon as possible.[64] Strelitzia pollen is 'even webby like flour in which mealworms have been active' so that 'Strings of it might soil a bird's breast and not just dust the feathers in the way Aloe-pollen dusts the foreheads of probing birds.'[65]

Hugh Chittenden

Adding to this 'puzzle of pollination',[66] Cape weavers have been seen perching to feed on *Strelitzia reginae* so that their feet unfold the flower segments to reveal the pollen, 'some of which doubtless adhered to the bird's feet.' Moving between flowers in a strelitzia clump:

> There can be little doubt that the bird's movements affected a transfer of pollen grains in such a way that probably both self-and cross-fertilization became possible.[67]

Ornithologists, birders and gardeners continue to observe sunbirds feeding on members of the strelitzia family. In the Pigeon Valley reserve in Durban, for instance, purple-banded sunbird (*Nectarinia bifasciata*) was recorded feeding on *Strelitzia nicolai* nectar.[68] In the coastal forest at Twinstreams Farm in Mtunzini, Zululand, one research team studied sunbird visits to *Strelitzia nicolai* over a year and found that 'No group of animals exploited *Strelitzia nicolai* nectar as assiduously as did the sunbirds.'[69] About one in 100 bird visits were from other species such as Cape white-eye (*Zosterops pallidus*) and spottedbacked and Cape weavers (*Ploceus cucullatus* and *Ploceus capensis*).[70]

Olive sunbird

Olive sunbirds were 'the most aggressive sunbird species' seen, with the males particularly 'appropriating the locally richest food sources', whether because of volume in the case of *Strelitzia nicolai* or rich nectar in the case of tree fuchsia (*Halleria lucida*) and wild dagga (*Leonotis leonurus*). Although *Strelitzia nicolai* flowers throughout the year, sunbirds visit it much more often in winter – eight or nine times an hour compared to two visits an hour during summer, when courtship, breeding and wet weather can all change feeding patterns. A wider range of five sunbirds – olive sunbird (*Nectarinia olivacea*), grey sunbird (*Nectarinia veroxii*), collared sunbird (*Anthreptes collaris*), black sunbird (*Nectarinia amethystina*) and whitebellied sunbird (*Nectarinia talatala*) – visited in winter compared to only two in summer. The sunbirds chose a range of perching positions including, crucially, the anther-sheath and stigma when the new flower 'was more or less horizontally aligned'.[71] Time-lapse photography revealed that each flower was visited an average of 27 times during its life cycle. On between about 15 percent and 40 percent of those visits:

> birds made contact with the anthers, and on at least one visit birds perched on the stigma. However, since most of the stigmas examined bore several feet marks these values probably represent an underestimation.[72]

Earlier, it had been suggested that the 'stringy, and slightly damp' clumps of strelitzia pollen could be a deterrent to nectar thieves such as sunbirds who want to enjoy the sweet exudate but come away clean. This seemed contradicted, however, by good seed set in the study areas and 'all of 270 seed pods examined contained mature seeds'.[73]

By the 2000s, more than a century after Scott-Elliot's work, bird pollination was seen as part of an 'evolutionary shift' that happened 'independently in many lineages of flowering plants' – about 65 families of flowering plants and particularly common in some plant orders, including the Zingiberales order, which include both the strelitzias and their distant relatives, the heliconias. It is considered especially common in 'tropical and subtropical shrublands, open woodland, and riverine communities' – all areas favoured by southern Africa's various strelitzias. From the birds' point of view, these almost season-less regions mean 'flowers and nectar are available year-round to support nectarivorous birds', usually sunbirds, white-eyes and sugar-birds in southern Africa. This is vital for birds to survive and thrive – being many times larger than insects, their food must give them much more energy. In turn, this is thought to be the reason why 'plants with bird-pollinated flowers tend to put more energy into nectar production and often produce larger flowers to accommodate their avian pollinators.' They also develop ways to prevent creatures that are not likely to pollinate it from consuming their nectar before 'legitimate pollinators' can reach it. The complex folds at the base of the strelitzia flower form a 'nectar barrier' making it harder for insects to steal nectar. But this is only moderately effective because sunbirds, 'the primary nectar thieves . . . can manipulate the nectar barrier with their beaks to gain access to the nectar without causing obvious damage to the flowers of *Strelitzia reginae*'. Weaver birds, though, may be more interested in feeding on pollen so they 'do not drink nectar from every flower they visit'.[74]

## *STRELITZIA NICOLAI* AT HOME

The impact of von Regel's gloating declaration seems to have been accepted by botanists only very slowly. Three decades after von Regel's publication in *Gartenflora*, another German publication, *Blumengarten*, included *Strelitzia augusta*, as *Strelitzia alba* was still known, but did not include *Strelitzia nicolai*. In 1915, John Medley Wood maintained that *Strelitzia augusta*, 'the only Natal species of the genus', was found in the Durban Botanic Gardens but *Strelitzia nicolai* – which in fact grew wild around the city – was not. Even in the 1950s when Dyer was reassessing the treelike strelitzias (see chapters 2 and 9), he noted in exasperation:

> This is the Natal *Strelitzia* or "wild banana," universally but erroneously referred to in botanical literature as *S. augusta*.
>
> The name *S. augusta*, which is a synonym of *S. alba* . . . is so firmly entrenched in botanical literature that it will take some time to remove it from common usage. Why it is that the Natal plants have so long been confused with the species endemic in the Knysna-Humansdorp Divisions of the Cape Province is hard to tell. The mistake, once made, seems to have been followed in all botanical literature.[75]

This can be seen in the original identification of South African herbarium holdings of *Strelitzia nicolai* specimens into the 1970s. It seems to have been a widespread problem as Dyer's comment was echoed in 1977 by Eve Palmer, who felt it necessary to make the point that *Strelitzia nicolai* was 'Sometimes wrongly called *S. augusta*' in her *Field guide to the trees of southern Africa*.[76]

The Drège brothers, Carl and Johann, had reached Port Natal as Durban was then known on 27 March 1832, having followed the coastal route from Port St Johns, along which *Strelitzia nicolai* still grows densely to this day. But it was another German, pharmacist and collector Christian Krauss (1812-1891), who in 1846 was first to publish a description of Natal's vegetation. He had arrived in Cape Town in 1838, travelled through the Cape and sailed from Port Elizabeth to Port Natal in June 1839 and collected inland at least as far as the Pietermaritzburg and Howick areas. His undated specimen of *Strelitzia nicolai* is held at the Natural History Museum in London.[77]

The French collector Adulphe Delegorgue (1814-1850) and the Swede Johann Wahlberg arrived on board the same ship as

Clinton Friedman

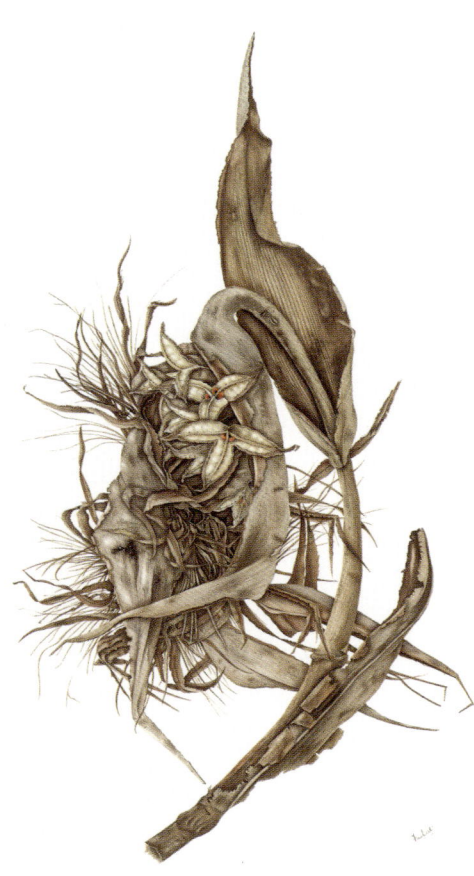

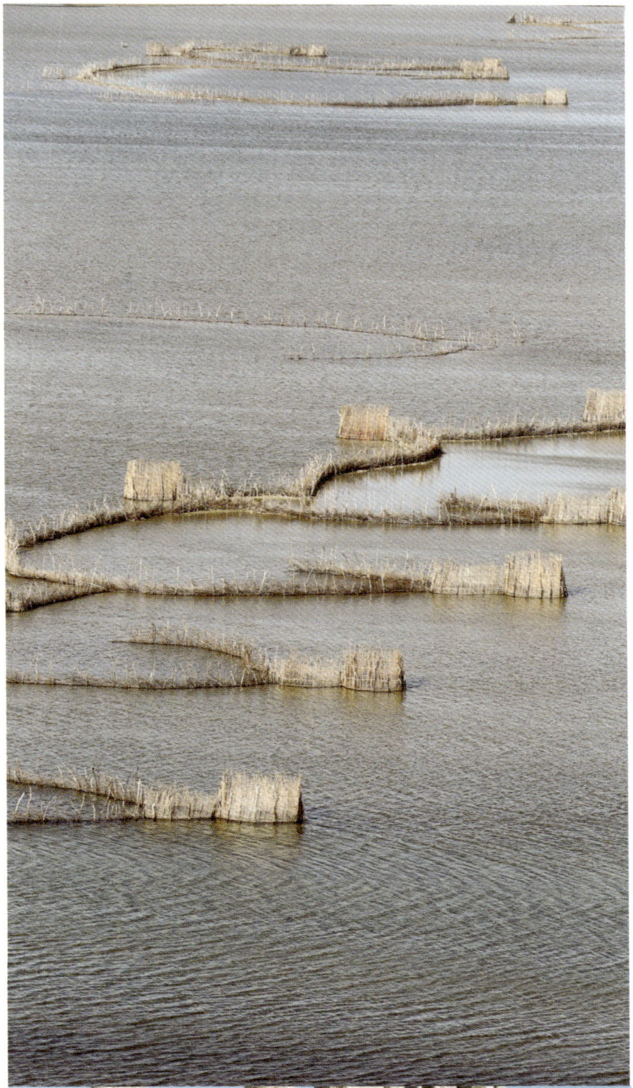

Ricky Taylor

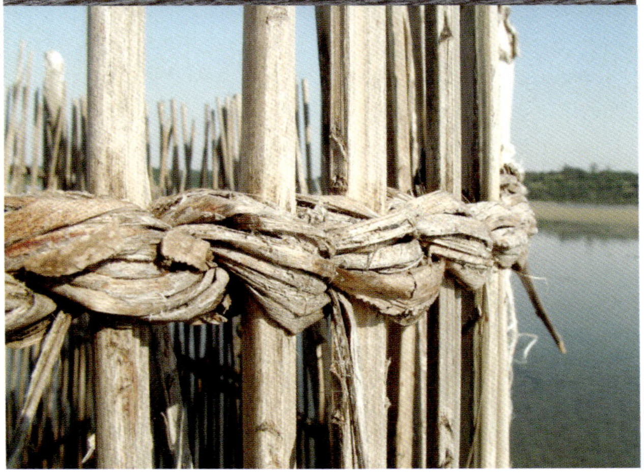

Krauss. Delegorgue wrote of the butterflies in the Durban forest 'with their rich, bright colours' and of the elephant herd that lived in the area. He collected the eastern bronze-naped pigeon (named *Columba delegorgeui* after him) in the forest on Durban's Berea, possibly the area now known as Pigeon Valley.[78] Wahlberg stayed in Congella, alongside Durban Bay, and saw the Africans using *Strelitzia nicolai* leaves to seal maize silos:

> When dry, the mealies are threshed with sticks and brought to holes in the earth, dug in a somewhat oval form, and so deep that a man can stand upright in them without being seen. They are smeared over with cow-dung and covered with the leaves of the wild banana; and then the hole is filled up, a flat stone placed over the opening, and earth shovelled over so that no sign of the hole can be seen on the surface.[79]

Dried leafstalks are still used 'as binding material for hut building and fish kraals'.[80]

By 1876, the region's 'beautiful and teeming vegetation' was an important feature of a book portraying the young Colony of Natal. 'Wild bananas (*Strelitzia angusta*) and wild date palms (*Phoenix reclinata*), give quite a tropical aspect to the scenery',

Farhat Iqbal, *Strelitzia nicolai*, watercolour, 2017 | Tsonga fishtraps bound with strelitzia fibre wrote author Henry Brooks, using a variant of *Strelitzia augusta* instead of von Regel's *Strelitzia nicolai*. Brooks also rhapsodised about *Strelitzia reginae*:

> These Strelitzias bear a remarkable brilliant-coloured flower, shaped like the gaping bill of a bird, with a projecting tongue that looks like a griffin's sting.[81]

Dyer and Hooker would probably have been kindred souls, frustrated by gaps in knowledge and sloppy usage. None of these early collectors had clarified key questions asked by Hooker by the time he described *Strelitzia nicolai* in 1889. He grumbled as much at the plant's misty history as at persistent confusion over the geographical ranges of *Strelitzia augusta* and *Strelitzia nicolai*:

> The date of introduction of this fine plant, which, seeing the stature it has attained, must have been cultivated in European Botanical Gardens for a great many years, is unknown; nor has its native locality in South Africa been ascertained . . . It would be as interesting to know the geographical area occupied by *S. Augusta* as to discover that of *S. Nicolai*.[82]

Even the renowned plant hunter John Medley Wood (1827-1915), director of the Durban Botanic Gardens, who maintained a regular correspondence with Kew and was a key figure in the development of botany in South Africa, made one of the 'mistakes' that must have irritated Dyer. Medley Wood's later lists of plants in the gardens include three species of strelitzia: '*Strelitzia Augusta*', '*Strelitzia juncea*' and '*Strelitzia regina*', although in 1883 he also included the now superseded name '*Strelitzia farinosa*'. It is unlikely, though just possible, that Medley Wood had obtained from the Cape a specimen of *Strelitzia augusta*, or *Strelitzia alba* as we now know it. This becomes even more unlikely given that *Strelitzia nicolai* grows naturally and easily in the Durban area and the gardens were cut into a section of Durban's Berea forest in 1851, making it more probable that an existing specimen of *Strelitzia nicolai* was incorporated into the gardens.[83] Medley Wood's text makes it clear that he believed this to be a local plant, even though he used the name *Strelitzia augusta*, which was potentially confusing and half a century out of date:

> This so-called "Wild Banana." A native plant found abundantly in the coast districts, and reaching to at least 2,000 feet [610 metres] above sea level. The young plants are much used in Europe for ornamental purposes. The leaf stalks and midribs of the leaves contain a useful fibre, but its extraction has not proved profitable, it has not been cultivated for this purpose, the fibre having been taken from the wild plants only. This is the only Natal specimen of the genus, but two Cape species [*Strelitzia juncea* and *Strelitzia reginae*] will be noticed in their place.[84]

## GIVING FOOD AND SHELTER

*Strelitzia nicolai* help support a wide range of creatures:[85]

- Birds probe for copious nectar.
  Monkeys eat soft parts of flowers.
- The tuft of orange aril on the seed is a favourite with monkeys and birds.
- Tree frogs make their homes at the leaf bases.
- Larvae of the banana tree nightfighter butterfly (*Moltena fiara*) feed on the leaves.
- Blue duiker shelter in clumps.

## BEES, BATS AND MONKEYS

Sunbirds are the renowned bird pollinators of southern African strelitzia, from double-collared sunbird and *Strelitzia reginae* in the Cape to grey sunbird and *Strelitzia nicolai* in Zululand, while redwinged starling have been seen enjoying eating the arils from the plant's seeds.[86] Although birds are a key strelitzia pollinator, other creatures feeding on the plants may help them reproduce to varying extents.

During the five-day lifespan of the *Strelitzia nicolai* flower, honey bees were seen feeding on the flowers' nectar particularly on days two and three, although also to some extent on day four. Although they might make contact with the pollen on the anther sheath, they are generally opportunistic and their contribution to pollinating the flowers is 'negligible'.[87]

During a study in Zululand, greater bushbabies (*Galago crassicaudatus*) were recorded visiting the flowers for just two hours in the evenings during October, possibly because fewer other food sources were available:

> Searching for nectar they went from clump to clump without stopping to feed in other vegetation on the way. In feeding from the flowers, bushbabies anchored their hind legs to the stem and then, in a standing position, held onto the upright white sepals and bract with their front paws. In this position they were able to drink nectar from the cup and lick pollen along the anther sheath.[88]

Bushbabies probably contribute to pollination but are not likely to be major pollinators because of the short period of the year in which they visit *Strelitzia nicolai*:

> Due to their large size and furry bodies, bushbabies probably collected pollen on their face and paws while positioning themselves or while feeding at the flowers. They could later transfer pollen to the stigma of the next flower.[89]

Vervet monkeys (*Cercopithecus aethiops*) 'were both infrequent and destructive visitors' when they arrive to feast on the plant's flowers, leaves and seeds: 'They usually pulled the entire flower out of the bract and chewed on the nectaries, thus acting solely as nectar thieves.'[90] *Strelitzia nicolai* seeds proved to be the monkeys' second favourite seed among the 29 species of trees and vines on which they were seen feeding in unmined areas around Richard's Bay.[91] But the thorough way in which the monkeys chew the *Strelitzia nicolai* seeds mean that they are 'destroyed' and it was not possible to identify even broken *Strelitzia nicolai* seeds in their faeces.[92] In such cases, the monkeys benefit from the seeds but do not help propagate the plant.

One mystery still to be solved is whether banana bats (*Neoromicia nana*) roost in the leaf tubes formed during the early development of leaves in the treelike strelitzias as they do in wild and cultivated bananas across sub-Saharan Africa. Banana bats spend much of their lives roosting and need to find new roosts every two to three days as the tubes where they roost unfurl into mature leaves.[93]

Plant hunters, naturalists and botanists had all noticed *Strelitzia nicolai* in KwaZulul-Natal – but if they took specimens, they have not survived. The earliest specimen at SANBI Herbarium in Pretoria dates back to June 1904. It had been collected at Port Shepstone on the KwaZulu-Natal south coast by Joseph Burtt Davy (1870-1940). He had been appointed to found the Transvaal Division of Botany within the new Transvaal Department of Agriculture the previous year. Burtt Davy had grown up in Derbyshire in England, worked at the Kew Herbarium as a technical assistant in the early 1890s and then spent a decade at various American botanical institutions before moving to South Africa for this new job. Over the next decade until his retirement in 1913, Burtt Davy continued collecting enthusiastically, with his collecting numbers eventually reaching above 14 000. Despite this pedigree and probably to Dyer's later annoyance, his predecessor's specimen not only used the outdated name but its label also featured a misspelling: 'Sterelitza august. Thunb.' This was later corrected to 'S. nicolai R & K'. *Strelitzia nicolai* specimens form the largest part of Kew Herbarium's holdings from southern Africa's treelike strelitzias. Of the 20 specimens, just over half were reclassified from *Strelitzia augusta* to *Strelitzia nicolai*, mainly at Dyer's instigation. By comparison, the Kew Herbarium has 10 *Strelitzia caudata* specimens and five of *Strelitzia alba*. Not surprisingly, SANBI's National Herbarium has at least twice as many of each: 40 of *Strelitzia nicolai*, 35 of *Strelitzia caudata*, Dyer's special project (see chapter 9), and 10 *Strelitzia alba*.[94]

Kew Herbarium's specimens are older than those in Pretoria, although both are from plants cultivated in gardens rather than from plant hunters out in the field. One Kew specimen was received from St Petersburg in 1867, probably sent with great satisfaction by von Regel, who also included a small envelope of 35 seeds for the herbarium. The other was taken in 1888 from a plant growing at the Royal Botanic Gardens, illustrated by Matilda Smith (1854-1926), Kew's Indian-born botanical artist, and described by Sir Joseph Hooker in the species' entry in *Curtis's Botanical Magazine*. This makes it a botanical landmark, internationally recognised as the type specimen. Similarly, both SANBI's Pretoria Herbarium and KwaZulu-Natal Herbarium hold specimens of the plant collected in September 1943 for the illustration in *Flowering Plants of Africa*, with Dyer pointing out that there should be no confusion in the field between *Strelitzia alba*, reaching about as far east as Humansdorp, and *Strelitzia nicolai*, growing in the wild from East London up the north-east coast, with 'a gap of about 250 miles' [400 kilometres] between them. The label observes that *Strelitzia nicolai* is a typical coastal plant that can withstand 'strong salt-laden winds' but not always frost when planted away from its subtropical home.[95]

## CONNECTING THE STRELITZIA FAMILY

Looking for differences and similarities between plants believed to be related is vital in helping botanists solve their key challenge of defining and naming plant species and how they relate to other plants and to other plant groups, or taxa: 'The history of the classification of the Zingiberales, like many tropical plant groups, is one of continual refinement and division.'[96]

*Strelitzia nicolai* seeds; adult female vervet monkey browsing on *Strelitzia nicolai*

The taxonomic method developed by Linnaeus aimed to use this comparative, scientific observation in 'determining, describing and naming of organisms, and classifying them in a way that attempts to reflect their inter-relationships'. The strelitzia family that we know today has very few members compared to the several hundred ericas, for example, but even so, the way strelitzias are believed to fit into their own family and the plant kingdom as a whole has varied as these plants have become known and understood in better detail. The process is rather like a genealogist starting with a bare outline of a family tree and filling in new members as they are found – as well as occasionally moving or removing some assumed members as it is discovered that they are not so closely related or even not related at all. Some agreement was reached about the overall relationships within the Zingiberales order in 1990 – nearly quarter of a millennium after Linnaeus named the first strelitzias he encountered as part of the heliconia family.[97]

This understanding of the relationships within the Zingiberales order was reached more than quarter of a century ago – after considerable shuffling and reshuffling of groups of plants around different levels of the family tree. The table opposite shows key landmarks in this research process.

Looking back over more than half a dozen taxonomic approaches to strelitzias and their broader relatives over the previous century and a half, botanical researchers still foresaw continuing debate:

> The changes in family concepts within the Zingiberales during the last hundred years can be attributed in part to an increased understanding over time of character distribution within the taxa. In the early classifications of the 1800s comparatively little was known about the number of taxa and the amount of character variation within each group. Hence, similarities among the taxa were stressed in devising classifications, as it was easier to fit new taxa into fewer categories. As more genera and species were discovered and described, discontinuities in character variation became more obvious and differences separating taxa became more apparent. The current recognition of eight families within the order is a direct result of the much larger data base of taxa and character distribution available today. An even further division of families may occur as additional data become available.[99]

But around this time, botanists began to discuss in much greater detail how they believed plants had evolved, particularly using the developing scientific approach of phylogeny, tracing the history of how organisms and their interrelation had been understood. New methods of plant analysis and new approaches to taxonomy would be harnessed to this transformation of botanical science in attempts to resolve these debates.

Edith Burges, *Strelitzia nicolai* in *Flowering Plants of Africa*, 1946

*Strelitzia nicolai* specimen from KwaZulu-Natal South Coast, collected 1950

## KEY LANDMARKS IN TRACING STRELITZIA FAMILY RELATIONSHIPS [98]

| DATE | BOTANIST | EVENT | FROM | TO |
|---|---|---|---|---|
| 1771 | Linneaus | Plants that would become *Strelitzia reginae* and *Strelitzia alba* allocated to heliconias | | *Heliconia bihai*; *Heliconia alba* |
| 1789 | Aiton & Banks | *Strelitzia* separated from *Heliconia* | *Heliconia bihai* | *Strelitzia reginae* |
| 1831 | Richard | Ravenala officially named for first time | | *Urania* |
| | | Includes *Urania* with heliconias in banana family | | Musaceae |
| 1880 | Bentham & Hooker | Place strelitzias and their broader Musaceae family within new order of plants | Musaceae | Scitamineae |
| 1889 | Petersen | Promotes four tribes to family status and adds subfamilies | Musaceae | Museae, including Musaeae and Heliconiaeae |
| 1900 | Schumann | Introduces new subfamily level between family and genus | Musaceae | 3 subfamilies: Musoideae, Strelitzoideae and Lowiodeae |
| | | Creates two tribes within Strelitzioideae, depending on characteristics such as flowers, position of leaves and presence or not of arils on seeds | Heliconiaceae | Strelitziaceae and Heliconiaceae |
| 1930 | Winkler | Removes ravenala from strelitzia family and places it in its own tribe | Strelitziaceae | Ravenaleae |
| 1934 | Hutchinson | Moves strelitzia family out of the banana family to a new family combining southern African strelitzias, ravenala and phenakospermum with heliconias | Musaceae | Strelitziaceae |
| 1948 | Nakai | Moves heliconias out of strelitzia family, within which he creates three new subfamilies | Strelitziaceae | Strelitzioidae, Ravenaloidae and Phenakospermoidae |
| | | Divides phenakospermum subfamily into 2 genera, though this is shortlived | Phenakospermum | Phenakospermum and Musidendron |
| 1959 | Hutchinson | Replaces heliconias within the strelitzia family | Strelitziaceae | |
| | | Renames Scitamineae order Zingiberales | Scitamineae | Zingiberales |
| 1985 | Dahlgren, Clifford & Yeo | Zingiberanae closely related to Bromelianae and Commelinanae | Zingiberales | Commelinadae |

*The Botanist's Repository, 1804-1805, Peter H. Raven Library/Missouri Botanical Garden*

# ART AND ARTIFICE
## *Strelitzia depicta to the citrine strelitzia*

CHAPTER FOUR

# 4

# ART AND ARTIFICE: *Strelitzia depicta* to the citrine strelitzia

The fashion for botany and botanical art boomed in Britain and Europe from the late 18th century well into the 19th, spurred on by the excitement of more and more exotic plants being introduced. Scientists at first controlled the descriptions and depictions of new plants published in scholarly texts. Fashionable enthusiasm for these new plants soon meant that luxury, limited-edition botanical volumes were created for the libraries and salons of the wealthy. These expanded into cheaper editions for the growing popular press and readership. Eventually, these new plants were admired not just for themselves but also for the unusual twists on nature that horticulturists could achieve by experimenting with plant hybridisation.

Banks's dedication of *Strelitzia reginae* to Queen Charlotte proved to be a master stroke in keeping the plant at the forefront of public attention for a century and a half. The plant's name, shape, colouration and unusual appearance saw it being celebrated in specially commissioned illustrations, in poetry and in visual allegory. In the language of flowers that Victorian Britain redeveloped from its traditional origins, Strelitzia reginae came to symbolise magnificence. Yet the flower's flamboyant image and celebrity by association with the royal court were at odds with the preference of King George III and Queen Charlotte for a lower-key lifestyle, which the king's recurrent mental-health problems would also necessitate.[1]

From 1775 onwards, Joseph Banks and William Aiton snr enjoyed using the royal gardens at Kew as a botanical workshop. The gardens also became an important part of the royal year, with George III spending about three months there from about mid-May. At Kew, he and Queen Charlotte pioneered the kind of informality that returned increasingly to the British royal family from the 1950s to the present day. For George III's seven surviving sons, this extended to learning the practicalities of gardening, agriculture and building. The six daughters joined their mother in gardening – she already looked after 'a small flower garden' at Buckingham House (now Palace) – walks, interior decoration, embroidery and painting. Princess Elizabeth (1770-1840) is believed to have depicted the plants that climb the walls in one room of the Queen's Cottage, a substantial summerhouse. Another part of the royal year was spent at Windsor, where Queen Charlotte also planned her garden as a 'little Paradise'. The queen had been interested in botany as a child and continued to study it with both Sir Joseph Banks and Sir James Edward Smith.[2] She became a royal patron of and leader of the fashion for botanical science and new plant discoveries. This had been sparked particularly by excitement over Banks's natural-history collection from the *Endeavour* voyage which was brought back in 1772, the year Banks met the royal couple and the year that they inherited Princess Augusta's property at Kew.

## ROYAL ARTISTRY

Proving that he took his wife's interest seriously, in 1788, King George paid 100 guineas to buy her the herbarium of mainly British plants collected by the late Reverend John Lightfoot (1735-1788). The plants were 'rearranged' by Smith and would eventually occupy '24 mahogany cabinets' at Windsor. Meanwhile, in Göttingen, Germany, Murray was preparing a 'collection of Hanoverian plants' for Queen Charlotte. The family even holidayed at Weymouth on the south Dorset coast, 'collecting, pressing and mounting marine plants'.[3]

Like well-to-do society in general, Queen Charlotte was captivated by the possibility that botanical painting offered the opportunity to use a ladylike skill in depicting beautiful or interesting plants and also to play some part in scientific development at its cutting edge. An elderly friend of the royal couple, Mary Delany (1700-1788), had charmed society and the discerning Sir Joseph Banks with her decorative 'paper mosaicks' depicting flowers. There seem to be several reasons why Mrs Delany did not depict *Strelitzia reginae*, the tribute to her royal friend who had arranged a grace-and-favour residence for her at Windsor after she was widowed. During the final years of Mrs Delany's life, her eyesight was not good enough to work with either her mosaics or her excellent embroidery.[4] Flowers of *Strelitzia reginae* appear to have created such a sensation because they were so extremely rare (see chapter 1). Finally, the flower was not formally dedicated to Queen Charlotte until *Hortus kewensis* was published in 1789, the year after Mrs Delany's death.

In March 1788, the month before Mrs Delany died, Queen Charlotte decided to combine two approaches, using dried plants to create a 'Herbal from impressions on black paper', she told Lord Bute. Soon Queen Charlotte was taking lessons of up to three hours in botanical painting from Margaret Meen (c.1752-1834), a Suffolk artist who exhibited at the Royal Academy from 1775 to 1785. In 1790, Meen dedicated to Queen Charlotte her *Exotic plants from the royal gardens at Kew*, including an illustration of *Strelitzia reginae* among the eight plants featured. Meen later sold various 'drawings' to the queen for £22 12 shillings. Meen has since become somewhat controversial for taking a creative rather than an accurate approach to botanical illustration and her *Strelitzia reginae* illustration has been dismissed as structurally 'inaccurate' by a taxonomist since 'the leaf is correct for *S. reginae* but the flower pedicel and number of floral parts do not represent any known form of *Strelitzia*.' Although one commentator considered her competent, another damned Meen with faint praise: 'In spite of all her immense industry and patience, however, she never quite rises above the level of a very highly gifted amateur.' It does not help Meen's reputation that she rarely signed her work or else marked it with 'initials . . . so small as to be scarcely legible'. This makes it more difficult to judge her whole output although Kew bought a large number of her paintings in 1933, noting that they 'provide ample evidence that Margaret Meen was a highly accomplished artist.'[5]

In the same year that Meen published her *Exotic plants*, editor William Curtis (1746-1799) enhanced the 'sensation' created by naming *Strelitzia reginae* after the queen when he experimentally splashed a depiction of the flower as a gatefold spread (quarto-sized) in his *Botanical Magazine* (see chapter 1). Curtis had founded the journal only in 1787 but it was already setting the pace for the growing market in fashionable botanical magazines, especially given the high quality of the botanical art produced by James Sowerby (1757-1822), Sydenham Edwards (1768-1819) and later Walter Hood Fitch (1817-1892).[6] Unfortunately, it is not known who was the artist or engraver for the plate of *Strelitzia reginae*. The splash given to the plant was an unprecedented honour in the magazine's early years but Curtis admitted that he used the gatefold format with trepidation:

> In order that we may give our readers an opportunity of seeing a coloured representation of one of the most scarce and magnificent plants introduced into this country, we

have in this number deviated from our usual plan with respect to plates, and though in so doing we shall have the pleasure of gratifying the warm wishes of many of our readers, we are not without apprehensions least others may not feel perfectly well satisfied; should it prove so, we wish such to rest assured that this is a deviation in which we shall very rarely indulge, and never but when something uncommonly beautiful or interesting presents itself[7]

There was one grudging comment from a subscriber, Mrs Wilson, who complained that the *Botanical Magazine* should be published as quarto size instead to accommodate large plates without the need for a gatefold format. Although sales kept climbing to 3 000 copies a month, Curtis had been scared off – and possibly found the cost difficult to justify. He never used a gatefold again and there was a gap of more than 600 plates before the new editor, John Sims featured a gatefold format once more.[8]

## BOTANIC PAINTER TO HIS MAJESTY

It was also 1790 when Banks, who was notoriously critical of botanical illustration, engraving and printing, finally found a botanical artist he could be happy with. He poached Franz Bauer (1758-1840) from the Schönbrunn Imperial Botanic Gardens outside Vienna, where he was illustrating the botanical writings of Baron Nikolaus Joseph von Jacquin and his son Baron Joseph Franz von Jacquin. Bauer was, in fact, visiting London with the younger Jacquin in 1788 when he first came to Banks's attention. Franz Bauer's brother Ferdinand (1760-1826) was already working in Oxford with the curator of the Oxford Botanic Gardens, Professor John Sibthorp (1758-1796), so Banks had probably at least already heard of, if not seen, the talent in Ferdinand Bauer's work. When he recognised the same level of skill in his brother Franz's work, Banks organised a post as a botanical illustrator at Kew. Banks sponsored the appointment by paying Franz Bauer a handsome salary of £300 himself. Ferdinand Bauer received little more than that, £310 a year, when he was contracted as a botanical illustrator for Flinders's pioneering, two-year circumnavigation of New Holland (Australia) from 1801.[9] Time would prove Banks's instinct about Franz Bauer correct as his memorial stated:

> In the delineation of plants he united the accuracy of a profound naturalist, with the skill of the accomplished artist, to a degree which has been only equalled by his brother Ferdinand.[10]

To add legitimacy to Franz Bauer's appointment, Banks also engineered the title of 'Botanick Painter to His Majesty' for him in 1790. Bauer's main purpose at Kew was to illustrate the hundreds of new plants arriving there each year but, like Meen and indeed his rival Pierre-Joseph Redouté (1759-1840), Bauer was soon also teaching botanical art at court. While Redouté, often called 'the Raphael of the rose', taught botanical painting to 'two Queens,

Margaret Meen, *Strelitzia reginae* in *Exotic plants from the royal gardens at Kew*, 1790

Franz Bauer, *Strelitzia reginae* from *Strelitzia depicta*, 1818

two Empresses and one other claimant to the throne of France', Franz Bauer taught Queen Charlotte, the Princess Royal and Princess Elizabeth, 'who meticulously coloured Bauer's *Erica* engravings, copied plates in botanical works, and tested the capabilities of pencil, pen-and-ink, crayon, watercolour and oils, usually in the intimacy of a family group.'[11]

For many observers, both Franz and Ferdinand Bauer represent 'the high point of late 18th and early 19th century botanical art', rivalled only by their predecessor Georg Ehret (1708-1770), who had illustrated the much-admired *Hortus Cliffortianus* compiled by Linnaeus snr, and by their contemporary Redouté. The Bauer brothers had developed a style that combined 'exacting observation' with graceful and engaging plant portraits that appeared natural, although the process was far from effortless. Franz Bauer was 'probably the first artist to draw detailed plant dissections for recording purposes at Kew'. In precisely recording differences and similarities, plant anatomy was a key identification tool: 'In microsopical drawing [Franz Bauer] was altogether unrivalled, and science will be ever indebted for his elaborate illustrations of animal and vegetable structures.' Franz Bauer seems to have been a kindred spirit to Banks in seeking perfect accuracy, exploiting his microscope to the full to draw even 'magnified pollen specimens with remarkable detail'.[12]

Bauer also adapted with alacrity from drawing for copper-plate engraving to the new, more economical lithography, a process that 'involves drawing with a greasy crayon on a limestone base which is then dampened and applied with ink. The ink will subsequently adhere only to the drawing.' When the new technology of the camera lucida was introduced, apparatus that assisted accurate drawing by reflecting an image of the subject onto the artist's paper, Bauer is believed to have used this too. This combination of art, science and craftsmanship represents a pinnacle of botanical art: 'The greatest flower painters have been those who have found beauty in truth: who have understood plants scientifically, but who have yet seen and described them with the eye and hand of the artist.'[13]

Bauer was a supreme exponent of the way botanical art 'unites function and aesthetics'. It is a 'hybrid genre constituted of two primary discourses – botany as science, and botany as visual art.' Bauer's combination of artistry with explicit diagram was so clear that Banks felt it eliminated the need for descriptive text in *Delineations of exotick plants*: 'each figure is intended to answer itself every question a Botanist can wish to ask, respecting the structure of the plant it represents.' As well as helping justify Banks's spending on botanical artists and his constant push for perfection in printing, this claim was part of a contemporary struggle in methodology. Did botanists or illustrators contribute more to the taxonomic record? Could a description or an illustration ever be complete proxies for each other? Some botanical writers supported the theory that a well-written plant description could be visualised by a well-trained botanist without any assistance from an illustration but such claims carried the danger of straying into the lesser territory of flower painting pure and simple, exemplified in the many popular florilegia published in the period.[14]

## A FINAL TRIBUTE

Queen Charlotte's strelitzia was one of the 486 plants portrayed by Redouté in his masterpiece collection *Les liliacées*. A massive collection of 10 000 plates published in a very limited edition of 200 copies between 1802 and 1816, with 18 prestige copies handfinished by Redouté himself. The publication was made possible by the patronage of the Empress Josephine, to whom it was dedicated, although she died before the collection was completely published. Only about half the plants illustrated were from the lily family, with the remainder being plants that were at least as 'resistan[t] to preservation in herbaria . . . it also had a special value to botanists in providing accurate drawings and descriptions of plants that would not otherwise be easily obtained for study.' Josephine had enthusiastically collected new

Pierre-Joseph Redouté, *Strelitzia reginae* in *Les Liliacées*, about 1816

species of plants for her garden at the Chateau de Malmaison, with nearly 200 new species flowering there between 1804 and 1814. It is very likely that the glamorous and intriguing queen's flower, 'Strelitzia de la reine', as Redouté called it, was one of these, especially as the Royal Botanic Gardens at Kew was one of Josephine's main plant sources. In his two plates, he depicted it in two very different ways. In one plate, a distinctly golden flower emerges from a softly bending clump of curved leaves. The other uses the style of botanical illustration to present a flowerstem with five small supplementary drawings detailing the flower parts. Curiously, this second plate makes the flower appear chunkier than usual and the spathe seems foreshortened.[15]

This publication – and possibly a competitive sense that he could do better – might have inspired Bauer, in his third decade at Kew, to undertake portraits of *Strelitzia reginae* and *Strelitzia augusta*, the only two members of the family known at the time. The project may have been encouraged by Banks as a consolation to the queen. Her own health was now poor and she was enduring the difficulties of her son's regency and her husband's final descent into permanent madness. Bauer rendered the plant's bold, structural flowers delicately and meticulously. Even in reproduction, their ethereal quality seems to float hauntingly on the page. Their contemplative mood contrasts strongly with the bright and vigorous illustrations common in botanical magazines of the time.

After completing his series of portraits so meticulous 'that you could be forgiven for thinking that the plate is an entirely original artwork', Bauer's 11 plates of the two plants were published as *Strelitzia depicta* in 1818 and became a final tribute to Queen Charlotte, who died on 17 November that year.[16] The volume was subtitled 'Coloured figures of the known species of the genus Strelitzia from the drawings in the Banksian Library', which suggests Bauer was working from an illustration or at best a specimen rather than a living, blooming plant. The later renowned Kew artist Walter Hood Fitch maintained:

Sketching living plants is merely a species of copying, but dried specimens test the artist's ability to the uttermost; and by drawings made from them would I be judged as a correct draughtsman.[17]

Coloured and uncoloured plates of Pierre-Joseph Redouté's *Strelitzia reginae* in *Les Liliacées*, about 1816

Bauer's strelitzia portraits are all vividly yet delicately hand coloured. To add to the publication's prestige, the format chosen was very large, allowing the flowers to be portrayed lifesize and intensifying their impact on the viewer. A print of one of the strelitzia plates was selected as part of the limited edition of 500 sets of seven of Kew's botanical treasures issued to mark the diamond jubilee of Queen Elizabeth II in 2012.[18]

By 1820, Banks was so confident that Bauer's work at Kew needed to be continued that he left Bauer an annuity of £300 'to continue unhampered his botanical art at Kew'. That allowed Bauer to continue producing his 'notably precise and scientifically accurate' illustrations for the new director of Kew, Sir William Jackson Hooker (1785-1865). Bauer died in 1840, just months after his rival Redouté. Bauer had served Kew for half a century, the warm tribute on his memorial concluded: 'The *works* of Francis Bauer are his best *monument*. Friendship inscribes this record on his honored tomb.'[19]

## REMEMBERED WITH HONOUR

The extensive patronage of botany and art by King George III and Queen Charlotte are not commemorated in Westminster Abbey or St Paul's Cathedral, or even the Royal Academy of Arts, but more personally and quietly in St Anne's Church on Kew Green. Queen Anne, George III and William IV were all among the church's benefactors.[20]

The area was popular among 18th-century artists and several society celebrity artists, such as Thomas Gainsborough and Joseph Zoffany, were buried at this elegant building. In 1962, these links were creatively marked when parishioners Mr and Mrs G.E. Cassidy initiated a series of commemorative embroidered kneelers. The 140 kneelers were mostly designed by Ursula Holttum, whose husband Professor Richard Holttum (1895-1990) worked in the Singapore Botanic Gardens and at Kew, and were embroidered by a group of about a dozen women from St Anne's congregation.[21] The kneelers commemorate many of those associated with Kew, including King George III, Queen Charlotte, both Princesses Augusta, the earl of Bute, Kew gardener William Aiton and botanical artists including Pierre-Joseph Redouté and Walter Hood Fitch. The kneeler commemorating Franz Bauer depicts a *Strelitzia reginae*.

In 1840, Bauer was buried at St Anne's next to Gainsborough. Bauer's service to the British royal family and his depiction of the queen's strelitzia also earned him a memorial in the church that was delicately sculpted with a frame of strelitzia flowers and leaves by Sir Richard Westmacott (1775-1856), then Britain's doyen of memorial sculpture.[22]

Memorial to Franz Bauer, St Anne's church, Kew (in nave to left of altar)

## THE WRECKED TEMPLE

While Franz Bauer created ever more delicate and detailed depictions of plants, a more vivid and more popular trend was emerging in botanical illustration. In contrast to the restraint of Franz Bauer's work for Kew was *Temple of Flora*, compiled by Robert Thornton (1768-1837), part of Thornton's *New illustration of the sexual system of Linnaeus*, published between 1799 and 1807. *Temple of Flora* celebrated the strelitzia among its 31 plates that combined romanticised classical allegory with often robust and vigorous depictions. It echoed 18th century fashion while looking forward to the increasingly populist botanical publishing of the 19th century just as its subtitle, 'Garden of the botanist, poet, painter and philosopher' bridged different cultural spheres. This work was dedicated to Queen Charlotte as 'The Patroness of Botany and Fine Arts' and prominently featured the 'Queen's flower', *Strelitzia reginae*. The publication has been called 'a romantic *tour de force*'. But this extravagant, atlas folio size volume was harshly criticised by the *Philosophical Magazine* and failed commercially.[23]

Thornton lavished care on the production, 'employ[ing] the best artists and engravers he could find . . . [to create] pictures . . . from aquatint, mezzotint or stipple engravings, partially printed in colours and finished by hand.'[24] This has created a bibliographical curiosity:

> No two copies . . . are alike. The delicate mezzotint or aquatint plates wore as each print was taken off them, and most of them were retouched, sometimes more than once, and in some cases fresh plates were substituted . . . The same number of prints are never found in two copies of the book.[25]

The narrative opens with the illustration 'Cupid inspiring plants with love', in which *Strelitzia reginae* features as the only flowering plant among a range of lushly leaved trees and is placed in the prominent righthand corner.[26] Elsewhere, the sturdy stems and bold colours dominating the portrait of *Strelitzia reginae* are offset by the pink blush on the flower's spathe and the soft-focus background of classical woodland and trees. In the 'Verses addressed to Dr Thornton, on Completion of his Temple of Flora', the Rev. Dr Maurice of the British Museum describes the plant's noble battle to survive:

> Though o'er her head the southern whirlwind rave
> Secure, behold! august STRELITZIA wave![27]

But this literary garden is a haven for *Strelitzia reginae* and other plants, declared Dr Shaw of the British Museum in his testimonial verses, 'Lines addressed to Doctor Thornton, on his Temple of Flora':

> Thy GARDEN of perpetual bloom
> No change of threat'ning skies can fear;
> Nor dashing rains, nor chilling blasts
> Can reach the lovely fav'rites here.[28]

It was a golden age for flower books but Thornton had misjudged the market. Despite lavishing money on 'the best artists and engravers', there were far too few subscribers for his great publication scheme. His *Temple of Flora* became a publication landmark for the wrong reasons: 'it was none the less a commercial failure and had to be stopped when half-completed because of lack of support.' This was not an uncommon dilemma for publishers at the time: 'Many fine books represent only a portion of what was planned.' Fierce letters were exchanged in the letters pages of the *Philosophical Magazine*. Publication ceased in 1807, although in 1811 Thornton was still trying to recoup some of the large sum that he had lost on the volume through a lottery for 'copies of the book, in which the plates are much worn and a pale shadow of the earlier ones, and individual prints were given as prizes'. The versified botany held echoes of Erasmus Darwin (1731-1802), the doctor, scientific philosopher and grandfather of Charles Darwin. Erasmus Darwin's verse epic on Linnaeus's then socially risqué new system of sexual classification of plants, *The loves of plants*, was slated by Coleridge when it was published in 1789. Thornton's *Temple of Flora* was not a great work of literature, either. To its contemporary audience, it was not even an artistic landmark although it is now seen as a classic representative of the trend for lavishly produced florilegia, perhaps the 'greatest of all English flower books'. Tastes were changing and later illustrated botanical works would still be decorative but would combine entertainment with useful information to suit the mood of Victorian England.[29]

## THE QUANDARY OF COLOUR

'What colour is that flower meant to be?' is a legitimate question when faced with some of the botanical painting from the 18th and 19th centuries. The quality of colour reproduction was such a sensitive subject that it could become part of a book's upmarket appeal. This contrasted with late medieval woodcuts whose heavy, inked lines inevitably detracted from the delicacy and elegance of plants. Even when newer techniques could create finer lines from the mid-16th century onwards, book buyers were expected to colour the illustrations themselves or commission another artist to do so.[30]

By the 19th century, colour printing was possible but usually crude. So publishing houses felt that the accuracy and nuance of the intended colour was better captured by hand colouring from their 'teams of painters' – usually women, as in the potteries. *Curtis's Botanical Magazine* was even published with hand-coloured plates until 1948. The magazine had been explicit about the challenges of colour reproduction when William Curtis founded it, promising no more than that its plates would be 'coloured as near to nature as the imperfection of colouring will admit'.[31] The limited range of artists' pigments available in the 18th century improved only gradually during the following century:

'The Queen' from Robert Thornton's *Temple of Flora*, 1807

*Henderson pinxt.*                                 *R. Cooper sculpt.*

*The Queen.*

London, Published Feby. 1, 1804, by Dr. Thornton.

Prussian blue was invented in 1704, but cobalt was not discovered until 1802, and French ultramarine about 1815 . . . The bright and durable cadmium and chromes did not come into use until late in the nineteenth century[32]

Crucially, the pigments available often had insufficient range to capture what the eye could see: 'it is impossible to mix anything but a very muddy green using only the totally safe and permanent true ultramarine with yellow ochre or the transparent "gumboge".' Ferdinand Bauer was so concerned about this that he developed three versions of his numbered colour chart, which also acted as a colour reference when he was processing specimens quickly during his travels. The colour-chart system dated back at least to the time of German painter and printmaker Albrecht Dürer (1471-1528) and Bauer probably adopted and adapted it from other draughtsmen who had been working for Jacquin in Vienna. Colour charts can also assist with reproduction given the perennial problem that all human beings perceive colour slightly differently. By the early 20th century, task-specific colour charts were being developed, for example, to help horticultural taxonomists describe slight differences in shade and tone in new cultivars and to assist designers and dyers in the textile industry. The Royal Horticultural Society's colour chart developed from standards set by the British Colour Council in the 1930s. These charts are considered a benchmark internationally and have been updated and extended several times since, valued for helping resolve some guesswork and assisting with any disputes over colour perception or colour blindness. A version is now available online.[33]

Even when artist, engraver and printer agreed on a colour, they could not prevent the decaying effect of time on pigments that were not fast:

> Carmine, made from cochineal and used also as a cosmetic, was apt to be fugitive; cinnabar or vermilion, a sulphide of mercury, was fairly reliable when pure, but darkened if mixed with white lead or adulterated, as it often was, with red lead. It is hard to find a picture of an orange or scarlet flower that has not been blackened by time.[34]

## THE QUESTION OF YELLOW STRELITZIAS

Could that darkening effect in contemporary orange pigment have been the reason why Suffolk artist Margaret Meen (fl.1775–1824) chose to colour the petals of her depiction *Strelitzia reginae* yellow? This is one of the earliest portraits of the flower and has often been featured in books and calendars of botanical illustrations, despite her depiction of the yellow flower now being broadly condemned as an aberration. There could be several reasons why Meen chose to use a downright dangerous pigment – like blue-green (verdigris), yellow (orpiment) was poisonous. She might have been creating from sketches and reference books rather than a live, flowering specimen, made all the more likely by the draughting inaccuracies in her portrayal. On the other hand, she might legitimately have been capturing an early yellow sport of the plant, a horticultural expression for a mutation in flower or form that often disappears as the plant reverts to type. There remains the possibility that Meen's yellow strelitzia was not an aberration to her. We do not know whether Meen painted the colour she saw or the colour that was reported to her, nor whether she was painting from life or a dry specimen. Interestingly, Redouté's first *Strelitzia reginae* painting also has a strong element of yellow in the flower.[35]

Although hindsight now tends to dismiss Meen's portrait particularly as fool's gold, early evidence of yellow *Strelitzia reginae* might have made this seem both more likely and more desirable to her and those in her circle. She might have been told of the rarer, yellow-flowered plants and decided that yellow was a good choice to emphasise rarity value and to make a golden offering to the queen.

Linnaeus jnr had included '*Flos* croceus' in his description, suggesting that he had seen or been told that the flower was a reddish gold. Thunberg, who saw the plant growing in the wild, had described the plant as '*Strelitsia* med gul bloma'. The Swedish *gul* means the flowers, as Thunberg saw them, were yellow, golden or amber. This was translated as yellow in the English edition on which the translator worked closely with Thunberg.[36]

Yellow strelitzias seem to have been discussed by Banks and Masson as a rarity around the time Meen's book containing her yellow *Strelitzia reginae* was published. In June 1790, Masson wrote to Banks: 'I have enclosed in this letter 4 Seeds of STRELITZIA – lutea on a presumption that you have never Seen the Seed of that Genus, one of the most beautiful in nature'. As if knowing that the name *Strelitzia* would catch Banks's attention, Masson wrote this in capital letters about twice the size of other capitals on the page. The botanical description *lutea* is defined as 'deep yellow, golden-yellow, buttercup-yellow . . . In general deeper than *flavus* and not verging to red as *croceus*'. Linnaeus jnr used '*croceus*' in his 1781 description based, it appears, on Miller's plate (see chapter 1), while the first *Hortus kewensis* in 1789 used '*lutea*'.[37] Masson continued to use the latter term, reiterating his excitement about '*Strelitzia lutea*' when he notified Banks in August 1790 that he had sent him a box of '200 sorts' of seeds with the *Southampton*, which had called into Cape Town en route back to England from Tierra del Fuego to pick up water. The ship was bound for its home port, Southampton, commanded by Captain Moorhead and the box to be collected care of Mr Thomas Potter in that town:

> In this collection you will find Seeds of *Strelitzia lutea* which perhaps you have never seen. [I]t is a very Singular Seed and m[a]y be wanting to compleat the character of the Genus.[38]

Although Banks had seen *Strelitzia reginae* in flower on more than one occasion, and although some of these blooms were described as '*lutea*' (golden yellow), the flower was depicted in

George Loddiges, *Strelitzia reginae* in *The Botanical Cabinet*, 1829

*Curtis's Botanical Magazine* in 1790 as a deep orange. At the very least Masson seems to believe that Banks has never seen the plant's seed – possibly that he was not familiar with golden-flowered strelitzias.

In May 1835, another traveller was captivated by the sight of golden strelitzias. Explorer and soldier Sir James Alexander (1803-1885), who had married the daughter of the Cape surveyor-general soon after arriving in Port Natal as aide-de-camp and private secretary to Sir Benjamin d'Urban, was in the Eastern Cape. He noted in his journal that near the source of the Nahoon River, he had seen:

> The splendid *Strelitzia regina* with its golden crest and red and green neck darted its pointed sheath, like the beak of a magnificent bird, among the vegetation by the river bank.[39]

Thunberg's description of the flower as yellow does not seem to be due to linguistic misunderstanding, although a *Strelitzia reginae* specimen linked to him in the Bergius Herbarium in Uppsala indicates that it was once called *Heliconia aurantiaca*. Although the description *aurantiaca* is derived from *aurum*, the Latin for gold, this stem became established in botanical Latin as meaning 'orange' as in *Citrus aurantium*, the Seville or marmalade orange. To complicate the question of colour perception even further, like Alexander, in 1890 British botanist Scott-Elliot described the *Strelitzia reginae* that he had seen flowering in Cape Town as having 'bright scarlet' sepals. He emphasised this by remarking that the 'red' sepals and blue petals were 'exactly the same as those on the breast of . . . its visitor', the greater double-collared sunbird (*Nectarinia afra*). Although a ruddy variety of strelitzia, *Strelitzia reginae* var. *rutilans*, had appeared in Belgium in 1826, this seems to have been a rusty orange-red rather than a scarlet red. It is not certain either whether this would have been imported back to the Cape where Scott-Elliot might have seen it.

By contrast, various strelitzia species were described as 'yellow' in a 'Select list of stove perennials' by Joseph Paxton in 1836. These were: *Strelitzia reginae, Strelitzia alba, Strelitzia juncea* and *Strelitzia ovata*. More than half a century later Thomas Baines (1823-1895), a gardener to the wealthy who later took up garden writing, described *Strelitzia reginae* as having 'purple and yellow flowers' in his 1894 book on greenhouse and stove plants. In 1921, William Fawcett (1851-1926), formerly director of public gardens and plantations in Jamaica, used the same description but limited this 'purple and yellow' combination to the species he called *Strelitzia parvifolia juncea* – *Strelitzia reginae*, he wrote, had 'orange and purple' flowers. This description of *Strelitzia reginae's* flowers was probably adopted from that given in the 1866 *Treasury of botany*, edited by renowned botanist John Lindley (1799-1865) and Thomas Moore, curator of the Chelsea Physic Garden (1821-1887).[40] Colour variations in some plants can be so unusual that this in itself can make a plant a coveted rarity. Blue roses, black tulips and golden camellias are all highly desirable for this reason.[41]

## OUTDOING NATURE

From the 1830s onwards, Britain's industrial revolution meant the middle and working classes increasingly had more money and time to collect and care for ornamental plants. A flourishing niche market in accessible and more affordable gardening books and magazines was made possible from about the 1840s as copper engraving was replaced by the cheaper but cruder reproduction techniques of chromolithography.

Lower production and sales prices helped boost the new *Gardeners' Chronicle*, founded in 1841 and running until 1986 when it was incorporated into *Horticulture Week*. The *Gardeners' Chronicle* was aimed mainly at professional horticulturists but was also read by enthusiastic amateurs. Gardening writer Jane Loudon (1807-1858) had successfully targeted her earlier books at women – for example, 1838's *Young lady's book of botany* and 1840's *Gardening for ladies* – and was able to capitalise on this by founding the *Ladies' Magazine of Gardening* in 1842. Gardening books and journals such as these tracked and encouraged a fresh direction in the excitement of the rarity – hybrids, which would create great rivalry among garden clubs in industrial areas and dreams of riches for nurserymen and horticulturists.[42]

In the 16th century, Huguenot immigrants from France and Flanders had brought to Britain both their weaving skills, their carnations, pinks and provins roses, and their passion for developing their own unusual varieties of bulbs and plants. They keenly fostered striped and variegated tulips, for instance, finding that 'whoever thinks of raising new varieties of Tulips from seed must be possessed of an ample fund of patience and perseverance'.

This marked the beginning of 'a remarkable new phase in horticulture'. Garden enthusiasts would now routinely 'set themselves the task of producing an ordered beauty, symmetry, and refinement in flowers hitherto grown for their natural charm'.[43]

At the beginning of the 19th century, the potential for new strelitzia varieties captured the attention of Belgian horticulturists. In 1810, *Strelitzia reginae* flowers had won prizes at the Ghent agricultural show for their growers: the Abbé Verdonck, a veteran member of the society; Baron d'Houdetot, then prefect of the *département* de l'Escaut; and the horticulturist F. Van Cassel, reported Belgian botanist Charles Morren (1807-1858), editor of the *Annales de la Société Royale de l'Agriculture de Gand*, in 1846.

Another *Strelitzia reginae* won a prize for Baron Dubois de Vroylande in 1816 but in 1819, Van Cassell again attracted attention with his *Strelitzia ovata*. This plant had seemed to be a new species to William Townsend Aiton, who recognised it in the 1814 edition of *Hortus kewensis*. Today, however, it is considered a synonym of *Strelitzia reginae*.[44]

A *'magnifique variété'* [magnificent variety], the *Strelitzia reginae* var. *rutilans* or ruddy strelitzia, was exhibited in 1826 from the collection of Auguste Vande Woestyne from Wondelghem, recorded Morren two decades later:[45]

*Elle se distingue par la côte rouge pourpre des feuilles, par le bord pourpre de ces mêmes organes, par la couleur empourprée et très prononcée de la spathe générale, par la couleur orange-rouge ou couleur de feu du périanthe et le violet foncé, presque noir des nectaires. Les lames des feuilles sont beaucoup plus grandes que sur les pieds ordinaires des strelitzioe reginae: elles mesurent près de quatre décimètres de longueur et le pétiole n'est pas plus long que cette lame*[46]

[It is recognised by the reddish purple side of the leaves, which have purple edges, by the pronounced purplish colour of the spathe generally, by the fiery, orange-red of the perianth and the deep violet, almost black of the nectaries. The leaf blades are bigger than on ordinary Strelitzia reginae plants: they measure almost 40 centimetres long and the petiole is not longer than this blade]

By the time of Morren's 1846 article, the plant was commercially available in Ghent and the University of Liège had acquired from Belgian horticulturist Ambroise Alexandre Verschaffelt (1825-1886) the best specimen seen by Morren. Despite this, the plant had not been previously featured published in Belgium, England or elsewhere in continental Europe.[47]

## A RARE JEWEL

It was a further half-century later, and almost a full century after *Strelitzia reginae* was named, before French botanist Professor Jules-Emile Planchon (1823-1888) of Montpellier University wrote an enraptured article celebrating his delight at finding a yellow-flowering strelitzia variety:

*Si l'on n'était un peu blasé sur les plus beaux objets qu'on voit tous les jours, l'admiration serait toujours fraîche devant les* Strelitzia *et notamment devant le* Strelitzia Reginae.[48]

[If one was not a touch blasé about the very beautiful objects one sees every day, then our admiration would be forever fresh seeing strelitzias and particularly seeing *Strelitzia reginae*.]

Planchon had worked at Kew after receiving his doctorate of science in 1844. He continued to publish occasionally in English in the *London Journal of Botany* after his return to France, where he taught in schools before being appointed head of the department of botanical sciences at Montpellier University in 1853. Planchon maintained his interest in botany into old age, particularly taxonomic questions, and wrote on the yellow-flowered strelitzia just two years before he died.

The original article was published in the 'Annales générales d'horticulture' section of the journal *Flores des Serres et des Jardins de l'Europe* [Greenhouse and garden flowers of Europe] in 1880, published alongside a dramatic illustration in vivid colour. This exciting curiosity was then splashed in botanical magazines and handbooks across continental Europe in the late 19th century.[49]

Planchon explained that *'ce joyau rare'* [this rare jewel] had appeared in Lille, near the French border with Belgium. An amateur gardener, Monsieur Lemoinier, gave a yellow-flowered strelitzia, to a M. Miellez, part of a well-established family nursery business specialising in exotics.[50] This led to the plant being named *Streliza reginae* var. *lemoinierii* Miellez ex Planch., although much later the World Checklist of Plants corrected the variety name to *lemoinieri*. Planchon emphasised the impact of *'une teinte d'or d'où se détache vivement l'azur de la langue en fer de lance, formée par deux des pétales* [a golden tint stands out vividly

Alex Lagarde, *Strelitzia reginae* var *rutilans* in Annales de la Société Royale d'Agriculture et de Botanique de Gand, 1846

against the azure of the blade-like tongue, formed by two of the petals]'. He commented that this softened ('*adoucit*') the original contrasting orange and blue combination, which had usually instead been heightened or reddened (*rutilans*) by hybridisation:

> *Si brillantes que soient ces fleurs [Strelitzia reginae] par elles-mêmes, les horticulteurs ne peuvaient manquer de provoquer, soit par le sample semis, soit par la voie des croisements, l'apparition de variétés ou d'hybrides modifiant le type primitif. Feu le professeur Lecoq, de Clermont, expert en fait d'hybridisation, avait obtenu par cette voie, avec le Strelitzia Reginae, pour porte-graines, des hybrides . . .*[51]

[These flowers [*Strelitzia reginae*] are so brilliant in themselves that horticulturists did not need to use seed selection or crossing to create varieties or hybrids that modify the original. Once Professor Lecoq, from Clermont, an expert in hybridisation, had obtained hybrids by pollination . . . ]

At Kew, a yellow-flowering strelitzia that appeared in the Palm House in December 1887 was considered a sport. A specimen of that flower, which had 'sepals citron yellow', can still be found in the Kew herbarium under the name of *Strelitzia reginae* var. *citrina*, or citrine bird of paradise.[52]

Accompanying the citrina specimen is a rushed and indignant complaint on headed notepaper from the Grahamstown Botanic Gardens, probably from the curator Edwin Tidmarsh (1831-1915). The letter, dated 3 February 1888 and addressed to 'Mr Watson', complained not about the flower colour but about the way the leaf shape was described:

> Have your notice of our 'Strelitzia regina var citrina' like the name well enough, but with us the plant is not of same foliage as S.regina being more oval in shape, thus our 'ovalifolia' not 'cordifolia' as you make us call it[53]

The addressee is probably former Kew curator William Watson (1858–1925), who was employed by Kew from 1879 onwards and was appointed curator in 1901. He had toured the Eastern Cape from Port Elizabeth to East London in February and March 1887. The visit had included a stop at Grahamstown, prompting the indignant retort from Tidmarsh, who was himself a regular plant collector.[54] Tidmarsh seems to have believed that Watson should have known better about this strelitzia or at least have consulted botanists in the area.

More yellow-flowered strelitzias followed. In Germany in 1896, botanists August Siebert (1854-1923) and Andreas Voss

*Strelitzia reginae* var. *lemoinierii* in *Flore des Serres et des Jardins de l'Europe*, 1880

Specimen of Lemoinier's yellow-flowering *Strelitzia reginae* variety sent by him to Kew in May 1870

(1857-1924) mentioned a '*forma flava (S. citrina)* [yellow form (S. citrina)]' among six forms of *Strelitzia reginae* known in cultivation in the Strelitzia entry in the third edition of *Vilmorin's Illustriete Blumengärtnerei*. No strelitzias had been included in the 1879 first edition of the compendium. Although this is not mentioned, it is possible that Siebert had seen the yellow form at the Frankfurt Palmengarten where he had been director since 1886.[55]

Another strelitzia blooming for the first time at Kew in 1910 produced the surprise of yellow flowers. This time it was put down to an accidental crossing of *Strelitzia reginae* and *Strelitzia augusta* (now *Strelitzia alba*) 11 years previously – technically, hybridisation between any two species in a genus is possible. Both parent species were in flower at the same time as their hybrid offspring, reported Kew's *Bulletin of Miscellaneous Information* that year. Named *Strelitzia* x *kewensis*, a specimen of this hybrid can also still be found in the Kew Herbarium. 'There is a wide difference between the two parents' but the hybrid had features of both, said the *Bulletin of Miscellaneous Information*.[56] All three plants grew in what was then the Mexican House and is now the south block of the Temperate House. The 1.5-metre *Strelitzia* x *kewensis* was somewhat taller than *Strelitzia reginae* and had '10 leaves shaped like *S. augusta*' but with leaf blades about 60 centimetres long and about 45 centimetres wide.

The general appearance of the flower in *S. kewensis* resembles that of *S. Reginae* more closely than that of *S. augusta*. The stamens are enclosed in the fused petal-sheath in the hybrid and in *S. Reginae* but are exposed in *S. augusta*. The small hooded petal of the hybrid, however, is much more like that of *S. augusta* and it is also interesting to notice that the lilac-pink patches at the base of the sepals in *S. augusta* are also present in a similar position in the hybrid.[57]

Just four years later, *The Garden* magazine featured on its letters page an illustration of a yellow-flowered strelitzia, for which it revived the name 'Strelitzia reginae citrina', together with a paragraph of historical and horticultural background. William Fawcett (1851-1926), director of public gardens and plantations in Jamaica, wrote of a *Strelitzia parvifolia juncea* '4 ft. high, with purple and yellow flowers'. After that, yellow-flowered strelitzias seemed to disappear from Kew's conservatories, although they still arrived at the herbarium and were better known in Africa.[58]

Yellow sports and unusually coloured cultivars of *Strelitzia reginae* kept botanical and horticultural taxonomists debating since the time of Banks, Aiton and Morren. In 1846, Morren had noted that the differing size of the ruddy strelitzia's leaf blades and petiole alone should mean that a botanist would separate it as a separate species, '*car le strelitzia humilis ne se distingue pas autrement du reginoe que par ce caractère de grandeur* [because the

# FROM CLASS TO SPECIES:
## THE STRELITZIAS IN THEIR GREATER CLAN OF GINGER, BANANA AND OTHER RELATIVES[60]

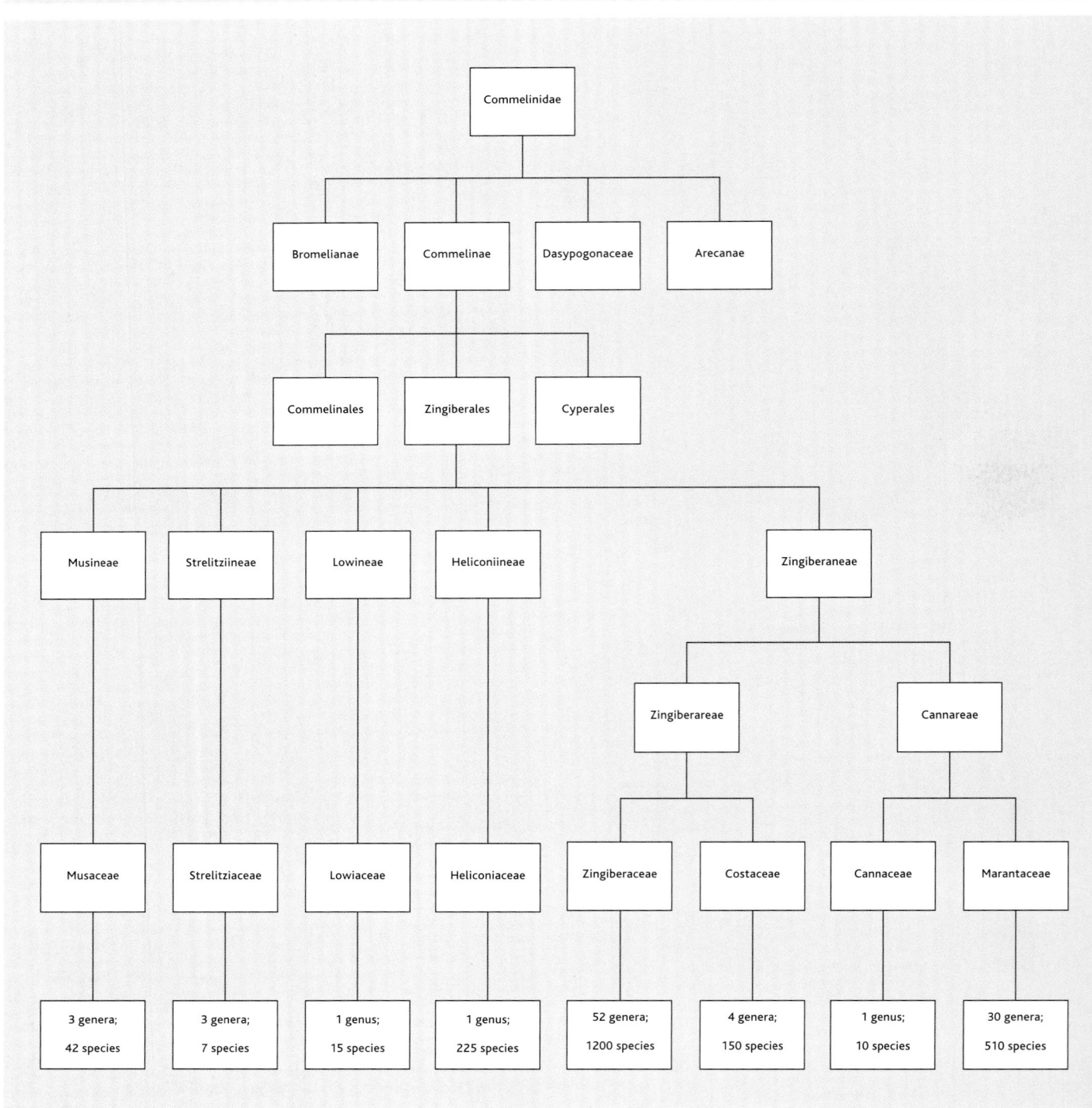

strelitzia humilis only distinguishes itself from the reginae by this characteristic of size]'. However, Morren also appreciated that a small size difference might be rather a spurious reason to declare a new species because he recommended that the whole strelitzia genus needed '*une révision sévère pour bien diagnostiquer les espèces* [severe revision for good species diagnostics]'. Once that was achieved, they could all be displayed together to great effect, he suggested:

> On ne saurait s'imaginer quel magnifique effet produit dans une collection de strelitzia et dans une serre chaude cette riche variété.[59]

> [One can but imagine the magnificent effect that this rich variety could produce in a strelitzia collection and in a hothouse.]

The yellow strelitzia had indeed been a rare novelty, though not completely unknown. From Lemoinier's plant, described with such delight by Planchon, to yellow cultivars that Yutaro Suzuki succeeded in breeding in Japan (see chapter 12), these occasional, dazzling appearances were feted. But achieving a stable yellow-flowering strelitzia cultivar ultimately required the kind of meticulous dedication that Franz Bauer had given to his strelitzia portraits (see chapter 13). It would not be until the end of the 20th century that such a plant would be launched commercially.

## TOOLS TO TRACE STRELITZIAS' ORIGINS

Discovering new species and subspecies of plants, whether *Strelitzia reginae* in the 18th century or its *mzimvubuensis* subspecies in the 21st century (see chapter 14), is one way of filling in the gaps in a plant family tree and mapping out about how the plants evolved from a common ancestor. Using plant systematics, the umbrella under which taxonomy and the study of plant evolution and genetics now tend to be grouped, botanical researchers try to add detail to the family trees of plants and unravel how the form and habits of distantly related plants, their families and orders appeared to have changed to grow closer or diverge further than might be expected. The graphic opposite shows recent consensus, reached by the beginning of the 21st century.

Botany has tried to make the most of newer scientific tools such as microscopes, first widely used to describe plant cells in the 19th century, chromosome analysis from the 1930s onwards, anatomy in the 1950s and then DNA and molecular analyses from the 1980s:

> molecular systematics has revolutionized our understanding of higher-level relationships within the angiosperms, shifting genera among families and families among orders and leading to the first coherent view of the phylogeny of flowering plants as a whole.

Such analyses provide the basis for rigorous studies of adaptive radiation.[61]

The new, hi-tech techniques have revitalised taxonomy and the broader study of how plants adapted and radiated across regions from their place of origin. This allows botanical researchers to test their detailed evidence against the broader theories of historical biogeographers as to how vegetation and climate have changed over the life of our planet.

The superorder Zingiberaneae constituted a particularly interesting candidate for the new techniques in plant systematics because its members – including the strelitzias – form a 'distinctive, isolated' group within the plant sub-kingdom of monocotyledons which have 'always been recognised' as descended from one common ancestor (monophyletic).[62]

***Anatomy***: This was, quite literally, the cutting-edge tool of the 1950s for those who wanted to find out more about plant systematics. Systematists dissect and observe plants, 'provid[ing] valuable evidence in taxonomic research' and steadily working through plant groups to enable better understanding, noted P. Barry Tomlinson. In 1955, he completed his doctorate at Kew's Jodrell Laboratory, using the systematic anatomy approach to the Zingiberaceae, consisting of one overarching order of plants, the Zingiberales or, broadly, the gingers and their relatives. Although he emphasised that he was never a plant collector, Tomlinson went on to become professor of botany at Harvard University in the USA and received both the Linnean Medal in 1999 and the Smithsonian Institute's José Cuatrecasas Medal for Excellence in Tropical Botany in 2002, as he pointed out when accepting the medal.'[63]

***Morphology***: Identifying, analysing and describing plant structures are key to this technique, which can be used to track evolution in a plant or relationships between plants. Irwin Lane compared heliconias and strelitzias,[64] revealing how strelitzia leaves are, in fact, arranged in a spiral but heliconia leaves appear on alternate sides of the vertical stem. Lane deduced that heliconias are a 'considerably more advanced genus' than strelitzias because heliconia flowers are more variable and the plant is better able to adapt 'to a less uniform climatic condition'. He believed that this meant heliconias were the strelitzia relatives that had separated most from the broader family, as well as changing most from their origin in the Musa group. But they remained part of this group and of the Musaceae family because they still resembled its members more than any other genus or family. Ravenala, he noted, was the family member that had developed least from its origins within the strelitzia family and the Musa group. There was a broader discussion in the 1970s and 1980s between half a dozen research teams, debating whether the similar flowerheads seen in members of ginger and bromeliad groups meant that they had evolved from a common ancestor and ultimately from the flowering plant group Commelinales.[65]

**Chromosome counts**: Some of the earliest work on strelitzia chromosomes dates back to the 1930s but in 1989, John Manning and Peter Goldblatt (of Kirstenbosch and Missouri Botanic Gardens respectively) used the technique to sketch a new picture of how the strelitzia family may have evolved. *Strelitzia reginae* and *Strelitzia juncea* had already been shown to be closely related – both have 28 chromosomes, half from a male and half from a female parent (diploid: 2n = 28). *Strelitzia nicolai*, however, had been shown to have 44 chromosomes, the same as *Strelitzia alba* (both diploid 2n = 44).[66] Manning and Goldblatt showed that ravenala and phenakospermum also had 44 chromosomes (both diploid, 2n = 44), the same as the treelike strelitzias.

This suggested to them that all treelike strelitzias have the same chromosome count (although *Strelitzia caudata*'s chromosome count is not yet known). The treelike strelitzias share their chromosome structure and count with more distant relatives, the banana and heliconia suborders (Musineae and Heliconineae). This meant, Manning and Goldblatt believed, that the treelike members of the strelitzias are more ancient and that the shrubby species such as *Strelitzia reginae* developed out of them through 'chromosome fusion'.[67]

**DNA**: Despite great progress in past decades, other major gaps remained in understanding both the Zingiberales order in general and the evolution of different characteristics across all the monocotyledons, Kress noted in 1990. Over the following five years, Kress tried to unravel relationships between the Zingiberales at family level using DNA. But he eventually concluded that better results were achieved by combining this with other available methods and data, especially molecular information and analyses of plant forms, particularly their flowers and seeds (morphology). He believed the strelitzia and heliconia families had both descended from the banana group (Musaceae) but that the Lowiaceae were the strelitzias' closest relatives.[68]

**Fossil analysis**: The fossils of an incomplete banana-like leaf about 16 centimetres long and two leaf fragments that had been found in eastern North Greenland were identified in 1992 as an ancestor of the modern families of Heliconiaceae, Musaceace and Strelitziaceae by Austin Boyd. They were believed to date from the early Tertiary period, up to 65 million years ago. Boyd analysed the vein pattern on the fossils, which is one of the vein forms seen in these three families, and the only vein pattern seen in the genera Strelitzia and Phenakospermum. These were 'the first known Arctic occurrence of fossil leaf material resembling this modern group of taxa.'[69]

Other fossil leaves, seeds and fruits suggest the Commelinales originated up to 100 million years ago during the early Cretaceous, with the ginger and banana group of Zingiberales developing by the following Tertiary period, 65 million years ago. Genetic and morphological analyses supported including the strelitzia family with the lowia family in a new suborder, the Strelitziineae, on a level with the banana, ginger and heliconia suborders.

Members of the Zingiberales diverged quickly from the original plant form and from each other, with 'direct fossil evidence shows members of the Zingiberaceae and Musaceae were present by the Late Cretaceous', American researcher John Benedict noted. He studied seeds, leaves, buds, adventitious roots and rhizomes from Zingiberales fossils found in North Dakota, USA. These did not include any strelitzia fossils, though he did point to a relationship, showing that 'anatomical characters of the seed coat in fossil and extant seeds provide the basis of a more accurate placement of the fossils and a better understanding of character evolution within the order.'[70]

**Phytoliths**: Recently, new technology has also allowed minute mineral deposits in plant cells, known as phytoliths, to be studied. Looking at phytoliths from fossils and from living plants and focusing on their shapes and markings (morphotypes) can help differentiate most Zingiberales, leading to insights into how the plants evolved and even allowing the vegetation of their ancient habitats to be reconstructed. Phytoliths occur in the strelitzia family mostly in a form known as druse, which is common to other families such as canna and maranta also living in similar areas, such as the edges of forest, disturbed ground or forest with clearings or an open canopy that allows through plenty of light.

Most studies have looked at the high-level groupings of Zingiberales such as orders, suborders and superfamilies. The first detailed study of the strelitzia family's evolution and how relationships developed within it found that 'Evolutionary trends are linked to changes in habitat and coevolution with pollinators' and intensified by climate change. Using a consensus of results from five different methods – taxon sampling, molecular characters, DNA extraction and sequence alignment, morphological characters and phylogenetic analyses – persuaded researchers to agree with previous theories that ravenala and phenakospermum are directly related, with southern African strelitzias descending from them. Within these southern African strelitzias, *Strelitzia nicolai* is a cousin of ravenala and phenakospermum. *Strelitzia alba* and *Strelitzia caudata* are directly related to each other, as are *Strelitzia reginae* and *Strelitzia juncea*, with both pairs being descended from *Strelitzia nicolai*. But cautioned the researchers, 'generic relationships within the Strelitziaceae remain equivocal.'[71] The debate continues . . .

Basia Hitchcock Swiel, *Strelitzia nicolai*, watercolour, 2016

# SECRETS OF THE HEAVENLY TREE
*Ravenala madagascariensis and its relatives*

CHAPTER FIVE

## 5 SECRETS OF THE HEAVENLY TREE: *Ravenala madagascariensis and its relatives*

Saviour of the traveller, emblem of the homesick, shelter for the poor, giver of hope to the sick – hoisted high by its lofty fan of leaves, ravenala, popularly known as the traveller's tree, carries this burden of fact and legend. Though long hidden from the outside world, in its home on the Indian Ocean island of Madagascar the tree was already an established contributor to daily life when European travellers first noticed it about four centuries ago. To this day, the tree is frequently described as 'breathtaking' by horticulturists and remains a favourite display specimen around the world. Its distinctive fan-shaped silhouette also makes it a popular visual subject, from botanical art and photography to modern use in a logo. Although the traveller's tree has in many ways become a tree of the world, its island home can never be forgotten as it is doubly commemorated in its scientific name, *Ravenala madagascarensis*. The genus name ravenala is simply derived from the tree's local name, *ravinala*, generally taken to mean 'leaf of the forest'.[1]

The ravenala was long hidden on Madagascar from the eyes of the wider world by the island's isolation and the fearsome cyclones that often assail it. Piracy and warfare in the region as European powers tussled for a strategic Indian Ocean base were also powerful deterrents for scientific investigators. Ultimately, even ravenala's unveiling in Europe was clouded by association with the intellectual piracy of a proven scientific fraudster.

To many people, ravenala has become the symbol of Madagascar. The island is about 1 600 kilometres by 575 kilometres, extending to nearly 600 000 square kilometres. This makes it half as big again as California, twice the size of Great Britain and the same size as France, Belgium and the Netherlands put together. Madagascar is understandably known as the 'Big Island', dwarfing the smaller islands off its coast, and sometimes even as the eighth continent. Madagascar finally adopted its dramatic and statuesque ravenala as the country's official emblem when independence from France was finalised in 1960. A mere quarter-century earlier, ravenala had been scientifically recognised as a member of the newly created Strelitziaceae family, where it joined its new cousins, the southern African strelitzias, and the South American phenakospermum (see chapters 3 and 7). The slow revelation of this botanical alignment seems to echo in the narrower, scientific domain the way that ravenala was slowly revealed to the world beyond the earth's oldest and fourth largest island. Relatively few fossils are found on Madagascar that would help piece together its biohistory, although dinosaur fossils of long-necked prosauropods have been found that are believed to be about 230 million years old. This period is well before the island separated first from Africa and South America about 120 to 135 million years ago and then from the Indian subcontinent about 80 to 95 million years ago. Fossils of the 10-metre dinosaur *Majungatholus atopus* found on Madagascar in 1996, for example, may be related to other creatures found in Argentina and India.[2]

Botanists believe, however, that ravenala was the only member of the strelitzia family to survive in the part of the Gondwana supercontinent that became Madagascar. The family's Gondwana distribution has long fascinated scientists and in 1911, pioneering biogeographer Alfred Russel Wallace (1823-1913), who had visited Brazil and the Amazon before the Indian Ocean islands, added to the second edition of his landmark book *Island life*:

> The celebrated "travellers' tree," *Ravenala madagascariensis* . . . has its nearest ally in a plant inhabiting N. Brazil and Guiana [phenakospermum, see chapter 7]. Echinolaena, a genus of grasses, has the same distribution.[3]

Ancient forms of ravenala probably developed in Madagascar's swampy forests which they still favour to this day. Researchers say the signs of this are the strongly anchored shoots which ravenala produce at the base, typical of larger plants in the group of monocotyledon plants (meaning one-leaf seed). The roots give traction to help the tree stabilise, whether in swampy areas, on steep slopes, or resisting serious waves of fire, either during dry periods or even during paleoclimatic episodes before any human beings arrived in Madagascar. Ravenala's characteristic crest of fanned leaves allows it to exploit gaps in the forest canopy without creating a large crown that could be a liability during one of the island's frequent devastating cyclones.[4]

Madagascar's comparative geographic isolation means highly specialised forms of plant, animal and invertebrate life have developed on the world's fourth largest island. With Madagascar lying 425 kilometres off the African coast, it is possible that visitors from Africa may have encountered ravenala as early as a couple of millennia before the current era, reaching the island at least to forage, possibly to settle. So far, the permanent early settlements found are later than this, dating from 500CE. It is not known when Madagascar's earliest settlers, from Polynesia and Africa, first took notice of ravenala although it seems strange that centuries apparently went by before ravenala, now one of the world's best-recognised trees, was mentioned in the records. This overlooks ravenala's quiet role in Madagascan society through the centuries when many everyday uses were found for it. Madagascar itself was first mentioned in the written record in the 7th century after Arab traders reached it, calling it Serandah. Swahili traders from East Africa used Madagascar as an outpost of the broad Indian Ocean trading network stretching from East Africa across to Arabia, Persia, India and Indonesia, through to China. It still tended to remain isolated, though, as prevailing and perilous

Overleaf: *Ravenala madagascariensis* in Nikolaus Jacquin's *Plantarum rariorum horti caesarei Schoenbrunnensis descriptiones et icones*, 1797

winds and weather patterns meant other Indian Ocean islands offered easier harbours.[5]

Particular mention of the ravenala does not yet seem to have come to light in the *sorabe*, Madagascar's earliest records. These were created in Malagasy using Arabic script after the 9th century when Arab traders had built a trading post or fort on the island's north coast. Such traders may not have tackled the more hazardous and swampy areas where ravenala would be found, though. So far no mention has been found either of the tree that would soon generate an international spiritual and utilitarian reputation in the extensive archives of the Merina kingdom, which flourished in the late pre-colonial period in the 16th century. Though ravenala may have remained in the shadows as far as the rest of the world was concerned, to the Malagasy at that time, its uses were clearly part of routine and ritual. There was no need to record them as being in any way out of the ordinary.[6]

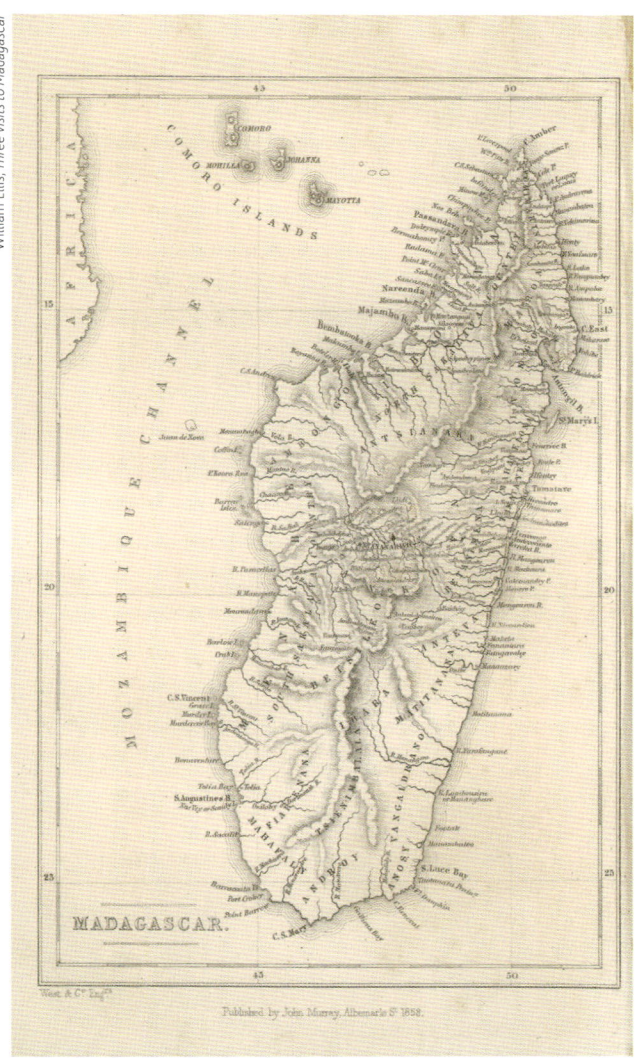

William Ellis, *Three visits to Madagascar*

Madagascar in 1850s

## IN THE SHADOWS OF PARADISE

Like the earlier Arab and Swahili traders, European adventurers of the early age of commercial navigation were still very much at the mercy of wind and weather. Marco Polo (1254-1324) and Pêro de Covilhã (1460-1526) had merely heard about the island. Diogo Dias (fl.1490s), brother of Bartolomeu Dias (c.1451-1500), was swept past it in a storm. In the same period as Europeans were beginning to consider exploring the southern Cape coast of Africa, Dom Francisco de Alameida (c.1450-1510), outgoing Portuguese governor and captain-general of the seas of India, on 1 February 1508 became the first European to land on Madagascar. A fort built in 1527 by Portuguese shipwreck survivors did not survive long and later Portuguese explorers were disappointed that the island would not produce trading staples, such as spices, ivory and gold. Madagascar was not the stereotypical 'paradise island' that poets and explorers depicted in Renaissance writing. Its terrain generally seemed daunting and there was no viable harbour. On the east, swamps flanked precipitous rainforests deluged by rainfall of up to 660 centimetres a year. Beyond jagged volcanic mountain ridges formed a 2 500m barrier from the western side of the island. This rain shadow created thirstlands of baobab on the west and deserts of spiny forest to the south-west.[7]

The Portuguese probably did not encounter the ravenala family of trees during their first century of passing acquaintance with Madagascar, particularly as they do not seem to have strayed far from the security of their preferred landing sites, so avoiding regular contact with the areas where the trees grow. The edges of swamps and forests were both renowned at the time for the danger of tropical fever-causing miasmas. Even if they did encounter the ravenala, though, it may not have warranted more than a passing glance at its eye-catching stature as its uses are more domestic than for the timber trade. As Goa-based Portuguese physician Garcia da Orta (1501-1568) observed in 1563:

> The Portuguese, who navigate over a greater part of the world, only procure a knowledge of how best to dispose of that merchandise, of what they bring there and what they shall take back. They are not desirous of knowing anything about the things in the countries they visit. If they know a product they do not seek to learn from what tree it comes... nor ask about its fruit or what it is like.[8]

A succession of commercial, colonial and missionary deterrents meant the Portuguese cut their losses in Madagascar. This must contribute to the fact that 'there is relatively little work on Madagascar in Portuguese scholarship'. Dutch, French and English writers still praised the riches and wonders of the island but of 16 attempts at settlement on Madagascar in the 16th to 18th centuries, only one had any longevity. Ironically, modern technology has finally shown ways to exploit the mineral wealth that earlier travellers believed and hoped would be found on the island. Sadly, this has been to the detriment of ravenala habitat.[9]

Beyond the possible resources it could yield, Madagascar had a strategic significance for the Portuguese, Dutch, French and English who fought out their intra-European power struggles around the globe. Each one vied for naval control of the western Indian Ocean and for sufficient dominance to exclude the others from lucrative commerce in spices, silks, china, ivory and gold. In turn, this attracted the attention of both pirates and privateers, essentially bounty hunters licensed by an imperial power, who harried naval vessels and preyed on merchant ships carrying rich cargos from the east to Europe. More than 1 000 pirates of various nationalities – English, French, Portuguese, Dutch and American – are said to have switched their attentions to the various East India Company fleets when authorities drove them out of the Caribbean in the 17th and 18th centuries. For about half a century from the 1670s, pirates flourished in bases on Madagascar, the smaller islands around it, particularly Nosy Baraha, or within the Mascarene archipelago. Between 1793 and 1810, privateers captured goods worth £2.5 million from British East India vessels and between 1807 and 1809 alone, the British East India Company lost at least 40 ships to the pirates in their fast Arab-style *boutres*.[10]

## RAVENALA REVEALED

France was the first would-be colonial power to gain a significant foothold on Madagascar. In 1642, in the south-east of the island, the French founded Fort Dauphin (also known as Faradofay and now called Tolagnaro). This was 10 years before the Dutch had landed Jan van Riebeeck (1619-1677) at the Cape of Good Hope to establish the Dutch East India Company's victualling post, including the Company garden. In 1658, the French consolidated their claim to Madagascar when Etienne de Flacourt (1607-1660), who styled himself governor of Madagascar and adjoining islands and was a director-general of the Compagnie des Indes (French East India Company), published an ambitious *Histoire de la Grande Isle Madagascar*.[12]

Whatever scientific advice was available to de Flacourt, it is unlikely anyone attempted to achieve much at that stage in the Madagascan woods or forests, by contrast with the forests of the Cape or India. Attempting to reach them was challenging in the utmost. Even two centuries later, the London Missionary Society's Charles T. Price (1847-1916) noted that the traveller on Madagascar who attempted to strike inland from the coast sometimes found themselves 'nearly up to their waists in soft sticky blood-red earth . . . [with] no rocks for footholds, nothing but . . . shifting, slimy soil.'[13]

## THE PLANT WORLD OF RAVENALA

Madagascar's flora is made up of as many as 12 000 different plant species, it has been estimated. It is as extraordinary for its great diversity as for its high proportion of plants found nowhere else on earth, known as endemics. Madagascar's vegetation was first categorised by Perrier de la Bâthie and then mapped in 1965 by Humbert and Cours-Daarne. The island's four main vegetation types were defined in a recent study as:

1. *Eastern rain forest*: This covers the east coast and eastern slopes of the eastern mountain range, running about 1 500 kilometres from north to south. It extends from the extreme south-east near Tolagnaro (Pic Saint Louis and Andohahela) to the extreme north at Montagne d'Ambre.

2. *Sambirano rain forest area*: This lowland rain forest covers the Tsaratanana and Manongarivo massifs and the Ambanja plain. It is also found on the neighbouring islands.

3. *Tropical deciduous forest area*: This is found through the western lowlands and in part of the north of the island.

4. *Semi-arid spiny bush area*: This occurs throughout the southwest and southern sections of the island.[11]

Vegetation in Madagascar with *Ravenala madagascarensis* (centre) in *The Garden* magazine, 1872

So it is hardly surprising that when ravenala made its first appearance in print in Europe in de Flacourt's 1658 book that this was because he recorded it as being found close to Fort Dauphin. But de Flacourt also spelled out the deterrent nature of the tree's preferred habitat, a lowland swamp area where the rivers were full of crocodiles. This seems appropriate for an island that proved stubbornly inhospitable to the successive waves of would-be imperial claimants who tried but failed to establish a presence throughout the 17th and 18th centuries.

De Flacourt's history was published at the peak of the 17th-century French effort to settle the island so interest was running high back home, both in political circles and in the salons of Parisian high society. Both missionaries and merchants were among the French colonists originally sponsored by the Société de l'Orient, founded by Cardinal Richelieu (1585-1642). French attempts to use divide-and-rule tactics with the small and often volatile Madagascan kingdoms backfired. A general uprising in 1672 ended with the French being driven back to La Réunion island, almost 1 000 kilometres away. Part of the reason for the uprising may have been retaliation against the growing trade in slaves, though indigenous leaders of the feudal-type chiefdoms at certain times cooperated. People were taken as slaves from Madagascar to Mauritius, Rodrigues, the Cape of Good Hope and into the western hemisphere. Later botanists and scientists tended to align themselves with enlightened voices that achieved the abolition of slavery in French possessions in 1794, although this was countermanded about a decade later by Napoleon Bonaparte (1769-1821).[14]

## A BLUE SURPRISE

Conditions might not have been particularly hospitable for naturalists but de Flacourt's privileged position gave him access to as reliable local sources as possible and also to as much protection as could be mustered. He was either personally knowledgeable or had well-informed assistance and categorically distinguished the ravenala with its fan of leaves and curving, almost banana-shaped fruit capsules – '*Ce graine … vient en forme d'un regime de banane*' [this seed comes in the shape of a type of banana] – from the banana tree itself. Even today the tree is recognised as appearing at first glance to be 'a curious mix of characters of both the banana family and the palm family.'[15]

De Flacourt's clear description, the first available written record, showcases the distinguishing ornamental features of ravenala:

> *La plante est fort belle à voir: car elle croist en forme d'un pennache, son fruict vient en façon d'un grand trochet, ainsi que les dattes qui est long comme un espi de bled de Turquie*[16]

> [The plant is really beautiful to see: for it grows in the form of a pennant, its fruit appears like a large cluster, while the dates are long like a Turkish sword]

He also carefully noted how the seeds with their unusual blue covering (aril) are enclosed in a hard casing. Like many ethnobotanists today, de Flacourt was particularly interested in pointing out the food value of the seeds, such as chewing the flour with slaked lime like betel. He described the seed as:

> *… enfermé dans une escorce dure, & chaque grain gros comme un poix, est enveloppé dans une certaine chair bleuë, dont ils font de l'huille, & du grain ils en font de la farine, qu'ils mangent auec du laict.*[17]

> [enclosed in a hard casing and each seed, the size of a pea, is enclosed in a kind of blue flesh, from which they make oil, and from the seed they make flour, which they eat with milk.]

Recent analysis has found that ravenala seeds contain 4 percent oil while the arils contain 68 percent. The fatty acids in

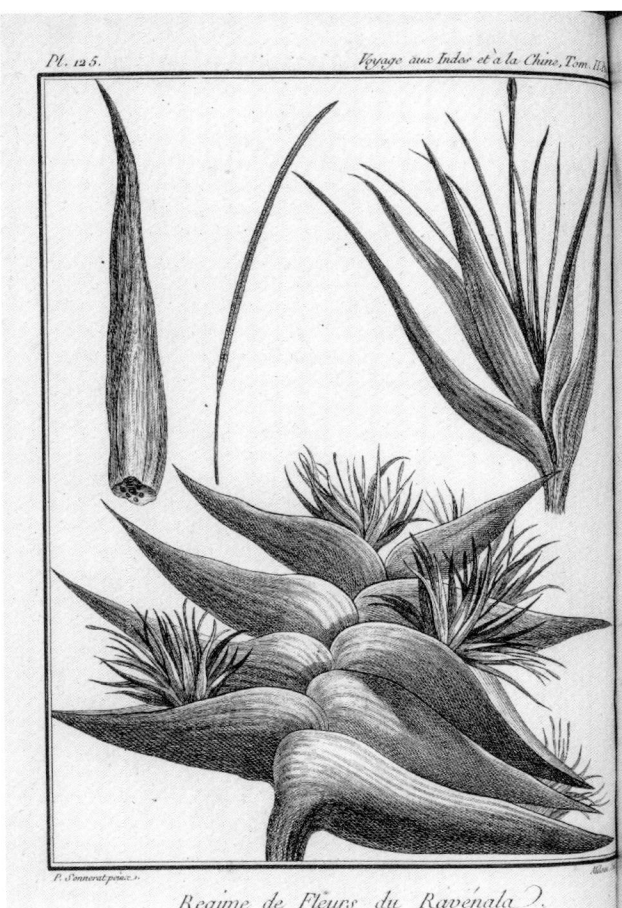

Pierre Sonnerat was first to depict *Ravenala madagascariensis* in his 1772 *Voyage aux Indes et à la Chine*

the oil fall between palm oil and cocoa butter, with their composition analysed at about 39 per cent oleic acid and 34 per cent to 42 per cent palmitic acid. Oil in the seed contains seven sterols, while the aril oil contains 12. The ravenala's blue jacket remained a surprise to botanists and horticulturists well into the 19th century, with French writer Charles Lemaire even admitting that at first it seemed that writers such as the missionary Rev. William Ellis had been mistaken when reporting the seed was enclosed in a blue or blueish-purple coat.[18]

De Flacourt referred to the tree as 'la plante du balizier'. This common name for the canna lily (Canna indica) is also used for both the South American heliconia family and for a South American strelitzia relative, phenakospermum, also familiarly known as 'la plante du balisier'. There has been a suggestion that de Flacourt believed the local name for the plant was Fontsi. Earlier, though, he gives this name as a local variant of Ontsi to bananas with fruit 'gros & long comme le bras [fat and long as one's arm]'. Others, he noted, are 'gros comme la moitié du bras [fat as half an arm]' and there were also Acondre, small bananas 'pas plus grosse qu'un bon pouce [no fatter than a good thumb]'. This indicates that he seems to have been able to distinguish clearly between Fontsi and the balizier. On the other hand, he gives no other local names for the tree, only for its fruits – voafontsi and voadourou.[19]

De Flacourt also highlighted the utilitarian value of ravenala's huge leaves:

> grande d'une brasse, & large de deux pieds. Il y en a quelques unes qui ont plus de huict & dix pieds de haut, sans la tige, qui a quelquefois plus de douze pieds.[20]

[as long as an arm and two feet wide [0.6 metres]. Some are eight or 10 feet high [2.4 to 3 metres], without the stem, which is sometimes more than 12 feet [3.8 metres].]

He focused first on use of ravenala leaves in house-building, then to make table settings:

> Ces fueilles estans seiches sont nommées rattes, des tiges de ces fueilles que l'on nomme estans seiches, Falaffes, l'on en fait les parois des maisons, les Falaffes, & les Rattes, durent six ans, sans se pourrir, les fueilles estans vertes, seruent de nappe, d'assiette, de cuillers & de gobelets à boire.[21]

[When these leaves are dried, they are called rattes [rattans], the dried stalks of the leaves are called Falaffes, and are used to make the walls of houses; the stems & the rattans last for six years without rotting; when the leaves are green, they are used as tablecloths, to make spoons and drinking goblets.]

Nearly two and a half centuries later, American missionary Belle McPherson Campbell observed similar uses:

> Quantities of fresh leaves of the "traveler's tree" are sold in the market every morning and take the place of plates and dishes. A kind of spoon is formed by twisting up a part of the leaf and tying it with tendrils of some climbing plant.

Ravenala and its leaves could be used for more than simply containing food or drink, claimed one French scientist – young leaves were made into a soup that was, he considered, 'très indigeste [very indigestible]', perhaps because it was so high in fibre.[22]

For all de Flacourt's concise and precise reporting on ravenala, it is clear that he found the tree interesting – but not significantly more so than several others among the 200 or so plants with useful seeds and fruit, including rice, manioc (cassava), the durian and Ravensara described in his 1658 book. Ravenala simply appears as number 23 in this account.

## A STRELITZIA FAMILY FIRST

Just over a century later, ravenala found its place in the scientific record when French botanist and academician Michel Adanson (1727-1806) included it under the one-word name Ravenala in his landmark Familles naturelles des plantes, published in 1763. This was a doubly interesting milestone as it would make ravenala the first member of the strelitzia family to receive any formal scientific name. The tree had captured the attention of botanists 10 years before its shrubby relative, Strelitzia reginae, reached Kew Gardens. Another 14 years later, in 1777, Austrian-Italian chemist and botanist Giovanni Antonio Scopoli (1723-1788) formally accepted the name ravenala in his Introductio ad historiam naturalam, a work aimed, in the ambitious contemporary style, at describing all world genera and species and

Ravenala leaves used for thatching in Madagascar

considered a masterpiece of its time. Nearly a decade and a half more would pass before *Strelitzia reginae* was first officially announced to the botanical world in *Curtis's Botanical Magazine* in 1790.[23]

Like many other plants, *Ravenala madagascariensis*, *Strelitzia reginae* and various members of their family carried different scientific names in the past, leaving a legacy of synonyms today. This is partly the result of the scientific naming wars that have broken out intermittently since the 18th century. Their aim was to develop the best way to catalogue scientifically the plants and creatures large and small that have been revealed to the scientific world, whether through scientific exploration, trade or dedicated use of new tools such as microscopes.

Ease of use and clarity helped ensure the triumph of the categorising system devised by the 18th-century Swedish pioneer Carl von Linné, generally known today as Linnaeus. But Adanson's book on the natural families of plants made him one of several contemporary scientists proposing alternatives to Linnaeus's system. Adanson's intelligence and flair for natural science saw this son of an equerry to the archbishop of Aix-en-Provence advance through his studies to be cataloguing plants at the Jardin du Roi in Paris by the age of 21. Adanson felt Linnaeus's method, focusing on certain selected reproductive characteristics as points of identification, was not sensitive enough to categorise every living thing. Instead, Adanson preferred to concentrate on family relationships, taking into account more than 60 characters.[24]

In his 1763 *Familles naturelles des plantes*, Adanson discusses 'Ravenala' as a member of the ginger family, Zingiberaceae. His compare-and-contrast style of describing the family, though, means he spares little particular description for ravenala itself. In his lifetime, Adanson found his system largely eclipsed by Linnaeus.

However, time would somewhat vindicate Adanson. The year 1789 'was almost as important in the history of science as it was in that of politics.' Already, only about 10 years after Linnaeus's death, Jardin du Roi professor of botany Antoine Laurent de Jussieu (1748-1836) had realigned many of Linnaeus's plant families, taking into account both the work of his own uncle, botanist Bernard de Jussieu (1699-1777), and that of Adanson

The 21st century's taxonomic developments relying on DNA rather than only visual evidence would have fascinated Adanson and, intriguingly, continue to support some of his insights. Though Adanson was renowned in his day, to concentrate on his work, he preferred a reclusive life, with no students and almost no friends. Even when the French Revolution left him poverty-stricken, Adanson continued with his work and managed to buy paper and ink. But he had to excuse himself from a meeting of the French Institut National, saying he had no shoes in which to appear before them.[25]

## HARD AS A ROCK

From Sonnerat onwards, ravenala's unusually bright blue aril enclosing the seed attracted the curiosity of scientists. But it was 1911 before French botanist and plant genetics specialist E. Decrock recorded that about a tenth of the vivid seed jacket in ravenala and the tufted reddish arils of its South American relative, now *Phenakospermum guyannense* (see chapters 7 & 8), consist of silicon dioxide ($SiO^2$). It was already well known that silicon dioxide, one of the most common minerals on earth, is present in almost all plants and particularly high in the grasses and horsetails. But the proportion in the ravenala or phenakospermum aril is between 10 and 100 times that of the seed coat of other plants, German plant biochemist Friedrich Czapek emphasised in 1925. It is the abundant fine grains of dark brown silicon dioxide rather than the structure of the ravenala and phenakospermum seed coverings that make them so strong: '*sur laquelle s'ébrèche le rasoir quand on essaie de l'entamer sur des matériaux secs* [that it blunts a razor when you try to use it on dry materials]'. Sulphuric, nitric and hydrochloric acid break down the cell walls but leave: '*petis prismes plus ou moins gonflés qui conservent la forme caractéristique des cellules. Or, il n'ya que la silice qui résiste, de cette manière, aux acides forts.* [little prisms of varying sizes which retain the characteristic cell shape. Now only silica resists strong acids in this way.]' All contemporary tests for silica were carried out and all were positive.[26]

Ravenala fruits showing the 'heavenly blue' of the arils

## A FRENCH BOTANICAL ENTREPRENEUR

Ravenala's ornamental potential was soon capitalised on even though a fuller scientific description would not appear for a quarter-century after Adanson when French naturalist and colonial administrator Pierre Sonnerat (1748-1814) published his *Voyages aux Indes & à la Chine* in 1782. The French had tried again unsuccessfully to take control of Madagascar in 1768 and 49-year-old Pierre Poivre (1719-1786), a horticulturist who had become administrator of Ile de France, as Mauritius was then known, remained interested in Madagascar as a source of plants of exotic, commercial or medicinal interest for his botanic gardens on Mauritius at Pamplemousses. Ravenala is believed to have been grown in Mauritius from about 1768. Certainly, in Sonnerat's 1782 account of his travels, he remarked: '*On l'a transporté à l'Ile de France, où il a très-bien réussi.* [It has been transported to Mauritius, where it has succeeded very well.]'[27]

After a Jesuit education in Paris, Poivre had gone to China as a missionary, become interested in the problems of colonialism and slavery and imprisoned by the British both in China and India – the second time having been captured by a British ship in an attack during which he lost one arm. Poivre used this time to study local cultural approaches to agriculture and land management and later worked on extracting spice plants from the East Indies to grow them in Mauritius and break the Dutch spice monopoly. By the time Poivre returned to Paris, he found a new economic philosophy of physiocracy was taking root. Its followers believed that agriculture and other natural resources were the cornerstone and driver of the economy. Poivre was won over by the opposition of leading thinker François Quesnay (1694-1774) to cash-crop plantations and by the radical ideas on a more equal distribution of wealth to encourage production developed by Irish-French economist Richard Cantillon (c.1680s-1734), called by some the father of modern economics.[28]

Agriculture was seen as the heart of the country's prosperity and discussion of the relationship between man and nature was both intellectual and practical. Poivre spent nearly a decade in academia and warned about imbalances creating deforestation, soil erosion and crop loss as he contributed his observations from his travels and studies. After the 1763 Peace of Paris ended the Seven Years' War between France, Britain, Spain and Portugal, Poivre became a key adviser to the French naval ministry, which had been forced by the bankruptcy of the Compagnie des Indes to take over supervision of the Mascarene island group. Mauritius, Réunion and Rodrigues are the largest of this archipelago of volcanic islands that lie about 700 to 1 500 kilometres east of Madagascar. By 1767, Poivre had given in to persuasion to accept the post of commissaire-general of the Mascarenes and had returned to make his headquarters in Mauritius as *intendant* [administrator].[29]

Poivre travelled with his superior, the new governor, Daniel Dumas (1721-1794). This does not seem to have helped the two establish a rapport and Poivre's five-year tenure proved stormy. He was determined to push policies agreed with the reforming Duc Etienne-François de Choiseul (1719-1785) even further in a radical direction. Choiseul, then minister for foreign affairs, may have quietly tolerated this, particularly because he expected Poivre to spy on Dumas, a senior army officer who seems to have been disillusioned by his experiences in Canada during the Seven Years War. By January 1769, Dumas was the loser in the running battles with Poivre and was recalled to Paris. Later, though, Dumas would not only be absolved of accusations of corruption but also found to have followed his usual pattern of slashing official expenses in half.[30]

## A SCIENTIFIC HOTHOUSE IN THE INDIAN OCEAN

During the final three months of this political drama, French doctor and acclaimed scientist Philibert Commerson (1727-1773) had been on Mauritius, being co-opted by Poivre into helping realise his ambitions in the field of natural history. Then 40 years old, Commerson was a physician and botanist of international standing, who had worked for Linnaeus, studied at Stockholm and Uppsala in Sweden and at the scientifically forward-thinking botanic gardens in Montpellier, France. In 1766, when King Louis XV gave French admiral, Comte Louis Antoine de Bougainville (1729-1811) permission to mount the first French expedition circumnavigating the globe. Commerson was nominated as botanist and naturalist to the king by Pierre-Isaac Poissonnier (1720-1798), royal doctor, diplomat and scientist, who had invented a way to desalinate sea-water that de Bougainville would use.[31]

The expedition had set off in November 1766 via Rio de Janeiro and Tierra del Fuego in South America. By March 1767, it had reached Otaheite, as Tahiti was then known. Commerson was enraptured by the island and de Bougainville similarly found himself recalling *Cythère* [Cythera], an ancient classical island-society idyll. This notion was already well known in France through the painting, 'Voyage to Cythera', by the French artist Jean-Antoine Watteau (1684-1721) and the more contemporary work of Jean-Jacques Rousseau (1712-1778). But Commerson's letter published in November 1769 in the French newspaper *Mercure de France* has been credited with truly popularising the cult of the tropical paradise and the noble savage. Overall, though, the appointment with de Bougainville proved rather a poisoned chalice to Commerson. As well as nursing a recurring leg ulcer, Commerson became seriously seasick as the circumnavigation wore on. When their ship – unfortunately named *La Boudeuse* [Sulker] – reached Mauritius after nearly two years at sea on 8 November 1768, Poivre saw this as an excellent opportunity for Commerson to convalesce on shore while becoming part of the scientific and intellectual team gathering around Poivre. Being released from his duties with the circumnavigation by Bougainville a week after their arrival in Mauritius would prove both a blessing and a curse for Commerson.[32]

Commerson was accompanied, as usual on his expeditions, by his female housekeeper, Jeanne Baret (1740-1807), who disguised herself as Commerson's manservant to work alongside him. On the *Etoile*, the pair had been fortunately allocated the privacy of the captain's cabin so that their large quantity of equipment could also be accommodated. Beyond their personal relationship, Baret was already helping manage Commerson's collections and papers. Some believe she covered for him in Rio de Janeiro when he was already suffering from the leg ulcers that would cut short Commerson's duties with de Bougainville. This included collecting for him – and in the process, she probably collected the specimen now familiar internationally as the tropical garden shrub bougainvillea, named to immortalise the expedition's captain.

The couple journeyed with de Bougainville and the crew on to Montevideo, through the Straits of Magellan to Tierra del Fuego. It seems to have been while they were in Tahiti, Samoa and the New Hebrides that local people saw through Baret's disguise and made public her true gender. She later married Commerson, becoming archivist of his work, a task she retained even after his death when she remarried.[33]

Poivre was already hosting Jacques-Henri Bernardin de Saint-Pierre (1737-1814), who had fought with the French in the Seven Years' War after completing his training as an engineer and who searched as much in his travels as in his writing for 'a lost golden age of human happiness'. Saint-Pierre's major influences were the Enlightenment philosopher Voltaire and, after his return from Mauritius in 1770, Jean-Jacques Rousseau. Saint-Pierre's visit to Mauritius was the sequel to a childhood dream for the boy who had read *Robinson Crusoe* at the age of 12 and sailed to the West Indies with his sea-captain uncle. Though Saint-Pierre would write a best-selling novel set on the island, *Paul et Virginie*, this period also became the prelude to the next, different chapter of his life. The botany that Saint-Pierre learned on Mauritius in the company of Poivre and Commerson meant that he eventually became *intendant* (administrator) of the Jardin des Plantes in Paris.[34]

Saint-Pierre would have become familiar with ravenala, for example, in Poivre's Mauritius garden and possibly also in the highly regarded garden of plant-loving engineer Jean-François Charpentier de Cossigny (1690-1780). His son, botanist Joseph-François Charpentier de Cossigny (1736-1809), who introduced the litchi to Mauritius and Réunion, was detailed to work with Commerson on Mauritius.

In addition, Mauritius was a port of call for about a third of all French shipping passing the Cape of Good Hope and was key to French trade and power in the Indian Ocean at the time. It is all the more likely that Poivre had a ravenala sent to the Jardin du Roi in Paris given that the tree was one of 52 species selected to be sent by the director of the Pamplemousses gardens, in early versions of the wardian case to the conservatory at Schönbrunn in Vienna, Austria, in June 1782. It was illustrated and described in

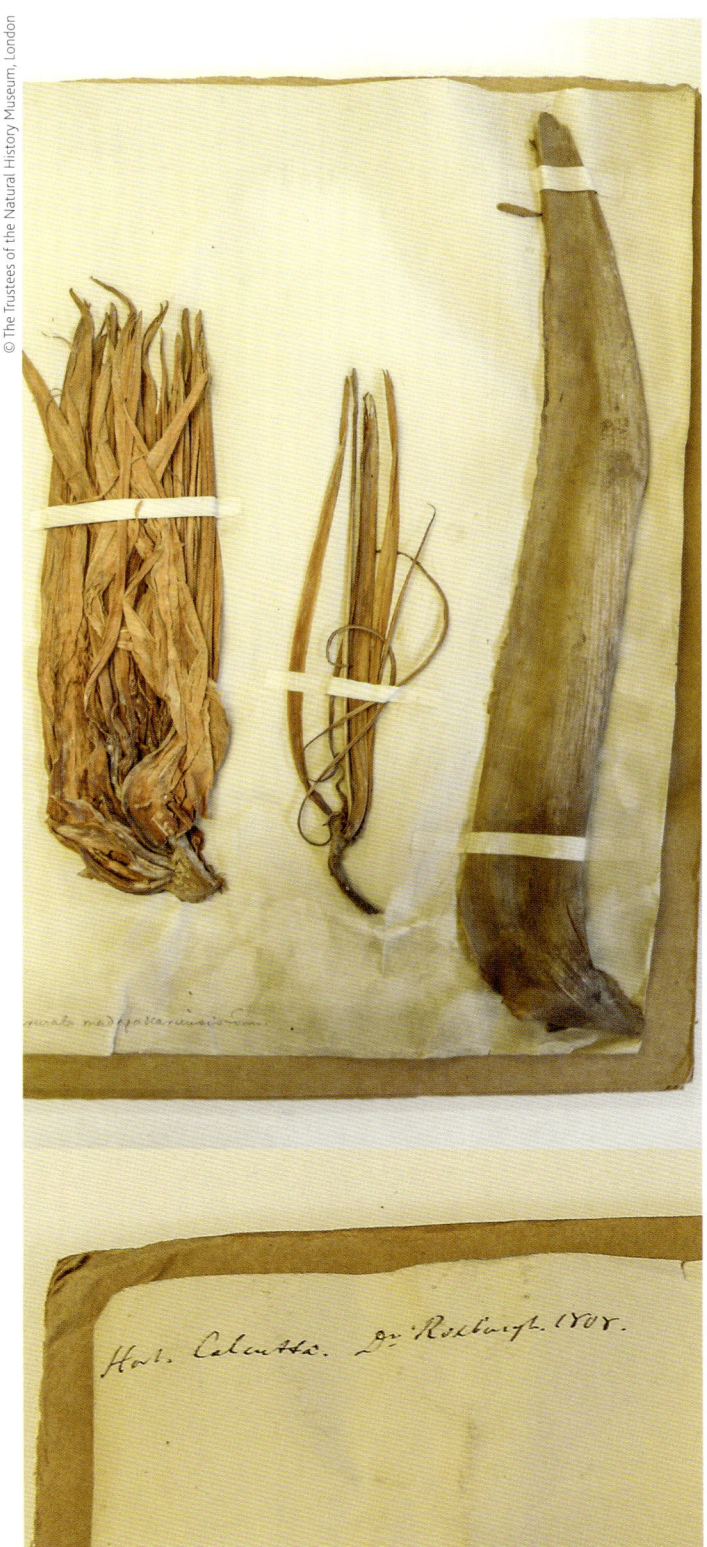
© The Trustees of the Natural History Museum, London

Specimen of *Ravenala madagascariensis* sent by William Roxburgh from Calcutta, early 1800s

Baron Nikolaus von Jacquin's four-volume celebration of the rare plants collected to enhance Emperor Franz Josef's palace. The Mauritian gift helped rebuild the collection after a disaster in 1780: 'the principal gardener, who was suffering from gout, left the care of the hothouses to his assiatants who, on a very cold winter night, forgot to fire the furnace.' Trying to revive his treasures, the gardener next morning set the furnaces roaring – only to see the precious rarities wither and die.[35]

In 1802, William Roxburgh recorded that 'three plants of this elegant tree were brought from the Island of Mauritius to the Botanic Garden in Calcutta.' All were planted 'in different soils, and situations' but the one that had been planted 'in a very moist place, and in a rich brownish black soil, throve more luxuriantly than the other two', flowering around the end of 1806 and producing ripe seed in November 1807. A specimen from Roxburgh in Calcutta, dated 1808, still exists in the Natural History Museum herbarium in London. In Bengal, the tree was known as the *panthopadop*, according to David Prain (1857-1944), superintendent of what had by then become the Calcutta Royal Botanic Gardens and botany professor at the Calcutta Medical College. The tree was also an early arrival at the Peradeniya Botanic Gardens, being introduced there before 1824.[36]

Poivre and Commerson struck up an excellent relationship, discussing land-use planning and rhapsodising together about Mauritius as a utopia. Poivre also put Commerson's skills to work on the island particularly and in the region generally. Officially, Commerson was supposed to survey the medicinal resources of Mauritius and eventually permission for this came through from Poissonnier in Paris. Unofficially, and more boldly, Poivre hoped that Commerson would survey the natural history of all the Mascarene islands. The ideas of a survey and of employing a professional staff naturalist were still a rarity but to entrust this to a man weakened by tropical sickness was a gamble. Even so, Commerson carried out a survey of the woodlands and forests on Mauritius, assisted by Joseph-François Charpentier de Cossigny.[37]

## DISCOVERING THE TRAVELLER'S FRIEND

Despite the opposition they had experienced, the French were still keen for observers such as Commerson to continue assessing Madagascar's economic potential. Poivre's brief to Commerson for his expedition twinned botanical exploration with a political request to scout for sites for a new French base on the island. Commerson was eventually fit enough to depart for Madagascar on 25 October 1770, only shortly before the French were due to pull out from Fort Dauphin. He was under no illusion that Poivre's broader interest was in the botanical findings and he himself was excited to make the expedition. The ravenala specimen in Poivre's Mauritian garden would itself have been an intriguing talking point and a foretaste of the novelties to expect on Madagascar. Commerson was able to spend about three weeks on Madagascar, collecting nearly 500 species for his survey of the island's flora and trying to rename ravenala.[38]

Commerson had come to Madagascar prepared, not only following the route de Flacourt had taken from Ile de France about a century earlier but also having studied his *Histoire de la Grande Isle Madagascar*. As well as Baret, Commerson was accompanied by Paul Sauguin de Jossigny (fl.1770s), the other illustrator whom he used in Mauritius as well as Sonnerat – though Sonnerat would later try to enhance his own prestige by claiming he and not de Jossigny had accompanied the scientist. Also part of the party was a young African slave who had been trained to collect plants and had a good instinct for novelties. Commerson was ambivalent about being involved with slavery, not least because the youngster reminded him of his son. Commerson wrote in a letter to his brother-in-law that: '*il lui fera oublier à jamais qu'il a été esclave*' [he would make him forget he had ever been a slave]'.[39]

Despite the political and military misgivings of the French on Madagascar and the hostile reception given to many other European landing-parties, Commerson wrote warmly of the inhabitants:

*Une forte preuve de la bonté, de la douceur et de l'humanité de ces insulaires, c'est que dans un temps où il se fallait se tenir respectivement sur ses gardes, j'ai parcouru toute la partie la moins bien famée de cette île, en caleçon et en veste, un jonc à la main, et j'ai trouvé partout un favorable acceuil.*[40]

[Strong evidence of the kindness and gentleness and humanity of these islanders is that in a time when both sides need to be on their guard, I have wandered around the worst-renowned parts of this island, in longjohns and waistcoat, a twig in my hand, and I found a good welcome everywhere.]

Commerson rhapsodised about Madagascar in this letter to his friend and former tutor, astronomer Jérôme Lalande (1737-1802).[41] Despite all the other wonders Commerson had seen with Bougainville, he wrote to Lalande:

*C'est à Madagascar que je puis annoncer aux naturalistes qu'est la veritable terre promise pour eux. C'est que la nature semble s'être retirée dans un sanctuaire particulier pour y travailler sur d'autres modèles que ceux auxquels elle est asservie ailleurs. Les formes les plus insolites y les plus merveilleuses s'y rencontrent à chaque pas.*[42]

[I must tell naturalists that it is in Madagascar they will find their real promised land. Here nature seems to have withdrawn into a special sanctuary to work on different models than the ones she uses elsewhere. The most unusual and marvellous forms are encountered at every step.]

Some of this hyperbole has been put down to Commerson's nose for publicity, with one tart comment noting: '*Commerson savait très bien que la lettre qu'il écrivait . . . ne finirait pas au fond d'un tiroir* [Commerson knew very well that his letter . . . would not end up at the bottom of a drawer]'.[43] Lalande not only shared Commerson's earlier correspondence from Tahiti with a Parisian salon but also published it. Although Baron Marc Antoine de Clugny (1741-1792), commander of the ship that had brought Commerson to Madagascar, was critical of the naturalist in other ways, he agreed with Commerson's view of: '*cette lande, qu'il appelle à juste titre, le paradis des Naturalistes* [this land, which he justly calls a naturalists' paradise]'.[44]

Commerson seems to have foreseen the depth of modern scientific research focus on Madagascar when he commented: '*Madagascar . . . mériterait à lui seul, non pas un observateur ambulant, mais des academies entières.* [Madagascar would merit to itself, not just a single roving observer, but whole academies of them.]' In fact, this also alludes to Commerson's 1770 suggestion that a scientific academy should be founded in Mauritius to study subjects from exotics to tropical diseases that were better studied on site than in Europe. The proposal seems far-sighted as it predates Europe's late 18th-century wave of founding learned societies, such as the Linnean Society, founded in London in 1788. Although the notion was supported, perhaps even fomented, by the ambitious Poivre, the academy was not then founded on Mauritius. The impact of Commerson's pioneering thinking on how migration contributes to species variation, however, has been traced through Bernardin de Saint-Pierre to the development of these theories by the better-known French naturalist Georges Cuvier (1769-1832) and Jean-Baptiste Lamarck (1744-1829).[45]

## MADAGASCAR'S MARTYRS OF BOTANY

Despite Commerson's enthusiasm for Madagascar and his efforts at collecting, fate was against him in several ways, including in his dealings with ravenala. The manuscript he never managed to publish shows that he wanted to rename the tree *Dalembertia uranoscopa*. *Dalembertia* was in tribute to mathematician Jean Le Rond d'Alembert (1717-1783) and his doctrine of natural philosophy to replace formal religion for enlightened scientists and scholars. *Uranoscopa* means 'heaven-seeing' for the vertical leaves which Commerson described as 'presenting to the sky their entire upper surface'. But unfortunately for Commerson's creative homage, the rules of science are firm and Michel Adanson's 1763 usage, ravenala, finally stood. Interestingly, Commerson also uses this himself, though in the sense of a common name: 'Ravenal resembles absolutely the banana'.[46]

Apart from this generalisation, Commerson's observations are usually more carefully detailed than de Flacourt. Commerson was also reputed to relish passing on an enjoyable story so it is

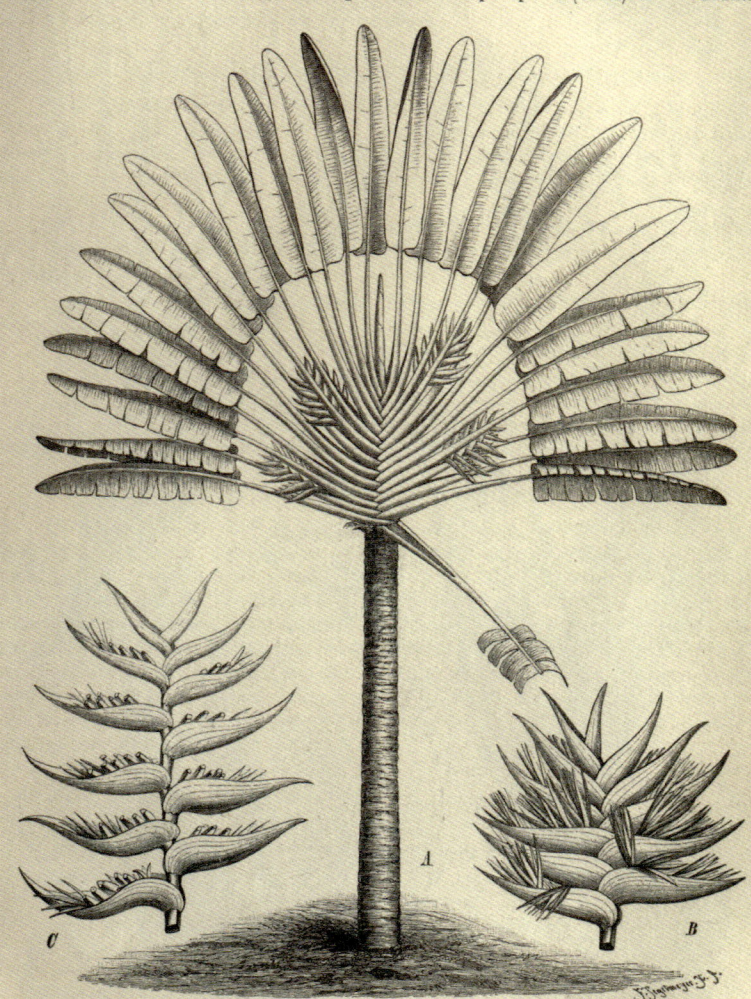

*Ravenala madagascariensis* depicted in *Das Pflanzenreich* magazine, 1900

hardly surprising that he is said to be the first to depict the tree as the traveller's friend. Despite some pragmatic and scientific attempts at debunking, this is the romanticised image attached to the tree ever since:

> These leaves, well suited by their immense size to receive a large quantity of rain water, transmit it by means of a channel, occupying the centre right up to the pedicel [stalk] of the same leaf, thus the pedicel is very flat and wide, and clasps the stem very closely, yet in a manner as to make a quite ample capsule where water is held in [a] reservoir for the need of the plant, giving it a particular flavour that is not at all without merit. This sort of reservoir can serve abundantly to quench the thirst of several travellers by piercing the [stalk] in the lowest part.[47]

Commerson's bad luck with his health culminated in his observations becoming a sidelight on ravenala's history. His collecting journey to Madagascar had been repeatedly delayed as he relapsed into sickness, including an attack of gout. He injured himself badly on his final day's collecting, however. The small fleet then endured a seasonal typhoon while leaving Madagascar on 4 December. This wrecked the accompanying ship *La Garonne* at the mouth of the Black River and caused so much damage to *L'Ambulante*, on which Commerson was sailing, that they had to put in at Réunion for repairs. Fortuitously, this allowed Commerson time to recover from his injury and do some botanical exploration of Réunion. But it may all have been too little, too late as far as Poissonnier and the officials in Paris were concerned. Once again at odds with the superior, this time Governor Julien Desroches, Poivre was recalled to Paris. Commerson's salary was also stopped but he was too sick to make the journey with his friend and administrative chief. Eventually, having relapsed again into fever after returning to Mauritius and then developing dysentery, Commerson died there aged 45 in March 1773. He has been nominated as one of Linnaeus's 'martyrs of botany'. One biography of Commerson is even entitled *Martyrologie*. Just a year later, Frans Oldenburg, who had been recommended to Governor van Plettenberg by Thunberg to join an expedition to Madagascar, died on the island, also of a fever. Ironically, Oldenburg had been nominally employed as surgeon's mate. So far, no Oldenburg ravenala specimens have come to light.[48]

Political considerations allowed Commerson's reputation largely to go into eclipse for a century or so, however. Ironically, part of Commerson's brief on Madagascar had also been to assess the suitability of the French military's new preferred outpost, the vividly named Foulepointe (now Mahavelona). Commerson pointed out that this area was fever-ridden from December to May, whereas the existing outpost of Fort Dauphin was fever-clear all year, adding: '*Toute la pointe n'est qu'un vaste cimetière de François* [The whole point is nothing but one vast cemetery of Frenchmen]'. He probably did not know that he was echoing Dutch seafarer and explorer Cornelius van Houtman (1565-1599), who had called Madagascar 'Coemiterium Batavorum', the graveyard of the Dutch.[49]

The lure of the island's rarities still outweighed the risks for many botanists, however. As English missionary Rev. William Ellis wrote in 1838:

> The productions of the west, north and southern coasts, and of all the interior, remain entirely unknown, and the slender documents that have been furnished as to the vegetation of the north-east by the French naturalists, most of whom have perished from the effects of the climate, serve rather to stimulate than to satisfy a botanist's curiosity.[50]

Sadly proving Ellis's point, in 1866, English plant hunter William Gerrard (c.1831-1866) died of yellow fever at Foulpointe – another ghost for Madagacar's imperial graveyard.[51]

*Ravenala madagascariensis* fruit in Pierre Sonnerat's 1772 *Voyage aux Indes et à la Chine*

## CLOUDING RAVENALA'S REPUTATION

In 1782, just over a decade after Commerson's visit to Madagascar, his protégé and Poivre's nephew, Pierre Sonnerat, published the first full description and series of illustrations of ravenala, as well as completing the tree's naming in Linnean terms as *Ravenala madagascariensis*. This form eventually won favour with the scientific and taxonomic community and ousted the claims for *Dalembertia uranoscopa*, the name put forward by Sonnerat's own mentor Commerson. In Sonnerat's lifetime, though, the name *Ravenala madagascariensis* was still disputed – and despite Sonnerat's talent and the apparent credibility of about 15 years spent journeying in the east, his ruthless ambition undermined his reputation and possibly immediate recognition of ravenala's scientific name.

Leaving Paris in 1768 at the age of 20, Sonnerat had gone first to Mauritius, initially to work as a secretary for Poivre, the island's administrator, who despatched him in 1769 as *écrivain* (scribe) on one of the French king's vessels sailing to India, the Philippines, the Moluccas and, according to the title of his later book, New Guinea. Before his departure, Sonnerat had benefitted from Poivre's scientific hothouse, being mentored as a naturalist by Commerson and through conversations with Commerson and Saint-Pierre being exposed to some of the most respected scientists and thinkers in France. Sonnerat was a talented illustrator who spent some time working alongside Paul Philippe Sanguin de Jossigny as an illustrator for Commerson. Sonnerat was later considered an 'able observer' by Messieurs de la Lande and Fougeroux of the Royal Academy of Sciences, who gave official approval for the publication of Sonnerat's book. Sonnerat's work was also praised by Parisian scientists such as Michel Adanson.[52]

In 1772, the year Poivre was recalled from Mauritius to Paris and while Commerson was heading to Madagascar, Sonnerat returned to the French capital to deposit his own natural-history collections.[53] These included the breadfruit (*Artocarpus altilis*) and birds of paradise (Paradisaeidae). Sonnerat was preparing the manuscript of his first book, *Voyage à la Nouvelle Guinée*, when news reached Paris of Commerson's death on Mauritius. That put in train a series of events which apparently left Sonnerat eager to put more excitement and rarities into his manuscript – but would leave a stain on his reputation, cloud his future credibility and may have sidetracked the scientific naming of ravenala for decades.

During 1771, while passing through the Cape of Good Hope, Sonnerat had met Joseph Banks and Daniel Solander, who were homeward bound after their voyage on the *Endeavour* with Captain James Cook and were staying there for a month while the surviving crew recovered from the ravages of malaria and dysentery (see chapter 1). Sonnerat's charm and enthusiasm struck a chord with Banks and to help the cause of international scientific exchange, Banks asked him to deliver some bird skins he had collected in the Antipodes and beyond to Philibert Commerson on Mauritius. Sonnerat gave the bird skins to de Jossigny to illustrate but when Commerson died shortly afterwards, Sonnerat was tempted to the deception that would be his scientific downfall. He included the de Jossigny illustrations as birds of New Guinea in his first book, eventually published in 1776, and did not shy away from the ambitious title *Voyage à la Nouvelle Guinée*. Sonnerat soon faced a threefold scandal. He had taken credit for the de Jossigny illustrations himself as well as claiming to have visited New Guinea without actually having done so. This was underlined by the fact that the birds he had chosen to identify with the island were an Australian kookaburra and gentoo and king penguins which do not occur there. Whether Sonnerat already had the illustrations or had retrieved them from Commerson's effects in Paris is not clear. But the fact of the deception and plagiarism was not unmasked until later in Sonnerat's life.[54]

Commerson's death on Mauritius in 1773 had left unpublished most of his manuscript observations on his travels in

Specimen of *Ravenala madagascariensis* from Pierre Sonnerat, 1770s

the Indies. Sonnerat set off on his second voyage the following year. He spent seven years, more than twice as long as his first voyage, travelling around India, China, the Moluccas and the Mascarenes. When Sonnerat returned to Paris in 1781, despite having no apparent competition, he rushed his new observations into print a year later. The three volumes of *Voyage aux Indes orientales & à la Chine* included his landmark description and illustrations of the ravenala. With a strong market nationally and internationally for travel books, many were promptly translated into English and other languages. Sonnerat's *Voyage à la Nouvelle Guinée* had appeared in English five years after its 1776 publication and his *Voyage aux Indes orientales & à la Chine* appeared in German two years after its 1782 publication, followed by Swedish and English. Sonnerat's godfather and mentor Pierre Poivre (who had been vindicated over his role in Mauritius by a commission of enquiry) probably assisted the process practically and could capitalise on support among the Parisian intellectual establishment. Whether Sonnerat could still make the most of that would be another question, although he highlighted his credentials on the title page: '*naturaliste pensionnaire du roi et correspondant de son cabinet* [the king's official naturalist and correspondent of his cabinet]', as well as commissioner and correspondent of the Royal Academy of Sciences in Paris and member of the Lyons Academy of Sciences. For added support, he dedicated the book to the Charles-Claude Flahaut de la Billaderie, Comte d'Angiviller, holder of a range of military and courtly honours, including overseer of Versailles and director of the king's buildings, gardens, arts, academies and royal factories – in other words, the director of royal patronage.[55]

Sonnerat also gave the first descriptions of key Madagascan animals such as the indri and the aye-aye but like Commerson, he was expected to extend beyond natural history to comment on the social and political situation as well.[56] Sonnerat appeared keen to strike a balance by appearing less effusive than Commerson in order to add credibility but did not shy away from familiar controversies, lacing the text with often tart observations.

William Ellis, *Three visits to Madagascar*

## REPORTING ON RAVENALA

By the fourth page of his general introductory chapter to Madagascar, Sonnerat had mentioned ravenala, though not with great fanfare. He was fascinated by how '*savans* [wise men]' reveal writing on ravenala leaves:

> *ils se servent du poinçon, à la manière des Indiens: les caractères tracés sur la feuille, n'y sont pas d'abord très-sensibles; mais à la mesure qu'elle séche, ils deviennent très-noirs.*[57]

[they use an awl, the way the Indians do: the characters traced on the leaf are not at first really visible, but as it dries they become very black.]

Another customary use of the ravenala which caught Sonnerat's interest was the elaborate boat blessing carried out when a voyage along the coast was to take place. Watched by his crew sitting on the shore, the pilot:

> *prend l'eau de mer dans un morceau de feuille de Ravénala, puis il addresse des prières à l'élément qui va le porter; il le conjure de ne point faire de mal à son navire, de la garantir au contraire de tous les écueils, & de le ramener promptement au port chargé de beaucoup d'esclaves: ensuite il se met dans l'eau, fait le tour de sa pirogue & l'asperge tout au tour … il revient sur le bord & fait un trou dans la terre, pour y déposer le morceau de feuille de Ravénala.*[58]

[takes sea water in a piece of ravenala leaf, then addresses prayers to the element [the sea] that will carry him; he abjures it not to harm his boat, to guarantee it on the contrary from all dangers, and to bring him promptly home to port loaded with many slaves: then he gets into the water, goes round the dugout, dowsing it with water all round …he returns to the shore and makes a hole in the ground, where he leaves the piece of ravenala leaf.]

Sonnerat questioned the existing slave trade and noted also the widespread use of rifles and pistols, supplied by both earlier traders and Europeans and used even to sound a volley of gunfire to warn evil spirits away from the bed of a seriously ill person. He elaborated on de Flacourt's earlier reference to ravenala as a staple for walling and thatching Madagascan homes. This prompted Ellis to call it the builder's tree, a name it still carries in parts of Asia.[59] Once the large supporting stakes have been driven into the ground, Sonnerat said:

> *les parois sont faites avec des côtes de la feuille de Ravénala jointes ensemble & liées contre des lattes de bamboo … Le toît est couvert de feuilles de Ravénala, dont les côtes sont rapprochées les unes à côté des autres, ce qui forme une couverture très-solide*

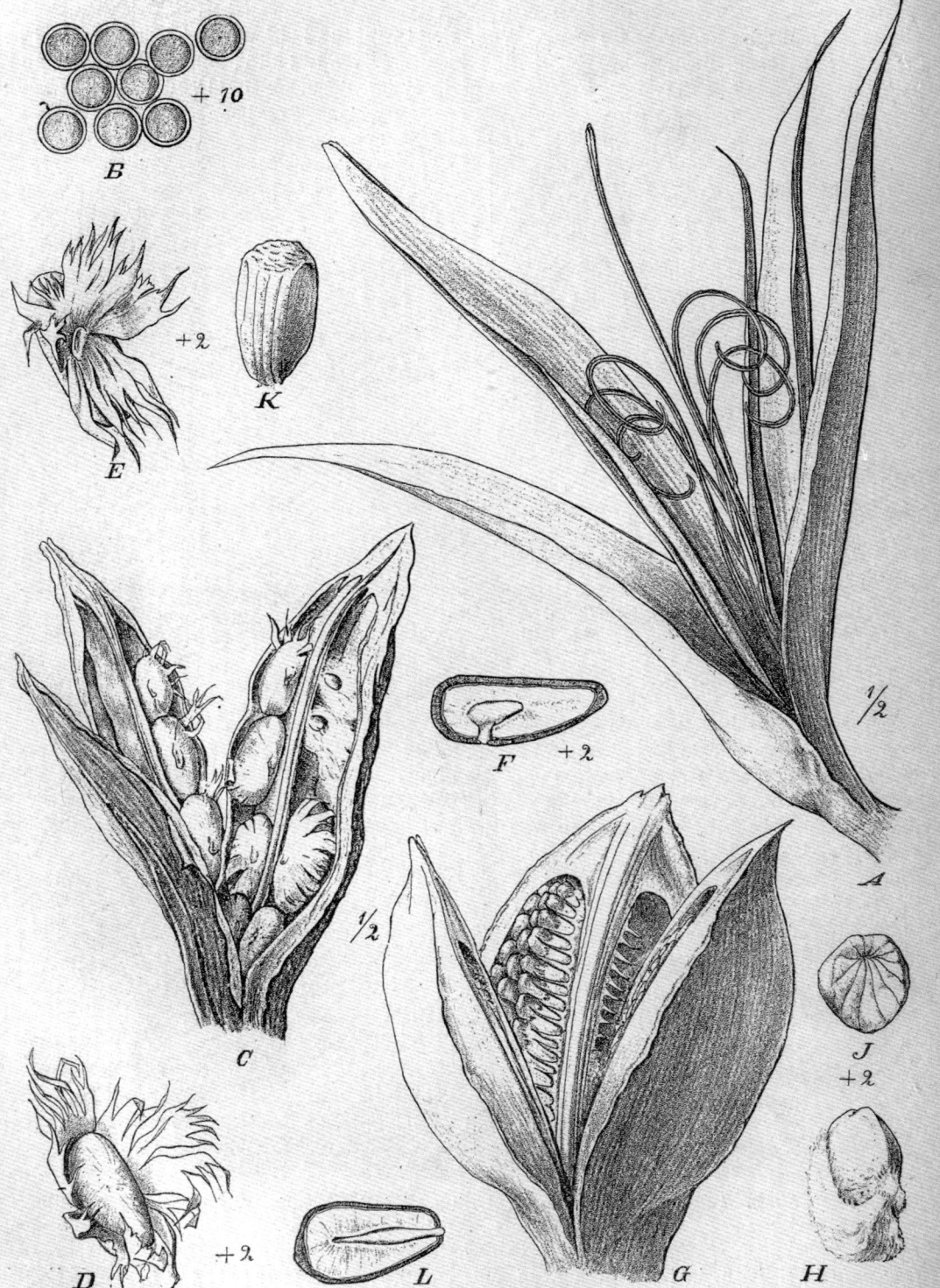

Fig. 7. *Ravenala madagascariensis* Sonn. *A* Flos. *B* Pollinis granula. *C* Capsula. *D* Semen. *E* Arillus. *F* Semen longitudinaliter sectum cum embryone. — *Ravenala guianensis* (L. C. Rich.) Benth. *G* Capsula. *H* Semen. *J* Idem ab apice visum. *K* Idem a latere visum. *L* Idem longitudinaliter sectum. — Icones originariae.

[the walls are made by joining the sides of the ravenala leaf together and tying them against bamboo battens … The roof is covered with ravenala leaves, with their edges close one to the other, so they form a very solid covering]

In the early 20th century, a French colonial official commented that one of the Betsimakara's most common practices was to use the upper bark of trunks to make winnowing baskets for threshed rice.[60]

## HONOURING THE ISLAND

When he named *Ravenala madagascariensis*, Sonnerat expanded Adanson's name to full Linnean two-word (binomial) format, choosing a specific name that referred to its island home. Despite pushing his connection with Commerson among others in Parisian intellectual society, Sonnerat did not use or reference the manuscripts of his ill-fated colleague which had been brought back to the Jardin du Roi. Nor did Sonnerat use or make reference to Commerson's manuscript proposal of the name *Dalembertia uranoscopa* for ravenala. Whether Sonnerat was trying to avoid another charge of plagiarism or simply trying to make his own mark is open to question. The type specimen at the Natural History Museum herbarium in London appears to be one collected by Sonnerat himself.[61]

Sonnerat's excitement at his travels and discoveries shines through at the end of his detailed botanical description while describing the ravenala aril as covered in a skin *d'un beau bleu de ciel* [a beautiful sky blue]'. He soon returned to the solemn tones of the scientist, though, prefiguring considerable scientific discussion by immediately recommending that the ravenala should be part of a separate genus within the banana family, even though its leaves look like those of a banana tree and it has banana characteristics. The deciding factor, he noted, is the tree's three-chambered fruit with multiple seeds (polysperm). He contrasts this with Linnaeus's heliconia, where the fruit chambers are single-seeded (monosperm) and the Musa 'or true banana', with only one one-chambered capsule.[62]

Sonnerat made three illustrations of ravenala. There is a study each of a flowerhead and of the fruit, as well as the now iconic image of the trunk '*couronné par un évantail parfait et superbe* [crowned by a perfect and superb fan]' reaching to the heavens. In his drawing as well as in his words, Sonnerat was careful to show the detail of how this fan of leaves is formed by the ravenala's long leaf stalks emerging from the trunk and crossing over each other near the base in the characteristic style of the treelike strelitzias like a classic French plait.[63]

Though Sonnerat's work contributed to the foundations of anthropology as well as natural history, several critics at the time reprimanded him for frivolous observations. Whether despite or because of this poor reception, Sonnerat promptly tried to exploit his connection with Sir Joseph Banks to achieve election to the Royal Society in London but did not succeed. He was honoured later, though, with election to the French Institut National in 1803.[64]

Even Sonnerat's enemies probably would not have wished his successful and comfortable later career as a colonial bureaucrat at the French enclave of Yanam, India, to end with 10 years in a British prison in Pondicherry (now Puducherry), India. But they might have smiled that even this side-effect of the impact of the French revolutionary wars in the Indian ocean was eventually resolved by Sonnerat's connections. Sir Joseph Banks and Antoine Laurent de Jussieu jointly interceded on his behalf. However, he survived only a few months before dying in Paris in March 1814. He had completed the manuscript of another book, 'Nouveau voyage aux Indes orientales', and de Jussieu was asked to edit it for publication. Whether Sonnerat ever salvaged his reputation sufficiently for a publisher to consider the manuscript is unknown. The book was not published, however, and the manuscript is lost.[65]

## CONFLICT OVER THE HEAVENLY TREE

Meanwhile, Sonnerat's scientific naming of ravenala was also derailed, with the tree being given another six name variations over the following half century. In 1789, German scientist Johann von Schreber (1739-1810) picked up on Commerson's '*uranoscopa*', used the name *Urania* in 1789 in his revision of the eighth edition of Linnaeus's *Genera plantarum*. Several other botanists followed the form given in this authoritative publication, although the name that Sonnerat preferred, *Ravenala madagascariensis*, was used in 1791 by Johannes F. Gmelin (1748-1804), who had taken over updating the benchmark reference work of Linnaeus's *Systema naturae* with the 13th edition following the systematist's death.[66]

Writing for Paulus Usteri's *Annalen der botanik* in 1796, Pierre Rémi Willemet (1735-1807) chose to hark back to Linnaeus's *Heliconia* (see chapter 1) and used the name *Heliconia ravenala*. The following year — and only a few years after von Schreber had become president of the German Academy of Science — for the third edition of *Nomenclator Botanicus omnes plantas*, German botanist Ernst Raeüschel (fl.1772-1797) reverted to *Urania*, combining it with the specific name *madagascariensis* favoured by Sonnerat and Gmelin. In 1799, German taxonomist Carl von Willdenow varied this as *Urania speciosa* in his updating of Linnaeus's *Species plantarum*. More than three decades later, French botanist and naval pharmacist Achille Richard (1794-1852) modified this again to *Urania ravenala* in 1831, attempting to create a compromise between the largely French support for ravenala from de Flacourt, Sonnerat, Adanson to Gmelin, against the heavyweight German support for *Urania* from von Schreber, Raueschel and Willdenow.[67]

While botanists took sides in this naming battle, another scientific tussle was developing over the tree's family identity (see chapter 3). A workable solution to this debate would not be reached until 1934 when the strelitzia family of ravenala, phenakospermum and the southern African strelitzias was created. But beyond the laboratories and the botanical debating chambers, ravenala had several more surprises to reveal, from its intriguing relationship with lemurs and bats to its own new variants, first noted by scientists in 1997 – nearly 340 years after Etienne de Flacourt announced ravenala to the world.

## RAVENALA AND THE HOMESICK PIRATE PRINCESS

Legend tells that ravenala may have reached Madagascar's neighbouring island, Mauritius, to please homesick Princess Betia, who also went by the European name of Marie Elizabeth Sobodie. She was the granddaughter of pirate Thomas Tew White and daughter of King Ratsimilao. At the time, pirates were the scourge of the Indian Ocean and in the 16th and 17th centuries were the only Europeans to succeed at all in creating a more permanent base in Madagascar. Their infamous stronghold was on the smaller offshore island of Nosy Boraha, or Isle Sainte-Marie, part of Ratsimilaho's kingdom.[68]

One romantic local legend tells how Compagnie des Indes corporal, Jean-Onésime (or Louis-Onésime) Filet, known as La Bigorne, was shipwrecked on the island shore after fleeing for his life from a fellow officer whom he had cuckolded. Among those who found him and helped nurse him back to health was Princess Betia (often known as Bety), 'sans contredit, l'une des plus belles femmes qu'on pût voir [without contradiction, one of the most beautiful women you could see]'.[69] The couple later married. Their 18th-century romance was made into a swashbuckling film in 1958, *La Bigorne, Caporal de France* – or more directly in English, *The amorous corporal*.[70]

The real-life story, though, seems to have become more mercenary. Filet may have pressured Betia into ceding the island to the Compagnie des Indes in 1750 after her father's death. Another version says that shortly before King Ratsimilaho died, he had been negotiating the island's cession to the French himself. After his death, Betia went ahead with this as part of a group of 30 local clan chiefs – indeed, some say her father had been a senior chief rather than the king.

But the local people rebelled, massacring the French settlers, possibly due to outrage after a Compagnie des Indes officer defiled Ratsimilaho's tomb or in protest at the land deal. For her own safety, Betia went into exile on Mauritius some time between 1751 and 1756, depending on the version of the story. Some accounts say this was complicated by the fact that in 1755, Princess Betia also married the Compagnie des Indes' Lieutenant Colonel René Forval de Grainville (1730-1815) who had led the negotiation for France.[71]

There are suggestions that ravenala first appeared on Mauritius in the 1750s, requested by the homesick pirate princess in exile. This is part of Malagasy culture, where taking plants and a bag or even a handful of soil when moving house is seen as a physical and metaphysical connection with the place of origin. What is known is that ravenala was one of the Madagascan trees chosen by Pierre Poivre to feature in his Pamplemousses garden in Mauritius by 1768 and was recorded about that time by Pierre Sonnerat for his book *Voyage aux Indes orientales & à la Chine*. It is now rampantly invasive in Mauritius, particularly in damp valleys. It is also present and invasive in Réunion and in Rodrigues, although cultivated but not invasive in Seychelles. A strelitzia leaf specimen presented by the Royal Society to the British Museum had been collected in Rodrigues during the 1874 expedition to observe the transit of the planet Venus. Although originally identified as *Canna indica*, it may be a young ravenala leaf as the expedition botanist Dr Isaac Balfour (1853-1922) noted that *Ravenala madagascariensis* was one of the trees often found 'In the vicinity of habitations or old plantations'.[72]

Betia lived comfortably in the Port Louis area of Mauritius and finally died on 14 October 1805. Apart from the ravenala legend, there are today no traces of Betia's long exile in Mauritius.

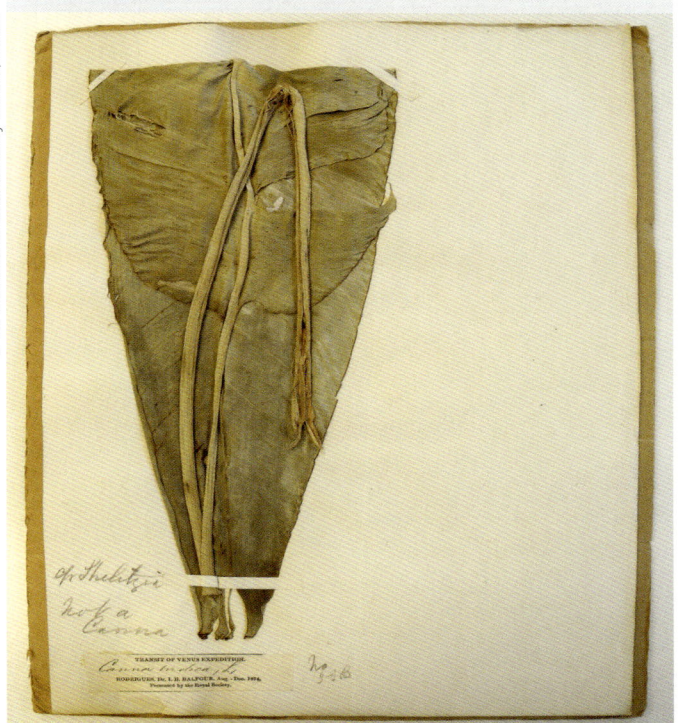

Probable leaf from a strelitzia family member collected on Rodrigues island during 1874 Transit of Venus expedition

# MYTH, TABOO AND REALITY
*Ravenala, the traveller's tree*

CHAPTER SIX

# MYTH, TABOO AND REALITY: *Ravenala, the traveller's tree*

A highway has been overlaid on the traditional route from the eastern Madagascan port of Toamasina to the island's capital, Antananarivo – but you can still hear the wind whisper between the tall, swaying ravenala leaves along the way. Or you can make your way by rail through the forest in a train aptly nicknamed after a forest snake, *fandrefiala*, glimpsing ravenala as you go.[1]

Following this route is now the adventure or heritage choice. It is impossible, though, to re-enact fully how a 19th-century traveller would have faced it, fed on the island's reputation as a death-trap, beset with fevers, incomprehensibly dangerous intertribal politics and astonishingly challenging terrain. Practicality and politics have dictated that most visitors have not arrived in Antananarivo along this old route for more than a century, since the island's key transport links were reorientated. The French had tried to take control of Madagascar from 1883, blockading the key route from the east coast to Antananarivo until 1885 during the first Franco-Malagasy war. When the French finally seized power in Madagascar in 1896, they made the practical and symbolic gesture of replacing the eastern route completely with an approach to Antananarivo from the west coast port of Mahajanga (still known to the French as Majunga), the side of the island where comparatively few ravenala are known. Then, in 1934, air services started flying direct into Antananarivo. To a small degree, these travellers still enjoy the ravenala experience during their journey – the national airline adopted the tree as its official symbol.[2]

In reality and eventually symbolically, the tree has overseen generations of outsiders trying to grapple with the challenges of learning about Madagascar or influencing its politics, religion or economics. Given ravenala's striking good looks, its social importance and the fact that it grows both on the forest edges and deep in the forest's heart, early 19th-century travellers continued the trend of identifying ravenala with Madagascar, rather than focusing on other common trees such as the bamboo and rofia palm (a Malagasy usage for raphia).

Plentiful ravenala loomed over the often tortuous quest to reach Antananarivo, perched on 12 hills about 1 400 metres above sea level.[3] The island's settlements had grown in size and influence by the 19th century and their geographical pattern reinforced the sense of being overseen by ravenala. Moving between settlements was distinctly and memorably challenging and it was impossible to avoid ravenala. Yet travellers seem to have had little energy or spirit to investigate the tree despite trekking through forests where ravenala thrived and would be hard to miss. The comparative rarity of early discussion about ravenala appears to have combined with the enormity of the task of many thousands of Madagascan plants also waiting for their first formal scientific identification. As a result, a team of French and Madagascan-based botanists succeeded in surprising the scientific world in 1997 when they announced four new ravenala variants more than 200 years after the tree was revealed to the outside world. They had concluded that four different names given to the tree locally were not simply local dialect but appeared to differentiate between at least four variants or forms, which may in time prove to be new subspecies. According to DNA testing, these may have evolved in marshy lowland forests from a plant similar to the present form *horonorono* (see panel, 'A family of four').[4]

## THE OLD ANTANANARIVO CHALLENGE

What had led to two centuries of oversight by naturalists? Ravenala grew near to the first, short-lived French Fort Dauphin (now Tolagnaro) in the south-east of the island, as Etienne de Flacourt had noted in 1668, and was still flourishing at 'lower levels of the south-east of the island' in 1939 when Olive Murray Chapman made an anthropological expedition. It also occurred along the eastern coast in areas explored by later travellers (see chapter 5). Today's capital Antananarivo was founded within the Merina kingdom in 1625, growing quickly to become a major centre although towards the end of that century, Merina power and the city's importance declined. By the time of the Merina kingdom's resurgence a century later, the challenges of visiting Madagacar had been intensified by a ban on foreigners visiting Antananarivo. By 1810, though, most of the island had been united under Merina rule and their king, Radama I (1793-1828), opened Antananarivo to foreigners and missionaries.[5]

Travellers found themselves tackling the swampy and challenging route through various forms of ravenala country up from the island's main port on the east coast, Toamasina (then known as Tamatave), to the capital on its mountain ridge. Far from being a straightforward hike or even pleasant sightseeing, this rugged route was by far the worst leg of the difficult journey which had already brought travellers across the Indian Ocean from Mauritius. As late as 1889, American missionary Belle McPherson Campbell of the Woman's Presbyterian Board of Missions of the Northwest wrote: 'It has heretofore been the policy of the Malagasy Government not to open roads, lest invasion by a foreign power be facilitated.' Even in the final decades of the 19th century, the increasing numbers of travellers, diplomats and plant collectors had 'only a few miles of public road . . . and no railroads' to help them. For most of the century, in most of Madagascar, travelling conditions remained extremely difficult. English artillery officer and geographer Samuel Pasfield Oliver (1838-1907) reported that the older, more direct but more demanding route

Overleaf: *Ravenala madagascariensis* in *L'Illustration Horticole* magazine, 1860

was only 118 miles (189 kilometres) from Toamasina to Antananarivo but the route used in Oliver's time was 126 miles (about 200 kilometres). The terrain meant that the fastest general travellers could manage the distance might be five days, although normally the uphill route would take from 11 to 17 days, or as little as 10 days downhill if the party of travellers could cover as much as 35 kilometres a day.[6] McPherson Campbell warned:

> Wheeled vehicles would be useless in a country full of marshes and dense forests of underbrush, tremendous rocks and gorges, and sloughs of adhesive clay . . . Frequently the bridge is a simple log, felled across a stream; often there are no bridges, and the stream is forded, though during the past few years a number of arched bridges have been built.[7]

Victorian travellers who did not want to walk could travel in relative comfort in a palanquin, a basic type of sedan chair or litter known in Madagascar as a *filanjàna* or *fitàkona* and carried by four bearers or *màromìta*, noted Rev. James Sibree (1836-1929) of the London Missionary Society, a prolific author in both English and Malagasy who also became a fellow of the Royal Geographical Society. In 1939, Olive Murray Chapman also had to use a *filanjàna* on her anthropological expedition through the island, supplementing this with a canoe or an 'ancient car', when her route took her near rivers or the 'newly constructed road':

> My porters were amazingly skilful, carrying me in my chair up steep mountain-sides and down again by precipitous rocky tracks, covered with loose stones, while I was tilted at times right back and at others forward to such an extent that I frequently found myself standing on the footrest, while I was forced to cling on to the arms of the chair for safety.[8]

Leaving the port of Toamasina, travellers would head through marshy country down the coast, south to Adevoranto, where the *horonorono* form of ravenala would probably be most common and where ravenala is believed to have first evolved. From day one, travellers would also be transported by canoe across the coastal lakes and rivers, 'permeated with miasma during the wet season'.

These canoes were like vast barrels and made of a single tree-trunk, four feet wide (1.2 metres) and often as much as 30 to 40 feet long (about nine to 12 metres). On the second day, travellers moved through 'thick woods' with 'plenty of lemurs' and by the fourth day, they reached Adevoranto and struck inland from the coast. Within a couple of hours, the broad estuary had narrowed into a river about 100 yards (90 metres) wide and the first hills had been reached. Perhaps it was relief at escaping these intense challenges on the route that allowed Sibree to be reminded by the woods and water in this area of 'some of the loveliest landscape that English river and lake scenery can present'. Whether Oliver wanted to make himself feel at home or make this distant Indian Ocean recognisable to his readers, he similarly recast the landscape as some English sylvan idyll with 'a path as broad as an English highway and covered with beautiful turf. The scenery altogether reminded me of a shrubbery or wilderness in some lovely park in England'.[9]

William Ellis, *Three visits to Madagascar*

Transport by *filanjàna*, Madagascar, about 1838

Another four miles upstream, the landscape changed dramatically:

> The inland journey commences, and a totally different country is traversed – the change of vegetation and soil being very apparent, whilst, instead of easy marching over almost level sand and firm turf, there is now a series of clay hills intersected by small streams with marsh and swamps between. The *ràvinala* (*Urania*), bamboo, and the *rofia* palm (*Sagus rofia* [now *Raphia farinifera*]) abound. The country is desolate and little inhabited. . . .[10]

Another traveller in 1920s Madagascar similarly commented on this '*terrain marécageux et inculte, couvert de* Ravinala madagascariensis [this swampy and uncultivated ground, covered with *Ravenala madagascariensis*]'. In this region, travellers would see the *bemavo* form of ravenala, with its particularly lofty and pronounced fanning crown of leaves. This is now the most widely distributed form of ravenala internationally, and generally the mainstay of traditional Malagasy house-building.[11]

The following day would be treacherous underfoot but the landscape rewarding:

> Leaving Rànomafàna, the country traversed has many beauties, and amidst a fine amphitheatre of hills there rise several lofty cones. The streams, of which several are crossed, run deep and strong over beds of quartz pebbles. The traveller's tree grows in enormous numbers, and large patches of rich black soil appear amidst the general masses of red clay.
>
> The course of the track is very tortuous and follows pretty nearly the course of the Fàrimbòngy, whose waters present a succession of deep pools, cascades, and small rapids. The track ascends and descends over most slippery and steep ground . . .[12]

## SLOGGING THROUGH THE RAVENALA FOREST

Day six was probably one of the most daunting of the route:

> The journey is most wearisome, continually ascending and descending over ridges and terraces, the track mostly leading along the spurs and round the shoulders of ravines between them . . . the only resting-place is reached at Màrozèvo, 7 miles, a poor village with no supplies . . . Now follows thick forest over slippery ground with narrow passes for some miles . . .[14]

Among that network of roots and fibres over which travellers clambered would be those of ravenala, creating a strong interwoven mat that can allow the tree to grow as high as 25 metres in the most precarious situations, whether on a precipitous mountain slope or in a swamp.

Emerging from this forest gave only brief relief, Oliver made clear: 'a clearing and flat open valley is reached where is situated a considerable valley . . . But the valley is swampy and the place notorious for fever . . .'[15] This gallery forest is one of the places where the *horonorono* form of ravenala thrives:

> *Un premier biotope est constitué par les ravins sinuant à la base des collines de basse altitude. Le fond de ces ravins est occupé par une forêt-galerie dans laquelle le 'horonorono' constitue l'élément émergent et le plus abundant. Ces forêt-galeries s'étendent sur quelques dizaines de mètres de part et d'autre du ruisseau central, et les 'horonorono' ne s'implantent pas*

### A FAMILY OF FOUR

The first illustration of ravenala, published by French adventurer Pierre Sonnerat in 1782, somewhat stylised this sturdy, evergreen tree, both generally and in terms of its distinctive flowerheads. It is simply portrayed as a single, ring-scarred trunk with a distinctive flowerhead that is rather more compact and regular than occurs in nature. The tree, with its grey to olive-green trunk, commonly grows to between 20 and 30 metres tall but it was more than two centuries later when it was noticed that one of the four different forms usually branches at the base. French botanists then focused on how the four different Malagasy names for ravenala might reflect four variants or subspecies:

1. ***The Immortal***: *Horonorono's* new stems simply spring from the base of older ones and from its own system of adventitious roots.

2. ***The Builder***: *Bemavo* is most frequently used for construction and most resistant to water stress.

3. ***The Lightshifter***: *Hiranirana* often grows in forests but as it becomes larger and its needs for photosynthesis greater, this form adapts the way it grows. It slightly shifts the flat plane of its leaf fan to give the blades as access to light and shade as needed.

4. ***The Twister***: While *malama* grows, this shade-loving form twists to get the most access to light in its forest home, producing a spiral-shaped rather than fan-shaped leaf bowl. When it reaches its full growth, this pattern tends to straighten out.

Other local names used for ravenala include: *honkodranto; maroanaka; tokampototra; ravimpotsy; bakabia* (fan-shaped); *akondrohazo* (banana tree). [13]

*plus au-delà de cette limite, où le sol perd ses caractères hydromorphes pour faire place à la savane qui subit les feux annuels. Dans cette forêt à 'horonorono' très peu d'arbres sont présents, mais d'autres plantes hygrophiles abondent dans le sous-bois créé par les couronnes des ravenalas, notamment des Cypéracées, Pandanacées et Taccacées.*[16]

[A primary biotope [habitat] is created among the ravines snaking along the base of the low-altitude hills. The ravine floor is occupied by gallery forest where *horonorono* is the emerging and most abundant element. These gallery forests extend for several dozen metres on either side of the central stream and the *horonorono* do not implant themselves beyond this limit, where the soil loses its hydromorphic characteristics, giving way to savannah which undergoes annual burns. In this *horonorono* forest, very few trees are present, but there are other hygrophilic plants in the undergrowth created below the ravenala crowns, particularly Cyperaceae, Pandanaceae and Taccaceae [sedges, screw-pines and relatives and yams].]

The typical prop-roots of screw-pines particularly could add to the instability underfoot when trying to negotiate this terrain. Beyond lay Madagascar's interior forest, said Oliver, marked by a turn to the north at a great wall of rock and 'many a climb and as many a descent' over red clay hills:

The coast tropical vegetation has disappeared; no longer do we meet with the *ràvinàla* and the *rofia*, but instead timber trees of enormous size, bamboos and tree-ferns, interlaced with a thick jungle of undergrowth and parasites . . . The track is absolutely frightful, requiring frequent detours on account of the gigantic trees which have fallen across the path.[17]

By this stage, though, travellers were reaching what Oliver had believed to be the 'Limit of Travellers Palms or Ravenale'. To Charles Lemaire of *L'Illustration Horticole*, writing in Belgium for the leisured owners of fashionably lavish hot houses, ravenala:

*par sa haute stature, ses grandes dimensions foliaires disposées en un gigantesque éventail, son majesteux ensemble, est un des plus nobles végétaux qui parent la surface de notre globe...*[18]

[with its lofty stature, its great leafy reach arranged in a gigantic fan, its overall majesty, is one of the most noble plants to share the surface of our globe . . .]

Lemaire added that as well as being one of the world's most noble plants, ravenala was also one of the most useful. Oliver was much more pragmatic, using ravenala trees he encountered functionally as a topographical marker on a map he had sketched

William Ellis, *Three visits to Madagascar*

Approach to Antananarivo, Madagascar, 1850s

in 1861 (and first published in 1865). These ravenala trees were probably mainly the *hiranirana*, the especially tall, single-stemmed trees, studded around the forest in gaps created by cyclones, stretching and rotating their leaves to the light. But late 20th-century collectors have found ravenala as high as 1 200 metres – and it might extend to an altitude of 1 600 metres. Within the forest itself, for example, would be the '*relativement discret* [relatively discreet]' *malama* form of ravenala, which can look more like a shade-loving tree fern or a large bird's nest fern (*Asplenium nidus*) as a young plant, sinuously curving to seek the light. When fully grown, it usually reverts to a straight posture.[19]

Early European visitors tended to be fascinated by the special character that ravenala gave the landscape but soon they often found this deep forest oppressive. Even as recently as the 1950s, British author Rupert Croft-Cooke (1903-1979) commented that it felt like:

> no forest such as I had known . . . none of the airy freedom of European woodlands, for there was a cathedral air about it . . . the gigantic trees, engaged for centuries in a life-and-death struggle to find light and air at another's expense.[20]

Oliver had considered also this area 'notorious for fever', a serious deterrent to outsiders. Not surprisingly, the Malagasy monarch Radama I believed his allies General Tazo (Fever), who had already harried newcomers through the coastal marshes, and General Hazo (Forest) joined forces here to help repel alien invaders.[21] From the beginning of this forest, it would be at least another three or four days before the travellers would at last reach the capital of Antananarivo.

## DRAMATISING 'THE BEAUTIFUL HEAVENLY ONE'

The term 'flagship species' was popularised in the 1980s, in part through the work on primates such as Madagascar's lemurs by Conservation International's Russell A. Mittermeier. Like lemurs, the threat to ravenala through logging and other habitat clearance flags it as one of the species to watch as landscapes struggle on the brink of degradation (see chapter 15). Ravenala's prominence in this debate was foreshadowed by the way the tree became an icon for Madagascar. Its dramatic and statuesque appearance, its varied botanical cycle and its range of daily and special uses quickly captured the imagination of European plant enthusiasts and travellers. This is commemorated to this day in the fact that most international common names for ravenala are variations of the traveller's tree.[22]

The tree's iconic status would be doubly poignant for renowned French scientist Philibert Commerson, who saw himself as a 'missionary naturalist' (see chapter 5). In 1770, he made the expedition to Madagascar that ultimately cost him his life. Commerson recorded in his trip notes how the convenient reservoir of water held at the heart of the ravenala was interpreted as offering the traveller the promise of survival. By the next century, ravenala would be deeply identified with Madagascar, from travellers' tales to French botanist Pierre Sonnerat's iconic illustrations. But how, after being witness to the misery of generations of travellers submitting themselves to the excruciating test of determination to reach Antananarivo through the fever-ridden forest along the eastern route, did ravenala continue to be portrayed as a friend of the traveller?

Missionaries who fed the fervent imaginations of their readers and followers were largely responsible. The tone was set by Rev. William Ellis (1794-1872), who had joined the London Missionary Society at the age of 20, serving first in Polynesia. After nearly 30 years back in England, he was sent out to Madagascar in 1853 and 1854 – but Queen Ranavalona refused to allow him to stay. In 1856, he succeeded in visiting for a month, the first English missionary to do so for more than 20 years. Given the queen's opposition to Christian missionary efforts, Ellis kept his visit low-key and used it for fact-finding. Even so, his 1858 book, *Three visits to Madagascar*, carries an image of the ravenala gilt-embossed on its front and back covers, the text since interpreted as repeatedly using the ravenala as a Christian image.[23]

Religious overtones in *Ravenala madagascariensis* depiction, cover of Rev. William Ellis's 1858 *Three visits to Madagascar*

Extracting water from the traveller's tree, Madagascar, in *L'Illustration Horticole* magazine, 1860

While praising the 'almost inconceivable magnificence' of the tree that he called 'the beautiful heavenly one', Ellis noted that it is 'even more celebrated for containing during the most arid season a large quantity of pure, fresh water, supplying to the *traveller* the place of wells in the desert'.[24] These biblical undertones are pursued as he continued:

> One of my bearers struck a spear four or five inches deep into the thick firm end of the stalk of the leaf, about six inches above its juncture with the trunk, and, on drawing it back, a stream of pure clear water gushed out, about a quart of which we caught in a pitcher, and all drank of it on the spot. It was cool, clear and perfectly sweet.[25]

## GIVING THE TRAVELLER'S TREE A MISSION

An echo of Christ's crucifixion is detected here by an ethnographical analyst, quoting both New Testament references to Christ's cross as the tree of life and specifically the account of St John's Gospel, taken to be a sign of Christ's humanity: 'But one of the soldiers with a spear pierced his side, and forthwith came there out blood and water.' Such hints could well be in keeping Ellis's view that he could not be outspoken as a missionary because local followers and potential converts were already in such danger. After a series of edicts banning Christian services and rites such as marriage and baptism on the island, Ellis's predecessors had to bury 70 Bibles and several boxes of books and tracts before leaving the country in 1836.[26]

So far, so symbolic – but nothing more, according to sceptics. One explained:

> Most symbolic explanations by anthropologists, psychoanalysts, art historians, or other sorts of theorists . . . treat symbolic associations as unconscious processes lying beyond the awareness of the individuals to whom they are attributed. Unfortunately, relegating symbolic associations to the unconscious places them outside the bounds of empirical test. Subjects can neither confirm nor deny the existence of something outside their own awareness. Researchers cannot use recordings or photographs or chemical analyses to detect associations that reside in a hypothetical realm, the unconscious, which is itself immune to direct measurement. In the end, belief in symbolic associations depends on the literary elegance and persuasiveness of the symbolic analyst's argument.[27]

But real life can provide circumstantial evidence that breaks through barriers of disbelief, admitted the researcher, if the person in question recognised the symbolic association attributed to them. Ellis, for example, circumspectly cloaked and confirmed his apparently mythologising observation by his remark that he had seen Sir William Hooker draw water in just this way from one of the ravenala specimens in the palm house at Kew. This seems irrefutable given Ellis's lifelong interest in plants and the fact that he began work aged 11 in a market garden, before taking up religion in his late teens. Ellis sent plants to Sir William Hooker at Kew and was touched when Hooker told him that he had enjoyed his book on Madagascar. Ellis's circumspect use of imagery was also probably deliberately diplomatic. He was still planning to return to minister in Madagascar – and eventually managed to spend four years there from 1861 to 1865.[28]

By the end of the century, William J. Townsend was prepared to be much more open. On the spine of his 1892 *Madagascar: Its Missionaries and martyrs*, fluid is shown filling a cup for a Malagasy traveller who has speared the side of a ravenala. This could also have implications of christian holy communion, given that he notes botanical imagery as key to the New Testament, linking it directly to the introduction of christianity to Madagascar and the issue of religious tolerance and martyrdom. He uses chapter titles such as 'The Gospel Rooting Itself' and creates such interpretations as 'the paradoxical secret of the bleeding palm, that death is the source of new life'. The 'banished ones joyfully mingled with their families again' are shown sheltering under the safety of a ravenala. The botanic imagery of martyrdom was widespread in England at that time and would also be taken up in the first major Malagasy account of the island's Christian martyrs, Pastor Rabary's 1910 *Ny Maritiora Malagasy*.[29]

Not everybody was convinced, though, and by 1914 Hugh Macmillan (1869-1948), the superintendent of the Perideniya Royal Botanic Gardens in Ceylon (now Sri Lanka) commented:

> The name "Travellers' Tree" is on account of the capacity of the tree for storing up water in the receptacles formed by the sheathing bases of the leaf-stalks, being thus supposed to be of service to travellers in deserts. The supposition, however, is rather discounted by the fact that the tree does not naturally grow in districts where water is scarce, and thrives only in regions where the rainfall is more or less abundant. Moreover, during the dry weather the water collected in the leaf-bases referred to becomes putrid and infested with the larvae of mosquitoes and other insects.[30]

## 'INSATIABLE CURIOSITY' FOR RAVENALA IN THE BELGIAN CONGO

If ravenala could be mythologised, it could also be demythologised – and was, deliberately or not, by Belgian hydrobiologist and botanist Dr Paul van Oye (1886-1966). Like many other tropical travellers, he had already been intrigued by and studied Java's pitcher-plants (nepenthes) and their apparently 'man-eating' qualities.[31]

Much later, van Oye collected plants in Iceland and took

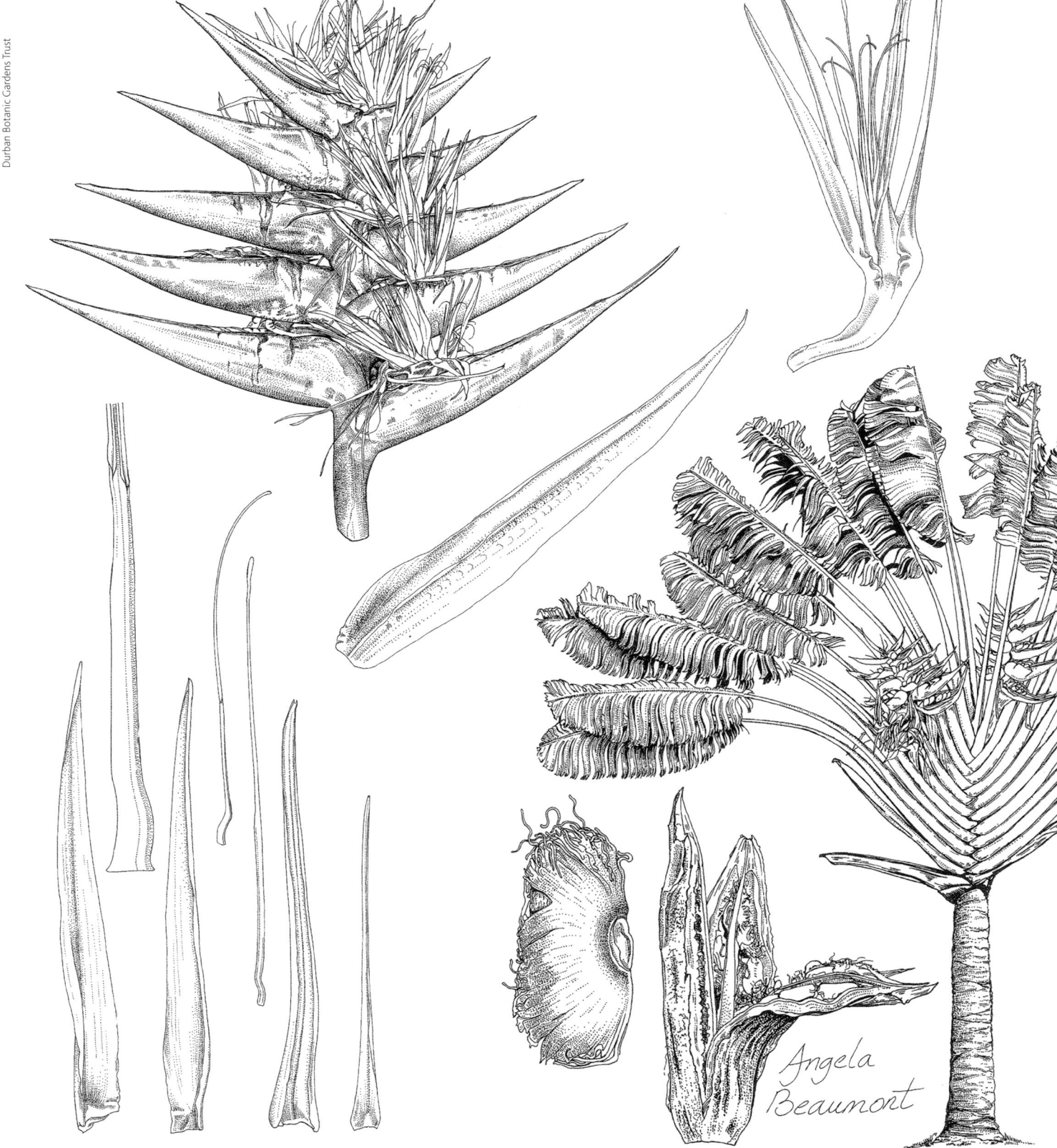

Angela Beaumont, *Ravenala madagascariensis*, pen and ink, 2002

over editing the authoritative Belgian flora. But in 1923 he found himself in Eala, five kilometres away from the equator in the then Belgian Congo (now Democratic Republic of Congo). Van Oye already held a doctorate in zoology, and the following year he would complete his doctor of medicine and doctor of tropical medicine degrees after a 12-year gap. At Eala's 371-hectare botanic gardens, created by decree of King Leopold of Belgium in 1900 and curated in the 1920s by V. Goosens, one thing that caught van Oye's eye was the ravenala.[32] He would not be alone in this. In 1925, the Eala botanic gardens attracted French writer and future Nobel Prize winner André Gide (1869-1951) – and its ravenala gave him pause for thought. In his travelogue, *Voyage au Congo*, Gide wrote:

> *Le 11, visite au jardin d'essai d'Eala, le vrai but de ce détour en Congo belge. M. Goosens, le directeur de ce jardin, présente à notre émerveillement les plus intéressants de ses élèves: cacaoyers, caféiers, arbres à pain, arbres à lait, arbres à bougies, arbres à pagnes, et cet étrange bananier de Madagascar, l'«arbre du voyageur», dont les larges feuilles laissent sourdre, à la base de leur pétiole qu'un coup de canif a crevé, un verre d'eau pure pour le voyageur altéré. Déjà nous avions passé à Eala, la veille, quelques heures exquises. Inépuisable science de M. Goosens, et complaisance inlassable à satisfaire notre insatiable curiosité.*[33]

> [11th we visited Eala experimental gardens, the real goal of this Belgian Congo detour. M. Goosens, the garden's director, presented his most interesting seedlings for us to marvel at: cocoa-palms, coffee trees, breadfruit trees, milk-trees, candle-trees, loin-cloth-trees and the extraordinary banana-type tree from Madagascar, the 'traveller's tree', from whose great leaves there flows a glass of pure water for the needy traveller at the slash of a penknife across the base of the petiole. We had already passed a few exquisite hours at Eala the day before. Inexhaustible knowledge from M. Goosen, and untiring accommodation of our insatiable curiosity.]

As at Eala, tourists in Córdoba, Mexico, were also introduced to ravenala's water reservoirs by a gardener slashing open the base of a leaf stalk (petiole) with a machete. At Córdoba, this was a daily performance, commented A-R. Proschavsky in a letter published in the June 1897 Parisian natural-science journal, *Bulletin de la Société National d'Acclimatation de France*. Prochavsky tasted the water, which he declared '*extrêmement pure et fraiche . . . sans y constater aucun gout particulier* [extremely pure and fresh . . . without any particular taste]'. During his Asian travels, August Daniel Frederickson echoed this, finding the water released in this way at Perideniya Royal Botanic Gardens in Ceylon 'cool and perfectly sweet'. The curiosity was often remarked on in gardening journals, dictionaries and encyclopaedias in Britain, France, Germany and Sweden. It gave the tree additional common names: in German *Quellenbaum* [waterspring tree] and in Dutch *piesang aijer* [water banana]. The economic museum at the New York Botanic Gardens even displayed a ravenala leaf base in its Beverages section.[34] In fact, though, the practice simply exposes the water gathered in this cranny, rather than the inner reservoirs that the insatiable curiosity of the determined Dr van Oye would reveal.

## UNMASKING THE RAVENALA

It rains a lot in the Belgian Congo and it is likely that van Oye sat on the veranda, pondering the rain sluicing down the ravenala, long enough to realise the mechanism was different from the easily filled pitcher plant. What was more, he saw something amiss with Commerson's original interpretation of the ravenala leaf's working like a glorified gutter to fill the tree's rainwater reservoir for the thirsty traveller (see chapter 5).[35]

By a series of detailed experiments establishing elements of basic science, van Oye swept aside 150 years of mistaken beliefs and clarified several elements of the ravenala's structure. He noted:

> *Nous savons que chez plusieurs espèces d'arbres, de familles très différentes, on peut rencontrer des collections d'eau sous des formes variées. En général on admet que l'eau emmagasinée sert de réserve.*[36]

> [We know that several tree species from very different families collect water in various ways. This stored water generally acts as a reserve.]

He was not thinking of how the desperate traveller could sip the rainwater caught in its supporting bracts – along with any insects and perhaps the occasional Malagasy frog. He was thinking more of succulents, which store water in this way.

These include the baobab, of which six of the world's nine species occur on Madagascar compared to just two species on the African mainland, as well as one in Australia. Yet though it is known that the baobab's water reservoir has been tapped by needy travellers in similar ways to depictions of use of the ravenala, it was the tall and elegant ravenala which instead became known as the traveller's friend or, in India, traveller's joy.[37]

Van Oye may also have been thinking of the ravenala's relative, the banana, because Professor P. Baccarini of the University of Catania in Italy had published the results of his research into water accumulation in banana leaf sheaths in 1904. Indeed, over in Jamaica, William Fawcett had compared the ravenala to 'wild pines of Jamaica', which he described as:

> Hold[ing] water, which is sometimes of service to hunters in the woods, so the sheaths of the leaf-stalks of the Traveller's

Tree store up water for the plants' own needs, and each one pierced from below will yield the thirsty traveller half-a-pint [about 230 ml] of refreshing water.[38]

The Bombay Natural History Society had heard in 1904 that 'It is alleged that a quart of water may be readily obtained in driest weather by piercing the leaf-stalk'. Van Oye, however, was looking well beyond this debate to internal reservoirs and he argued that fluid such as rainwater does not flow into the ravenala from outside – it is created by the plant itself.

He pointed to the large quantity of stomata, cells on the underside of the leaf that allow transpiration of water and oxygen. On the stalk (petiole), cells of the outer wall are small, very close together and quadrangular, he noted. But on the stalk's inner wall, he found bigger epithelial cells to be '*allongées et lâches*' [enlarged and flabby] – implying they were ready to be swollen by water.[39]

He also noted that the ravenala leaf structure was designed to allow water to run off, rather than channelling it down a central gutter as Commerson had suggested – and proved this with a series of back-to-basics experiments:[40]

- He sprayed water from a pipette over a leaf to track the direction of water – it did not reach the petiole.

- He put colourant on a leaf in different places but during rain the colourant stayed on the leaf and did not reach the gutter or the reservoir.

- During rain, he observed the path of rainfall across a leaf – the result was the same as in the lab, proving rain did not prompt either the leaf blade or the stalk to move to allow water into the reservoirs.

Then van Oye considered the purpose of the fluid. Like pitcher plants, did the fluid stored by ravenala predigest creatures on which the plant feeds? His next series of experiments involved observing egg-white in ravenala fluid till it stank of rotten eggs after a fortnight. Then he looked at corpses of termites, flies and ants – which started smelling after nine days, indicating putrefaction, he noted.[41]

Angela Beaumont, *Ravenala madagascariensis*, pen, ink and watercolour, in *Flowering Plants of Africa*, 2002

Finally, van Oye placed some live ants in ravenala fluid and others in filtered water – only to find that they both died at the same rate. This was a long shot. He admitted ant corpses are exceptionally rare finds in the fluid of ravenala reservoirs in the wild, while they are found almost all the time in wild pitcher-plants. He concluded ants entered the ravenala reservoir by chance, rather than being lured in. Once there, they become prey to any micro-organisms present, rather than the plant itself preying on them, he argued.[42]

Van Oye then began calculating how much fluid could be accumulated in a ravenala reservoir. He extracted a total of 1,3 litres of fluid from one of these reservoirs and calculated that up to 150 millilitres remained inside, slightly less from another reservoir. So he concluded that a reservoir might hold up to 1,5 litres of fluid. Why would the tree accumulate so much fluid? This is not a reserve of fluid as we normally understand it, he suggested, because levels in three experimental plants did not drop during three weeks without rain – nor did they rise during periods of heavy rain.[43]

The key, he believed, lies in the construction of the cells making up the interior of the stalks, which were like 'os des squelettes, c'est-à-dire qu'ils offrent le plus de résistance avec la plus petite quantité de matière [skeleton bones, that is to say they offer as much resistance as possible with the least amount of weight]'. Given the regular cyclone conditions in Madagascar, these parts of the plant must be able to resist great pressures up to two kilograms per square metre when a wind is blowing at four metres a second, he estimated. The ravenala's fluid reservoirs reinforce the plant's strength – and he found 35 reservoirs in the ravenala he examined. Van Oye conservatively estimated that a single ravenala could be carrying up to 40 litres of fluid in total in its reservoirs. In turn, if the reservoirs were equally distributed and the fluid was pure water, this would create a remarkable, 40-kilogram ballast. Importantly, this ballast is sealed and cannot evaporate in younger plants, helping them resist the wind and protect growth. In older plants, on the other hand, which no longer resist the wind and whose leaves are torn, the reservoir is open so evaporation is possible.[44]

## INSIDE RAVENALA'S WORLD

Given that van Oye went on to become an acclaimed hydrobiologist and found the journal *Hydrobiologia*, it is not surprising that he concluded his ravenala study with identification of the 10 different organisms who had made their home in the tree's reservoirs: '*des petits étangs en miniature, dans lesquels une association aquatique s'est constituée* [miniature ponds in which aquatic life has established]'.[45] This world-renowned scientist and charter member of the World Academy of Art & Science was, however, just one of several people who has interested themselves as much in ravenala as in the ecosphere the tree creates. On Madagascar, an intriguing range of creatures feed and live in this ravenala ecosphere, a few even becoming complete ravenala specialists.

Along with plants such as bromeliads and pitcher plants, ravenalas provide a particular microhabitat that because of its abundance on the island has been called the 'most remarkable of these niches in Madagascar'. Scientifically known as phytotelms, these mini-homes are created by small amounts of water collected in places on plants such as ravenala leaf joints (axils), open fluid reservoirs or crevices or holes in tree-trunks. Phytotelms are found in other forest areas round the world but on Madagascar, they are both especially common and often surprisingly large – particularly if an open ravenala reservoir is involved, offering a small pond containing a litre or more of water.[46]

Mosquitoes find the ravenalas' leaf sheaths particularly welcoming microhabitats and several make their home or breed there. Many of these mosquitoes are endemic to Madagascar, with their most diversified genera known so far being those that house their larvae on plants such as ravenala. Researchers have suggested this species development is thanks to Madagascar's highly diverse flora providing a wide range of breeding sites, 'rather than the variety of prey of these blood-sucking insects'. One subgenus of the *Mimomyia* mosquito was named *Ravenalites* because it had adapted its life-cycle to exploit the small pockets of water held in ravenala leaf sheaths. It has since been renamed *Ingramia*. Most of its members are found along Madagascar's eastern slopes.[47]

Two species of the *Toxorynchite*s mosquito are found in eastern Madagascar and one of them, *Toxorynchites pauliani* is endemic to the island. Both it and *Toxorynchites brevipalpis* (also found in sub-Saharan Africa) lay their eggs in ravenalas' trunk crevices and leaf sheaths. Their larvae are carnivorous – which is risky for the eight out of 12 species of the 'beautifully marked' *Orthopodomyia* mosquitoes, which are endemic to Madagascar and use ravenala leaf sheaths and joints (axils) as sites to develop their larvae.[48]

*Orthopodomyia* seem to carry avian arboviruses rather than have any human medical implications. Similarly, travellers may be relieved that *Uranotaenia* mosquitoes are 'not particularly attracted to humans and are of little importance from a medical perspective' – but it seems there are a lot of them and researchers suspect they will discover even more. So far 28 species have been found on the island in this 'very poorly known genus', some shared with Africa and some endemic to Madagascar. It is expected that the number of known endemic species of this genus will at least double. So far the larvae of at least five of them have been found developing in water in the nooks and crannies of ravenalas' trunks and in leaf joints.[49]

## SHELTERING FROGS, CRABS AND BATS

Other creatures, though, simply use ravenala as a shelter. Endemic Malagasy frogs belonging to the family Mantellidae make their homes in water caught in ravenala leaf joints. The freshwater, tree-climbing crab *Malagasya goodmani* uses these small water spaces as its microhabitat. Researchers were struck by this because:

> It is still a relatively rare occurrence to find any crab in a phytotelmic microhabitat. To date only some six out of the almost 1 000 species of true freshwater crabs ... are known to be phytotelmic.[50]

*Malagasya goodmani* made a seventh. Seven species of sesarmid crabs, a family known for its tree-climbing, also live in these tiny phytotelms. Crabs turning into tree-dwellers is rare because of the way the creature must adapt to its environment. It had to evolve to survive out of water for long periods, breathe air and conserve water. These crabs usually develop long, slim legs that help them carry their weight as they climb trees and they often adapt their reproduction patterns, laying larger eggs.[51]

Reptiles and amphibians, or the herpetofauna, are among some of Madagascar's lesser-known creatures. They include one of the skink genus *Amphioglossus*, several of which specialise in particular areas of the humid eastern forests of the island. One species, *Amphioglossus punctatus*, is believed to give birth to its young in the leaf axils of ravenala. The *Amphioglossus* belong to a group known as the Scinidae, which appears to have established itself there after rafting over from Africa in some previous era. On Madagascar, a Merina proverb relating to Scinidae could apply more broadly regarding the daily battle for survival in the ravenala ecosphere: '*Androngo anaty asa, samy miaro ny rambony tsy ho tapaka*' [Like lizards in an agricultural field, each one takes care not to lose his tail]'.[52]

For some of the herpetofauna, that survival battle may come down to how much nectar can be found from ravenala trees in the area. It has been noted, for example, that reproduction rates are reduced in female day geckos such as *Phelsuma dubia* when less nectar is available from favoured sources such as ravenala and the coconut palm (*Cocos nucifera*). Sometimes, researchers have found, these geckos milk planthoppers for nectar instead of going to the trouble of seeking out the nectar directly themselves.[53]

But it seems that a relative of *Phelsuma dubia* at some point became so focused and adapted to ravenala that it will now live nowhere else. While conducting a survey of day geckos in the area of Mananjary, a town more or less at sea level in eastern Madagascar, a team of researchers spotted a new species of this creature. The gecko seemed to live only on ravenala trunks or leaf stalks, whether growing in plantations, grassland or gardens. Sometimes as many as five were found on one tree, but never with any other species of day gecko. Because of its slimmer body than *Phelsuma dubia*, it could take refuge in ravenala leaf joints, usually about eight metres above the ground, and was often seen licking ravenala nectar. 'Intensive searches' of other trees in the area, such as coconut and screw palms and banana plants, turned up other *Phelsuma* relations but not this different species. Though they noted a similarity to *Phelsuma berghofi*, this *Phelsuma* seemed clearly to be a ravenala specialist. So they named this new species *Phelsuma ravenala*.[54]

Many Madagascan bats are engaged in a battle for survival as their habitat is threatened and that includes some that make themselves at home in ravenala. Various kinds of bats are found all over the island, except the high mountain region. Insect-eating bats mostly favour the western region's dry forest and spiny bush, with 27 of the 30 species found here. The other three species – *Myzopoda aurita*, *Miniopterus fraterculus* and *Tadarida pumila* – prefer the eastern humid forests or the moist, central mountain forests. Because ravenala has thrived in some areas where ground has been disturbed, this has also proved an advantage for the Old World sucker-footed bat (*Myzopoda aurita*) which likes to choose this species of tree for roosting, particularly where there is also marshland. Researchers observed that the specialised suckers on the bat's thumbs and soles helped it move smoothly across ravenala leaves at the roost.[55]

Commerson's leaf-nosed bat · Madagascan flying fox

## CREATING RAVENALA'S SEED RAIN

Ravenala's relationship with fruit bats and with birds such as parrots that enjoy its fruit is more mutually beneficial than with insect-eating bats or creatures such as tree-climbing crabs or geckos. Three species of parrots are found on Madagascar, two of which are endemic to this island and the Comoros – the greater and lesser Vasa parrots (*Coracopsis vasa* and *Coracopsis nigra*). They skim rapidly among the treetops, sometimes in flocks of more than 100, making them particularly difficult to tell apart in the field.[56]

Despite there being relatively little research on these parrots, the greater Vasa parrot has attracted some attention for its breeding system. The female advertises extensively and copulation often lasts more than 100 minutes. There is usually a range of likely paternity options and a generally cooperative approach between the genders to raising chicks. By comparison, lesser Vasa parrot females are more social and less individualistic, flying, displaying and mating in flocks.[57]

Both species face the same kind of risks, though. Feeding in flocks has made them into agricultural pests if they target flowering or fruiting orchards and or crops such as maize, cassava and rice. They are hunted for meat and trapped for sale as caged birds. The Vasa parrot population is declining somewhat, partly because forest clearances have curtailed the habitat where they nest in tree-holes as well as targeting trees such as the ravenala, where they feed on its seeds.[58]

Birds such as parrots help disperse the digested ravenala seeds in their droppings, though often this may not be far from the tree where they feasted. More helpful to the ravenala in the long run are Madagascar's fruit bats, part of the Pteropodidae family. Having eaten the ravenala's oily arils, they can scatter the tree's seeds long distances from the original trees because they often range more than 30 kilometres away. This is really a hop, skip and a flutter for a fruit bat, which are quite far-ranging. Interestingly, a recent study found that despite roosting up to 500 kilometres apart, individual fruit bats are 'very similar genetically'.[59]

Only a third of the trees where Madagascar's fruit bats prefer to source their food are endemic to the island. Ravenala and baobab are among their particular favourites and they also seem to pollinate these trees when they feed on their flowers. Seeds recovered from bat faeces were 80 percent more likely to germinate within four to six weeks than those of the same plants taken straight from intact fruit, researchers found. Bats' other major benefit to ravenala is the quantity of their 'seed rain' as they move in flocks, dropping plenty of the same fruit-seeds in the same area. This eventually gives new young trees a better chance of pollinating each other. In addition, because bats scatter the seeds they eat far from the trees on which the fruit grows, this has given hope for efforts to regenerate deforested areas and to conserve Madagascar's fragmented forests.[60]

Only three species of fruit bats exist in Madagascar: the Madagascan flying fox (*Pteropus rufus*); the Madagascan straw-coloured fruit bat (*Eidolon dupraenum*); and the Madagascan rousette (*Rousettus madagascariensis*). This small range of fruit bats is considered 'rather limited in comparison with other Old World tropical regions', against Africa's 25 species and south-eastern Asia's 82 species. Unusually for Madagascan vertebrates, the species do not seem to specialise in a particular habitat but share access to the same habitats and, to some extent, the same food resources.

Their preferred menu is known to stretch to 18 different tree species, though the Madagascan flying fox has become notorious for damaging litchi and mango crops. Blaming and tracking the flying fox, the island's largest fruit bat with a wingspan of up to 125 centimetres and weighing up to 750 grams, might have been too much of an assumption simply because it is so obvious. Researchers believe another perpetrator has simply been overlooked. This is the Madagascan rousette, the island's smallest and least studied fruit bat. It is less than half the size of a flying fox, with a wingspan of up to 52 centimetres and weighing only as much as 80 grams. But this taste for forbidden fruit puts both bats at greater risk from man, with the Madagascan flying fox already on the Red Data List as vulnerable and the rousette as near threatened.[61]

## CAUGHT IN THE CROSSFIRE

The rousette roosts in its thousands in caves – where fights occasionally break out. The middle-sized straw-coloured fruit bat, which has a wingspan of up to 95 centimetres and weighs up to 340 grams, is equally at home roosting in caves or trees. This makes it somewhat less vulnerable than the flying fox, which roost in trees in large, noisy roosting colonies, or 'camps', numbering anything from 10 to 5 000 bats. The trees where flying foxes make their camps are closer to humans and regularly face the risk of land-clearance and deforestation. During crop gluts, orchards now pose a major danger to flying foxes, potentially luring them to be shot for the pot. Hunting may, in turn, cause these bats to desert an established roost. Ironically, though, it is now widely understood that bats do not cause as much damage to fruit crops as first believed.[62]

This cycle is at its most deadly in the south-east of Madagascar, where a particular hunting method may be used that destroys both the host ravenala and the bats. On the first day of the two-day hunt, a hunter selects a ravenala with a flying-fox roost and cuts a preliminary slice into the trunk that still leaves the tree standing. Next day, he can fell the tree quickly when he returns. Any bats left clinging to the fronds of the felled ravenala are then beaten to death with sticks.[63]

Taboos against eating bats among the Mahafaly and Antandroy people, as well as among Muslims in the north, help protect the creatures. While 'traditional beliefs limit human pressures in some areas', in others human pressures are so great that taboos have broken down and the respect for not hunting bats in sacred sites, for example, has been lost.[64]

In a sense, fruit bats seem to have been under pressure in Madagascar ever since they arrived about 20 million years ago, probably with the Gondwana dispersal of species. This timing may have left them second in the evolutionary feeding queue as they arrived well after the lemurs had begun colonising Madagascar. Certain plants that would also suit the fruit bats appear to have already adapted to be consumed and have their seeds dispersed mainly by lemurs. Even so, ravenala has some adaptations that make it irresistible to flying foxes and other Madagascan fruit bats, such as its white flowers and its typically strong, musky scent. In fact, globally these factors are so characteristic of plants that are orientated towards bat-feeding and pollination that they are known as 'bat syndrome'. Baobabs are some of the few other endemic Malagasy plants to show this. Most species attracting bats on Madagascar have been introduced and are cultivated. Generally, though, it is believed that fruit bats had to fit into a Madagascan ecological niche that had already been shaped by co-evolution between lemurs and the ravenala in the cretaceous period, about 150 million to 65 million years ago, rather than shaping it more themselves. In turn, that might have given them greater opportunities to adapt and diversify their own species.[65]

William Ellis, *Three visits to Madagascar*

## MADAGASCAR'S ICONS IN MUTUAL BENEFIT

Lemurs even beat fruit bats to ravenala – recent studies confirm that some species of lemurs feed on ravenala nectar and in the process often carry pollen from flower to flower and tree to tree. Lemurs were first reported enjoying visiting ravenala, sipping its nectar and eating its flower petals, in 1910. They were observed by French-based botanist and naturalist Henri Jumelle (1866-1935), director of the Marseille Botanic Gardens and the town's colonial museum, and by Henri Perrier de la Bâthie (1873-1958), a French botanist who spent much of his working life in Madagascar. Ravenala flowers usually open at night and about 60 per cent of the nectar is produced when the flower is first opened. The nectar has a sucrose content of about 14 per cent – nearly twice as high as honey. Back in 1822, Samuel Copland (d.1876), then a missionary in Madagascar but later a leader writer for the English agricultural press and also known as 'the Old Norfolk Farmer', described ravenala nectar as 'of an exquisite flavour which may be termed natural honey.'[66]

The ravenala flower secretes nectar continuously for about 24 hours, dripping it in sticky streams. When the lemurs can obtain this nectar, they seem to enjoy it as their main food although overall it ranks as the lemurs' second most important foodstuff as it is not available year round. Through the year, each ravenala produces about eight sterile leaves and four leaves with flowerheads.[67]

Black and white ruffed lemurs (*Varecia variegata variegata*) were observed by one research team clambering from one ravenala flower to another, if necessary opening the beak-shaped flower to reach the real prize – the cache of nectar inside. This prospect of nectar tempted ruffed lemurs to visit repeatedly between dawn and sunset, with each flower visited at least once an hour and with the lemur spending on average seven minutes on a plant with four flowers. Various lemurs were also seen feeding on the nectar, including aye-aye (*Daubetonia madagascariensis*), with white-fronted lemur (*Eulemur fulvus albifrons*) visiting occasionally during the day and the greater dwarf lemur (*Cheirogaleus major*) at night. Whichever species was involved, the lemur usually foraged alone – and aggression flared occasionally if two lemurs headed for the same flower at the same time. Research on Madagascar's biodiversity has moved fast in recent decades and this lemur research project took place before the 1997 proposal of ravenala variants or possibly subspecies, which might be used to narrow down food choices by various lemur species.[68]

The black lemur (*Eulemur macaco*) was observed pollinating ravenala, as well as *Parkia madagascariensis*, by another research team. They described the lemurs reaching the crown of the ravenala tree from neighbouring plants, then climbing to the flowers using the huge petioles like steps of a ladder. To extract the nectar they supported themselves on the bracts, pulled apart the [flower] using their hands, pushed their snouts into the centre of the flower, and lapped nectar from the nectar chamber . . . Occasionally, an animal plunged a hand into the flower, withdrew it, and licked the nectar adhering to it. Lemurs were also seen holding the style and stamens and licking pollen from their surface . . . and left with their snouts dusted with pollen.[69]

Earlier researchers had in the 1970s discussed lemurs destroying the flowers but the black lemur researchers did not see lemurs either doing this or eating the flowers. On several occasions, researchers saw female black lemurs displacing males feeding at these flowers. This is considered fairly common behaviour among lemurs, though unusual among primates where males usually dominate.

Detecting exactly how the ruffed lemur, a versatile fruit-eater and ravenala-pollinator, succeeds in pollinating the tree's flowers was the goal of a 1994 research team. The ruffed lemur's 'pointed snout and long tongue' make it a well-designed nectar-seeker in flowers such as ravenala, they noted. This was interpreted as more evidence of ancient systems of plant and pollinator evolving together which seem to have survived only in parts of the world that have no or few flying pollinators such as birds and bats – Madagascar qualifies in this sense, they point out, given that it has only three species each of birds and bats that are known nectar-feeders.

But the feeding-pollination relationship between lemurs and ravenala is not absolute. The lemur can and does feed on other plants and the ravenala can and does self-pollinate, reproducing by seed and by suckers. This meant the researchers could not classify it as a tightly co-evolved relationship. But they did conclude, 'the pollination system may be ancient, at least in the Strelitzaceae.' The most primitive of the three genera in the Strelitzaceae family, they believed, is ravenala because it has six fertile anthers, considered to be the ancestral, or pleisomorphic, state for the order.[70]

But this ancient ravenala-lemur partnership is now in danger. Deliberately or not, humans put double pressure on ravenala. They make its reproduction more difficult by reducing the lemur population through hunting for bushmeat, while their domestic dogs also sometimes prey on lemurs. In addition, ravenala are being felled in forest clearance or simply under increased pressure of traditional *tavy*, slash-and-burn extension of agriculture and homesteads.

## THE BUILDER'S TREE

Ravenala has acquired an image as the saviour of the thirsty traveller but in Madagascar ravenala is more commonly thought of as a builder's tree, its common name in several parts of Asia. This is because it is still commonly felled so various parts can be used to build traditional tranogasy homes, which remain popular because of the advantage of using locally available materials that are '*renouvelables et bon marché* [renewable and cheap]'.[71]

It is the most common *bemavo* type of ravenala, which often forms forests at altitudes between 300 metres and 600 metres in eastern Madagascar, which is traditionally used for building homes. Ravenala is also common in the northern-western area of the island and used for housing there, too. Floors built from ravenala trunks and carpeted with ravenala bark were noted in a 1990s survey. Ravenala leaf stalks were harvested for walls and the leaves themselves for thatch roofing as well as some basketry. House frames were built from other trees in the coastal forest, such as *Faucherea thouvenotii* or *Terminalia ombrophila*. Almost everybody interviewed used the ravenala which they found within about two kilometres of their homes in some way, including making household implements or using its fibres.[72]

Even where more contemporary materials are now used for the main body of a house, the advantages of thatching with ravenala leaves are still recognised and this is often done: '*elles présentent une bonne étanchéité et un coefficient d'isolation*

'Builder's tree' used in Madagascar

Specimen from *Ravenala madagascariensis* growing in Durban Botanic Gardens, 1939

*thermique supérieur à celui de beaucoup de matériaux d'importation . . .* [they offer good water-proofing and better thermal insulation than most imported materials . . .]'. Ravenala thatching lasts between three and five years depending on the area's microclimate and on the ravenala variant used. Where the plant has become established in southern India, similar uses have developed with the ravenala stem being used for construction, midribs and stalks for walls and leaves for packing material.[73] In the so-called famine period between rice harvests, ravenala is not sought after by the poorest for its water stores but as a source of food. About one in eight said they eat the palm-heart, usually taken from a young *bemavo* ravenala with a trunk no more than a metre high as other forms are more bitter. In the 17th century, de Flacourt had noted that oil was pressed from the blue ravenala arils and the seeds ground to make flour, eaten with milk or chewed with slaked lime, rather like betel. But these days, it is considered difficult to collect a large enough quantity of seeds to be useful to humans so they tend to be left to bats and birds, which ultimately helps regenerate the ravenala population.[74]

## TAKEN FAR AND WIDE

The travels of the traveller's tree started soon after ravenala came to the notice of the world beyond Madagascar. It was widely grown in the tropics, usually the more common *bemavo* form. The labels on cultivated specimens at the Kew Herbarium evoke intriguing snapshots of time and place, tracking changes in international and national politics as much as in the development of society, politics and botany:[75]

- *Malaysia, 1830*: Collected by Colonel George Warren Walker (d.1844) of the British army, and simply marked 'Singapore, Penang etc', this also bears the stamp Kew 1867, indicating it was part of Sir William Hooker's herbarium, which came to Kew after he had died in 1865.

- *Mauritius, 1864*: Collected by Carmichael – probably J.R. Carmichael, a doctor in charge of the London Missionary Society hospital in Canton, China, in 1862 so possibly collected when calling at the island.

- *Philippines, n.d.*: A duplicate from the Philippines Herbarium, collected by August Loher (1874-1930), a German botanist who worked in the Phillipines from 1888 to about 1909 and in Madagascar in 1911.

- *Madras, India, 24 May 1900*: Part of the herbarium of Sir Alfred Gibbs Bourne and Lady Bourne, presented to Kew in July 1915. Sir Alfred (1859-1940) collected plants in India and what were then Burma and Siam. Born in Suffolk, Bourne had obtained a DSc at London University, then gone to Madras in 1885, where he was botanist to the Government of Madras for two years from 1898, as well as managing and teaching in tertiary education in Madras and Bangalore.

- *Oahu, Hawaii, 28 October 1937*: Collected by Albert F. Judd (1874-1939) at Kamehameha Girls' School in Honolulu for the Flora Hawaiiensis. Judd's father had been a senior judge in the Kingdom of Hawaii during its transition to the United States and his brother became governor of the Territory of Hawaii. Judd himself was a lawyer who served in the Hawaiian territorial senate and helped compile the territory's laws, as well as serving as a trustee of the Kamehameha schools, where he collected this specimen. He delivered a series of lectures on trees and plants in ancient Hawaiian civilisation at the school that was published in 1933 and included a list of local plants with botanical and vernacular names and uses.

- *Durban Botanic Gardens, Natal, May 1939*: Collected by the German-born botanist Herold Georg Schweickerdt (1903-1977), while working at the Natal Herbarium, South Africa's oldest herbarium building, constructed in 1902 and now part of the SA National Biodiversity Institute. Schweickerdt had been South African liaison officer at the Kew Herbarium and later became professor of general botany at Pretoria University. The specimen, a duplicate marked 'Natal Herbarium, Union of South Africa', was collected on the eve of the declaration of the Second World War and just nine years before the National Party finally took power in South Africa, legally instituting apartheid and then making the country a republic in 1960. Another duplicate from the same collector was received via the National Herbarium, Pretoria, its label categorised in Latin.

- *Hong Kong, 1969*: Collected for Plants of Hong Kong.

- *Lae Sub-Province, New Guinea, 1982:* As part of the Flora of New Guinea, collectors from the New Guinea Department of Forests sent to Kew a duplicate specimen from the ravenala cultivated beside a house in Morobe.

## THE TRIUMPH OF THE TRAVELLER'S TREE

Back on its island home, ravenala emerged only quite slowly as Madagascar's own official botanical symbol. This was despite its central role in Malagasy life and the fact that having a tree as a national symbol would chime with Madagascar's long-standing traditions and practices connected with ancestral trees (*togny*), around which everything from royal centres to provincial capitals and villages have been planned.[76]

Another likely reason for ravenala's slow climb to symbolic ascendancy among the Malagasy is probably its relatively limited natural range on the island. It prefers swampy flats or cloud forest so it is not found in the dry, barren west, nor in the south-western spiny desert. Though records point to shifts in its growth pattern, even historically it seems to have grown mainly only across the east and south-east of the island. Findings from several mid-20th century expeditions were not published but plant audits at the end of the 20th century also showed ravenala occurring in both the Parc National de Marojejy, in the north-east of the island, and in the Réserve Naturelle Intégrale d'Andohahela, Madagascar's 'most southerly example of . . . humid montaine forest.' It is 'not strictly tropical' as it lies just south of the tropic of Capricorn. At Marojejy, ravenala was found at 800 metres but at Andohahela, it was found at about 300 metres altitude where the forest was 'disturbed', a similar setting to raverala found in the Réserve Naturelle Intégrale d'Andringitra, also in the south-east of the island. At Andohahela, ravenala was found to be 'more common in light gaps along the river than in tall forest . . . Above this altitude genera such as *Ravenala* tended to drop out, being replaced by *Cyathea* in their narrower streambeds with their lower filtering light values.'[77]

Local Madagascan political changes, though, branded ravenala on the memory of 19th-century foreign travellers – and missionaries helped do the rest. However, the oldest known Madagascan heraldic symbol is a red bull. This was the emblem of King Andrianahifotsi (1610-1685), whose kingdom, Menabe, in the west of the island meant 'Great Red'. The mango appears to have been the tree first associated both with Sakalava royalty in western Madagascar as well as later with French imperial interests, particularly by Caporal Jean-Onésime Filet, La Bigorne, of the Compagnie des Indes in the 1750s (see chapter 5). In 1895, the replanning of the western port of Mahajanga (then Majunga) took into account sensitivities around its 'ancient landmark', a giant baobab planted by Muslim Arabs. By the time this new port and the modern airport were established, however, ravenala had triumphed over other tree contenders. The fact that the 'traveller's friend' had in fact been witness to the misery of generations of travellers submitting themselves to the excruciating test of tenacity on the eastern route through the fever-ridden forest to Antananarivo was barely remembered. Instead, the tree's visual appeal to the imaginations of visitors, particularly missionaries targeting the island, was triumphant.

In 1960, ravenala was chosen to feature in the design for the newly independent country's coat-of-arms. The tree had the twin virtues of being endemic to Madagascar and well known internationally. Yet for all this global interest in the tree and the interpretative focus around 'the traveller's friend', a key factor is often overlooked. Ravenala already had its own local heraldic pedigree. Queen Ranavalona II (1829-1883) was the first known to feature a ravenala palm on her coat of arms, combining it with zebu cattle among other motifs. In this sense, enshrining it once again as the island's emblem could be seen as a post-colonial triumph. It repossesses the island's icon made familiar to the world from early commentaries and depicted over the centuries for its striking appeal, multipurpose uses from house-building to food. Ravenala is also a dramatic example of the biological diversity which was and remains one of Madagascar's great attractions.[78]

## OVERWHELMED BY MADAGASCAR

English artillery officer and geographer Samuel Pasfield Oliver, who visited Madagascar in 1861 and 1863, felt so overwhelmed by the island's plantlife that he invited John Baker (1839-1920), then assistant keeper of the Kew Herbarium, to contribute 'the particularly interesting notes on the insular flora' to his two-volume, 1886 book simply entitled *Madagascar*. At that time, Baker noted, about 700 genera of plants and trees were then known on the island and about 100 had already been distributed to various parts of the world or collections.[79] Of these, about 80 were

> supposed to be endemic, as far as present knowledge extends. . . Several of these are represented by a single species only, and none of them by more than five or six. Many of them are little known, and of several of them we have no authentically-named specimen in the Kew Herbarium.[80]

It is significant that following his general overview, the first plant that Baker dealt with individually was ravenala:

> The endemic type that influences most the general physiognomy of the vegetation [on the east coast – S.P.O.] is the 'travellers' tree', *Ravenala madagascariensis* [Footnote: Also called *Akòndrohàzo, Ràvinàla, Ràvimpòtsy.*] It is allied to Heliconia and the banana, and has a tall simple woody trunk . . . [and] a blue pulpy arillus.[81]

This entry of 10 lines, including a careful botanical description, is at least double the other longest entry. Most entries are just between one and three lines long.

Madagascan national airline logo interpretation of *Ravenala madagascariensis*

# EMERGING FROM RAVENALA'S SHADOW
## *Phenakospermum guyannense*

CHAPTER SEVEN

## 7    EMERGING FROM RAVENALA'S SHADOW:  *Phenakospermum guyannense*

Every family has its likenesses and its surprises. At first, it might have appeared to western botanists that strelitzias were a family found naturally only in south-eastern Africa and on a remote though huge island in the Indian Ocean. However, the breakup of the supercontinent of Gondwana, when South America separated from Africa more than 110 million years ago, led to strelitzias also having a western-hemisphere relative from South America. The only neotropical member of the strelitzia family, *Phenakospermum guyannense*, often commonly known as the big palulu, is found scattered across the edges of marshy Amazonian forest in the north-eastern corner of this subcontinent.

At first glance, phenakospermum could be overlooked or even thought to be a ravenala with particularly shaggy leaves, especially if you do not know the two trees well or have no reference available. This may explain why phenakospermum was left in ravenala's shadow for a long time. Ravenala was a popular prestige plant in the USA from the second half of the 19th century but phenakospermum was not introduced there until 1932 as a specimen from a village near Paramaribo in Surinam, according to the US Department of Agriculture. Even in 1958, an American horticultural dictionary still firmly listed ravenala and phenakospermum as one and the same species, treating the two as variants of each other and pinpointing the key to telling the two apart as the length of the leaf-stalk – longer than the leaves in ravenala and the same length as the leaves in what we now know as phenakospermum.[1] The fact that this western strelitzia cousin looks superficially similar to ravenala may have flattered some botanists only to deceive. This eventually seems to have been wryly acknowledged in its botanical genus name, whose Greek root means 'cheat' – or perhaps more politely, 'mimic' – both of which might suggest a plant with an identity crisis.

The tree occurs east of the Andes in tropical northern South America and central South America across a wide range of countries – French Guyana (Guyane Française), Guyana (previously British Guiana), Surinam (previously Dutch Guiana), Venezuela, Colombia, Bolivia, Ecuador, Peru and northern and west-central Brazil. Yet phenakospermum has been 'little collected and poorly studied'. The Smithsonian Institution's Biological Diversity of the Guyanas Program collected only two specimens of phenakospermum when documenting the botany of Guyana from 1989 to 1991. As late as 1993, phenakospermum was noted as being common in the Reserva Florestal Alfredo Ducke in the Amazônia Central, Brazil 'but not yet collected'.[2]

Although phenakospermum is often described as 'sparse' but 'locally common' in the wild, it would seem quite hard to miss. Descriptions repeatedly comment on its size, about six times the height of a grown man, using variations of 'massive', 'gigantic' or 'vast' when discussing what has been defined in botanical terms as a tree-sized herb. Each shoot of phenakospermum can grow up to 12 metres tall and is topped off by a huge flowerhead of nearly four metres, or slightly more than a third as high again as the stem. The vast flower-head remains eye-catching when the seed-heads split open to reveal its vivid red-tufted fruits. But these seeds can be a relatively rare treat. Firstly, each flower in that dramatic flower-head lasts for only one night; and secondly, it often takes a long time for the tree to mature sufficiently to bloom and set fruit. It took 23 years for the big palulu to produce its first flowers at the Hawaii Tropical Botanic Gardens in December 2013. Apart from being elusive, phenakospermum was also overshadowed in the broader region by its immigrant cousin, *Ravenala madagascariensis* with its more defined, showier outline, which appeared in South American civic planting and even attracted British traveller and artist Marianne North to paint it.[3]

Scientists' research and debates around the botanical affiliations of the strelitzia family have been rife with mystery and confusion for about three centuries (see chapter 3). Despite some significant scientific discussion during the 19th century, by 1921 William Fawcett, formerly director of public gardens and plantations in Jamaica and co-author of the early volumes of *Flora of Jamaica*, still summed up South America's phenakospermum as 'not so well known' compared to its Madagascar's ravenala cousin. He was familiar with *Ravenala madagascariensis* from the Castleton Botanic Gardens, where it flowered in April and fruited in September – and had led his 1896 list of plants installed in the gardens' new tropical African and Madagascar section with this prize. This may have coloured his view of phenakospermum, which he referred to in his 1921 discussion by a long superseded botanical name, *Ravenala guianensis*, as if proving his own point.[4]

Phenakospermum has also shadowed ravenala with a debate over the number of species. In the case of phenakospermum, though, this debate is intermittent and simply considers a second species, rather than three or more potential additional variants like ravenala. Again, though, the phenakospermum species division was first sparked by habitat – in this case, Amazonia versus the Guyanas. Fawcett does not leave a clue as to whether he was passing judgement in this debate or simply following the precedent that was familiar to him. In 1911, Dutch botanist Johannes Lotsy (1867-1931), who had served as director of the Rijksherbarium in Leiden from 1906 to 1909, had believed phenakospermum occurred as far as northern Brazil, which seems plausible as the evidence of specimens shows. Fawcett, though, placed the tree as a 'native of Guiana' only, as given in 1912 by Hubert Winkler (1875-1941), a botany professor at Breslau University.[5]

Overleaf: *Phenakospermum guyannense* in Miquel's *Stirpes surinamenses selectae*, 1850

## SEEKING OUT THE BIG PALULU

Welsh-born Alfred Russel Wallace (1823-1913) originally trained as a lawyer but became one of the most distinguished plant collectors and scientists of his day.[6] He frankly admitted his overwhelmed awe when first confronting the Amazon forests:

> Huge trees with buttressed stems, tangled climbers of fantastic forms, and strange parasitical plants everywhere meet the admiring gaze of the naturalist fresh from the meadows and heaths of Europe. . . . I first endeavoured to familiarise myself with the aspect of each species and to learn to know it by its native name; but even this was not a very easy matter, for I was often unable to see any difference between trees which the Indians assured me were quite distinct, and had widely different properties and uses.[7]

In a similar but more focused way, getting to know and understand phenakospermum has been a major hurdle for generations of botanists. First they had to encounter phenakospermum because it is distinctly fussy about the habitats it prefers despite its deceptively wide range across a huge swathe of northern South America. One research group rated phenakospermum as common across the Guyanas, in other words from Guyana in the north and Surinam in the west, through to French Guyana, Venezuela and Brazil's Amapá. But in a study of plantlife diversity in the Sierra Maigualida in Venezuelan Guayana, another group found phenakospermum in only one out of four separate one-hectare plots. Importantly, that particular plot, which had four phenakospermum trees, had the greatest plant diversity of all four study areas. This was due mainly to being surrounded by 'a mosaic of diverse eco-vegetational zones' – evolutionary development drives the species occurring in a region but local ecological patterns of climate and soil productivity drive whether they thrive in numbers. This research also underlined how even those watching out for phenakospermum might overlook it among the vast variety of trees in the region's rain forests and swamps.[8]

Phenakospermum chooses far-flung or highly competitive situations, thriving in tropical habitats, for instance, where either its feet or its head will be kept damp – swampy situations or rain or cloud forests. The great range of species competing for space in these habitats probably contributes to phenakospermum's 'sparse' representation, although it can then become 'locally common' when it gains a foothold. In lowland forests, phenakospermum appears in damp patches, particularly the edges of seasonally flooded forests, known in the Amazon as *várzea*. Outside swamp forests, it seems to prefer clayish soils. Interestingly, rather like the *bemavo* form of ravenala thriving in slash-and-burn areas in Madagascar, phenakospermum can grow well in terrain with unstable soil and often flourishes in numbers at the top of low ridges. Both ravenala and phenakospermum seem to favour high-risk strategies, relying on seasonal climate fluctuations or on the benign side-effects of human intervention on the landscape.

To the west of the Andes, a rain shadow creates the Atacama desert, notorious as the driest place in the world. On the eastern side, though, enough lush tropical forest grows to allow phenakospermum to thrive here and there – but exactly how and why is not yet clear. As well as looking at height above sea level, for instance, researchers have compared water available from ground sources and naturally occurring rain and fog – making a distinction between rain forests' vertical precipitation and cloud forests' misty horizontal precipitation. So far, though, general understanding of the local patterns of plant diversity in this tropical Andean region is considered no more than 'fragmentary'.[9]

Detail of *Phenakospermum guyannense* flowerhead and comparative size of tree, Surinam

## THE CHALLENGE OF HERBARIUM SPECIMENS

Despite its own more recent expedition specimens, Kew selected a cultivated specimen for its DNA bank voucher in 1988.[10] A voucher herbarium specimen differs from specimens that record plants collected during an expedition – it is a representative pressed sample of a plant deposited in carefully maintained herbarium conditions to assist future study. Voucher specimens are benchmarks allowing researchers to see the material on which their predecessors based their conclusions. Researchers will refer to it when cross-checking the identity of a plant, discussing the range of variation within a plant species or when proposing changes in plant classification due to new scientific discoveries.

Using a cultivated plant in controlled conditions should give the best results for a voucher specimen. Attempting to press plant specimens in the midst of an expedition is often a major professional challenge for a collector, especially in an unfamiliar area where the climate may prove even more of a hurdle than anticipated. Kew's John Whitehead recalled an ill-fated attempt during an exploration of Mount Kenya:

> On the way down I strapped an impressive fungus in the plant press, only to find the next day that it had gone, leaving just an outline of the toadstool and a mass of maggots on the paper![11]

Renowned tropical botanist Richard Holttum despaired: 'The very large tropical monocots cannot be adequately represented in herbaria and few people have taken the trouble to make good observations on the living plant.'[12]

Whether preparing a voucher specimen or a survey record, the process begins by pressing the specimen in a plant press. This wooden frame holds firm the papery layers which help press out as much moisture as possible as quickly as possible from the plant, while ensuring that it remains recognisable, in one piece and its key diagnostic features are clear.

The plant is carefully positioned to achieve this on folded paper, often newsprint. Each time a specimen is collected, it is given its own unique number by the collector or herbarium. This is clearly marked on this inner paper and listed in the herbarium's own records.

The package is then enclosed within blotting paper to absorb moisture. Finally, corrugated cardboard is placed on either side to allow air to circulate and prevent mould. This plant parcel is secured inside a wooden-framed plant press. It is held tightly closed with buckled straps so the plant shrinks as little as possible and remains flat rather than wrinkling up as it dries. For this reason, the straps may have to be tightened further during drying, particularly with larger specimens.

Modern methods use an electric drier that should give steady heat from below and ideally allow specimens to dry rapidly so they retain their colours as much as possible. But blackened, discoloured or brittle specimens can equally emerge from drying specimens for too long or at too high a temperature. There remains an art to producing a good specimen.[13]

## FIRST SHARPEN YOUR MACHETE . . .

It takes a tough botanist with a sharp machete and a strong backup team to take good, herbarium-quality phenakospermum specimens. Although it is said that the tree can be felled with one strong blow, carefully bringing down the huge flower-head to avoid causing personal injury and securing some of the tree's long leaves are time-consuming challenges. The process ends three or four weeks later after repeated turning, pressing and changing of sheets always hoping and trying, in an already damp environment, to produce a dry and fungus-free specimen.[14] The challenge for amateurs of taking good specimens of these lofty, fleshy plants would be correspondingly even greater. Few casual collectors seem to have been tempted out into the field to collect phenakospermum, perhaps because they felt more restricted by language and political issues under Spanish and Portuguese colonial and post-colonial rule in South America compared to areas such as southern Africa. The challenges of collecting phenakospermum have helped trap it in a scientific vicious circle. With relatively little work being conducted on the plant, there has been little reason for professional botanists to add collecting this awkward and over-sized specimen to the challenges of coping in the kind of terrain it favours. The basic logistics and time required to grapple with phenakospermum's 'fleshy nature and massive size of the inflorescences and flowers' could well also put researchers off collecting the plant if their research focus was other botanical families.

A botanical naming battle raged over phenakospermum from the early 1830s until the 1850s. Yet there were relatively few specimens available then of this 'sparse' plant – so where had the botanists acquired them? Much of the early history of phenakospermum's recognition remains sketchy, even jinxed. In the 1500s, the Dutch occupied what they called de Wilde Kust (the Wild Coast) – known in English as the Guyanas and to the Dutch as Surinam – and the Dutch West India Company was established in 1621. This area has now become Surinam (formerly Nederlands-Guiana or Dutch Guyana); Guyana (formerly British Guyana); the Guayana region of Venezuela; Guyane Française (French Guyana, now an overseas department of France); and Amapá (a state in Brazil). Several early European botanists in this vast region could have collected the plant, perhaps they should have collected the plant. But having motive and opportunity are never enough, especially in cases where the outcomes were stymied by overheating international rivalries between botanists and catastrophes conspiring against them. Despite gallant efforts by collectors, it seems that early specimens did not reach Linnaeus or his team as they do not seem to have been catalogued and are not cited in the early classification discussions.

In 1699 pioneering female botanical illustrator Maria Sibylla Merian (1647-1717) received a grant from the city of Amsterdam to travel to the area now known as Surinam and work on its natural history. Exploiting the wonders of one of the newly developed microscopes, she studied the region's insects, including butterflies, as well as their host plants. She based herself at Paramaribo but also spent some time on a plantation as well as enlisting local guides to help her travel in the interior, where she noted local plant knowledge as well as insects and plants. Three factors hastened Merian's return to Europe after two years. She had been suffering from malaria and was 53 years old. She was also critical of how Dutch planters treated both their slaves and the free indigenous people. Although Surinam is one of the more likely places to find *Phenakospermum guyannense*, Merian had focused on somewhat smaller and more accessible plants. She published a first collection of engravings of Surinam's plants and animals, followed by another on its insects, which made her a renowned pioneer of documenting butterfly metamorphosis.[15]

## THE FATE OF THE GREAT VULTURE

When Nikolaus von Jacquin followed in Merian's footsteps about half a century later, he accumulated a spectacular collection of specimens from this Caribbean region that he featured in one of natural history's landmark publications, *Selectarum stirpium americanarum historia*, published in Vienna in 1763. Although the title underlines that he selected the most interesting of his collection, it is unlikely that he encountered phenakospermum as he used the expedition's ships as mobile collecting bases and so kept within easy range of the coast – and out of range of phenakospermum.[16]

Phenakospermum similarly eluded the succession of French botanists and plant-collectors who worked in French Guyana. Pierre Barrère (1690-1755), a physician who became professor of botany at the University of Perpignan, visited Cayenne for five years from 1722 but focused on medico-pharmaceutical plants in the books he published in 1741 and 1743. As a pharmacist, Jean-Baptiste Fusée-Aublet (1720-1778) had similar interests and although he surveyed a significant number of the region's plants, his 1775 four-volume flora – featuring nearly 400 engravings – did not include phenakospermum. Joseph Martin (fl.1788-1826), who had been trained by the renowned André Thouin (1746-1824) at the Paris Jardin du Roi, became director of spice-plant cultivation at L'Almirante acclimatisation garden, 12 kilometres outside Cayenne. But in May 1803, while trying to sail back to safety in France during the Napoleonic wars, Martin and his herbarium collection were on board *L'Union* when it was captured by British privateers. Martin's herbarium was auctioned in England after both the ship and its contents were sold for prize money by the privateers. Some decades later, a proportion of Martin's specimens found their way to the British Museum so it is not certain whether early phenakospermum specimens might have been dispersed among the lost specimens.[17]

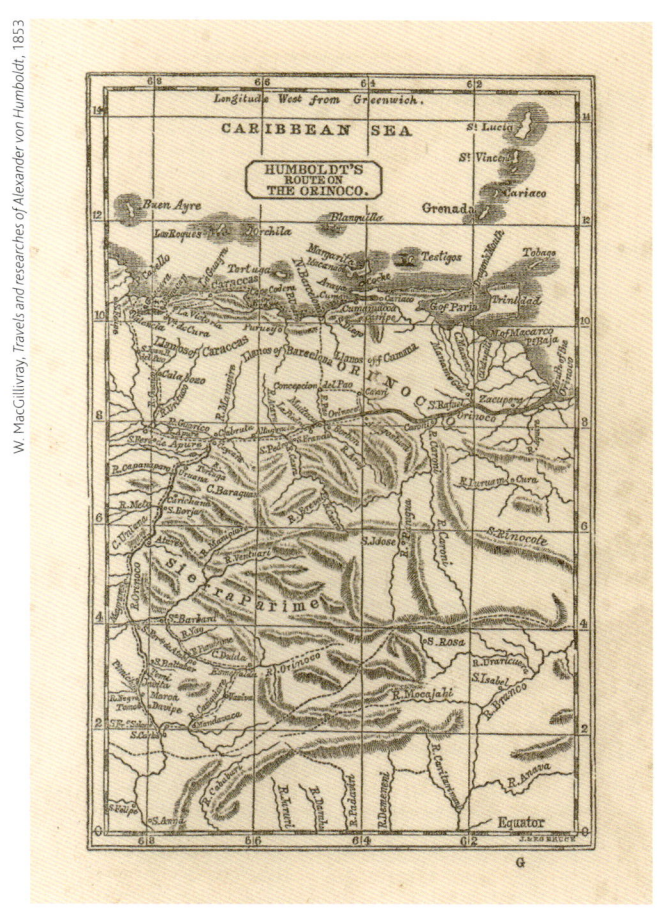

W. MacGillivray, Travels and researches of Alexander von Humboldt, 1853

All these botanists survived their expeditions to phenakospermum territory but Swedish botanist Pehr Löfling (1729-1756) was not so lucky. His mentor Linnaeus was proposing that Spain establish a botanic gardens which would also help broaden his own international plant exchange network. Linnaeus encouraged Löfling, his 'dearest and best pupil', nicknamed the Great Vulture for his keen eye for detail, to travel to Spain to study the country's natural history and economic plants. Löfling arrived there in 1751 and collected about 1 400 plants over the next three years. His hard work and Linnaeus's

Map of Orinoco region, about 1830

influence meant that in 1753 he was appointed to lead the natural history team on the Spanish Expedición de Límites al Orinoco. The expedition was led by Basque naval commander José de Iturriaga (1699-1767), who was also a long-standing member of the Real Companía Guipuzcoana de Caracas (Royal Guipuzcoan Company of Caracas), a Basque merchant company that held the monopoly of Venezuelan trade for most of the 18th century. The expedition's immediate purpose was to map the 1750 Treaty of Madrid boundary between Spanish and Portuguese territory, with contested border territory already prone to Portuguese incursions and Spanish smuggling. Its ambitious brief was then to travel on to Bogotá, Quito, Lima and Buenos Aires, ending in Patagonia.[18]

They finally set off in 1754, their two ships escorted by three naval escort frigates because of the risk of attack by Algerian pirates. The convoy reached the mouth of the Manzanares River and began work first at Cumaná on the coast, next in the Guayana missions region, beyond what is now Ciudad Guayana and about 50 kilometres westwards along the banks of the Unara River, and Puerto Piritu, now in Venezuela. Löfling and his two illustrators were expected to describe as much as possible of the area's natural history, as well as help in prospecting for cinnamon, quinine, minerals and other potential economic assets.

Löfling was then sent with a party tracking the source of the Orinoco river overland, travelling deep into the Caroní region, one of the Orinoco tributaries. Altagracia, an area still 'sparsely populated' in the 21st century, about 15 metres above sea level and covered with regularly flooded broadleaved forest, was what would now be regarded as one of the best prospects for phenakospermum territory on the expedition so far. After four months of journeying through the rigours of the rainy season, though, Löfling was fighting off fever. Finally, he died at Murucuri on 22 February 1756, aged just 27. He was buried amid the headwaters of the Orinoco – another martyr to botany. Linnaeus lamented to Abraham Bäck (1713-1795), 'The Great Vulture is dead'. Löfling's manuscripts were carefully protected and brought home by his colleagues. The expedition had travelled in regions where phenakospermum and many other plants occur but Löfling's notes sometimes proved difficult to decipher, especially in terms of location. Linnaeus edited what he could as *Iter hispanicum*, which was published in 1758. In the preface, he wrote: 'Löfling gave his life for Flora and her lovers, they mourn his loss.'[19]

## THE LANGUAGE OF FLOWERS TURNS HOSTILE

From the early 1500s, plants from the Americas brought back by Spanish soldiers and merchants created a lively interest in the subcontinent. Iturriaga's expedition was wrapped up in 1761 and he chose to stay on in Spanish Guayana as commander of new settlements. The expedition had been a political damp squib but it helped set a new post-Conquista template for almost 60 Spanish scientific expeditions within Spain's empire. These were boosted further when Carlos III (1716-1788) succeeded to the throne and the Real Jardín Botánico was established in Madrid in 1775 with Linnaeus's friend and correspondent, Casimiro Gómez Ortega (1741-1818), as founding director. Spain's botanical expeditions were of a scale or duration unrivalled by any other global power of the time. The global circumnavigation from 1789 to 1794 by Italian nobleman turned Spanish naval officer Alessandro Malaspina (1754-1810) brought back 30 000 specimens, for example. But this was ultimately self-defeating. Much of the overwhelming amount of material that they sent back became stuck in a processing bottleneck in Madrid. However, as no record of any of these expeditions having collected a phenakospermum specimen or its fruits seems to have survived, it seems the jinx on collecting phenakospermum had struck again.[20]

The expeditions that were most likely to have brought back to Europe specimens of phenakospermum were the 1777 to 1788 expedition to Peru and Chile and the 1783 to 1816 expedition to Nueva Granada, which included regions that are now part of Colombia, Venezuela, northern Brazil and western Guyana. The Peruvian expedition reached Huánaco, at the junction of the Higueras and Huallaga Rivers, known as the gateway to the Amazon. The British captured the ship transporting back to France specimens from experienced French collector Joseph Dombey (1742-1794) who was participating in the Spanish Peruvian expedition. These trophies of war were eventually lodged with the British Museum where they remain to this day, despite later French claims for their return. The expedition was riven by rows, which eventually pushed Dombey to withdraw. All pretence of international cooperation was dropped and he found himself imprisoned on arrival back in Cádiz in February 1785 and forced to give up half the collections he had brought back before he was freed. Dombey's luck went from bad to disastrous as on a later expedition as he was captured by pirates and died in prison on the Caribbean island of Montserrat at the age of 52.[21]

The remaining members of the expedition fared little better. In 1786, they lost a consignment of specimens when the ship transporting them went down. Just a year later, fire destroyed yet more specimens before they could be shipped. After the team arrived back in Spain in 1788, it took 10 years before the first volume of their new plant finds was published, *La flora peruviana et chilensis, prodromus*. In all, 12 volumes were gradually published before sections of the expedition's specimens were sold by surviving expedition botanist José Antonio Pavón Jiménez to British botanists Philip Barker Webb (1793-1854) and Aylmer Bourke Lambert (1761-1842), both of whom had interests in South American flora. The rest languished in Spanish archives, where about 3 000 specimens and 2 264 drawings survive to this day.[22]

About twice as many specimens and drawings survived from the Nueva Granada expedition but many times more were lost. José Celestino Mutis y Bosio (1732-1808), a priest and doctor

Los volcancitos, Colombia, about 1830

who had studied botany for three years in Madrid, led an expedition that should have been a colossal contribution to science, with more than 60 artists alone working on illustrations of plants collected. Instead, it became something of a frustrating footnote to botanical history. Among the areas that Mutis and his team explored were low, swampy valleys north-west of Bogotá. Renowned as a perfectionist, he encouraged the naturalists and artists to revise their observations as new material was found so by the time he died in 1808, no specimens or illustrations had yet been sent back to the Real Jardín Botánico in Madrid. A shipment of Mutis's 105 boxes was eventually sent on in 1817 but at some point in transit, the material seems to have been simply tipped into different boxes so that collation between the plates and the appropriate notes went awry. In Madrid, the collections that had been so admired by German geographer and naturalist Alexander von Humboldt (1769-1859) when he met Mutis during his own American expedition were stored in a tool-house at the Real Jardín Botánico. The gardens was perhaps overwhelmed by specimens and illustrations that had arrived from four major, concurrent expeditions.[23]

Despite not sending his material back to Spain, Mutis did send some specimens to European scientists with whom he corresponded. As a result, 306 families, genera or species were attributed to him by the scientists who published his finds. These include Linnaeus and, following his death in 1778, his son Carl Linnaeus jnr, who died in 1783; Humboldt and Aimé Bonpland (1773-1858), who would make their own expedition through South America; and Antonio José Cavanilles (1745-1804), another clergyman-botanist and one of the first Spanish botanists to use Linnaeus's classification system. Some 20th-century botanists also worked on Mutis's specimens.[24] So far no phenakospermum specimen or drawing seems to have come to light from Mutis.

Even the ambitious expedition from 1799 to 1804 by Humboldt and Bonpland through a region including what is now Colombia, Venezuela, Ecuador, Peru, Cuba and Mexico does not seem to have collected the plant either. This is sadly ironic given that Humboldt was one of the earliest scientists to suggest that the land forming the continents of South America and Africa had once been joined. Landing first in Cumaná on 16 July, 45 years after Löfling and the Iturriaga expedition, Humboldt and Bonpland's objectives included trying to track the source of the Amazon from the Peruvian side. On their return, the first volume of *Plantes equinoxiales* was published in 1808 but Bonpland was then sidetracked by French imperial duties. Napoleon (1769-1821) had already given Bonpland an official pension but Bonpland's gift of plants and seeds to the Empress Josephine (1763-1814) so impressed her that she appointed him superintendent of her gardens at Malmaison and in 1813, he published *Description des plantes rares cultivées à Malmaison et à Navarre* (*Description of rare plants grown at Malmaison and Navarre*). Josephine's death in 1814 followed by Napoleon's defeat at Waterloo in 1815 meant that Bonpland's comfortable niche at the French imperial court disappeared. Meanwhile, though, Humboldt had co-opted German botanist Carl Kunth (1788-1850) into the project to ensure publication of the seven-volume *Nova genera et species plantarum*, featuring another 4 500 plants, published from 1815 to 1825. Publication of the remaining volumes of *Plantes equinoxiales* went ahead and Bonpland returned to settle in South America, while Humboldt became a celebrated intellectual throughout Europe.[25]

## PHENAKOSPERMUM'S FIRST PORTRAIT

While earlier collectors overlooked or bypassed phenakospermum, the scientist who eventually showed the world phenakospermum's portrait also unleashed a naming debate around the tree, whether intentionally or not. Two Germans, botanist Carl von Martius (1794-1868) and biologist Johann Baptist von Spix (1781-1826), had been assigned by King Maximilian Josef I of Bavaria to join a group of Austrian scientists on an expedition to Brazil from 1817 to 1820. They started out as part of the entourage of Maria Leopoldina of Austria (1797-1826) as she travelled across Europe and the Atlantic to become the wife of Pedro de Bragança (1798-1834) and the first empress consort of Brazil.

From meteorites to fossilised fish, from blue and yellow macaws to plants and people, Martius and Spix observed as much as they could of scientific, cultural and economic interest, travelling together from Rio de Janeiro south-west to São Paulo and through the state of Minas Gerais to the Salvador coast. They battled with disease and in north-eastern Catinga they nearly died of thirst. Finally, they reached the Amazon, with Spix investigating up to the Peruvian border and along the Negro River and Martius tackling the Yupurá River.

When Martius returned to Europe in 1820, he was appointed conservator of the Munich botanic gardens. After Spix died six years later, it is believed of a tropical disease, Martius continued publishing their findings. Martius's real passion was for palms and he produced three folio volumes on this subject between 1823 and 1850, with one volume given over to all the palm species he had discovered himself. Ironically, Martius may have taken notice of phenakospermum only because without its flower-head or fruit, it might at first glance be overlooked as apparently just one more palm. The Natural History Museum in Munich was founded around the collections of Martius and Spix and to this day holds in its herbarium collection a landmark phenakospermum specimen simply numbered 'Martius 3316' and marked 'Martius's Brazilian journey'. No date or more exact place of collection is given.

Aligning his specimen with ravenala, Martius adopted the genus name *Urania*, which had been coined by his professor Johann von Schreber (see chapter 5), and named the tree *Urania amazonica* when it was illustrated and captioned in volume three of Spix and Martius's *Reise in Brasilien*, published in 1831. Spix and Martius simply captioned this illustration as, '*Species caulescens – Crescit ad flumen Amazonicum (Martius)* [Species with trunk – Grows by the River Amazon (Martius)]', referring to it in the main text as '*Pacova Sororoca* (*Urania amazonica*, M.)'. However, without a full description, Martius's claim to name the tree constitutes what is known to botanists as a *nomen nudum*, or naked name. Martius noted that phenakospermum grew among a jungle of gigantic forest trees ('*den gigantischen Formen der Urwaldbäume*'). He went on to reflect more broadly that the rich abundance of vegetation ('*dieser Fülle des Pflanzenwuchses*') seemed proof of a life force ('*das Maas aller vegativen Bildungskraft*').[26]

Less philosophically, while discussing the wider banana or Musaceae family in a further publication that same year, *Die pflanzen und tiere des tropischen America*, Martius recorded:

> *In den heissen und feuchten Gründen dem Amazonenstrome entlang tritt ein malerischer Repräsentant der Musaceen zwischen dem dichten Urwalde hervor: die sogenannte*

W. MacGillivray, *Travels and researches of Alexander von Humboldt*, 1853

Pacova Sororoca, *a.i. Banane zum Dachdecken* (Urania amazonica, Mart. *Tab.1. vi.2.*)²⁷

[In the hot, humid area on the banks of the Amazon River, there is picturesque dense virgin forest, among which is found a member of the Musaceae family: the so-called *Pacova Sororoca*, i.e. Roofing Banana (*Urania amazonica*, Mart. *Tab.1. vi.2.*).]

In 1860, on behalf of the publisher the Belgian magazine *L'Illustration Horticole*, horticulturist Ambroise Alexandre Verschaffelt, editor Charles Lemaire swiped at other plantsmen:

*Nous trouvons en outre dans quelques catalogues marchands récemment lancées, sous les noms impropres de* Ravenala *ou d'*Urania amazonica, *une plante qui semblerait synonymiquement la même* . . .²⁸

[We find in some recently launched merchants' catalogues, under the improper names of *Ravenala* or of *Urania amazonica*, a plant that seems to be a synonym [of *Phenakospermum guyannense*]

## A FORGOTTEN SPECIMEN

The earliest phenakospermum specimen at Kew came from English naturalist William Burchell (1781-1863), a collector with an obsessive style, a restless soul and, it was said, a broken heart. He had trained at Kew and followed a business opportunity to the island of St Helena in 1805 to avoid joining his father in the family business. But the trading partnership broke up in under a year, so he took the post of the island schoolmaster. Soon, though, Burchell's skills meant he was also appointed official botanist in 1807. It appeared to be the personal breakthrough for which he had been waiting since he had faced family opposition to becoming engaged to his sweetheart, Lucia Green, before he had left London. She set out to join him in St Helena but it seems that her shipboard romance with the vessel's captain made her break off their engagement. To add to his misery, Burchell gradually became unhappy enough with his conditions of service as a teacher and as curator of the St Helena botanic gardens to resign and decamp to Cape Town in 1810. However, Burchell's time in Africa made his name in scientific circles, with various species being named after him, including a zebra.²⁹

Martius's specimen of *Phenakospermum guyannense* from his 1817-1820 Brazilian expedition

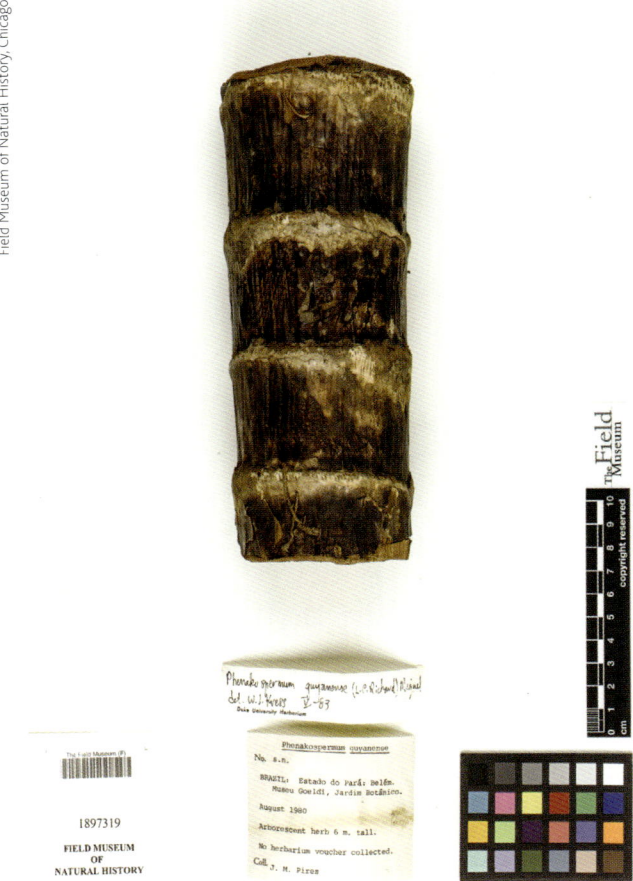

Section of *Phenakospermum guyannense* bark, collected by João Pires (1980) in the Museu Goeldi botanic gardens, Belém, Brazil, confirmed by John Kress (1983)

Burchell became renowned for collecting vast quantities of natural-history specimens and during his first southern Africa expedition, for example, he collected at a rate of about seven specimens per kilometre. Since he stayed for five years, covering about 7 000 kilometres, this amounted to more than 50 000 specimens. Later, from 1825 to 1830, Burchell also collected widely in Brazil, including 20 000 insects alone. Like Bonpland after Humboldt's expedition, Burchell may have been partly overwhelmed by the sheer volume of material to be worked through and did not begin unpacking and organising his 132 Brazilian herbarium packages until February 1847, 17 years after he had arrived back in London. The task took him three years. In addition, at some point the journals from Burchell's expedition were lost, though his field notebooks survive. Burchell's plants and the surviving field notebooks did not reach Kew until after the 82-year-old's death in 1863, possibly as long as 40 years after he had collected the specimens.[30]

Today, Burchell's Brazilian phenakospermum collection in Kew's herbarium bears neat printed labels attached to each specimen simply noting: 'Burchell Catalogus Geographicus Plantarum Brasilae Tropicae No.9691'. They form part of a group of specimens collected towards what proved to be the end of his Brazilian journey. He had reached the Belém area of the state of Pará, which forms part of the Amazonian region and within the decade would become the centre of the rubber boom. He had originally planned to travel on north-westwards to Peru but in early 1828, while in Goiás in central Brazil, he received letters warning him that his father was seriously ill. He decided instead to sail back from Belém and reached home in March 1830, only to discover that his father had died the previous year.[31]

While Burchell was still grieving his father's loss, Martius's 1831 publication appeared, which Burchell might have felt pre-empted his own efforts. More broadly, though, Martius's publication also unleashed a polite but sharp scientific dispute over the naming of what would become phenakospermum. This is not untypical of the cut-and-thrust of debate throughout the scientific world to this day as issues of naming continue to cause

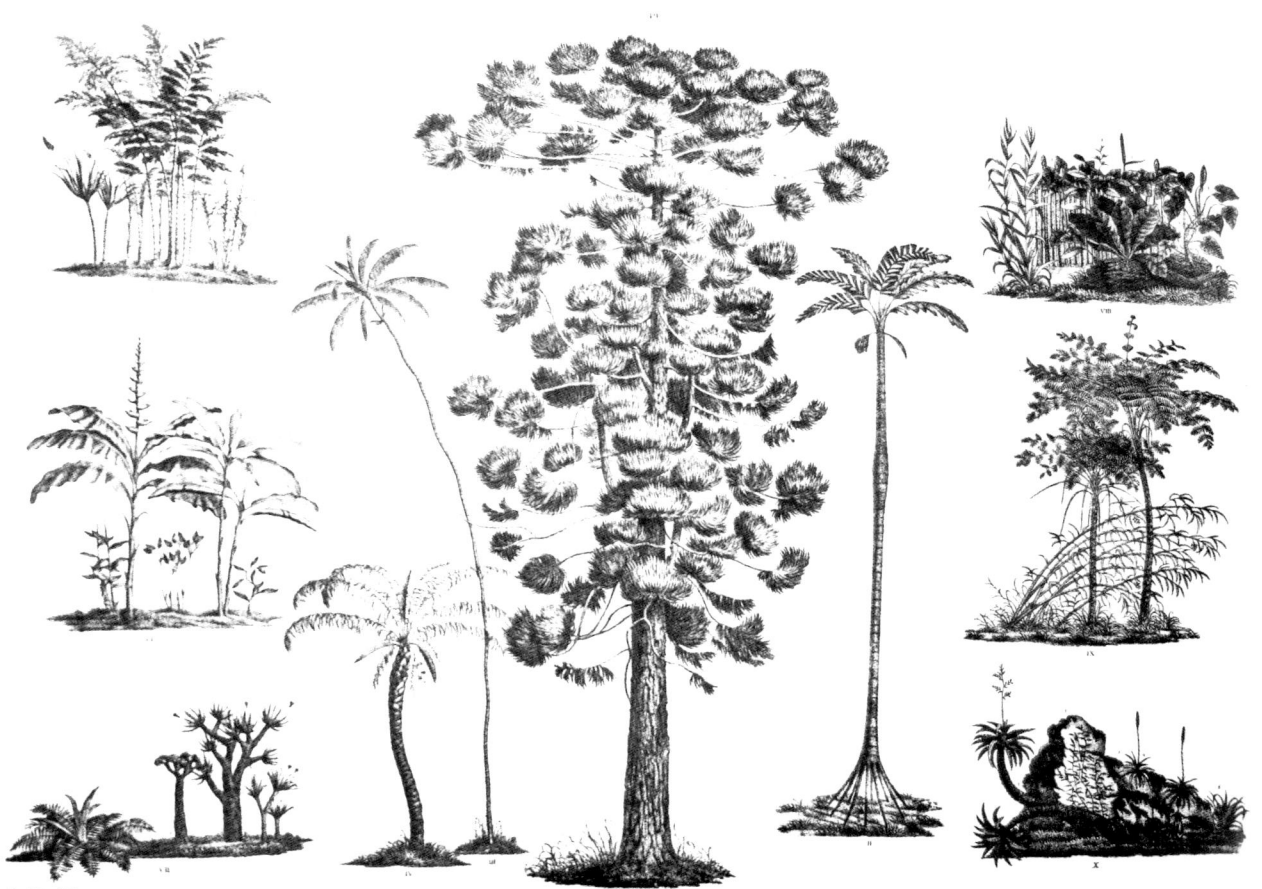

Martius included *Phenakospermum guyannense* (centre left) in a group of palm-like tropical American plants

fierce debate between constituencies, sometimes fed by academic or global politics. Anybody who asks 'What's in a name?' if they become exasperated about botanical systems can look to the confusion around phenakospermum as an example of how continuing development of botanical taxonomy, backed by modern communications technology, tries to pinpoint exactly which plant is being discussed and from where.

## DEFENDING A FATHER'S RIGHTS

The first battle in the phenakospermum naming war opened when hostilities were declared against Martius by Achille Richard (1794-1852), a French naval pharmacist and botanist who was also protecting the work of his father – esteemed French botanist Louis-Claude Richard (1754-1821), who had spent time in French Guyana.[32] Richard jnr was committed to ensuring his father's work was published and the first fruit of this was a journal article that appeared in 1831, the same year as Martius published his book illustrating the tree and naming it *Urania amazonica*. Richard, however, marked the territory in geographical and attribution terms, naming the tree *Urania guyannensis*. As part of a detailed description, Richard snr had noted:

*FLORES non vidi. Color, si qua fides incolis, luteus.*

[FLOWERS I have not seen. Colour, trusting to hearsay, yellow.][33]

Martius's specimen had showed the flower but his illustration, like Richard's, concentrated on the fruit.

Within months, Richard jnr followed this article with a book analysing the banana family, *De musaceis commentatio botanica*. This included commentary on heliconias, under which he included two members of the urania genus, the trees we know today as ravenala and phenakospermum. He noted that ravenala was the only one of the two to have been described previously, which he attributed to Jussieu, not Sonnerat (see chapter 5). Whether this was a ploy to distract attention from Martius's book naming the tree as *Urania amazonica* is not clear, though in both his earlier article and the Musaceae book Richard claimed *Urania guyannensis* as a new species with the notation '*Nob*', short for *nobis* meaning ours. Using the term '*Nob*' to underline the new claim to extend the species in the genus was not unusual – but it would have an added sting if Richard jnr was, in fact, aware of Martius's rival publication. Richard jnr could even have been hinting that Martius's submission should be considered inadequate. He might not have seen Martius's specimen, illustrations and brief reference as trumping Richard snr's illustrations and careful description, despite his father not having seen and so being unable to describe the flower. In neither article does Richard jnr make reference to having seen a specimen of the tree himself.

It is quite possible that Richard jnr did intend such emphasis. Perhaps his dying father had extracted a pledge from his son to go ahead with the work. At the very least that Richard jnr may have felt guilty about letting 10 years elapse before he completed the book for publication. He certainly emphasises ownership in the book:[34]

*L'une, la seule que l'on ait décrite jusqu'à present, est l'Urania Ravenala . . . l'autre a été trouvée par mon père à la Guyane . . .*

[One, the only one described until the present, is *Urania Ravenala* . . . the other was found by my father in Guyana.]

## THE MIMIC'S SEARCH FOR IDENTITY

Six years later, in 1837, Austrian Stephan Endlicher (1804-1849), director of the Vienna botanic gardens, sidestepped any debate by upgrading the tree to a genus in its own right instead of a species within the urania genus. In the process, Endlicher used the name *Phenakospermum* for the first time. The tree's elusive essence as a 'clever cheat' had been recognised for the first time.

This would prove short lived, initially, however. Martius stubbornly repeated his original name, *Urania amazonica*, in his *Flora Brasiliensis* in 1840. In 1841, though, when German botanist Ernst von Steudel (1783-1856) was attempting to compile a complete name index of all plants known to botanists at the time, he took the debate forward. Von Steudel gave only a brief listing in his *Nomenclator botanicus* but chose to acknowledge Richard jnr's name partially and to adapt it as the newer *Ravenala guyannensis*.[35]

By 1845, the battle had been decided except for a few final skirmishes. Friedrich Miquel (1811-1871) – a young Dutch specialist on flora from the Australian and Dutch colonies who headed in turn the botanic gardens at Rotterdam, Amsterdam and Utrecht, then directed the Rijksherbarium at Leiden from 1862 – used the plant's full and now-established name, *Phenakospermum guyannense*, in Berlin's *Botanische Zeitung*. From Richard jnr's point of view, the name preserved his father's insistence on referring to Guyana in the epithet. Just as pleasing, Miquel referred to the comments by the Dutch botanist Frederik Splitgerber (1801-1845), who collected in Surinam from 1837 to 1838 and later in Brazil, on the quality of Richard snr's illustration of the capsule: '*surtout à cause de l'excellente figure de la capsule de Richard* [mainly because of Richard's excellent drawing of the capsule.]'[36] Otherwise, like other botanical colleagues, Miquel was not interested in Splitgerber's claim two years earlier for a new genus of '*Urania* non Schreber nec Richard [neither Schreber nor Richard]'. Splitgerber, though, had gone further, recording his own investigations into Martius's manuscripts, his visits to Martius and the information that Martius had searched in vain for an ovary for his '*Urania amazonica*'. Splitgerber effectively divided the credit between Martius and Richard by dividing the genus into *Urania guianensis* and *Urania amazonica*.

When German botanist Richard Schomburgk (1804-1865) came to list the plants of Guyana in 1848, following an expedition he had made with his explorer older brother Robert from 1840 to 1844, he also split the genus – but differently again under the 'Tribus Uranieae'. With a nod to Miquel, he varied the name only slightly to '*Phenakospermum guianense*', commenting that it was: '*Ueber die ganze Region verbreitet in feuchtes Waldugen. Blüht fast das ganze Jahr hindurch*. [Found throughout entire region, usually in damp woodlands. Blooms almost all year round.]' Meanwhile, Schomburgk attributed '*Ravenala guianensis*' to Richard snr and described it as: '*Standort und Blühtbezeit wie vorige* [Location and flowering habit as before].'[37]

By the time, Miquel was publishing his assessment of the naming question, Splitgerber could no longer defend his friend and colleague, Martius. Splitgerber died in 1845, after being paralysed for three years and unable to work. And by 1849, Miquel's other opponent, Richard Schomburgk, had been back from South America for five years and still out of work. With his other brother, Alfred, Richard Schomburgk sailed for Australia, where he eventually became curator of Adelaide Botanic Gardens. So when Miquel returned to the battle more directly in 1851, nobody was left to carry on the fight.[38]

## SECRETS LINGER IN THE AMAZON

Miquel damned Martius with faint praise as, in *Stirpes surinamenses selectae*, he gave what has so far remained the final casting vote in 1851:

> Stirpes, in omni Guiana ut videtur haud rara, incolis Banana sylvestris dicta, ab auctoribus laudatis descripta, nunquam autem accuratiore partium analysi illustrata fuit, nam ill. RICHARD fructum tantum cognovit, SPLITGERBERUSQUE solam floris adumbratrionem exhibuit.
>
> ... Adnotatio. Phenakospermi amazonici, speciei caulescentis nondum accurate descriptae, imago (sub *Urania*) extat in MART. Reise in Brasilien . . .[39]

> [Plant rarely seen throughout Guyana, called by residents the wild banana, described by renowned authors, although the parts of the fruit have never been accurately analysed and illustrated, RICHARD knew the fruit, and SPLITGERBER sketched only the flower.
>
> Note. *Phenakospermum amazonicum*, an image of this not yet accurately described caulescent species (under the name *Urania*) exists in MART. Reise in Brasilien . . .]

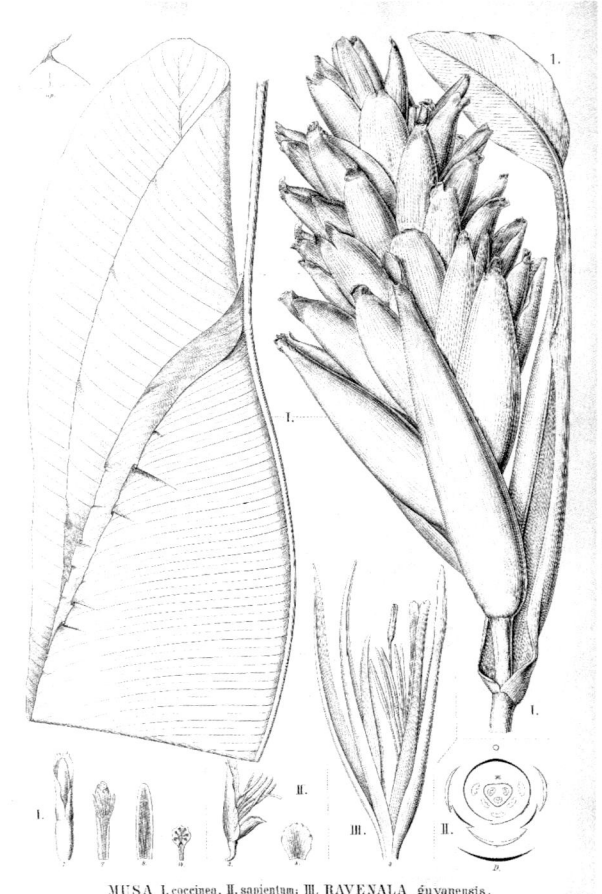

Richard jnr died the following year feeling that his father's claims had been vindicated. By 1867, the International Committee of Botanical Nomenclature entered the fray with its regulations on the hierarchy and nature of botanical names, essentially taking forward the work of Linnaeus and Alphonse de Candolle.[40] Yet even 80 years after Miquel's last, 1851 broadside, not every botanist believed that phenakospermum's description was finally resolved and that each one of its clever cheats had been revealed. Some botanists had for several decades also accepted a second species, *Phenakospermum amazonicum* – intriguingly reminiscent of recent proposals to divide the ravenala species. When four phenakospermum trees flowered spectacularly in 1945 in Buitenzorg Botanic Gardens in Java (now Bogor in Indonesia), they triggered the interest of Japanese botanist Takenosin Nakai (1882-1952). He went even further, though, and attempted to separate von Martius's *Phenakospermum amazonicum* into a new genus as *Musidendron amazonicum*. This eventually failed comprehensively when Nakai's proposal was overruled. So far, *Phenakospermum amazonicum* is not recognised as a separate species.

*Phenakospermum guyannense* (as *Ravenala guyanensis*) shown with banana family relatives, Carl Martius & Stephan Endlicher's *Flora brasiliensis*, begun 1840

The honour of the first scientific description still rests with Richard snr and Miquel's name *Phenakospermum guyannense* still holds despite classification of plants and other living creatures evolving as new information, new species discoveries and new diagnostic techniques come to light. Within the phenakospermum's extended family, the order of Zingiberales, three new species and one new subspecies of Ecuadorian heliconia were recognised by the Smithsonian's John Kress in 1991, for example. William T. Stearn (1911-2001), sometimes called 'the modern Linnaeus', wisely chose his words carefully when he came down firmly on the fence in the debate between these 'splitters' and 'lumpers', writing:

> *Ravenala (Musaceae)* has one species, *R. Madagascariensis* Sonnerat, the well-known 'Traveller's Tree' in Madagascar; its only close ally is a species of tropical America (Brazil, Guyana), *R. guianensis* (L.C. Rich.) Petersen; the latter having five stamens (instead of six) can be put in a genus by itself as *Phenakospermum guianense* (L.C. Rich.) Miq., but, whether the two are treated as forming one genus with two species or as two monotypic genera, their affinity and their distinctness from all other genera remain beyond dispute.[41]

## EXPLORING A BOTANICAL CABINET

The phenakospermum specimens in Kew's herbarium seem a handful compared to specimens of its strelitzia cousin, the southern African bird-of-paradise relatives. These bulge plumply on their stacked shelves, filling one cabinet and overflowing to another. But Kew's slimmer phenakospermum files still mark most of the key points in the tree's story as we know it so far, as well as providing more than straightforward botanical evidence. They also offer insight into different eras of the working lives of plant-hunters and practices of botanical networking. The labelling styles alone are snapshots demonstrating how botanical methods and technology have changed in a comparatively brief space of time.

After contributions received from Burchell, Kew's earliest phenakospermum specimens date from after the main naming war. They originated in French Guiana in 1857 and despite coming from the bequest of the herbarium of the former director of the Royal Botanic Gardens, Kew, Sir William Hooker (1785-1865), the specimen sheet carries its own mysteries. The sheet bears the stamp 'Herbarium Hookerianum 1867' alongside the label from the Herb. Sag. where Hooker originally acquired it. This is marked with the original herbarium's own number 578. In the handwriting of the original herbarium botanist, both the names '*Urania amazonica*' and '*Phenakospermum guianense miq*' are given on this label. Someone, probably a member of the Kew herbarium staff, has written the institution's preferred name at that time directly on the specimen sheet: '*Ravenala guyanensis*, Benth.', referring the researcher to 'Petersen, Mort. Fl. Bras III.pt3, p.6.'

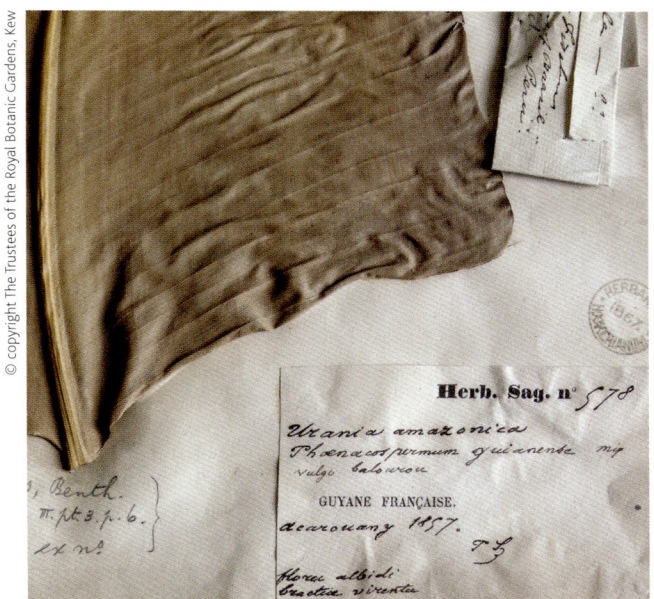

© copyright The Trustees of the Royal Botanic Gardens, Kew

The original label gives a couple of perfunctory notes on what the now discoloured flower-head specimen might look like in real life: '*florae albidi bracteae virenta* [white flowers green bracts]'. It also lists the specimen as collected at 'Acarouany' – the name of both a river in north-west French Guyana and of a small town, just over 180 kilometres north-west of the capital Cayenne. Just four years earlier, the French had established penal colonies in French Guyana at Cayenne, Kourou and St Laurent, as well as three islands of the Salut group, the most notorious being *l'Ile du Diable* [Devil's Island], where Captain Dreyfus would be infamously imprisoned in the 1890s.

Curiously, the makeshift paper envelope marked 'Ravenala – ??' seeds also stuck on the sheet is marked in a third hand with the name 'Dr Tisalune' (possibly, as it is rather illegible) and the place of collection, 'Border of Brasil & Peru'. Either this location is given incorrectly – French Guyana has no border with both Brazil and Peru, only with Brazil and Surinam; and the Aracouany River runs nowhere near French Guyana's Brazilian border – or a second collection was attached to the sheets for convenience sake. Such conundrums make working with old specimens particularly difficult for botanists and botanical historians and have contributed to an international movement over the years towards much fuller label descriptions.[42]

## THE KEWITE TRADITION

Seven decades after the Hooker bequest brought the Aracouany specimen to Kew, another phenakospermum specimen arrived from French Guyana. It was a duplicate specimen collected by Walter Broadway (1863-1935) as part of the Plants of French

*Phenakospermum guyannense* collected in French Guyana in about 1857

Guiana project, jointly run by Harvard University's Gray Herbarium, the United States National Museum and New York Botanic Garden. The specimen was supplied to Kew by one of these institutions as part of the centuries-old botanical tradition of exchange of duplicate live and herbarium specimens, instituted by the earliest botanical gardens and zealously pursued at Kew under the Hookers. On average, each year in the decade from 1871 to 1880, 32 wardian cases were received at Kew and 39 despatched; 2 632 packets of seeds received and 3 615 despatched. In 1899 alone, about 14 500 letters were sent from the small staff that ran the director's office, curator's office and museum.[43]

The specimen label on the first sheet is not completely handwritten, instead a basic label has been used that was printed to create uniformity among the project's specimens although it was also produced separately for each collector, showing as well the increased sophistication of botanical record-keeping. Later sheets for this specimen carry a different, less formal label. The specimen is taped down and the label contains a much fuller handwritten note than earlier specimens.

Much of this information might have appeared previously only in the botanist's field notebook. It describes the plant as growing in wet places and savannahs and having a 'flower stem 30 ft. [10 metres] tall'. It concludes: 'Leaves utilized for many domestic purposes.' This label is also clearly part of the development away from previously typically 'scanty' notes on habitat and other key features when collecting natural-history specimens.[44]

Broadway's specimen might have shown methodological progress and originated with a US project but Broadway himself was a quintessial Kewite in his day. Following in his father's footsteps to become a gardener, Broadway had started working in a local Hampshire garden in 1879 at the age of 16, then worked a further five years in a Devon garden before obtaining a gardener's post at Kew, where he reached the level of sub-foreman. His aptitude fast-tracked him into Kew's well-established programme of supplying keen young gardeners to posts in the imperial botanic gardens network and he arrived to fill the newly created post of assistant superintendent at the Royal Botanic Gardens in Trinidad in 1887 highly recommended by Kew director, William Thiselton-Dyer (1843-1928).[45]

Within less than 18 months of Broadway's arrival, his 'agreeable disposition and youthful enthusiasm' had captivated the solicitor general's eldest daughter, Ethel Philip, and the couple were married on New Year's Day 1890. About a year later, their daughter Elsie was born. Unfortunately, Broadway's boss, John Hart (1847-1911), was less captivated.[46] He insisted on the protocol of Broadway sending all his plant specimens through himself to Kew and the British Museum, rather than Broadway dealing with these institutions directly. By 1891, when Broadway was one of the six co-founders of the Trinidad Field Naturalists' Club, Hart was also trying to exercise strict control over his subordinate's off-duty hours as well.

Hart has been portrayed as 'paranoid and vindictive'; Broadway as 'ambitious, willing to comply with his seniors to secure and promote his career [but] . . . not prepared to sacrifice his free time entirely to the garden when he could be out collecting plants and insects.' Eventually, the row went as far as the colonial secretary, Henry Fowler, who asked the two men to settle their differences. The standoff was finally resolved by promoting Broadway out of Trinidad to become curator of Grenada Botanic Gardens in 1894. Here Broadway was able to devote himself seriously to plant-collecting and was listed as a Kew collector on the island for two years.[47]

It may also be possible, however, that some of the rigorous Hart's worst fears were realised as well. Within a couple of years, Broadway was selling unmounted specimens to institutions such as the Berlin herbarium, the Paris Institute, the New York Botanic Gardens, Harvard's Gray Herbarium and the British Museum. In 1904, a decade after his arrival, Broadway was forced to resign 'having given way to drink'.

At first, Broadway remained in Grenada, continuing with his trade in herbarium specimens as well as running his own nursery but in 1906, he moved his herbarium business back to Trinidad. He might have received a tip from friends because in 1908, Hart was forced into early retirement and Broadway given the post of acting curator at the botanic station in Tobago. This posting was particularly valuable to Broadway as Tobago's flora was previously underexploited. In 1915, he was transferred back to Trinidad as horticulturist and assistant to the government botanist, William Freeman (1874-c.1930). Freeman had written several entries for the 1911 *Encyclopaedia Britannica* as well as co-authoring a book on the world's commercial plants, published in 1908. He and his son William had discovered a white variety of purple wreath (*Petrea volubilis*), well timed for Freeman's next publication, *Useful and ornamental plants of Trinidad and Tobago*.[48]

Not surprisingly, plant-collecting once again became an important part of Broadway's official duties. In 1920 and 1921, he was one of the local guides assisting a study of the flora and plant products of northern South America by the US National Herbarium, the Gray Herbarium and the New York Botanical Gardens. As a result, these institutions jointly sponsored his visit to French Guyana later in 1921, when he collected the phenakospermum specimen now at Kew. They also sponsored his trip to Venezuela in 1922 and 1923. Broadway retired in September 1923 but continued selling specimens to overseas herbaria; these later British-held specimens are in the British Museum rather than at the Kew Herbarium. They include a specimen of *Ravenala madagascariensis* which seems to have reached London via the Missouri Botanic Gardens. Broadway had

Controversial collector Walter Broadway was accused of corruptly selling official botanical specimens

137

collected it at the Tobago botanic station where he had served Kew from 1908 until 1915. Broadway also served as secretary of the local Horticulture Society, which he co-founded and which still awards the W.E. Broadway Memorial Trophy for the best foliage plant exhibit at the annual show. He died in 1935, aged 72.[49]

## THE CHAMPION PLANT-COLLECTOR

For an American to be collecting phenakospermum in Venezuela in August 1944, with the second world war raging in Europe and the Pacific, could seem to be putting the maxim 'plants have no ideology' to the test. But Julian A. Steyermark (1909-1988) was not only out in the field collecting for the Plants of Venezuela project, he was also hunting for new botanical sources of cinchona, the raw material of quinine needed to treat soldiers who had contracted malaria in the Pacific, where Asian sources had largely fallen under Japanese control.

In the 10 years since Steyermark had obtained his doctorate at Harvard University at the age of 24, he had worked as a research assistant at Missouri Botanical Gardens, tasted the excitement of a three-month South American field trip to Panama but then had to survive the worst of the 1930s depression, as a biology teacher, later in the US Forest Service. His move to the Chicago Field Museum to assist in producing a flora of Guatemala helped place him in the right spotlight to use his botanical skills to serve his country. For Steyermark himself, it was also almost as if he was bringing childhood dreams to life. He had first discovered the plant world through the Boy Scouts, high-school nature trips and reading Arthur Conan Doyle's *The lost world*.

In 1943, by which time Steyermark had been collecting seriously for nearly 20 years, he was on a two-year assignment to Ecuador and Venezuela, looking for cinchona and collecting other herbarium specimens along the way. This brought him to Venezuela's Chimanta mountain area and what he would always fondly call his 'Lost World', a plateau of exotic plants and animals that resembled his memories of Conan Doyle's writing. Steyermark was transported by dugout canoe and obtained meat from indigenous Indians who had hunted animals with blowdarts. Whatever the conditions and however scarce the water supplies, Steyermark was renowned for sticking to his 'fastidious' shaving routine on all his field trips.

It was during that first visit to his 'Lost World' that Steyermark collected his phenakospermum specimen. His printed labels place him in the forest area on the south-eastern slopes of Cerro Duida, along a river called Caño Negro, a tributary of Caño Iguapo, at an altitude of about 260 metres. Only the specimen name and its number needed to be added to the label – not by hand this time, but by typewriter. Curiously, though, Steyermark calls the plant *Ravenala guianensis*, citing as his reference '(L.C.

Rich.) Benth'.[50] Steyermark later moved to Venezuela in 1959, becoming curator of the Herbarium Instituto Botánico in Caracas and a Venezuelan citizen. He eventually spent 25 years exploring his once 'Lost World', which he still had to reach by helicopter for his final field-trip.

As longtime friend and fellow collector Virginia L. Beatty reminisced after Steyermark's death:

> Some collectors stay within five meters of the road. Others go for only the beautiful stuff, the flashy flowers. He went on foot and got on a mule and traveled across the country. He was willing to do the hard work of finding the plants and then the necessary research and writing.[51]

Steyermark was renowned for his '"eagle eye" for spotting plants', according to Bruce K. Hoist, project coordinator for the Flora of Venezuelan Guayana, Steyermark's final major project:

> . . . and if you didn't see something that he did, he would let you know it. He was just so excited about collecting plants, it was impossible not to be caught up by his enthusiasm. He

*Phenakospermum guyannense* on the road near Hannover, Surinam

always likened collecting to a 'treasure hunt'! Even with his two artificial hips, he would plow through brush, crawl over and under logs, fall down, laugh about it and push on . . .[52]

Steyermark was driven by his belief he was 'contributing to the sum total of the world's knowledge.' He wrote landmark flora of Guatemala, the Guayana highlands and Venezuela's Ptara-tepui among other major works and was grateful for having been given time and space to do so: 'Whatever I have achieved in my field of plant and taxonomy and phytogeography has been the result of [my] interest and enthusiasm combined with hard work without worrying about the time spent on achieving my objectives.'

When Steyermark returned to the USA in 1984 at the age of 75, it was to his home town of St Louis and the post of curator of the Missouri Botanic Gardens. He later became curator of the Field Museum of Natural History. Steyermark received several official honours from the government of Venezuela and Guatemala, including Guatemala's Order of the Quetzal. Partly because he never took holidays, always preferring to go on field trips instead, Steyermark is reputed to have collected more than 138 000 specimens. The final number in his collecting book was 132 223.

As a result, Steyermark's collecting colleague Gerrit Davidse ensured he was listed in the 1986 *Guinness book of world records* as a 'champion plant collector'. Steyermark also described more than 1 000 new species. Although bromeliads and orchids were an enduring special interest for him, at his death in 1988, Steyermark was believed to have 'discovered and described more plant species than any other botanist of recent times.'[53]

Despite all this collecting activity, until at least the mid-20th century, the *Phenakospermum guyannense* tree remained shadowy and little known beyond the basics of later taxonomic corrections. Botanists are still closing in on its mysteries.

## PHENAKOSPERMUM MOVES NORTH AND EAST

Even in North America, the African members of the strelitzia family, such as ravenala and bird-of-paradise (*Strelitzia reginae*), are far better known and more widely seen, both in landscaping and for commercial growing, than the plant that originates from just across the Gulf of Mexico. Phenakospermum may have proved slightly less appealing than ravenala as an ornamental plant for landscaping and street-planting because it is not different enough in size and shape and in overall appearance, it also has a rather more windswept look like *Strelitzia nicolai*.

*Phenakospermum guyannense* was named within 80 years of its strelitzia cousins but unlike *Strelitzia reginae* and *Strelitzia nicolai*, its first descriptions were taxonomically oriented and did not give any details of how the tree was introduced to the world beyond its South American home. When it was featured in *L'Illustration Horticole* in 1860, Lemaire deliberately left this vague:

> *On sait, à ce qu'il paraît*, bien peu de choses *au sujet de l'histoire de cette belle plante, introduite vraisemblablement en Europe, il y a peu d'années*[54]

> [Little is known, it seems, *very little* about the history of this beautiful plant, probably introduced into Europe a few years ago . . .]

Durban botanist John Medley Wood believed *Phenakospermum guyannense* (which he called *Ravenala guianensis*) had been introduced into cultivation in 1848, though he did not specify where or how. He had planted the tree in the Durban Botanic Gardens in 1894. There is also a record of phenakospermum being offered for sale the previous year in southern Florida.

Interestingly, it was apparently phenakospermum and not ravenala that was taken to Java (now Indonesia) to add to the collections of exotic plants there. It was grown in Buitenzorg (now Bogor) Botanic Gardens, where a specimen was collected in 1923 that is now in the Natural History Museum in London. It is not clear whether the *Ravenala madagascariensis* referred to in the gardens' 1844 catalogue, again by Miquel in 1855 and also in the 1893 gardens' catalogue was this plant or had been assumed to be ravenala when it was, in fact, its lookalike *Phenakospermum guyannense*. In the Singapore Botanic Gardens, 'a small grove' of phenakospermum also formed a feature at the head of the lake. Its southern African relative, *Strelitzia nicolai*, was also present in Singapore as Henry Ridley (1855-1956), director of the gardens from 1888 to 1911, sent a specimen to London in 1909.[55]

Interestingly, neither these nor any other African members of the strelitzia family are mentioned in some 1946 correspondence between South Africa and Singapore. Murray Ross Henderson, the Scottish curator of the herbarium and acting curator at Singapore Botanic Gardens was evacuated from Singapore in 1941 and spent the rest of World War II at Kirstenbosch. After his return to Singapore, he stayed in touch with Sussex-born Frank Thorns, curator of Kirstenbosch Botanic Gardens, previously superintendent of the Government Gardens in Khartoum, Sudan, and curator of the Durban Botanic Gardens. Henderson believed South African plants from the Western Cape would grow well in Singapore and help improve the inevitable neglect of the gardens' collection and appearance during the war and he hoped Kirstenbosch would help by donating some plants.[56]

*Phenakospermum guyannense* in *L'Illustration Horticole* magazine, 1860

# AMAZONIA'S STAFF OF LIFE
*Phenakospermum, the big palulu*

CHAPTER EIGHT

# AMAZONIA'S STAFF OF LIFE: *Phenakospermum, the big palulu*

In the decade or so before US champion plant collector Julian Steyermark died in 1988, Kew's reawakening to South America's plant riches led to their being able to motivate for and fund their own collectors instead of simply piggybacking onto the Americans' expeditions and in the process acquiring duplicates. Their participation over the next dozen years in research into the plants of Colombia, Venezuela, Guyana and Brazil is reflected in Kew's herbarium collection.

This, however, does not reflect the full story of difficulties overcome in that region to collect scientific specimens. Zoologist Adrian N. Warren (1949-2011) was intrepid and later turned wildlife film-maker, yet 'several misfortunes . . . befell' his first expedition to collect zoological specimens in Guyana in 1968. Even three years later, the Venezuelan government refused to grant permission for a zoological expedition to enter the country. Warren compromised by climbing Mt Roraima from the Guyana side and his tenacity was commemorated when a local tree frog, *Hyla warreni*, was named after him.[1]

## KEW'S NEW GENERATION

David Philcox (1926-2003) was known to many Kewites in the 1960s as one of 'our helpful herbarium heroes'. His own herbarium labels for two phenakospermum specimens found in wet forest near the collectors' base camp in Mato Grosso, Brazil, are intriguing. Completely handwritten, they succinctly detail the two specimens – one with a yellow aril capping the seed, the other with orange – both collected from wet forest near the team's base camp. Map coordinates of 51°46'W 12°49'S on the Xavantina to Cachimbo road are given. They are signed simply 'D. Philcox' but undated. The Harvard Herbarium database shows Philcox was part of a collecting team in Amazonian Brazil led by (now) Sir Ghillean Prance during 1967 and 1968. Research in this area would be a major focus for Oxford-educated Prance, who spent the first 25 years of his career based at New York Botanical Gardens before becoming director of the Royal Botanic Gardens, Kew, from 1988 to 1999. Another team member with Philcox was Prance's first graduate student, the Colombian Enrique González Forero. By Forero's retirement in 2003, he had been a full professor at the National University of Colombia for 20 years, was also head curator of the National Herbarium of Colombia and had collected in Madagascar as well as in North America. Brazilian collectors on the team included Lois F. Coelho (fl.1955-1987), L.G. Farias (fl.1962-1968) and A. Fereira (fl.1967-1968), as well as Belém-based tree-climber and collector José Ferreira Ramos (fl.1966-2001).[2]

The next phenakospermum specimen marked only with the surname Lister bears a simple typed label but this runs to 75 words, recording a wealth of information surrounding its collection in December 1975. From notes that the local name is *daweyo* to the local habits of cutting the stem to drink the sap or using the edible fruit 'as a cure for abundant excretion', many of the observations are fresh and vivid. That includes descriptions of the setting: 'low-ground forest with abundant shrubs, ferns and herbaceous plants' at Santa Rosa de Tencua, Rio Ventuari, at an altitude of about 450 feet (about 135 metres) in the Venezuelan Amazon border lands. Notes such as 'alluvial soil' and the 'distinct seasonal variation' with a dry season from December to May are important environmental and horticultural pointers. Lister notes the six-metre-high tree has white flowers in November and large leaves in a fan shape up to two metres long. Both the typed herbarium specimen label and the handwritten endocarp specimen label for Lister bear the collector number 96. He is remembered by one former Kewite as 'an adventurous collector in Venezuela who came from Oxford or Cambridge' – but he is not mentioned in the institution's list of collectors or in the *Kew Guild* and so far Kew has not been able to trace any further details on him.[3]

Zoologist turned ethnologist Marcus Colchester collected phenakospermum fruit in February 1976 in another part of Venezuelan Amazonia. He noted this was typically flooded for four months a year and appropriately called Caño Mosquito. Colchester also recorded more local names, Ta Né and Platanillo Grande (from the Spanish literally meaning great little banana).[4]

*Gardeners' Chronicle*, 1893

John Gilbert Baker, keeper of the Kew Herbarium from 1890 to 1899

*Phenakospermum guyannense* collected for regional flora projects by James Zarucchi (Colombia, 1976); & by Paulus Maas, Hiltje Maas & Rufus Boyan (Guyana, 1977)

## DRIVEN BY ENTHUSIASMS

Across in Mitú, Colombia, in May of that same year, 1976, 27-year-old American James L. Zarucchi needed to go only as far as the road out of town to come across 10-metre-high specimens of the plant among the secondary growth. He described the pale yellow' flowers as fragrant and noted the bracts were pale green with 'a white waxy covering'. Zarucchi is now in his 60s, curator of Missouri Botanic Gardens and a legume specialist, with a particular interest in the dogbane and bean families. He is editorial director of the Flora of North America and an adjunct associate professor at the University of Missouri. He maintains his interest in the flora of north-western South America and also takes an interest in the flora of China.[5]

The carefully sawn cross-section of a bract features in the sheets of specimens collected in August 1977 five kilometres west of Timehri airport, Guyana, by Paulus and Hillegonda Maas with Rufus Boyan. Compared to Zarucchi's specimens, these collectors describe their target plant as a 'small tree', six metres high growing on a wet, open place on Adrian Thompson's farm. Paulus and Hillegonda Maas are a Dutch husband-and-wife team of botanists; she is also known as Hiltje and was formerly van de Kamer. They have collected together in Brazil, Costa Rica and Surinam, as well as French Guyana in South America, and in the Netherlands, France and Spain in Europe. Boyan collected with them only on the 1977 trip but his collecting career ran from 1960 to 1986, with a particular interest in spermatophytes.

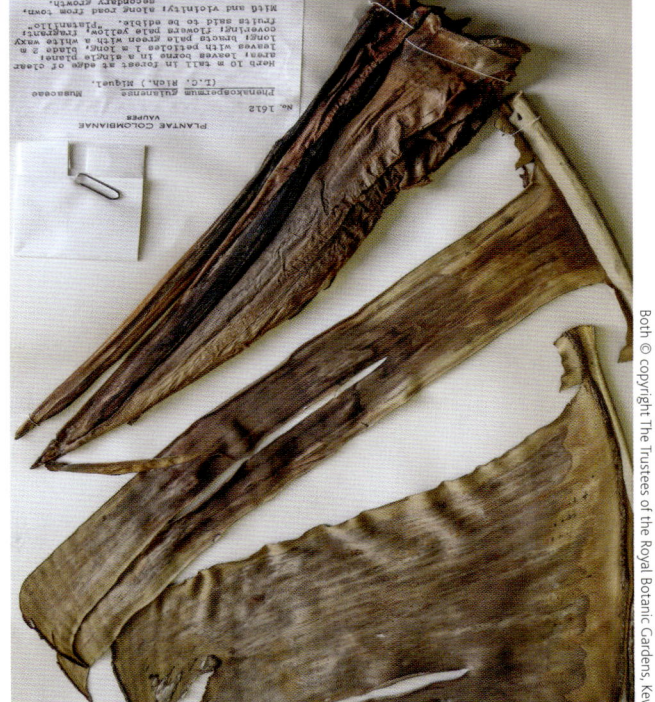

## THE ILL-FATED ETHNOBOTANIST

A label and a half was filled with 156 words of highly detailed botanical description of a phenakospermum on the only specimen in this collection of the often pioneering ethnobotanical work in South America by American Tim Plowman (1944-1989). He was a protégé of Richard Evans Schultes (1915-2001), who is considered to have shaped the modern study of ethnobotany. Plowman eventually collected more than 15 000 specimens. Many of these are housed at Chicago's Field Museum of Natural History, where he was first employed in 1978. Plowman specialised in erythroxylum, a genus of more than 200 species.

Paulus Maas studied biology at Utrecht University from 1957 to 1964 but fell in love with the neotropics during a postgraduate trip to Surinam to catalogue the plantlife of the Bakhuis mountains and has since travelled to more than 20 tropical American countries. He remained based at the University of Utrecht and was also appointed curator of flowering plants at the National Herbarium of the Netherlands in 1974. Between 1972 and 1984, he wrote five volumes on phenakospermum's broader grouping of relations, the order of Zingiberales or ginger plants, for *Flora neotropica*. While Paulus Maas has collected more than 10 000 specimens and described more than 250 taxa, he is also renowned for his devotion to soccer and is said to have tracked down a television in the jungle so he could watch the Netherlands win the 1988 European Cup.[6]

*Phenakospermum guyannense* leaf sections collected by: J.J. Strudwick & G.L. Sobel (Brazil, 1981); & William Milliken (Brazil, 1987)

## REDISCOVERING AMAZONIAN RICHES

By the 1980s, Kew was once again having its collection supplemented by receiving duplicates from New York Botanical Gardens, thanks to its ongoing collaboration in projects in Brazilian Amazonia. This includes a phenakospermum specimen collected in Pará by a team of six as part of the Flora of Brazil project with the Instituto Nacional de Pesquisas da Amazonia and Museo Paraense Emilio Goeldi. Once again, this specimen was found among secondary vegetation and (perhaps conveniently) alongside an airstrip at Macau, which is noted as upstream from Lageira airstrip at an altitude of about 800 feet (just under 250 metres) on the Maicuru River. For the first time in this grouping, full map coordinates are given: 0°55'S, 54°26'W. Also supplied are details of how the specimen has been prepared, particularly challenging with large, fleshy plants such as phenakospermum. In this case, it was pressed with alcohol.[8]

In December 1987 phenakospermum was collected among riverside vegetation from Sema ecological reserve on Maracá Island, Roraima, at 3°21'N, 61°27'W. The team of four collecting for Flora of Brazil also recorded the vernacular names *sorococa* and *banana brava* for the tree. Leading this team as part of the Royal Geographical Society's two-year Maracá Rainforest Project was William Milliken, now head of tropical botany at Kew. Comparing recent fieldwork in Brazil in the footsteps of earlier botanists such as Martius, he wrote in 2014:

While their challenges were primarily linked to transport, survival in remote regions, language difficulties and losses of shipments, our recently-gained ability to record plant data in a much more precise manner (including GPS coordinates,

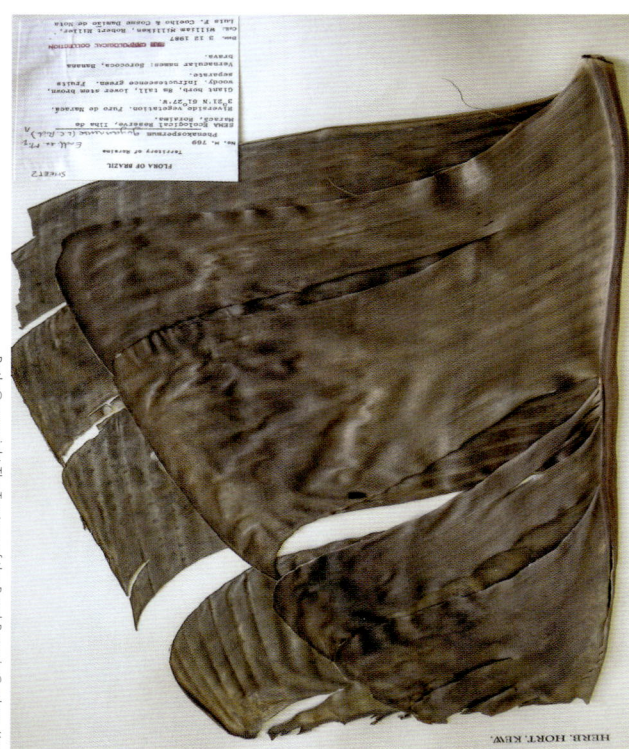

many of which contain coca, the basis of cocaine. Early in his career in South America, Plowman collected phenakospermum on an expedition into the Venezuelan Federal Amazon Territory in February 1979. This specimen was collected about 35 kilometres south of Puerto Ayacucho in a 'forest immediately behind "El Tobogán de La Selva" camping area', at an altitude of 85 metres. Plowman describes it as coming from a 'gigantic herb', 10 to 16 metres high. He also suggested that phenakospermum trees all flower at the same time (synchronous flowering).

Plowman became curator of the Field Museum in 1988 and his own scientific output was prolific, publishing more than 80 scientific papers. His interdisciplinary work included examining archaeological data for use of coca leaves as a stimulant, medicine and in religious practices in pre-Colombian civilisations. From 1981 to 1987, he was coprincipal investigator of the Flora Amazonica project and his contribution to ethnobotany was memorialised by fellow Schultes alumnus Wade Davis in his 1997 book on the Amazon, *One river*. Plowman collected his phenakospermum specimen in February 1979. Ten years later, aged 45, he died from AIDS said to be contracted from pre-trip vaccinations.[7]

images, databases etc.) creates a proportionally much larger volume of information. The old botanists had a collecting book and perhaps a travel diary, while we are now laptop-dependent expeditioners. But when faced with these amazing plants, we get every bit as excited as they did 200 years ago.[9]

Phenakospermum often proved marginal to the researches of these collectors who brought back the specimens. But they nevertheless contributed to two botanical and broader scientific principles – laying down the bones historically and geographically of a collection of specimens that would be a treasure trove for other researchers near and far. This would help create a body of knowledge, creating greater understanding of phenakospermum particularly and, more generally, both the flora of regions of South America and of the strelitzia family.

Many of these later field botanists were collecting for national flora, as was becoming the trend, and also making observations that formed part of the wider interdisciplinary study of the valleys, forests and marshes where they collected, seeing the areas more broadly geographically, ecologically and ultimately even sociologically. Kew's comparatively few specimens, mostly from geographically scattered locations, form part of the attempt to solve the mysteries of this relatively little-known plant. Questions being studied range from how humans have found uses for phenakospermum to its habitat range, how its life cycle might differ from its relatives and checking whether it has any

Woolly opossum (left) and white-necked jacobin (right)

sub-species in South America. These strategies are seen as part of enabling a balanced way of living within the environment without destroying 'the biodiversity that has accumulated over millennia.'[10]

Though superficially, phenakospermum looks quite similar to ravenala, phenakospermum is less finely formed and has so far proved much less in demand for landscaping and horticulture. Phenakospermum may have been eclipsed horticulturally by ravenala both in North America and in its own native South America but in its natural habitat, its attractions bring it a legion of followers, keen to make the most of its offerings – right down to hosting a specialised fungus, *Meliola musae*.[11]

## QUEEN FOR A NIGHT

Within the strelitzia family, phenakospermum most resembles its Madagascan cousin, ravenala – but there is an intriguing difference in their floral biology. While each ravenala flower opens for 24 hours, phenakospermum opens for just the 12 hours of night. So it is particularly creatures of the night that phenakospermum attracts.

Phenakospermum flowers slowly but steadily from the end of the rainy season through to the middle of the dry season – generally from about June to November. Usually only one bisexual flower appears every four nights in a flowerhead on one of the

145

canoe-shaped bracts, a form of modified leaf bract. Other surrounding flowers may be waiting their turn to open or starting to produce fruit following their cycle of opening and pollination that ultimately culminates in the trunk's leaves dying back.

Often nectar flows so abundantly during the night that enough remains to attract daytime nectar-seekers. Among those spotted trying to take nectar during the day in French Guyana have been small, dark, sting-less trigonid bees and a hummingbird species, the white-necked jacobin (*Florisuga mellivora*). The jacobin often struggles to pierce the nectar chamber of unopened flowers. Woolly opossum (*Caluromys philander*) have also been observed visiting for a late-evening snack, generally about 40 minutes after sunset. This is deliberately well timed as the flowers start opening about an hour before sunset, though about one in 20 open during the night.

But though such creatures feed, or attempt to feed on the nectar, neither the jacobin nor the opossum seems to collect pollen as well. An important difference between temperate regions and the tropics is the vital role that bats play in pollinating plants. In fact, within a region where bats are key to pollinating flowers and dispersing fruit, phenakospermum seems to have evolved as one of the plants specialised for bat pollination. This applies to more than 1 861 plant species from the central region of French Guyana alone, for instance. Among the region's 37 nectar-feeding bat species, eight feed exclusively or mainly on nectar, and a further 29 are known or suspected to feed on nectar when they find it. Phenakospermum is considered a typical example of bat-pollination syndrome, its flowers are creamy white to greenish white and its bracts green with a yellow tinge. This makes it less attractive to birds and insects, who usually have good colour vision. In addition, at night the huge flowerheads at the end of its lofty stems send out wafts of fragrance aimed at potential visitors flying past. Similarly, the bright flowers of the distantly related South American heliconia family in South America are mostly pollinated by humming birds. By contrast, its South Pacific cousin, *Heliconia solomonensis*, has greenish-white flowers that open in the evening and are pollinated by bats.[12]

## BELOVED OF BATS

About a third of all mammals are said to be herbivores because they consume some part of green plants. This includes specialist fruit and nectar-feeders occurring mainly in tropical habitats. Because fruit, pollen and nectar often have only brief seasons compared with other food resources, it appears that relatively few mammal families compete with each other for the food resources within this niche. Only about eight mammal families, containing about 460 species, are mainly fruit-eaters. Though in French Guyana, for example, as many as 52 bat species eat fruit, they tend to favour smaller, softer fruits over

phenakospermum. Its large capsule, armoured with a tough shell, would be impossible for them to tackle and even the hard seeds inside are unappealing.

Fewest and most specialised of all are the mammals whose diet is mainly nectar – just two families of bats, the New World leaf-nosed bats (Phyllostomidae) and the fruit bats or flying foxes (Pteropodidae). Significantly, phenakospermum meets not just one but three out of four factors that make a plant flowerhead appeal to bats. First, access is quite direct because it is not hidden in foliage. Second, it produces rich nectar, with a sugar content of about 17 per cent. This is slightly higher than the concentration of 14 per cent calculated for ravenala and about the same concentration as other bat-pollinated plants in Costa Rica, for example. Phenakospermum, though, produces a slightly higher volume of nectar. Finally, the flower's scent also becomes stronger as the night progresses, but generally is relatively light and cannot be detected by humans more than 30 centimetres away from the flower.

Two bat species returned night after night to feed on the phenakospermum flowers in French Guyana, researchers observed. Visiting singly between 30 and 100 minutes after sunset were the greater spear-nosed bats (*Phyllostomus hastatus*). Chattering groups of smaller bats arriving between 45 and 60 minutes after sunset were their cousins, the pale spear-nosed bats

*Phenakospermum guyannense* (left) flowers for about 12 hours, mostly at night, releasing large quantities of nectar; white-lipped peccary (right)

*Phyllostomus discolor*). Both bat species gather nectar and act as pollinators identically. The ways in which they feed and pollinate complement the structure of phenakospermum's flowers, underlining the theory that plant and bat have evolved to live in mutually useful relationship.

The opening approach of circling round the flowers is the same for the bats as for bees and humming-birds. But as a phenakospermum flower prepares to open at sunset, it first rises out of the supporting bract, making it easier for the bats to seize the flower's closed petals with their teeth and/or claws. This triggers the petals' opening, allowing the bats to push their heads into the flower and feed. It also allows them to access to the reproductive organs, the stamens and stigma, so they collect fresh pollen and deposit pollen from elsewhere. The flower seems to have evolved a means of preventing self-pollination. The bat has contact with the stigma only when it opens the flower while still carrying pollen from a different flower. As the flower opens, the slim stalk-like style carrying the stigma moves safely out sideways. As the bat or bats return to feed repeatedly on the nectar, they can then remain in good contact with the stamens to collect the flower's pollen but unable to deposit it back on the flower's own stigma.[13]

The highly mobile bats appear to be effective pollinators – but there are questions about how efficient they are. By introducing pollen from different plants (outcrossing), they should promote strong gene flow and variation. Yet the genetic variation in phenakospermum has proved surprisingly low. The question of whether bats pollinate efficiently as well as effectively has been linked to 'a series of bottlenecks' across phenakospermum's range, which in turn creates the low genetic variation. Fruit bats have also been seen inflicting more damage on male flowers than female flowers in the vine *Freycinetia reineckei*, for example, though this behaviour has not been observed so far with phenakospermum. Ultimately, this could influence breeding systems in ancestrally bisexual plants, such as phenakospermum. Dangerously low fertility levels could possibly result from phenakospermum's specialised pollination system – in one study, only about a third of flowers was successfully pollinated. Yet in other reproductive areas, phenakospermum seems remarkably successful: 'fruit set might be lower than expected ... [but] a single infructescence can produce over 18 000 seeds per season'.[14]

Some botanists have noted that phenakospermum's distribution is 'sparse' but 'locally common'. High concentrations of phenakospermum have been found, for example, in the territory of the Nukak people in Colombia's Guaviare department around the Amazon basin. These trees particularly grow on low ridges of disturbed ground but do not seem to be associated with human activity because the Nukak completely grind and eat the

Black-necked aracari feasts on *Phenakospermum guyannense* seeds

plant's seeds. Instead, it may suggest that phenakospermum thrives less well in prime tropical rain forest where competition for sunlight and nutrient is typically very high.[15]

## PERFECT PARTNERSHIPS

The phenakospermum of the South American rainforest demonstrates a great natural multi-level partnership. As phenakospermum's large leaves curl inwards and die, for instance, another bat, Spix's disk-winged bat (*Thyroptera tricolor*) chooses to gather there in small roosts.

The leaf formation around phenakospermum's main stem base allows a water reservoir to form, as with ravenala. Even in these usually swampy areas, water can sometimes be scarce and during dry periods, passing pale-fronted capuchin monkeys (*Cebus albifrons*), for example, are known to bite open the petiole bases to quench their thirst from the water reservoirs. Small aquatic animals such as tree frogs (*Hyla* species) breed in water reservoirs and joints between phenakospermum leaf stalks and the trunk or hide in the unfurling new leaves. On Maracá Island Ecological Reserve in Brazil's Roraima State, part of a 'mosaic of habitat types', white-lipped peccaries (*Tayassu peccari*) have been seen in riverine forest feeding on phenakospermum leaf bases and probably taking water from them as well. These animals usually feed or palm pulp, seeds, seedlings as well as invertebrates from swamps of mauritia palms.[16]

In addition to the design of phenakospermum flowers attracting bats, researchers have suggested the phenakospermum fruit is an example of a possible co-evolution. A study in a Colombian neotropical forest, about two-thirds of the plants were found to have either large, protected, dull-coloured fruit, which would be preyed on by mammals. Those with small, unprotected, bright-coloured fruit were found to be mainly preyed on by birds. The fruit capsule, or phenocarp, of phenakospermum typically measures about 15 centimetres long and about eight centimetres wide. This sturdy case contains about 400 black seeds with a bright topping of red tufts (aril). Each seed measures about eight by 10 millimetres. Monkeys such as the pale-fronted capuchin (*Cebus albifrons*) usually break open this fruit capsule and so feast on it first. Surinam's brown capuchins (*Cebus apella*) have also been seen working with particular skill to extract the seeds (see panel 'The capuchin surprise').[17]

Some plants develop husks that limit birds' direct access to the pulp, researchers have suggested. Generally only after monkeys have crunched the phenakospermum fruit capsules are the birds attracted to the remaining scarlet-plumed seeds. Birds taking their share include the Cayenne jay (*Cynocorax cayanus*) and toucans such as the black-necked and green aracaris (*Pteroglossus aracari* and *Pteroglossus viridis*). Birds cannot cope with the fruit capsule until the hard outer shell has been opened or removed so they look for fruit capsules that have split open spontaneously or else scavenge the monkeys' leftovers. As several of these birds are more territorial, phenakospermum seeds may not be dispersed as far by birds as different, softer seeds are by fruit-eating bats.

The hard phenakospermum seed itself passes right through the birds' guts, enabling it to be scattered through the forest. As birds taking phenakospermum seeds tend to be 'gulpers', swallowing them whole, this increases the plant's chances of having its seeds dispersed somewhat further away. 'Mashers', on the other hand, would chew up fruits and seeds but tend to drop all larger seeds around the plant. The taste of the fruit matters less to a gulper as well, since bitter flavours in fruit pulp are not tasted when the fruit is swallowed down whole without chewing. In either case, with bird or bat, there remains a risk that the seed could be dispersed into a habitat that is not suitable for phenakospermum to germinate or grow.

The bright red aril tufts on the seed are more than just an attractive, colourful topping to attract attention from fruit-eating birds and to give the plant a second chance beyond the monkeys to disperse its seeds. This aril, rather than the seed, is what the birds digest, benefiting from its rich oils and other secondary compounds. Research in laboratories is showing that for human beings, those secondary compounds could even prove to be key medicines for our future (see also chapter 15).[18]

## THE STAFF OF LIFE

From food to medicine, from domestic utensils to human shelters, across the Amazon forests, phenakospermum has emerged as one of the region's most useful plants. When Brazil's Parakana live as nomads, they make a phenakospermum gruel or bread from the fruit they call *banana brava* and which is available year round. For variety, they alternate it with the *babaçu* palm nut (*Orbignya* sp.). Phenakospermum's versatility compares well to another of the region's major multipurpose plants, seje (*Oenocarpus bataua*), and it can be seen as one of the two cornerstones of life for the Colombian Nukak.[19] Collectors and other researchers had gradually became aware of phenakospermum's important role in the everyday lives of the indigenous inhabitants of the Amazon states.

From termite 'fishers' to baking trays, phenakospermum's large, pliable and relatively lightweight leaves are in daily demand because they can conveniently be made into a variety of packaging and utensils, rather like the wide range of local uses for their ravenala cousins in Madagascar. Venezuela's Makiritare Amerindians use phenakospermum leaf-blades – known as *cucuruciu* – as large leaf wraps to secure their booty of termite soldiers (*Syntermes* genus). Using a filed-down midrib of a palm-leaf, this high-protein, high-fat foodstuff is 'fished' from its burrows, often three metres deep into the earth and inhabited by at least 30 000 termites. The termites come out at night to forage on the rainforest floor for litter to eat and take back into their nests. Termites attack the 'fishing stick' and are brought to the surface clinging to it with their strong mandibles. At swarming time, queens are collected more easily by blocking nest entrances with leaves so they are forced to exit through small holes where they are picked off.

The termites may be eaten raw or after dipping in almost boiling water, often when wrapped in leaves. They are usually served with cassava and hot pepper. Rather like the western eating of snails, for example, termites have traditionally been a prized part of the menu in most parts of the Amazon since before the first western soldiers and travellers visited. As a result, Amazonian people have a far greater range of names and understanding of these creatures than western taxonomy has yet achieved. Though not all Amazonian people eat termites, generally they are considered flavourful and nutritious.

In South America, the termites which are eaten are mostly syntermes, a genus that occurs only on that continent. About 23 species of the rainforest's biggest termites are members of the syntermes, with several of the species also found in some parts of towns and cities, such as parks and well-greened grounds and large gardens. They are the dominant leaf-feeding termites in South America and can weigh up to 13 milligrams each.

Compared to fishing and hunting, collecting termites from known insect colonies is also considered to be much more efficient in time and energy than searching for other prey. Total consumption of invertebrates per year, including caterpillars, grasshoppers, palm grubs and earthworms as well as termites, probably amounts to about two kilograms per person per year. By contrast, each day the Nukak, for example, bring about 2.54 kilograms per person of wild fruit into camp in the dry season and 2.8 kilograms per person in the wet season. Among the Tukanaons of Colombia, it has been estimated that termites formed about 40 per cent of the invertebrates eaten.

Fondness for eating termites is also found elsewhere in the world. In Africa, for instance, members of the macrotermes genus are as big as South America's biggest syntermes. Insects may even be served as curiosities in some restaurants in countries such as Mexico and South Africa. Termites' high protein content – often double that of beef by weight – means that interest among policymakers such as the United Nations Food and Agriculture Organisation is increasing in how termites could play a greater role in helping combat the hunger that affects more than 2 billion people worldwide rather than only being considered for eating when and where poverty means better food cannot be afforded.[20]

The Yanomami people of the north-west Amazon and southern Orinoco have been popularised as a warlike, 'noble savage' people, although they have had 'contact with religious missions and government outposts since the 1960s' and 'more intense contact with the surrounding society after the mid-1980s when their territory was invaded by gold miners'. It has been suggested searching for metal tools and goods has been one of the main drivers of conflicts. However, one of the main uses to which they put phenakospermum is taking a fresh leaf as a baking tray to allow them to cook cassava pulp over a small fire. Among the Tatuyo, living in Colombia's Vaupes area, phenakospermum leaves can also be used as part of a means of preserving and storing pulses. This can be either by lining pits that are then filled with prepared, mashed pulses or by creating packets that are filled with uncooked pulses. More than half the harvest of Monopteryx seeds, for example, is often stored like this for about six months, allowing them to ferment, 'developing a sharp, pleasant taste.'[21]

In the Amazon region, salt is widely derived from tree ash and vegetable salt. In the north-western Amazon, Colombia's Witoto people prepare vegetable salts known as *iyoberi* or *uiyoberi* by burning parts of various plants, including phenakospermum. These are also used for rituals that confer shamanic properties on the salts. Each one becomes 'the spirit of an animal', according to the elders, emphasising humanity being at one with nature – the phenakospermum being associated with the earthworm. The salts

*Phenakospermum guyannense* leaves used to collect and bake termites

are used to prepare *ambil*, a tobacco paste which is licked or spread on the gums, and also often have medicinal purposes. Phenakospermum salts are considered part of the phlegm group and specifically associated with the stomach and teeth, seen as a remedy for parasites as well as protecting the teeth.

The salts are extracted from the ash by filtering them, rather like drip-filtering coffee. The ashes are placed on a lining of fern leaves or moss inside a piece of heliconia bark, with the water slowly filtering through it. This can be tasted from time to time to check the concentration and probably also the flavour. When the liquid extract is ready, it is reduced by boiling until the salts crystallise. These mainly contain potassium (33.4 per cent) and carbonate (27.4 per cent) but also contain much smaller amounts of substances including sulphate (8 per cent), chloride (3.3 per cent) and sodium (0.2 per cent). Often a vegetable slime is added to the final tobacco paste.[22]

## SURVIVING WITH INDIGENOUS KNOWLEDGE

The practical or food use for which a plant is renowned in one area may be unknown elsewhere, a phenomenon noted broadly about medicinal uses of plants in the Amazon area. Harvesting various parts of the plant seems to allow different groups of humans to overlap in the forests, reducing pressure on resources. The Arawete, for instance, classify honey into 45 different types. Their favourite food is tortoise liver but they dislike tapir, a preferred meat for Parakana. They have adopted the cassava (*Manihot esculenta*) as their basic crop, even though phenakospermum fruit and *Orbignya babaçu* palm nuts are available to them throughout the year. Instead, they eat a gruel or bread made from phenakospermum only during periods of nomadic living.[23]

Falling back on the old traditional ways when on the move may be only temporary for the Parakana, but for the Nukak, being on the move and relying on such indigenous knowledge are a way of life. For them, the phenakospermum, which they call the *juná* and which is regionally known as the *platanillo*, is a plant for all seasons and almost all reasons. Though broad uses as tools for construction and food might be similar, specific uses observed in a detailed study did not overlap with the practices of Makaritare, Yanomami or Witoto.[24]

'The last primitive Indians of the Amazon' is how the Nukak were described when they first made regular contact with Colombian colonists in 1988. Like the Bará and the Jupdu, the Nukak are a subgroup of the Makú who occupy the north-western Amazon basin. Most Makú subgroups have adopted a more settled life and subsist on farming cassava or manioc and yucca. But the Nukak follow an existence that is largely detached from modern society, though with certain exceptions. For help treating some illnesses, they may choose to make use of a regional clinic that provides western pharmaceuticals (allopathic medicines). They also use metal pans, for example. Otherwise, phenakospermum and *seje* are the key staples of their daily lives. Phenakospermum provides everything from food to shelter, as well as baskets, fibres for making darts for hunting animals and beds for breeding palm-grubs. In fact, phenakospermum also has a central part in their worldview. According to Nukak mythology, the sun, moon, clouds, rivers, fish, some birds and some trees existed before they arrived on earth. But until then there were no mountains and few animals. Among the trees that the Nukak believe came to earth with them or after their arrival was phenakospermum.[25]

Today the Nukak currently remain a significantly nomadic, hunter-gatherer society. In fact, moving up to 80 times a year makes them one of the world's most mobile forager societies. It is probable that no more than 2 000 Nukak live in an area of approximately 10 000 square kilometres between the Guaviare and the Inirida Rivers in the Colombian Amazon. Significantly, it is common to find phenakospermum among the secondary species in forests on alluvial plains and 'patches' of phenakospermum in forests on low slopes.[26]

Each Nukak band of four or five couples or families has a territory of a few hundred square kilometres, most of them alongside the floodplain of the Guaviare and Inirida Rivers. Moving anything from a kilometre to 10 kilometres every four to five days means they have streamlined the art of building effective temporary shelters. They move in small groups, each couple or family building their own shelter – usually the man does so unless he is absent. Building a more complex, new winter camp usually takes two and a half hours at the most. Generally, the shelters are

W. MacGillivray, Travels and researches of Alexander von Humboldt, 1853

JAGUAR, OR AMERICAN TIGER.

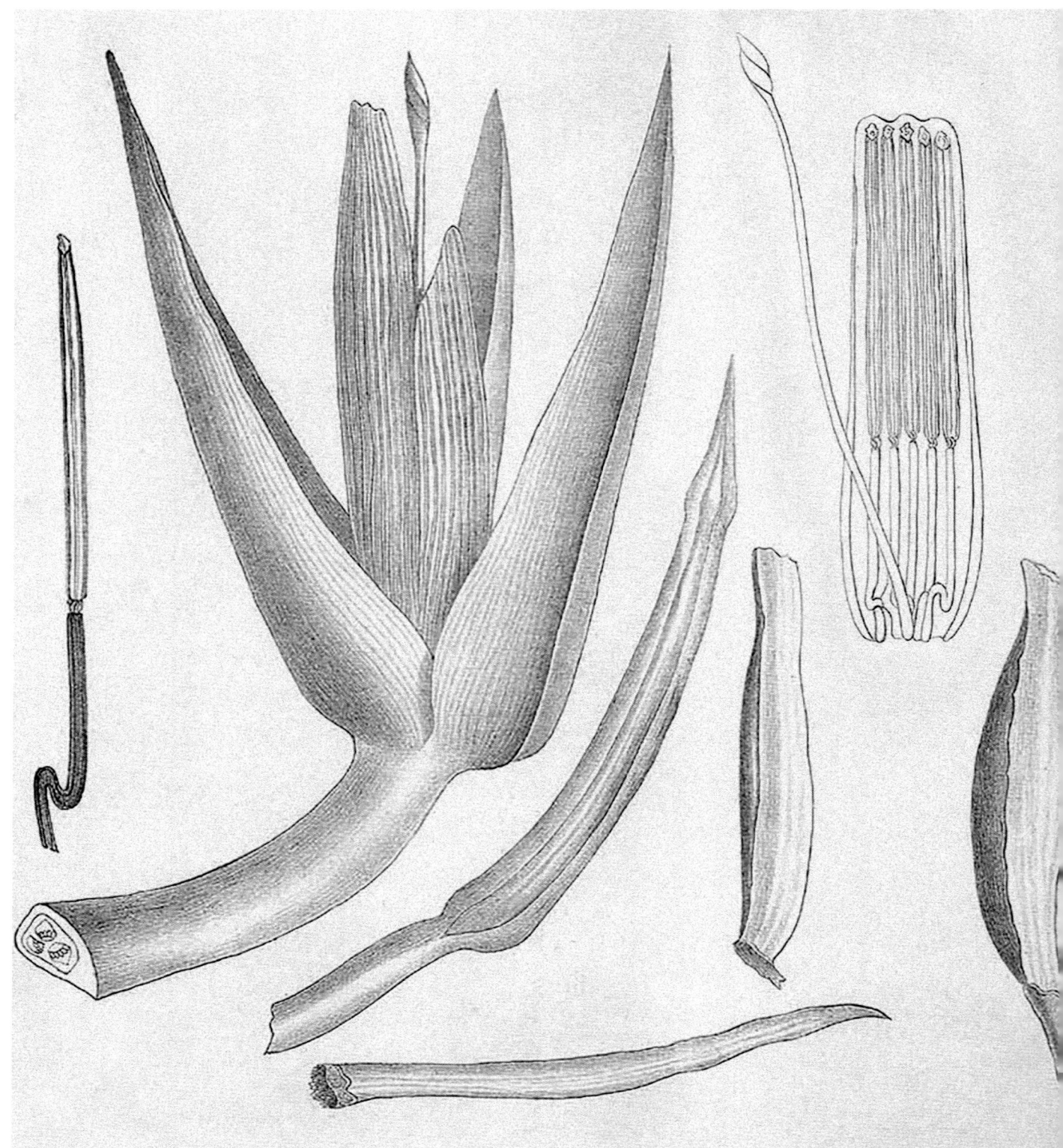

*Phenakospermum guyannense* in Miquel's *Stirpes surinamenses selectae,* 1850

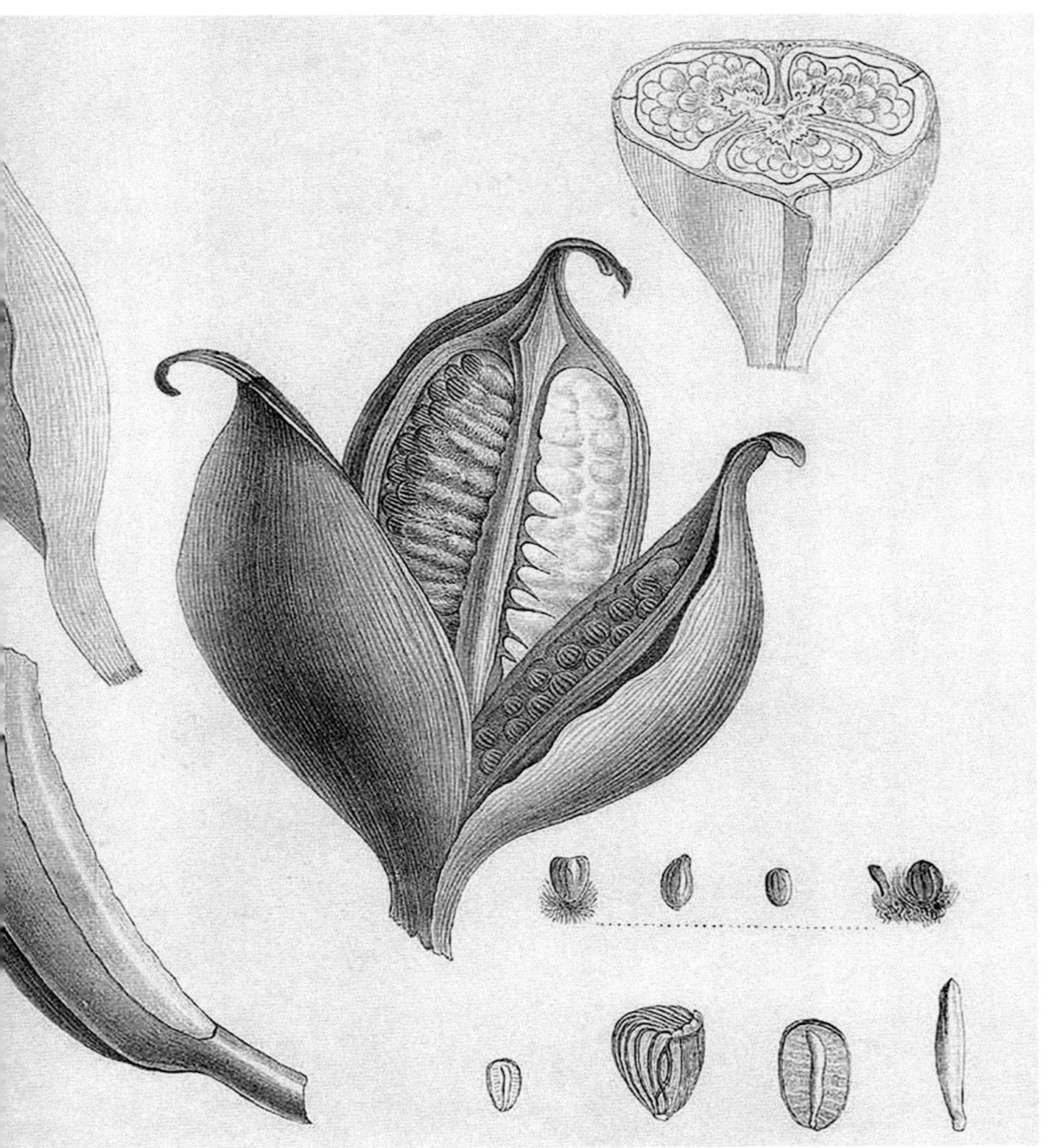

built close together to help protect against evil spirits. They are grouped around a central clearing, which is sometimes man-made. Here children might play or spar and communal goods, from axes to firewood, might be stored. Refuse may be abandoned here, but also between the shelters or on the edge of the camp.[27]

In Surinam, phenakospermum leaves are woven into mats used for thatching roofs after they have been air-dried and fermented so they are less attractive to insects. But the Nukak use phenakospermum differently – it is both the prop of their diets and the prop of their homes. Each Nukak shelter is built using a tree-trunk as the main support, with two posts erected to form a triangle. Cross-beams are placed about 1.6m to 1.9m high between the three points of this triangle. Rain is, of course, a regular feature with annual rainfall of between 2 500 and 3 000 millimetres. There is a short drier season during January and February when the monthly rainfall averages 50 millimetres to 100 millimetres. Otherwise, rainfall usually peaks from June to August at about 400 millimetres a month. During the rainy season, the roofs of the shelters are covered with phenakospermum leaves, with driplines carefully calculated at front and back to be well out of the way of the main activities. The arched roof is built by propping both phenakospermum and *seje* leaves up on their stalks (petioles) at an angle of between 55 and 70 degrees. The finished roof is usually between five and seven metres long and 2.5 metres to 4.3 metres wide. It has been calculated this design ensures that about 80 per cent of the rain that falls on the shelter runs off the strongest element, the stems. If the group intends to occupy the camp for longer than one night, a row of *seje* leaves is used to reinforce the low arch of shelters' roofs.

Drifting smoke gathers under the roof that arches low above the ground – about 90 centimetres to 1.1 metres – making the shelter warmer for sleeping and deterring insect life. Air circulating through gaps between the leaf stems helps keep the fire burning on the hearth in the front section of the shelter, despite the often damp firewood and the very high humidity. Hammocks are slung from the cross-beam, with the man's above and woman's below. For play, a simple shelter of one bent phenakospermum leaf may be created for a child. Separate shelters are also sometimes built to be used during menstruation or rituals.[28]

## THATCHING AND WEAVING

To prepare phenakospermum leaves for roofing their homes, the indigenous people of Surinam first dry them to deter insect infestations, according to Dr Paulus Maas. They then fold the leaves together along the main vein and weave them together, tying them at the petiole with makatité, fibre from *Smilax cumanensis*. The finished mats are then thatched onto the roof.

By comparison, in Indonesia, introduced ravenala are used similarly although leaves for making the roof mats are not first dried before weaving. During the weaving process, the leaf petioles are tied with maranta cane (*Ischnosiphon gracilis*), also used in basket-making. The completed mats are then rolled up to dry before being used for thatching.

Plants and fruits form a major part of the Nukak diet and are generally collected every day by foraging within an hour or two of the camp. Phenakospermum's leaves also have their uses during Nukak food-production. The leaves are spread as a carpet when the Nukak are shelling seje palm fruits which can be mashed to form a fruit 'milk'. Phenakospermum itself produces abundant and multipurpose capsules that are collected by the whole branch or individually and carried back to camp in *búrup* (bags) usually woven from heliconia leaves. The phenakospermum fruits are shelled at the camp and the shells make useful kindling on the fires. The red aril tufts topping the seeds are chewed immediately to enjoy the oily liquid they contain, while the seeds themselves are made into platanillo flour, a carbohydrate-rich staple within

Nukak people use *Phenakospermum guyannense* trunks, stalks & leaves to make shelters & gather seeds to grind for meal in heliconia-leaf bags (centre)

Clayton Burne

Adam Riley

the diet of the hunter-gatherers. This flour is made by grinding phenakospermum seeds in a mortar using a pestle made from a stick about 1.2 metres to 1.5 metres long and five to 10 centimetres thick, then pushing the flour through a sieve (*werep*). The flour is steamed in a wrapping of green leaves, making a form of potbread.

Both men and women can work together in choosing the pieces of tree trunk suitable for making mortars, usually pieces between 24 centimetres and 30 centimetres wide and 50 centimetres and 70 centimetres long are chosen. But it is a woman or women who make the mortars by burning out the central part of the trunk and blowing through a hollow cane when necessary to keep the embers alive. When enough of the tree heart has been burnt away, the charred sections are scraped out with a machete to clean and shape the mortar.

Not all Nukak make sieves and among those who do, the sieves are made by specialists in the group, usually two people, who could be men or women. The mesh of the sieves is woven from the green stems of *bórop* (*Ischnosiphon arouma*) and the frame from a thick vine.[29]

Monkeys are usually hunted every two out of three days for bushmeat using blowpipe guns with curare-poisoned darts. When available, herds of white-lipped peccary are also hunted with a wooden spear. The meat is eaten only by men as it is taboo to women. Several other larger animals are taboo to all Nukak, such as tapir, deer and jaguar, but tortoises and birds are regularly hunted and other creatures on the menu included caiman, armadillo, and frogs. Birds' eggs may be taken.[30]

## MOVING TO PRODUCE

Today Yanomami tend to stay longer in their camps, moving less frequently – game supplies running low in the area is the most common prompt. Though the Yanomami will sometimes need to move up to 10 kilometres, they prefer to move less than a kilometre, 'the limit that leads to building a new house':

> because of the effort require[d]... The fact that mobility strategies are maintained suggests how important mobility is to the Yanomami people, despite contact posts now being widespread in the TIY [Terra Indígena Yanomami] with resulting pressure for sedenterization.[31]

The Nukak move very much more frequently, about every three days in the dry season, when they also move further, and about every five days in the winter. They do not harvest or hunt to the maximum potential of the area but instead move '*before* there is an observable decline in the available resources'. Insects beginning to gather around the camp, partly due to refuse, can be a factor, though. If the Nukak choose to delay their departure by a day, they might escape troublesome insects by hanging their hammocks in the open near a smoking fire 15 metres to 30 metres outside the camp. Occasionally, too, longer journeys might be planned within the territory of the regional group, which can range between 1 000 and 2 000 square kilometres. Planning rituals, marriages, meeetings with other groups to trade, treatment of illness or men needing a particular type of cane for blowpipes – one area was called Cerro de las Cerbatanas (Blow Pipe Hill) – all may prompt journeys. But tensions can also arise between bands of people, causing territories to be redrawn. Or there may be the kind of journey in search of ancestral territories which first brought the Nukak into the western focus.[32]

The foods that can be harvested or hunted at different seasons of the year may influence the areas to be targeted. During the dry season, for example, when fish are caught using a poison called barbasco and sometimes honey is collected from wild beehives, bands move towards areas where both fish and bees' nests are plentiful. In winter, certain palms are particularly attractive and in January and February fruit of the chontaduro or peach palm (*Bactris gasipaes*) is harvested. Phenakospermum, however, is a year-round staple.

Caiman, lowland tapir and nine-banded armadillo

The group studied 'almost never reoccupied an abandoned camp, even when it was in good condition', though they did sometimes build again close by. Useful items the group might not wish to carry when moving on may be stored in the camp, for example, wooden mortars placed on platforms above the ground to help protect them from rotting, canes for blowpipes with their ends sealed with leaves to prevent insects infesting them. Sometimes when relocating once again or on a foraging trip, these might be collected from a former camp. A feature of abandoned Nukak camps is often several holes each about 50 centimetres across. This is because the wooden mortars have steadily been forced down into the earth by the pressure of grinding.

Nukak living closer to the north-eastern edge of the rainforest seem to have adapted this aspect of their lives, though, with about 19 per cent of them known to reoccupy a camp. This is generally considered 'exceptional', however. Camps may even become places to be avoided because of the spirits of those buried at the camp or illnesses which could be caught by occupying another group's abandoned camp. It has been estimated that each nomadic Nukak group alters about 6 400 square metres of the forest per year.[33]

## MAGIC OF THE DARK EARTH

Kitchen and other middens and charcoal from fires give organic enrichment back to the soil, encouraging this process. They create the *terra preta* or 'Amazonian dark earth' with its 'fertile anthropogenic soil patches', as opposed to the natural *terra mulata* ('brown earth'). One researcher believes that produce from *terra preta* patches becomes so important that these areas can be considered to be invested with a form of magic.[34]

The oldest *terra preta* has been dated back to 4 500 BP in the Jamari river area (near the Ji-Parana river at the source of the Madeira). Some social change seems to have sparked rapid increase in the number of these areas in the 500 years from 2500 BP. They are now scattered widely across Amazonia, occupying about 15 500 to 20 700 square kilometres or just 0.1 per cent to 0.3 per cent of forested Amazonia. Fish scraps seem to contribute strongly to the process. The largest patches of *terra preta* – up to 500 hectares – are found near confluences of rivers richest in fish, white-water rivers such as the Amazon, Ji-Parana, Madeira, Purús, Tocantins and Uatumã. Black-water rivers drain from nutrient-poor (oligotrophic) soils, tending to attract fewer fish. This ground is suitable for growing cassava, though, which in turn encourages a more settled way of life. Even so, there is human habitation in these areas and also in the *varzea* (seasonally flooded) and *terra firme* (interfluvial) forest areas where phenakospermum also typically grow. The two have similar flora and it seems that *terra firme* develops out of older *varzea* forest with more diverse microhabitats and allowing trees like phenakospermum growing into spectacular specimens.[35]

Climate change from 18000 BCE had encouraged grasses such as the ancestor of maize to survive on the savannahs. Later, when modern climate conditions set in about 9 000 years ago, it became possible to farm maize, which has a short maturity, on seasonal floodplains. Plants and other foodstuffs have been traded across regions since pre-Columbian times so everyone could exploit the produce from the different ecosystems:

> People from different regions utilized a variety of ecological niches . . . smoked game meat and forest fruits from upland groups were traded in exchange for fish and agricultural products from the riverine inhabitants.[36]

Most commonly today, the camp area becomes a 'future resource patch', regenerating in a form of silviculture. As with other Amazonian groupings, including more settled groups such as the Kayapó from eastern Amazonia and the Ka'apor from the middle Xingu river, the large quantity of seeds from popular Nukak fruits and herbs becomes a swidden, or an informal garden or wild orchard where the Nukak can find a high concentration of their favourite fruit and other useful plants. Phenakospermum tend to be the exception, though, because the Nukak usually completely consume its seeds. Generally, this so-called virtuous circle tends to make the area more and more useful to Nukak, who can be seen as 'sophisticated managers of their landscape . . . [who] move to produce.'[37]

## THE CAPUCHIN SURPRISE

Even before his South American expedition, Alexander von Humboldt (1769-1859) had never wanted to write just another travel journal. Yet his publications are still remembered as much for his vivid descriptions as for his scientific leaps of insight. The Orinoco section, for example, is probably best remembered for Humboldt and Bonpland's encounter with electric eels, which gave them significant electric shocks. But near Maipures (now Maipuri) on the upper Orinoco River, they also met and named regular members of the phenakospermum world, the white-fronted capuchin monkey (*Simia albifrons*, now *Cebus albifrons*), which is familiarly named after the Capuchin monks who were one of the main orders of missionary monks working in the region. Their relatives, the brown capuchins (*Cebus apella*) forage for husked fruits such as phenakospermum – but their behaviour also surprised 21st-century researchers in Raleighvallen, a rain forest in Surinam.[38]

Bearded sakis (*Chiropotes*), sakis (*Pithecia*) and uakaris (*Cacajao*) are considered the 'New World seed specialists' but brown and other capuchin species, as well as other primates, had already been seen attempting to break open large phenakospermum pods to extract their seeds in Colombia, Costa Rica, Peru and elsewhere. Instead of showing the same

'inept, sloppy and uncertain movements' seen in most wild monkeys, however, the Surinam troop had a persistent, precise approach to pounding open the large husks, which seemed to demonstrate reasoning and learning skills. This reminded researchers of a brown capuchin seen in a Brazilian mangrove swamp 'using a chunk of oyster bed to break open a closed oyster shell' to eat the flesh inside and a capuchin in Costa Rica 'using a large branch to repeatedly club a venomous snake, eventually killing it after more than 50 blows'.[39]

The brown capuchins were seen using their fingers, feet and teeth to force open pods with thick, hard husks to retrieve the seeds inside. They have developed with 'large teeth, thick jaws and powerful jaw musculature' which help them in 'vigorous and forceful processing' of food and objects such as husks containing food. The rewards were clearly worthwhile as they spent about 15 percent of their foraging time on this task, sometimes having to labour over one single husk for more than 30 minutes. Phenakospermum pods were particularly high on their target list as they are 'an abundant and predictable resource at this site during the wet season'. Even though a human armed with a machete struggles to open phenakospermum's large pods, it was clearly the favourite target for this kind of foraging and was documented 29 times compared to 27 for all 10 other species observed combined.[40]

Young brown capuchins learned the pounding technique by watching and imitating their elders and practising it and were sometimes seen standing upright on two back feet on a phenakospermum leaf, wrestling with a pod. Young capuchins struggled at first to break open the husks, mostly because they pounded the pod in the wrong way – on the long 'and extremely resilient' side, for instance. The trick the youngsters needed to learn was to position, even wedge, the phenakospermum pod so it could be hit at the best target area, the convex apex. This makes the effort behind the blow far more effective because the entire downward force is focused on this fulcrum point instead of being weakened by being distributed across the pod's whole surface. The brown capuchins were often seen rolling over the pod to check for signs of progress in breaking it open. If it could see a gap appearing, they would often 'gnaw with its teeth and tear with its fingers at the opening' in the phenakospermum pod to help the process along. Youngsters had become fully adept in this technique by between two and three years old, researchers observed. This process of learning skills reminded the research team of how wild chimpanzees in Guinea picked up the use of stone tools although, by contrast, chimpanzees take about a decade to perfect their foraging and tool use techniques.[41]

Foraging for closed phenakospermum pods is very hard work for brown capuchins but being able to do this gives them an advantage over competitors that can only take fruit and arils from pods that are already split open. This must be well rewarded because even though they avoid using exposed granite outcrops as anvils, the sound of their pounding carries through the forest and alerts predators – including human poachers of bush meat.[42]

(From top left) white-fronted capuchin monkey; *Phenakospermum guyannense* seed case split open; brown capuchin monkey

*Strelitzia nicolai*, Fransiolemtru Pretorius, watercolour, 20

# SPLITTING THE DIFFERENCE
*Strelitzia caudata & Strelitzia juncea*

CHAPTER NINE

# SPLITTING THE DIFFERENCE: *Strelitzia caudata* & *Strelitzia juncea*

From about 1800, when strelitzias were at a peak of fame in Europe and North America, they seem conspicuous by their absence from most of southern Africa's early modern writing, art and records for at least a century and a half. They made only a fleeting appearance in early Victorian botanical art from the Cape. Arabella Roupell (1817–1914), the wife of a British East India Company official who took a year's long leave at the Cape from 1843 to 1844, depicted *Strelitzia reginae* rather unconventionally for a period that preferred botanical perfection, with a leaf that seems to be withered offsetting a glowing bloom. Thanks to Danish botanist Nathaniel Wallich (1786-1854) persuading Roupell to allow him to show her paintings to Sir William Hooker at the Royal Botanic Gardens, Kew, they were first published anonymously in 1849. Although entitled simply *Specimens of the flora of South Africa* by a lady, Roupell's work was already acclaimed by high society. Queen Victoria topped the subscribers' list but despite *Strelitzia reginae*'s royal connections, the selection of 10 plates did not include Roupell's unconventional portrait of the flower. Instead, it is glimpsed only in one plate where its unusual shape and bright colours jar rather awkwardly among the other selected flowers that look more conventional and delicate with their softer, pretty tints. The volume's introduction was written by William Harvey (1811-1866), colonial treasurer and then treasurer general at the Cape, who joked that he spent so much time collecting plants that he might risk being retitled 'Her Majesty's pleasurer general'.[1]

Getty Research Institute

The bird of paradise also put in only a brief appearance in the collection of South African plants painted and scientifically identified by Maria Holland (1836–1878). Harvey particularly praised Holland's work for its accuracy when she supplied it to him for use in his *Thesaurus Capensis*, one of several pioneering books on Cape flora which he published after returning to Ireland, where he became curator of the herbarium at Trinity College Dublin and professor of botany at the Royal Dublin Society. Holland's background was in the Eastern Cape: she had been brought up in Uitenhage and married in Port Elizabeth. This good access to strelitzias probably contributed to her interest in the plants, reinforced by their appearance in European botanical magazines. The white bird of paradise, *Strelitzia alba*, which had been Auge's pride and joy, was probably not featured as its limited range around what is now known as the Garden Route between George and Humansdorp means it occurs at least 80 kilometres west of Uitenhage and 100 kilometres west of Port Elizabeth.[2]

Holland's inclusion of the dramatic bird of paradise in her collection of about 150 paintings meant little, if anything, in southern Africa as Harvey's book had very limited circulation in the region and the painting itself is now in London's Natural History Museum. The heartlands of strelitzias' popularity remained Europe and what would become its adoptive homeland of California (see chapter 10). Being generally a slow-maturing plant may have made strelitzias more challenging, but they would not become more common even in Western Cape gardens until they were swept in by the wave of national botanical pride in the mid-20th century.

## THE MYSTERY OF MISSING STRELITZIAS

The apparently mysterious neglect of bird of paradise flowers is echoed by a nagging gap in their history. The legacy of shipwrecks is more than death and the destruction of hopes – often they can also foster lingering mysteries. Around the time that Harvey was compiling his *Thesaurus Capensis* in Dublin, circumstance and shipwreck created one of the great mysteries of southern African botanical science. It even contrived to erase possible attention to the strelitzia by Harry Bolus (1834-1911) who is considered 'one of the founding fathers of South African botany'. To this day, nobody knows exactly which plant specimens were lost from Bolus's first landmark collection of specimens when the *Windsor Castle* went down in Table Bay in 1876 after hitting a reef off Dassen Island.[3]

Bolus had originally been lured to the Eastern Cape from his birthplace of Nottingham in the English midlands at the age of just 15 by the promise of getting to know exciting new plants. As

S.T. Edwards, *Strelitzia reginae* in *The new botanic gardens*, 1812

Maria Holland, *Strelitzia reginae*, about 1860

a pupil, he had become fascinated by the plants sent with letters from Grahamstown to his headmaster George Herbert. These were from Herbert's friend William Kensit, a trader in the Eastern Cape, who later requested Herbert find him an apprentice for his trading store from among the pupils. Bolus, who had already read contemporary works on southern Africa by great naturalists, such as Burchell's *Travels* and Thunberg's *Voyages*, was selected. It was the late 1840s and European settlement in the Eastern Cape was barely three decades old. Bolus's ambition was to be a plant collector – but like several others, he found collecting offered a precarious income.[4]

After spending two years in Grahamstown, Bolus moved to Port Elizabeth on the Eastern Cape's southern coast for a further three years. When he was about 20, he moved again to work in Graaff-Reinet for the Board of Executors, South Africa's longest-established stockbrokers. This gave him the means by 1857 to marry his former employer's sister, Sophia Kensit. It also led Bolus, through the family tragedy of the death of the couple's eldest son in 1864, to begin a serious botanical project. In an era when biogeography was in its infancy, Bolus had become interested by differences between the flora around Grahamstown, Port Elizabeth and Graaff-Reinet. A local teacher, Francis Guthrie (1831-1899), offered to give him private botany tuition, believing it would help ease his grief. Starting in 1865, Bolus collected systematically in the region for 10 years and then later around Cape Town, having moved there to go into partnership with his brother Walter.[5]

Tantalisingly, the *Windsor Castle* shipwreck meant that we do not know which strelitzias Bolus encountered where and what he made of them. It was an era which marked one of the region's peaks in botanical exploration as settlers turned citizen botanists. They became actively involved in collecting and observing the new flora they found around them, with Bolus ranking as one of the most active and knowledgeable: 'the indigenous flora was prized. It was collected . . . in the colonial herbaria. It was shipped for posterity in the wardian case to London, Berlin, Paris and Vienna.'[6] Although the citizen botanists competed with each other to find rarities and send them to Kew through their local botanic gardens, there had been no great wave of excitement over strelitzias at the Cape similar to the one that swept through London society between the 1780s and 1830s. Throughout the 19th century and well into the 20th century, in one of southern Africa's main centres of botanical endeavour, the Cape peninsula and its hinterland, the strelitzia would remain a botanical curiosity.

Often these citizen botanists lobbied Kew for the honour of being commemorated in the name of a 'novelty', as plants new to science were called. Bolus, though, was altogether more

Sir Joseph Hooker (shown at home) had collected widely across Indian subcontinent before becoming Kew's director from 1865 to 1885

discerning and demanding. Botany had captivated him and he made a bolder plan. He was determined to enlist Kew's help in naming the various specimens in his collection. Fortunately for science and for posterity, during his visit to London, Bolus had donated half the collection to Sir Joseph Hooker at Kew, where it safely remained. But no strelitzia specimen appears in Kew's collection either under Bolus's name or accessioned around that time. This may be because Bolus donated only specimens for which he had duplicates. Or it is always possible that Bolus had not collected strelitzia. While in the Eastern Cape, he was still a novice collector and it is demanding to produce an adequate specimen of large, fleshy plants such as strelitzias.[7]

Despite the shipwreck disaster, Bolus was inspired by the assistance and enthusiasm he and Guthrie had enjoyed at Kew. Bolus resolved to rebuild his collection and regularly returned to Kew to check his specimen naming. Eventually, his collection became the foundation of the world-famous Bolus Herbarium at the University of Cape Town. Bolus contributed a landmark overview of southern African plants, 'A sketch of the flora of South Africa' to the 1886 Cape of Good Hope *Official handbook* but while discussing the broad group of monocotyledons in the south-eastern and eastern 'Tropical' region, he simply commented: 'The *Strelitzia* are found as far north as Natal, and may occur beyond that country.' In practice, Bolus was more particularly interested in orchids, on which he published a three-volume survey, and in ericas, on which he worked with his lifelong friend Guthrie until Guthrie died in 1899. This would be Bolus's final contribution for *Flora Capensis*, too, published in 1905, six years before he himself died. Guthrie's daughter, Louise (1879-1966), also worked at the Bolus Herbarium for a while.[8]

## THE ELUSIVE STRELITZIA

This new focus on indigenous plants would have pleased William Burchell had it prevailed when he visited Africa nearly a hundred years earlier (see chapters 1 and 7). He had arrived fired by the Victorian interest in the Cape flora but was dismayed by the local settlers' disdain for the wayside local plants which were cosseted and coveted a hemisphere away. Burchell's disbelief was understandable given that in 1802 alone, about four out of every five plants published in London in *Curtis's Botanical Magazine* came from the Cape. The botanical magazines of the early 19th century popularised Cape plants across Europe and North America. In Britain, titles such as the *Botanical Register*, *Botanical Cabinet* and *Maund's Botanical Garden* often competed to feature the southern African plants which were so popular with readers.[9] In southern Africa itself, though, instead of cultivating gardens of local veld plants, European immigrants were likely to be nostalgic about familiar plants with which they had grown up. They challenged themselves trying to grow their favourites in these new, less hospitable settings, combined with dazzling, exotic curiosities from South America, Australia and Asia that had been revealed by imperial travellers and botanists.

In Britain, the craze for Cape bulbs was followed by another for heaths, which grew well in the stove houses installed by well-off Georgian gardeners. Though 'few in Britain or Europe would have thought of the geranium as an exotic African plant', by the end of the 19th-century, the Cape Floral Kingdom's great supply of new species (novelties) was no longer enough to hold the attention of trend-hungry Victorian gardeners:

> That this enthusiasm passed, in part reflected the changing fashion in plants and in particular the advance in popularity of the South American orchid. But it also reflected a decline in interest in things botanical at the Cape itself. Whilst the colony could boast . . . Bolus and . . . MacOwan as world-class botanists, the fact remained that the craze for indigenous flora that existed in the sister Colony of Natal no longer was to be found at the Cape of Good Hope. Indeed, by the time the Union of South Africa was created in 1910 the Cape had no surviving botanic gardens.[10]

Patricia McCracken

## THE WEB OF BOTANICAL NATIONALISM

Innovative plans in the early 20th century to renew Cape Town's botanic gardens also promoted the notion of demonstrating pride in the region's flowers and plants – and resulted another century later in a new strelitzia creating greater excitement than ever round the world. Over the years, attention to the city's Company Garden had waned.

By 1910, the year when the region would become the Union of South Africa, there was a push to restructure the botanic gardens hierarchy in the region. Pearson had used his 1910

Arderne gardens, Cape Town

presidential address to the South African Association for the Advancement of Science to give a strong voice to debate over the need for a national botanic gardens that would conserve indigenous plants and serve as a symbol of national pride:

> The inheritors of one of the most remarkable and most beautiful of existing floras, we take so little interest in it that we have not yet been at the trouble to bring its treasures into cultivation! Such as we do find room for, we grow because the enterprise of European horticulturists has made them popular and, as a rule, we are content to be ignorant that we have but brought them back to their own country. This is surely not in harmony with the traditions of South African patriotism! A tardy recognition of a national duty has given rise to legislation designed to protect some of our more attractive and rarer plants from a threatened extinction, but we cannot stop here. The public taste must be stimulated to a proper appreciation of the aesthetic value of one of the most striking of the products of the country, and our duty as custodians of a unique vegetation — many of whose constituents have already disappeared, and others can with difficulty be saved— must be realised.[11]

In his crusading call, Pearson had picked up Burchell's surprise nearly a century earlier and reshaped it for a contemporary audience. Instead of the Cape Colony being diminished as one of four provinces in the new Union of South Africa, Pearson was pioneering an environmentally conscious patriotism centred on the Western Cape. The Royal Botanic Gardens at Kew, which had been such a major force in the botanical exploration of the Cape, had expressed its regret when the Company garden was taken over by the Cape Town municipality in 1891:

> It is to be hoped, however, that botanical enterprise at the Cape has not so entirely died out and that it may not be possible at some future time to establish a Botanical Garden, under scientific control worthy of the Colony and of its vast and valuable resources. The Cape Flora is one of the most interesting in the world. A large number of very interesting and highly-valuable plants belonging to this Flora are gradually becoming extinct. The opportunity for preserving them for observation and investigation will soon pass away.[12]

## SIDELINING THE SUBTROPICAL

Pearson foresaw a new emphasis on indigenous plants to supplement the tradition of imperial botanic gardens at Kew and throughout its global network, which 'sought to display the flora of the world not the as yet unendangered flora beyond their gates'. Both approaches, though, regarded botanic gardens as reflecting botanical riches in an informative and educational way.[13]

In the first of Pearson's five conditions for choosing a site for the new flagship gardens, he firmly stamped on potential subtropical rivals, such as Durban, founded in 1847 and which is now the oldest surviving botanic gardens on the African continent. Durban had a strong history of plant collecting and scientific botany thanks to curators such as Wilhelm Keit, Mark McKen and John Medley Wood. Instead, Pearson established strong territorial rights:

> Its climate must not be sub-tropical. This follows from the fact that only a small proportion of the country whose interests are to be considered possesses a wet sub-tropical climate. A garden placed in such conditions would be to a large extent out of touch with the sub-ordinate establishments. It is hardly less important to avoid a hot damp climate, in order that the conditions may be as favourable as possible for work.[14]

His second essential factor stamped on a possible bid from the rival province of the Transvaal, as he maintained: 'It must be near the sea . . . Plants fresh from a sea voyage are not in a condition to withstand a long railway journey over the hot, dry plains of South, Central Africa.' He concluded that the new national botanic gardens must 'be an expression of the intellectual and artistic aspirations of the New Nation'.[15]

While welcoming Pearson's success in opening the new national botanic gardens at Kirstenbosch in 1913, Kew sounded a warning note at the territorial rivalries which Pearson had fed and stressed the value and heritage of the Durban Botanic Gardens. But Pearson was triumphant. Durban Botanic Gardens remained apart from the new national network as its management had been taken over that same year by the Durban municipality.[16] Pearson's successors in the South African botanical world developed the trend for wildflower reserves, while Professor Brian Rycroft (1918-1990) began realising the vision of a National Botanic Gardens network.

*Strelitzia reginae*, emblem of SA National Biodiversity Institute, at Kirstenbosch National Botanic Gardens

THE HERBARIUM AND LIBRARY.

W.J. Bean, *The Royal Botanic Gardens, Kew*, 1908

The Karoo Botanic Gardens became a Kirstenbosch satellite in 1956 and by 1983, when Rycroft retired, five additional, established botanic gardens had become part of the network. Only smaller institutions in Port Elizabeth, Grahamstown and Nelspruit were left within the National Botanic Gardens network to safeguard interest in strelitzias. Despite its restricted foothold in the strelitzia hinterland, when this network merged in 1989 with the 86-year-old Botanical Research Institute, the new National Botanical Institute and its successor the SA Biodiversity Institute (from 2004) both retained the Botanical Research Institute's strelitzia emblem and entitled its publication series *Strelitzia*.[17]

## DOCUMENTING SOUTH AFRICA'S WILDFLOWERS

The year 1913 had marked both the foundation of the new National Botanic Gardens at Kirstenbosch and, at the other side of the new country, the year when Welsh-born Dr I.B. (Illtyd Buller) Pole Evans (1879-1968) took over from Burtt Davy (see chapter 3) and amalgamated the Transvaal Department of Agriculture Divisions of Botany and Plant Pathology under his control. Pole Evans had been founding head of the Transvaal Division of Mycology and Plant Pathology in 1912. As director of the newly amalgamated Divisions of Botany and Plant Pathology, although still then within the Department of Agriculture, Pole Evans energetically built on Burtt Davy's work. The herbarium that Burtt Davy had founded in Pretoria in 1903 along the model established at Kew was not named the National Herbarium until 1918, however. The expansion driven by Pole Evans created a powerful northern counterpoint in the region to Pearson's ambitions for Kirstenbosch. Within five years, Pole Evans's recommendation for a Botanical Survey Advisory Committee was accepted, soon followed by various publications documenting the work of their contributors: *Memoirs of the Botanical Survey of South Africa* (founded 1919); *The flowering plants of South Africa* (founded 1920); and *Bothalia* (founded 1921).[18]

The first officer appointed to join the new botanical survey team is considered to be Pietermaritzburg-born R.A. (Robert Allen) Dyer (1900–1987), who became assistant to German-born Professor Selmar Schonland (1860-1940) in Grahamstown in 1925. Schonland served as the first professor of botany at Rhodes University for 21 years from 1905 and expanded the Albany Museum herbarium 100-fold from about 1 000 sheets to about 100 000. Schonland had also married MacOwan's daughter in

Angela Beaumont, *Strelitzia caudata*, pen and ink, 2009

1896 and was a fellow of prestigious academic societies such as the Linnean Society of London and the Royal Society of South Africa.[19] As Schonland retired from the survey the year after Dyer's appointment, this appears to have been succession planning.

## STRELITZIAS RETURN TO THE SPOTLIGHT

Dyer took the opportunity to build on Schonland's work, producing a detailed survey of the vegetation of the Albany and Bathurst districts. With its wide range of succulents, the region inspired Dyer to specialise in this family — but his wide-ranging curiosity meant he did not exclude other interests. After three years as South African botanical liaison officer at Kew from 1931 to 1934 and a decade at the National Herbarium, Pretoria, Dyer was appointed chief of the Division of Botany and director of the Botanical Survey in 1944. He also became editor of *Flowering plants of South Africa*, which he promptly extended and renamed as *Flowering plants of Africa* from volume 25, the 1945-6 edition. The following year, Dyer published his reappraisal of the tree-like strelitzias in that journal, giving the first scientific description of the mountain banana, *Strelitzia caudata*, differentiating it from the related and similar *Strelitzia alba* and *Strelitzia nicolai*, which he described in preceding plates.[20]

Dyer would have known *Strelitzia nicolai* when growing up in Natal and *Strelitzia alba* when working in the Eastern Cape. At Westphalia in the Pietersburg district of the Transvaal in 1906, his own predecessor, Burtt Davy, had collected what he had identified as *Strelitzia nicolai*. This was one of the strelitzia specimens that Dyer later reassessed and described as *Strelitzia caudata*. More specimens were added from the same district in September 1935 by forester A. E. (Ted) Grewcock (fl.1920-1940). He collected a specimen of mountain strelitzia on the edge of the Limpopo escarpment at Haenertsberg, or Mudallo in sePedi, between Polokwane and Tzaneen. He noted that 'the wild Banana is known as Mugamaka' — in siSwati, it is known as *inkhamango* — and that it was indigenous rather than cultivated. Grewcock's label emphasised the mountainous terrain where the plants were found, 'at a lower and warmer altitude than woodbush Forest Stn. It favours open rocky situations, where moisture is present.'[21]

## CHARTING THE MOUNTAIN STRELITZIA

That same year additional specimens had already been sent from the Soutpansberg by Ernest Galpin (1858-1941), to whom the 1935 edition of *Flowering Plants of South Africa* was dedicated. Galpin had collected his strelitzia specimens in July at Pisang Hoek, or Banana Coombe, named after the Transvaal wild banana (*Ensete ventricosum*), ironically, and not one of the banana-like strelitzias. Galpin was the fifth of seven sons of the renowned Grahamstown watchmaker Henry Galpin (1820-1886) and had been encouraged by his mother Georgina to make collections of

Treelike strelitzias specimens reassessed by Dyer when describing *Strelitzia caudata*: collected by Ted Grewcock near Tzaneen (1935; top) & by Joseph Burtt Davy near Pietersburg (now Polokwane; 1906; bottom)

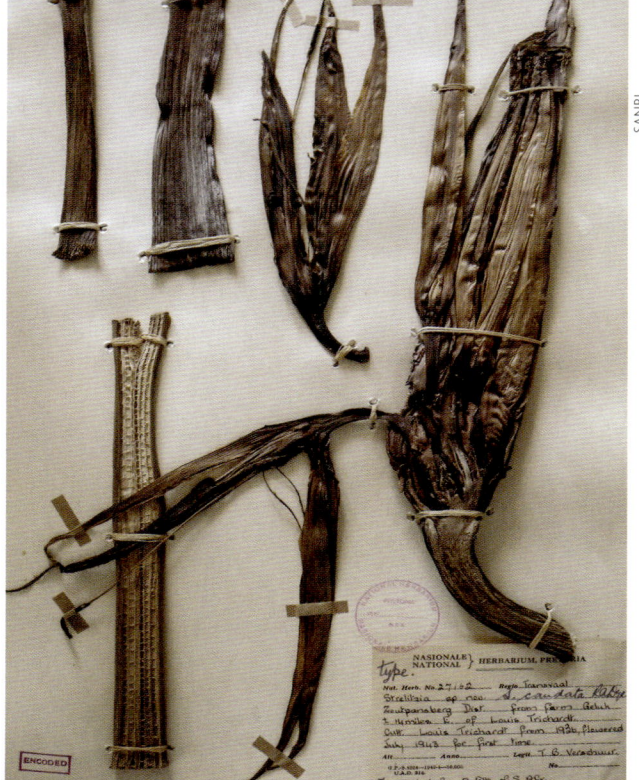

natural history specimens. After fighting on the Cape frontier in 1877 and 1878, Ernest Galpin became a banker and in 1883 began studying botany seriously, comparing finds from his walks with reference books such as Harvey's *Genera* and *Flora capensis*. Although he did not start his own herbarium until 1888, it was his transfer to Barberton the following year that prompted him to take his collecting even more seriously and distribute duplicates to the botanic gardens at Kew and Zurich, to Bolus and MacOwan in Cape Town, and to Medley Wood in Durban. Like Grewcock, Galpin noted the mountainous terrain in which he found his strelitzia specimen:

> Strelitzia sp.
>
> Like plant from Zululand.
>
> Amongst rock in thick bush on summit & on rocks or bush covered krans on mountain side.

Galpin died six years later, too soon to see this strelitzia specimen identified as *Strelitzia caudata* by Dyer after Galpin's collections were received at the National Herbarium.[22]

In 1936, T.B Verschuur sent specimens to the Botanical Research Institute from his farm Geluk, nearly 25 kilometres east of Louis Trichardt in the Soutpansberg. The plants, which he had nurtured from seed since 1936, had grown to more than 2.1 metres tall but did not bloom for the first time until July 1943. Verschuur was able to send these flower specimens and two months later, in September, he sent 'fruiting material'. These new specimens confirmed that, unlike a hybrid, the plant could reproduce effectively. Their arrival seems to have accelerated re-examination of the National Herbarium specimens, a process that Dyer would have overseen following his mentor Pole Evans's retirement in 1939. Ultimately, Verschuur's July 1943 specimens were illustrated to accompany Dyer's description in volume 25-26 (1945-1946) of *Flowering Plants of Africa*.

## RECOGNISING A NEW STRELITZIA

Dyer, a small and precise man whose major leisure activity was playing bowls, did not embark lightly on discussion of a possible new species of treelike strelitzia. Describing a new species from existing specimens can be a controversial botanical venture. It revives the grumbling, sometimes erupting, taxonomic feud which has smouldered since at least the 1840s between the 'lumpers' and 'splitters' (see chapter 2). On 20 October 1943, to assist in reviewing Verschuur's fresh specimens from Geluk, Dyer requested new specimens of the Eastern Cape *Strelitzia alba* from J.H. Keet, district forest officer at Storms River, west of Port Elizabeth. On 5 November, Keet posted to Dyer three specimens of *Strelitzia alba* from 'the Groot River Nature Reserve, which is portion of the Salt River Demarcated Forest in the Division of Knysna'. This was the only site where he knew strelitzias trees to grow in the area, although he undertook to doublecheck this with his senior, Dr Henri Fourcade (see chapter 2).[23]

*Strelitzia caudata* 1943 type specimen *(top left)*; *Strelitzia alba* *(above)*

About 15 years after Dyer published his description and discussion of the new species, *Strelitzia caudata*, further specimens of the mountain strelitzia confirmed his demarcation. One was collected in June 1957 among rocks at Millers Falls, near Mbabane, Swaziland, at about 1 300 metres. The plant's flower was noted as white and the plant about six metres high. In July 1958, a specimen of 'Wilde piesang [wild banana]' arrived from an altitude of about 2 700 metres on the southern side of Piesangkop [Banana Hill] near Letaba, Transvaal. The plant was described as having:

> Unbranched slender stems to 80' [feet; 24 metres] tall leaves on young stemless specimens up to +- 10' [3 metres] long with a 4' [1.2 metres] long and more and blades to 5' [1.5 metres] long and more. This very variable in size. Waxy white and blue.[25]

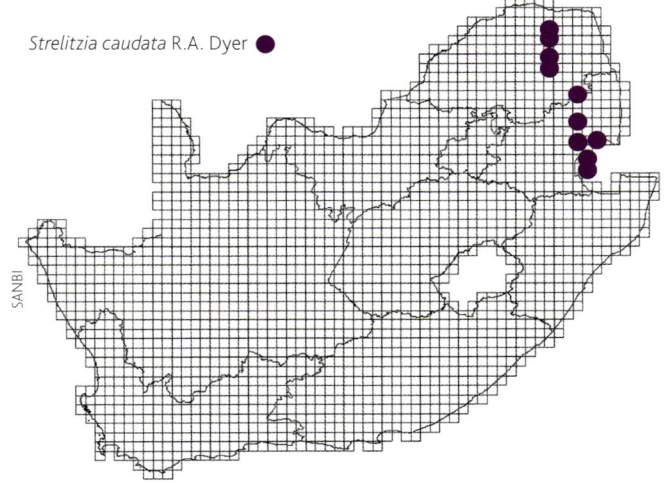

*Strelitzia caudata* R.A. Dyer ●

Keet had already observed that *Strelitzia alba* grew both at the bottom of the forest kloofs and on rocky ledges on these cliffsides. The apparently unexpected preference of some strelitzia trees for rocky heights instead of low-lying forest edges would later become an important characteristic, even opening the possibility of more specimens in suitable habitats outside its first recognised range. But for the time being, Dyer's description concentrated squarely on the plants' form, or morphology to botanists. Dyer admitted that 'confusion is almost inevitable' with other treelike strelitzia species, whether in the field or in the herbarium, 'unless particular attention is directed to the characters of the inflorescence'. Comparing Verschuur's specimens to Keet's, it was the delicate short extension of about 1.5cm to 2.5cm to the flower's lowest sepal that prompted Dyer to name it *caudata*, or tailed. Otherwise, Dyer noted, *Strelitzia caudata*'s flower is simpler than that of *Strelitzia nicolai*, with a different-shaped base to the petals. The lower petals also have more pronounced lobes than *Strelitzia alba*. Examining Keet's specimens alongside existing herbarium specimens considered to be *Strelitzia alba* or *Strelitzia nicolai*, Dyer concluded that by his new definition, it was more likely that *Strelitzia alba* would be confused with *Strelitzia caudata*. *Strelitzia nicolai* was less similar physically but in terms of both latitude and altitude can be found growing much closer to *Strelitzia caudata*'s preferred habitat. How close is not yet sure as 'Careful observations are needed' of treelike strelitzias growing on the KwaZulu-Natal Drakensberg escarpment, for example, to clarify whether they are *Strelitzia caudata* or *Strelitzia nicolai*.[24]

## CROSSING BORDERS

Until Dyer's careful re-examination of treelike strelitzia specimens, it had been believed that only *Strelitzia nicolai* reached beyond the borders of South Africa, extending from the coastal province of KwaZulu-Natal north into Mozambique.[26] *Strelitzia caudata* proved to be another species ranging across regional political borders and prompting reassessment of specimens already catalogued in herbaria locally and internationally.

Swaziland resident Denis Heenan confirmed this in September 1971 when he collected a specimen from Piggs Peak where the plants were 'Plentiful in some kloofs – stems up to 7 m. tall.' Heenan's particular enthusiasm was cycads and the woolly cycad (*Encephalartos heenanii*) was named after him by Dyer.[27] Like Keet, Heenan described the plant as 'Branching from base', adding; 'perianth, segs white, keel with or without "cauda".'

Edith Burges of SANBI, *Strelitzia alba*, watercolour, 1943 for *Flowering Plants of Africa* 1946; distribution of *Strelitzia caudata*

The species' range was originally given as 'Transvaal, Swaziland' but it became clear that it is also found as far north as Zimbabwe's Chimanimani mountains and Mozambique's Mount Gorongosa. Strelitzia did not appear in two early Southern Rhodesian floras, 1898's *Flora of tropical Africa* and F. Eyles's 1915 'Record of plants collected in Southern Rhodesia'.[28] But Albert and Lucy Crook, collecting in Southern Rhodesia in the 1950s, reminded Kew of an acute observation by Indian-born, English-educated Charles Swynnerton (1877-1938), who had arrived in the country in 1897, working briefly in a trading store before farming first near Melsetter (now Chimanimani) then in the Chipinga district and later becoming the first game warden in Tanzania in 1919. Swynnerton had notified the June 1910 meeting of the Linnean Society that he had seen treelike strelitzia in Southern Rhodesia's Melsetter district, between the Chimanimani mountains and Mount Pene. The society's September 1911 journal recorded Swynnerton's comment:

> There are clumps of fine Strelitzias at the heads of many of the glens, and the wood covering its eastern face is, especially towards the summit, one of the best examples of mountain forest I have seen . . . [29]

In fact, the collection at the Natural History Museum in London shows that in 1908 Swynnerton had also seen the plant across the border. He was at an altitude of 7 000 feet (about 2 100 metres) on Mount Pene itself in Gazaland (now Gaza and Inhambane provinces in Mozambique). He described it on his label as, 'A handsome tree with straight slender herbaceous stem and bluish leaves.' In a note included on the herbarium sheet, he added:

> This banana is very common in the glens and on the higher slopes of Mount Pene and in portions of the Chimanimani at from 5,000 to 7,000 feet. In an extensive thicket, completely burnt out by the exceptionally severe fires of 1906, amongst the crags of the latter range, I noted that the bananas, present in great numbers and themselves badly scorched, were the only living vegetation left, a cut with a knife producing a fair flow of watery sap – a great advantage to the tree in such an emergency. This sap did not produce a darkening of the blade as does that of the cultivated banana being evidently only slightly if at all acid.[30]

It is a botanical might-have-been. Perhaps if Dyer had had access to Swynnerton's observation and the Crooks' later Melsetter herbarium specimens, he would have chosen *Strelitzia swynnertonii* instead of *Strelitzia caudata* as the name of the new species, given that it later became reasonably clear that the strelitzia Swynnerton had observed was in fact what Dyer described as *Strelitzia caudata*. This was emphasised by the additional 1954 specimens collected by Albert Crook in the belief and hope of a possible new species

Top: *Strelitzia caudata* collected at Miller's Falls, Mbabane, Swaziland (R.H. Compton, 1956); bottom: *Strelitzia caudata* growing near Mbabane (2017)

('sp. nov.'). Crook also supplied six photographs of the plants in their habitat at an altitude of 1 200 to 1 500 metres near Melsetter. As the label put it:

> Wild Banana, Mtsoro. 3 ½ miles N of Melsetter on road to Sawercombi (farm) near top of steep side of ravine, in rich moist soil. Abundant. Alt 4 – 5 000 [feet]

As with the other treelike strelitzias, though, the reclassification took some time to become common and more than half a century later, Braam van Wyk and Piet van Wyk found it necessary to state firmly in their *Field guide to trees of southern Africa*: 'The plants that grow in the Eastern Highlands of Zimbabwe are *S. caudata* and not, as has been claimed in literature, *S. nicolai*.' The species' range may still need to be extended, too (see chapter 3).[31]

In late 1954, when Dyer visited Kew Herbarium, he reclassified these specimens as *Strelitzia caudata* during a review of its strelitzia holdings. But, intriguingly, Crook's specimens were relabelled *Strelitzia alba* in May 2005 for *Flora Zambesiaca*, then returned to *Strelitzia caudata* in September 2009.[32] Underlining Dyer's cautious reference to confusion between the species, specimens collected by Albert

(From top left) *Strelitzia caudata* growing in Chimanimani Mountains, Zimbabwe (2016); at Kirstenbosch Botanic Gardens; collected by Albert Crook (Melsetter district, 1950)

and Lucy Crook in September 1951 at Umtali and at 'Tarka Forest Reserve 3 m[iles] S. of Melsetter small colony on banks of Chanabuca River. Leaves not much glaucous petals blue mauve, filaments not spirally twisted as in Natal plant' remained classified as *Strelitzia nicolai*.

Mozambican specimens collected in 1943 and 1944 do not seem to have been reviewed by Dyer in 1954 and might have reached the Kew collection only as part of the Flora Zambesiaca project in the early 21st century. These specimens also reflect the confusion between the three species that Dyer highlighted but in this case between *Strelitzia caudata* and *Strelitzia alba*. One set of specimens was collected in Gorongosa at 1 500 metres in September 1943 for the Herbário do Centro de Botânica, Instituto de Investigação Científica Tropical in Lisbon, Portugal, by A.R. Torre, who noted:

*Erva com as folhas dispostas no mesmo plano mas simetricamente.*

*Abundante nos lugares rochosos.*

*No cume da serra da Gorongosa, associada a Aloe sp, Erica, etc.*

[Plant with leaves arranged in same plane but symmetrically.

Abundant in rocky places.

At summit of Mt Gorongosa, associated with Aloe sp, Erica etc.]

The second set of specimens in October 1944, when the herbarium had become part of the Junta de Investigações do Ultramar, were collected by Federico de Ascenção Mendonça who noted the vernacular name as 'Kô Kumovo' and:

*Manica e Sofala, Serra da Gorongosa.*

*Lugares rochosos, formacôes arbustivas.*

*Caule de medulla comestível.*

[Province of Manica & Sofala, Gorongosa Mountains.

Rocky places, shrubby vegetation.

Stalk of medulla edible.]

This second set of specimens was also named *Strelitzia alba* in 2005 and reclassified as *Strelitzia caudata* in 2009. A third set, collected at 1 600 metres in December 1965 by Torre and M.F. Correia, remains classified as *Strelitzia alba* like the first set. The collectors' label noted:

*Espique c. 4 m; folhas ca. 1.5m.*

*Savana herbosa junto à floresta densa de nevoeiro, no plateau; solos argilosos.*

*MANICA E SOFALA: Báruè, serra de Choa, no km. 28 de Vila Gouveia, pionda nova para a fronteira (Di).*

[Spike c. 4 m; leaves c. 1.5m.

Grassland next to dense mist forest, on plateau; clay soils.

[Province of] Manica & Sofala: Báruè, Choa mountains, 28 km from Vila Gouveia, new border post (Di).]

## NEGOTIATING FAMILY MEMBERSHIP

As well as reclassifying the treelike strelitzias, Dyer was also determined to put right what he seemed to believe was an injustice to the plant we now know as *Strelitzia juncea*, or rush-like strelitzia. For more than a century and a half, its botanical status fluctuated as botanists differed in whether they considered it a full species or simply a subspecies or even a variety of *Strelitzia reginae*. This bumpy road to scientific acceptance for *Strelitzia*

*Strelitzia nicolai* (Mozambique coast near Maputo, 1968)

*juncea* was more like establishing a difficult stepchild in a blended family than revealing a long-lost near relation and it was ultimately Dyer's support for South African scientific enthusiasm in the 1970s that confirmed it as a species in its own right.

The first written record of the plant we now know as *Strelitzia juncea* appears to be from Sir John Barrow (1764-1848), who served as Lord Macartney's private secretary in the Cape from 1797 to 1803 before becoming secretary to the admiralty and co-founder of the Royal Geographical Society. Barrow was tenacious about broadening his knowledge and before setting off with Macartney for the Cape had 'spent three days a week at Kew studying Cape plants.' During his tours of the Cape Colony, Barrow spent his evenings recording the day's events and his observations of the people and country, as well as organising his plant and rock specimens. In mid-August 1797, Barrow was heading directly south to the coast from Graaff-Reinet through thick bush and had reached the Zwart Ruggens area of the Camdeboo plain.[33] Interesting plants included a cycad which reminded Barrow of 'a species of *zamia* . . . described by Mr. Masson' (see chapter 1). Even more exciting, what seemed to Barrow a new species of strelitzia spangled the forest edge:

> On the slope of a hill, towards the southern verge of the forest, I distinguished among the clumps of frutescent plants several flowers of a strelitzia, which I took to be the reginae, but on a nearer approach it turned out to be a new species differing remarkably in the foliage from the two already known. Instead of the broad plantain-like leaves of these, those of the new species were round, a little compressed, half an inch in diameter at the base, tapering to a point at the top, and from six to ten feet [two to three metres] high: the flowers appeared to be the same as those of the reginae, the colors perhaps a little deeper, particularly that of the nectarium. I procured half a dozen roots, which I sent down to the botanic garden at the Cape; and the plant is now in England, and likely to become as common as the other species.[34]

Barrow's determination to introduce his new species is confirmed by his correspondence. In a letter to Sir Joseph Banks in December 1800, he mentions that he has 'sent some roots of *Strelitzia*, one of a new species'. To Lady Anne Barnard, wife of Macartney's official secretary Andrew Barnard, and with whom he had climbed Table Mountain before leaving on his tour, Barrow wrote, 'I found a new Strelitzia, a curious plant which will be a very acceptable thing in England', and sent her a bulb. Although Barrow did not suggest a name for this 'new species', he referred to it later as *teretifolia* while noting the abundance of 'beautiful' *Strelitzia reginae* near the Great Fish River:

> The cerulean blue nectarium of the reginae seemed to be uniformly faded, and it lost its colour by a short exposure to the weather, which did not appear to be the case with that of violet blue of the *teretifolia*.[35]

Unfortunately for Barrow, this passing reference with a suggested name – but without a formal botanical description – does not fulfil the conditions needed for valid botanical publication. In addition, there is no indication whether any plants ever grew from the 'roots' that he sent. If he took field specimens, they do not seem to have survived. Kew's earliest specimens of the plant that became known as *Strelitzia juncea* date from a decade or more after Barrow's journey. One was collected by William Burchell some time between 1811 and 1815 but not identified by him and received at Kew only in 1865, though it was also stamped as part of the herbarium bought from William Hooker's estate in 1867. The other was supplied by 'Villett', a father-and-son business of natural-history dealers in Long Street, Cape Town. This specimen is completely undated and also came via Hooker's herbarium. It is marked both 'Strelitzia juncea' and 'Strelitzia parvifolia var. juncea'. Two specimens from the German collector Carl Drège (1791-1867), both stamped 'Herbarium Benthamianum', are dated 1840. Both are labelled as 'Strelitzia parvifolia var. juncea' but one is also labelled 'Strelitzia juncea Ait.' and the other 'Strel. juncea forma angustifol'.[36]

This range of varying and partially duplicated names was due to a naming conundrum that arose within a few years of Barrow publishing his travel memoir. Heinrich Link (1767-1851), the renowned curator of the Berlin herbarium and director of the

*Strelitzia juncea* collected by Carl Drège (Cape Colony, about 1840)

Berlin botanic gardens from 1815, had followed William Aiton's 1789 *Hortus kewensis* in including an oval-leaved and a narrow-leaved strelitzia species (*Strelitzia ovata* and *Strelitzia angustifolia*) and added the rush-like *Strelitzia juncea*. The difference between the rush-like and the narrow-leaved species, Link believed, was that *Strelitizia juncea* had a very long petiole ('*petiolo longissimo*') and had developed so that the leafblade was obliterated ('*lamina obliterata*'). But that same year, John Ker Gawler (1765-1842) saw the plant differently – as a variety of a new species that he called *Strelitzia parvifolia* var. *juncea*. In both the 1913 and 1948 editions of *Flora Capensis*, *Strelitzia juncea* was still categorised as an 'imperfectly known species' and it was noted that it might be 'the same as *S. parvifolia* var *juncea*'. Later, French botanist René Maire (1878-1949) supported both *Strelitzia parvifolia* and *Strelitzia latifolia* as separate species, names that focused on the size and shape of the plants' leaves, meaning small-leaved and broad-leaved respectively.[37]

But botanists of the 'lumper' inclination did not see that varying leaf shapes were enough to justify separate species for these shrubby southern African strelitzias. Marc Weiller (fl.1940), updating Maire's *Flore de l'Afrique du Nord*, discarded *latifolia* and *parvifolia* in 1960. Several botanists disputed whether they should be separate species on the grounds that *Strelitzia reginae* and *Strelitzia juncea* hybridise quite easily in the wild and that 'no structural difference' had been found between flowers of the two plants. In 1970, the American Harold Moore (1917-80), professor of botany at the Bailey Hortorium at Cornell University, declared that *Strelitzia juncea* must be treated as a variety of *Strelitzia reginae*. He had debated the issue between hemispheres with Dyer and the two had agreed in their letters with 'mutual regret that ou[r] views do not coincide.' Moore's reservations were based on the fact that two early descriptions of the plant in 1821 were based on 'plants in cultivation', though he did not rule out the possibility that field work would prove Dyer's belief that *Strelitzia juncea* should be recognised as a species in its own right: 'We agree with Dr Dyer that a careful investigation needs to be made and are prepared to have such an investigation show our reduction of *S. juncea* to *S. reginae* to be wrong.'

Dyer's latest information had come from one of his contacts in the Eastern Cape, the botanist and conservationist Noel Urton (1917-2002). This left him 'back at square 1', he admitted. After having studied the local members of the strelitzia family around the *Strelitzia juncea* stronghold of Uitenhage for 30 years, Urton had noticed a range of leaf shapes. By far 'the great majority' had rush-like leaves with 'practically no' leafblade. But some had a 'small' blade and 'a few' had a somewhat larger blade 'up to 15 cm long and 3.75 cm broad' – conspicuously smaller than the leafblade of a *Strelitzia reginae* but present nevertheless.[38]

Eventually, new evidence from wild South African plants did indeed call into question Moore's 'lumping'. At the University of Port Elizabeth, Albertus (H.A.) van de Venter carefully catalogued wild populations of stemless strelitzias for a 1974 doctoral thesis on the distribution, leaf morphology, embryology and germination of *Strelitzia* Ait. This revealed that '*S. juncea* was found in abundance north-west of Uitenhage and in two other relatively small populations, the one north of Port Elizabeth and the other near Patensie'. Crucially, at Patensie, *Strelitzia juncea* overlapped with *Strelitzia reginae*. Dyer believed this 'help[ed] explain the intermediate forms which had reached Europe in the early days before any artificial hybridization had been possible in gardens.' Having also grown both *Strelitzia reginae* and *Strelitzia juncea* from seed to observe leaf development, Van de Venter was convinced that the two had 'genetically governed' different leaf shapes and sizes and that the 'intermediate forms' found at Patensie were hybrids between the two. Dyer speculated that *Strelitzia juncea* had developed from a *Strelitzia reginae* mutation – but maintained that it was now 'worthy of specific status'.[39]

After nearly two centuries of debate, *Strelitzia juncea* is now accepted internationally as a valid species in its own right. Meanwhile enthusiasm for indigenous plants steadily took root in several parts of the world, intensifying in South Africa the web of protective laws and regulations into which strelitzia would have to fit. Later, the plants also rode high on increasing botanical nationalism which would eventually slot into early waves of homegrown pride summed up in the South African catchphrase 'local is lekker' [great]. The newer species, *Strelitzia caudata* and *Strelitzia juncea*, would be followed by a new subspecies as well as by much-desired new cultivars. Strelitzias had largely emerged from their mysterious shadows and consolidated their slow but steady path from the shadows into the horticultural and floricultural spotlight.

Throughout much of this time, the dramatic plants which had once attracted the attention of artists such as the legendary Redouté and Bauer lived on quietly in private gardens. Eclipsed by Western Cape floral enthusiasms and botanical politics, even the developing local flower industry would for decades sideline the strelitzia in favour of proteas, chincherinchees and everlastings. Commercial growing of strelitzias was founded not in South Africa but halfway across the world.

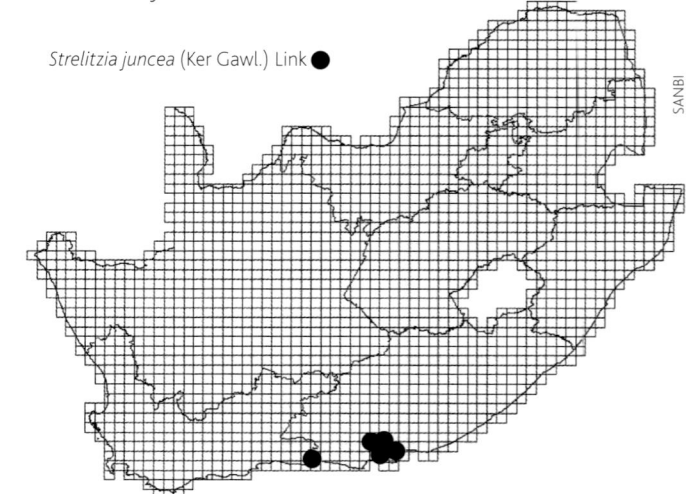

*Strelitzia juncea* (Ker Gawl.) Link ●

Distribution of *Strelitzia juncea*

Basia Hitchcock Swiel, *Strelitzia juncea*, watercolour, 2007

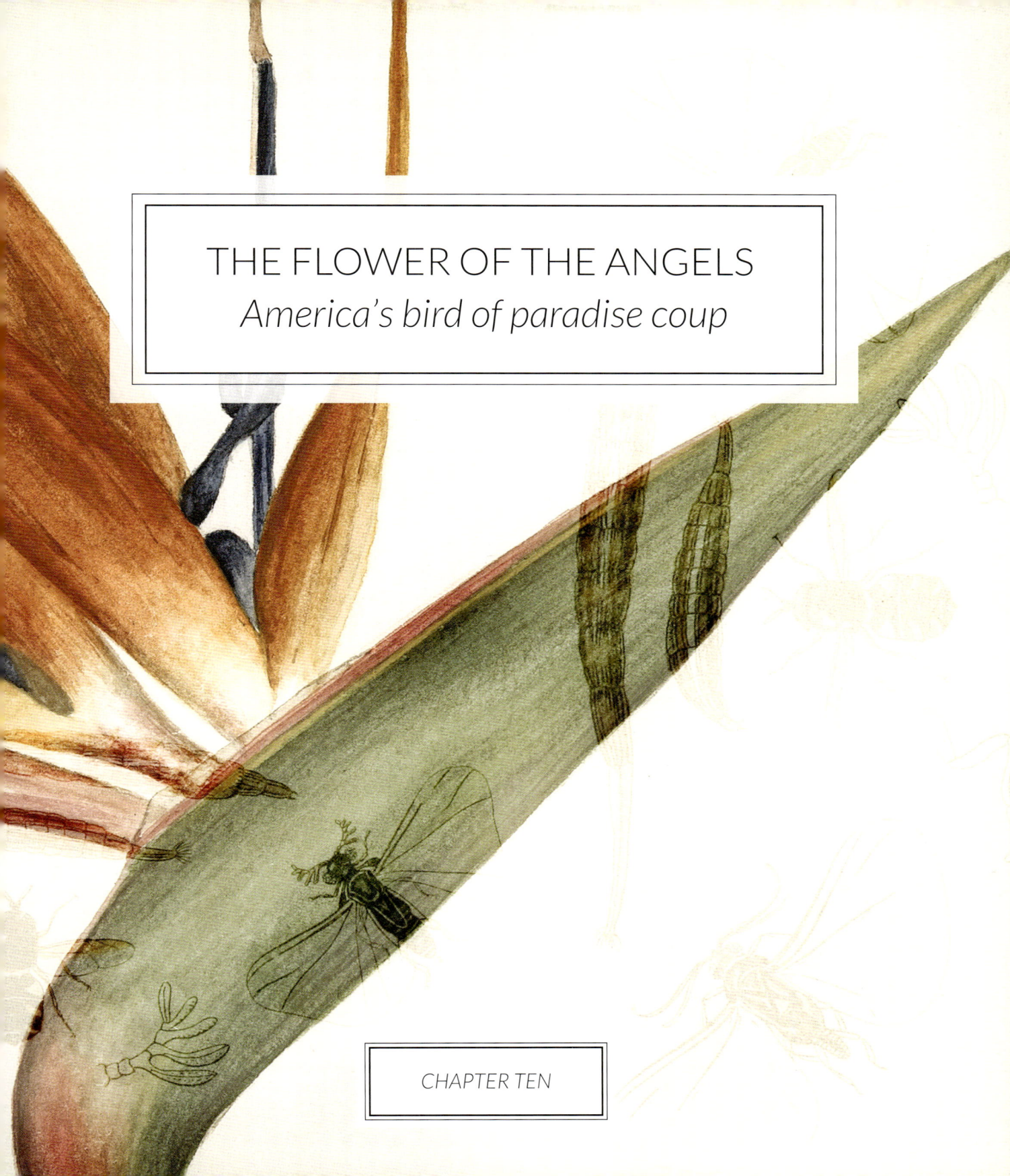

# THE FLOWER OF THE ANGELS
## America's bird of paradise coup

CHAPTER TEN

# THE FLOWER OF THE ANGELS: *America's bird of paradise coup*

The discovery of *Strelitzia reginae's* strange, glowing flowers on the South African veld had thrilled the worlds of botany and horticulture. Royalty, aristocrats and Europe's wealthy had clamoured to possess this fascinating rarity. But it was on a third continent, America, that the bird of paradise's future as a top bloom in the modern cut-flower industry first developed. Like many great American immigration stories, the early days of the bird of paradise flowering in California's floral fields began with hope and prayer. The bird of paradise, with its dramatic, sculptural blooms, was an exciting but untried prospect for floriculture and floristry. However, it would sweep into the American public's consciousness, leading to milestones such as the 'legendary bird of paradise coup' and the flower being adopted as the emblem of the American city of Los Angeles. Once again, the bird of paradise would capture the imagination and become an icon – this time making a fortune for growers and florists in both the cut-flower and the foliage-plant industries.[1]

Through its celebrity as a flower by royal appointment and a luxury for conservatories, *Strelitzia reginae* was a status symbol. It appeared rarely, being comparatively more demanding to grow than many of the popular and voluptuous orchids that might at first glance seem to overshadow it. With far fewer species within the family than either orchids or heaths, both of which sparked long-running collectors' crazes, the strelitzia was instead a statement plant. It has been and remains in demand by American gardeners, whether wanting to recreate a luxuriant tropical jungle effect and, especially more recently, by those looking for a showy, eye-catching specimen within water-conscious garden planting.

The bird of paradise preferred a drier heat to the steamy orchid hothouse, at first making it an either-or choice for all but the wealthiest of plant enthusiasts. Strelitzia also demanded extended patience, usually slowly developing their first flowers over at least three years of settling in. They also required much more space to flourish than the comparatively dainty orchids. Mid-19th century gardeners preferred headier perfumes and more voluptuous flower forms. Relatively few were prepared to put in the effort to succeed with the crane flower, or crane lily as the bird of paradise was also sometimes known, possibly to highlight its gawky quirkiness. As the century turned, though, the strelitzias' often languidly bending leaves meant they fitted the flowing naturalist shapes of art nouveau, born in the 1890s. Their angled flowers then gave it longevity through the swerves and curves of art deco in the 1920s and 1930s and it became an ideal, eye-catching flower to feature, for instance, in window displays.[2]

When the new dawn came for the bird of paradise, it was in one of the newest parts of the new world where climate allowed the plants to be grown in the open air in private and public gardens. The birth of the region's floriculture meant that as commercial flower growers and markets were established and developed, bird of paradise flowers became more and more widely available. Today they are part of the total global flower trade now estimated at about US$40 billion. These flowers are grown in more than 80 countries, though about 54 per cent come from the Netherlands. In the United Kingdom alone, the cut-flower industry is believed to be worth £1.5 billion a year, which closely rivals Germany as the world's largest importer of cut flowers. The United States is not far behind and Japan's market has been increasing significantly (see chapter 12).

## NEW GROWTH AT THE TRAIL'S END

It was in California, at the far end of the trails across the legendary Wild West, that the bird of paradise began to conquer a new domain and a new era. The area was first seen by Iberian explorers in the 1540s but it was not until 1769 that the Spanish moved up here from South America. They began to establish missions in the area they called Alta [Upper] California, with what is today Mexico known to them as Baja [Lower] California. As part of this expansion of the territories Spain had acquired in central America, Alta California was included in the ambitious Expedición al Virreinato de Nueva España (Expedition to the Viceroyalty of New Spain). Lasting for 16 years from 1787, this swept from Mexico and Guatemala, via California north to Vancouver Island, as well as east to Cuba and Puerto Rico. Under the leadership of Spanish doctor-botanist Martin de Sessé y Lacasta (1751-1804), a friend of the founding director of the Madrid Royal Botanic Gardens, Casimiro Gómez Ortega (1741-1818), the expedition collected about 1 335 drawings and about 3 500 specimens in all.

Its collections of American indigenous flora may not initially have attracted much more attention than those of Juan Mutís's expedition in Colombia and Peru (see chapter 7). But its reports from both the expedition and the missionaries of the agricultural potential meant most of the missions founded by the Spanish were established as agricultural centres to take advantage of areas of temperate climate and sheltered, well-watered productivity. This in turn encouraged more daring Spanish settlers and others followed later.

As many traditional farmers were dispossessed in this process, they came to join the flocks of settlers searching for fortune and salvation in San Francisco after the 1848 discovery of gold. Immigrants poured in, moving on from the eastern United States or taking long, often gruelling passage across oceans. The region's population rose more than fivefold in 1849 alone, from about 20 000 to more than 100 000.[3]

Overleaf: *Strelitzia reginae*: Detail from *Dittionaire pittoresque d'histoire naturelle et des phénomènes de la nature*, 1839; Helen Sharp, *Water color sketches of the fruits and flowers of Bermuda* (1892-1903)

California officially became a state in 1850 but was still considered largely unknown frontier territory for the daring by the time of the American civil war (1860-4). The journey from the east coast to Oregon alone was about 2 000 miles (3 200km) long and usually took four months. The main Mormon and Oregon trails led through Missouri into brutal weather over the challenging Rocky Mountains. The determination and desperation of would-be settlers could be measured by the number of crosses marking graves and by the rusting, abandoned stoves – legend had it that 'cowards never started and weaklings died along the way'.[4]

Those who could still face pressing south to California might be lucky to find the 'land of milk and honey' promised in powerful contemporary accounts such as William Brewer's *Up and down California in 1860-1864*. Brewer and other travellers depicted fruit orchards and vineyards flourishing in lush farming areas around mission villages such as Los Angeles. In fact, there was upheaval on the land after the new States Land Commission began reviewing all Mexican land grants in 1851. Many *rancheros* of Spanish-American descent found they could not prove title. Now dispossessed in their turn, they often also found digging for gold was not only a way of trying their luck, but a means of survival. In time, as the reality of hard life with often poor rewards on the diggings hit home, some reused their agricultural skills to form the backbone of the gardening and flower industries.

Early Los Angeles residents were so enthusiastic to beautify their gardens that they often sent for seeds and supplies to be mailed from the eastern states, where the Missouri Botanic Gardens, for example, already had *Strelitzia reginae* specimens as early as 1827. By 1853, Col Warren was planting *Strelitzia reginae* in Sacramento, making it 'one of the earliest plants introduced into California'. The first nurseries were established to supply such enthusiasms. In March 1855, William B. Osborne advertised for sale a shipment of roses, lilies and shrubs. Two years later, Ozro W. Childs and W. Huber founded the Los Angeles Nursery and Fruit Garden, located just east of Main Street. By 1870, Joseph Sexton was selling *Strelitzia reginae* in his Montecito nursery.[5]

Gardeners found the contrasting climate of a torrential rainy season and a parched dry season challenging. From 1874, gardens were sufficiently flourishing and local residents sufficiently settled for the commercial flower trade to be pioneered. These early entrepreneurs were considered simply housewives, earning extra money by selling flowers from their gardens but, in fact, were shrewdly capitalising on a countrywide trend:

> Homemakers were becoming increasingly interested in creating pleasant surroundings for their families both inside the home and in the garden. Popular women's magazines featured articles about flower care and display. Flowers and flowering plants, with all their color and image, were decorating dining tables and living rooms everywhere.[6]

This was due to more than the sense of triumphing over dusty new frontiers and building lots. The trend trickled down from the fashionable and the wealthy, in turn sparked by a renaissance of flower painting in the international art world:

> It helped that by 1875, the Impressionist style of painting had become the rage for collectors of art and those who cultivated an appreciation of beauty, and that many Impressionist painters chose flowers as their subjects. Refreshingly, painting styles had evolved from dark, somber and depressing depictions of dramatic events to happy scenes with color and loose, almost casual brushstrokes. Paintings of roses, which dominated the art scene from 1840 to 1900, were bold and full of color, inspiring those who saw them to grow or buy flowers so they could duplicate the scene for themselves.[7]

## PIONEERS AMONG THE PIONEERS

By the final quarter of the 19th century, strelitzias and other exotic flowers were in demand for Californian gardens and parks. The Californians were keen to follow the fashions of Europe and the eastern seaboard of the United States and also realised that the region's modified Mediterranean climate meant their displays

P. Badenhorst, *Strelitzia reginae* in *Flowering Plants of South Africa*, 1925

of exotic blooms from Africa and Australasia would compete strongly alongside the more familiar inhabitants of the flower beds.

On the state's southern border in San Diego, about 120 miles (200 kilometres) south east of Los Angeles along the Pacific coast, 1 400 acres (570 hectares) had been set aside by the city fathers as early as 1868 for a park, though it remained mesa scrub for at least 20 years. Today this might be seen as an unparalleled opportunity to preserve and showcase indigenous vegetation. But as with the new parks and botanic gardens in regions such as Australia, India and southern Africa, at that time planners felt there was plenty of wild, indigenous vegetation to be seen beyond the city limits. Instead, they wanted their landscaping to demonstrate their ability to subdue the natural surroundings and to show off their understanding of current international gardening trends and their excitement at California's flourishing prospects.

They accepted a proposal from female nursery-keeper Kate Sessions (1857-1940). She was far from the earliest Californian to establish herself in the nursery trade, though she was unusually a woman in an industry that was generally male dominated. In exchange for a plum 32-acre (13-hectare) position inside the park boundary where she would operate her nursery, Sessions offered to plant 100 trees a year inside the park, as well as donating trees and shrubs to be planted around the city. She particularly favoured highlighting the exotic impact of imported Australian, South African and South American plants. Thanks to her – and possibly to her two-month visit to the luxuriant growth of Hawaii just before she began a science degree at the University of California, Berkeley – well-known features of this park to this day include its queen palms, poinsettias and birds of paradise.[8]

The park was renamed in 1910 after Spanish explorer Vasco Nuñez de Balboa. San Diego first hosted the World's Fair in 1915-16 and the Panama-California Exposition celebrated the opening of the Panama Canal with 'Art and Culture, Gardens and Spanish-Renaissance Architecture' as its theme. The bird of paradise plantings played a resplendent part in portraying the theme colour of gold, which was also reproduced in many floral displays across the city and neighbouring hillsides.[9]

## THE CUT-FLOWER BOOM

The Washington political scene was a factor in sparking the 20th-century sales boom in flowers. Some said Edgar Wells (1883–1958) was a socialite, others that he was a playboy – but he is generally credited with spotting a marketing opportunity for his homeland, Colombia, in undercutting high fresh flower prices in Washington and elsewhere in the United States. The initial aim had been to take a piece of the market in fresh flowers often bought by guests for gifts to Washington's elite political hostesses. But in the process, Wells and others such as David Cheever and John Vaughan who pioneered the Colombian flower industry, helped turn the USA from a major flower grower into a major flower importer.[10]

The UK cut-flower industry was partly galvanised by cheap flower imports from the Netherlands that enabled supermarkets and then garage stores to introduce flower sales from the 1990s. Those who might not have a florist in their area could now make cut flowers a regular buy or part of the weekly shop. By 2004, UK supermarket chain Tesco was named Flower Retailer of the Year at the Retail Industry Awards and reported £200 million in flower sales. As many as seven out of every 10 cut flowers in the UK are now believed to be bought at a supermarket. Similar flower booms have been seen elsewhere. This led, for example, to a 44 per cent increase in the value of global exports of cut flowers according to International Trade Centre figures, notes SA economist Danie Jordaan, with a further 30 per cent increase expected by 2014.[11]

This would not be out of line with the dramatic rate of recent historic expansion, Jordaan has demonstrated by realigning figures for three decades at 2010 values. This showed that in 1980 the world import market for cut flowers was SA R5.8 billion. By 2002, the market had grown almost fivefold to over R27.3 billion. In just six more years to 2008, it grew almost nine times bigger again to R241.4 billion.[12]

Bunching bird-of-paradise stems for packing

## THE BIRDS FIND A NEW PARADISE

By 1935 and the time of the Second World Fair, or California Pacific International Exposition, birds of paradise were well on their way to becoming popular exotics in the gardens of southern California particularly, as well as being promoted by the California Agricultural Extension Service as an indoor house plant. Sessions was by then recognised as 'the Mother of Balboa Park' and its displays and the World Fair spotlight had drawn attention. But birds of paradise had also been made more generally available. *Strelitzia reginae* were also among the exotic plants such as anthurium, cattleya and other orchids exhibited by Siebrecht's House of Flowers in Pasadena, south of Los Angeles, in October 1920, for instance.[13]

Lowell Swisher, an enthusiastic amateur flower breeder, member of the Los Angeles Men's Garden Club and a Hollywood insurance executive, was looking for an eyecatching flower that would turn an equally eyecatching profit. Cornering the market in this exotic new tropical-looking bloom seemed like a good prospect to him so he planted hundreds of birds of paradise to harvest their seeds. Given strelitzias' notorious years of settling in before blooming, as well as his own need to learn how to hand-pollinate, this was a painstaking process. But he realised this method would in time outstrip the earlier method of dividing rootstock and waiting for the clumps to resettle and bloom again, helping him more than double the yield.

Swisher's strelitzia seed business thrived for some time, supplying 'the majority of local commercial bird growers'. Eventually, though, he was undercut by cheaper seed imported from Hawaii. However, like any shrewd businessman, Swisher recognised that the competition was outpacing him. He needed a new line and moved on to experimenting with orchids. As a well-rounded plants enthusiast, he was also a dahlia fancier and featured as a dahlia judge at the September 1927 San Diego Floral Association's autumn flower show.[14]

Swisher's initial enthusiasm for birds of paradise helped supply and grow the demand for the plants and flowers. After seeing these striking exotics in parks and featured in some gardens of the rich and famous, the general public wanted to introduce them to their own gardens. The angular flowers also added a fashionable note to art deco settings, creating a demand for them in local flower markets. Their theatrical showiness was even borrowed by the expanding film world, bracketed with 1930s sex symbol Dolores del Rio (1904-1983) for the 1932 film *Bird of paradise*, which became renowned for the leading lady's skimpy costumes and sensual dancing. Having bought the rights to a Broadway play, media mogul David Selznick (1902-1965) decreed it should be shot in Hawaii but told his director, King Vidor (1894-1982), 'I don't care what you use so long as we call it "Bird of Paradise" and Del Rio jumps into a volcano at the finish.' A stylised *Strelitzia reginae* appeared in the background of the movie poster.[15]

Pearl de Chalain, *Strelitzia reginae*, colour pencil, 2017

## FINDING THE BIRD OF PARADISE GOLDMINE

With housing and city projects booming and commerce thriving in California from the mid-19th century onwards, pursuing the nursery trade in an apparently attractive climate had long seemed a good proposition to a range of the region's settlers. Combining flower sales with the nursery trade was a different proposition, though. It entailed having the resources to tend the stock as well as going out into town to sell the flowers. Even when flowers were sold on street corners, this was a demanding task. But when California's flower markets were founded in the first decades of the 20th century, the trade was pushed onto the more professional footing that higher turnovers required. Yet even before formal markets were founded, some growers and dealers were taking advantage of the refrigerated rail cars introduced in the 1890s to ship blooms to the booming east coast. Eventually, California and Colorado would become the prime local flower suppliers to the US market.

In San Francisco, for instance, the California Flower Growers Association was founded in 1906, spearheaded by the Italian and Japanese growers. They quickly graduated from an open-air to a covered market three years later, drawing other immigrant networks such as the Greeks along with them. In Los Angeles, growers had also grouped into informal markets but these reflected lingering divisions in the community that culminated in the 1913 Aliens' Land Act, dispossessing California's Asian community. As a result, Los Angeles' European immigrant flower-growers gathered to sell their harvest in one area and Japanese-American flower growers elsewhere.

Santa Monica carnation-grower Edwin James Vawter (1848-1914) had begun his flower business in 1899 with just two acres (0.8 hectares), which had expanded to 20 acres (eight hectares) by 1905. He was less fortunate in his 1905 attempt to establish a flower market, though, as this lasted only three years. Japanese-American growers organised their own flower market in 1912. Since European immigrants could not operate on this market, they founded their own market about 1917. The two legacy markets became the Southern California Flower Market and the Los Angeles Flower Market. They are now both based in the Los Angeles Flower District but still remain separate, each with their own warehouses, though more and more competition has entered the area. The most significant is from the California Flower Mall which allows the public to buy at any time, while other markets are aimed at wholesalers and give only restricted public access.[16]

America's Japanese immigrants had often come from an agricultural or floricultural background and, like European immigrants, were among the most successful and prominent in the new Californian flower trade – which culminated in the bird of paradise becoming a common sight in the fields of Japanese flower farmers in southern California. Several of these farmers could trace their expertise in their adopted country back to the Domoto brothers, Kanetaro 'Thomas' and Motonoshin 'Henry', who arrived in America in 1883 and founded a nursery in the San Francisco area within a couple of years. By 1912, their premises was rated 'the largest nursery in the state' at 500 000 square feet (46 500 square metres) and so renowned for training and mentoring generations of Japanese immigrant nurserymen and gardeners that it was known as Domoto College.[17]

Kanetaro Domoto died in a World War II American internment camp but his son Toichi (1902-2001) eventually became president of the California Horticultural Society and received the California Association of Nurserymen's highest individual award, Pacific Coast Nurseryman of the Year, but closed the family nursery in 1988. Ito Uneka became president of the California Association of Nurserymen and later the first second-generation member of the Japanese community to be president of the American Association of Nurserymen. Although the bird of paradise flower had been known in Japan since the late 19th century, in Oaklands and elsewhere around San Francisco, the Domoto family tended to grow more easily marketed flowers such as violets and roses and to use their homeland contacts to develop their key ranges of the popular flowers from the east, such as chrysanthemums, camellias and peonies. Growers in the Los Angeles area and southern California generally, though, were eager to capitalise on the growing vogue for birds of paradise – eventually engineering not just one but two bird of paradise coups.[18]

## THE BIRD OF PARADISE COUPS

Commercial Los Angeles flower growers such as Manfred Meyberg (1886-1956), president of Germain's Seed & Plant Company, found birds of paradise a thriving addition to their business and sang their praises to colleagues. Gladioli grower Donald F. Briggs, for example, was 'talked into planting [birds] on his expansive acreage' by Meyberg, resulting in the renowned California Birds, co-owned with brothers Elmer and Clinton Pedley at Carlsbad, north of San Diego. It was little surprise to fellow growers that Meyberg became one of the leaders of the 1952 campaign by local bird growers to install the bird of paradise as the official flower of Los Angeles.

As both a prominent businessman and a civic leader who had founded the Keep Los Angeles Beautiful organisation and the Men's Garden Club of Los Angeles, Meyberg was able to leverage several constituencies for his campaign. Ultimately, the bird of paradise succeeded in ousting the challenge of the calla lily, bougainvillea and geranium, which scented many fields far and wide across the Valley of the Flowers in the Lompoc area northwest of Santa Barbara. The bird of paradise was doubly appropriate – showy enough for an already glitzy city and with a

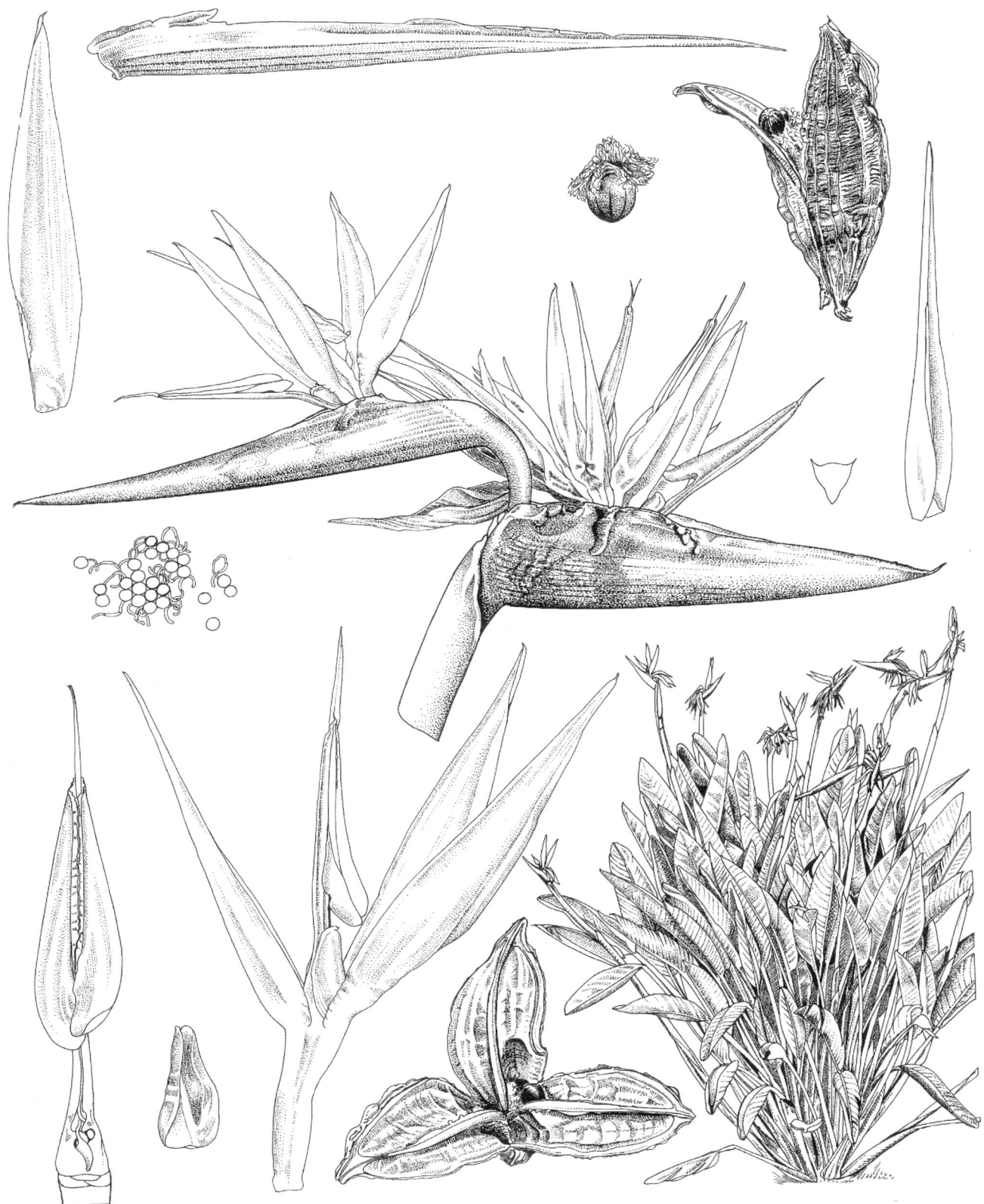

Angela Beaumont, *Strelitzia reginae*, pen and ink, 2011

common name that resonated with the underlying meaning of the city's own name, City of Angels. Within the mainland United States, cultivation of the bird of paradise was seen as unique to California, thanks to both climate and the early development of the nursery trade around San Francisco and Los Angeles. The plant tends to flower in the Californian winter, doing best in areas with milder winters – although planting against a sun-drenched wall or as a container under a covered patio can compensate – and with little if any watering in summer.[19]

Combining the nursery trade with floriculture meant higher risks but also economies of scale. Among the flower-growers who saw such opportunities was Waandert J. vander Bruggen (1887-1974), who had decided to emigrate to the USA from Sassenheim, Netherlands, where his family grew iris, calla lilies and bulb stock. In 1907, he arrived first at a nursery in New Jersey, moving to Chicago and Washington before settling in California in 1914, ultimately joining a large group of Dutch immigrants in Newark, as Montebello was then known. First, he worked as a park curator in Santa Ana until he was able to set up his own landscaping and nursery business. In 1923, this developed into growing bulbs and flowers and he acquired a licence to sell flowers at the American Florists' Exchange in 1925. Finally, early in the 1930s, vander Bruggen was able to buy land for his own flower farm in the Montebello area, east of Los Angeles.

It took more than two decades before vander Bruggen 'realised the potential goldmine that cultivating birds represented'. Another cycle of the flower's popularity had emerged, partly influenced by the fashion for tropical decor and moods and scenery, highlighted by the media and film industry following the introduction of new, more convenient air flights to South America and elsewhere. Vander Bruggen was also well positioned to take advantage of a local bird of paradise boom when it was established as the official flower of Los Angeles in 1952.[20]

Birds of paradise came to be the speciality of vander Bruggen's business and floricultural development. For decades, he worked hard to propagate and improve them, both for cut-flower production and as garden ornamentals. Shrewdly, he worked equally hard in promoting the flowers. Together, he and fellow Montebello grower Cornelius J. Groen, who had immigrated to California in 1912 and founded the Groen Rose Company, engineered the 'bird of paradise coup'. The incident has become legendary in the American cut-flower industry and proves the observation: 'Small, well-placed groups of traditional florists, freelance floral designers, and traditional wholesale florists shape taste, push innovation, and communicate changing consumer culture.' Essentially they proved that they never missed an opportunity to market their product, especially if that meant breaking into new regions.

Ostensibly, the two growers were sailing back from New York to the Netherlands in 1956 for a holiday with family. In reality, they despatched a thousand bird-of-paradise blooms to the liner from their flower farms. Given that many of those living on the United States east coast had never seen these flowers before, the sight of the liner's salons ablaze with these extraordinary blooms not only dazzled many passengers but also New York florists, invited on board by vander Bruggen and Groen to admire the display.[21] Even more profitable publicity was generated by media reports about the bold new flower and the California growers who farmed it:

> Newspapers blossomed with stories about the exotic African flower that came from California. By the time the bird growers returned, their business had taken a decided upswing.[22]

As outgoing president of the Southern California Floral Association, vander Bruggen ensured about 300 heads of the flower were splashed in a cover photograph for the November 1958 *Bloomin' News*. This marked the opening of National Flower Week, depicting vander Bruggen himself, Walter Swartz of the Southern Californian Floral Association and various Los Angeles city officials and civic leaders.

## COMPETITION FROM HAWAII

California's first competitor in bird of paradise production was Hawaii, which became a territory of the United States in 1898. In the Los Angeles region, the astute Lowell Swisher had felt forced to change his ambitions to orchids when imported Hawaiian strelitzia seed began to undercut market prices.

Six decades later, in 1959, access to markets became even smoother for Hawaiian growers after the territory became the 50th state of the United States. Though strelitzia seed production became gradually less viable in Hawaii, too, refrigerated air transport and refrigerated storage facilities became more common, making it possible to airlift birds of paradise as cut flowers to the Californian markets. Production peaked during the early 1990s, as well as at the beginning of the 21st century, Hawaii Department of Agriculture figures show. Currently, however, production is 'at a reduced level from its best years', though Hawaii is still considered a major supplier to the domestic US market.[23]

Competition between the Californian and Hawaiian growers had remained tight but fairly evenly matched for nearly three-quarters of a century. Where Hawaii might win on climate, they lost on air transport costs, for example. Other input costs from labour to agricultural products were fairly similar for the two. But the nature of the competition then changed dramatically for both of them (see chapter 12).

## A STRELITZIA DYNASTY

Meanwhile, in Los Angeles, in 1926 at the age of just 22, Giovanni Mellano (1903-1972) had become the third partner of Nick Gandolfo and John Rainero in the Los Angeles Evergreen Company at the American Florists' Exchange. During his first four years in the USA, Giovanni Mellano had foraged with other Italian families in the Santa Cruz mountains of northern California for wild ferns, acacia and branches of evergreens which were railed to the San Francisco and Los Angeles flower markets. By 1934, he was ready to set up his own wholesale business and to bring back from Italy a wife, Maria. Giovanni Mellano soon became a leader in the Italian-American community at the Los Angeles Flower Market while Maria remained on the farm, six and a quarter acres (2.5 hectares) they had bought in Artesia, south-eastern Los Angeles County, in 1936. Between raising their children Johnny, Michael, Harry Michael and Rose Marie, Maria Mellano ran the house in the morning and in the afternoon supervised the field workers, picking flowers alongside them.[24]

By the time of the Second World War, Mellano & Company was a major grower and wholesaler of both cut flowers and evergreens. The Japanese-Americans, including the Domoto family, were interned – Toichi Domoto remembered how Japanese and Japanese-American internees were transported off to the internment camps bravely and defiantly holding flowers. Both his father and uncle died a month apart in 1943.[25] Italian immigrants were not required to be interned but the period was still politically fraught for them:

When America entered the Second World War in December 1941, Maria had not yet received her final citizenship papers. Johnny Mellano recalls a scary evening in the war's early years when the FBI came to their house with bright lights and a bullhorn. A suspicious neighbor had accused Maria of being an enemy spy. The investigation was short-lived and Maria was allowed to stay with her family, although she retained 'enemy alien' status and was subject to travel restrictions. That meant she could not travel more than ten miles from home in daylight hours and not at all when it was dark – a curfew that lasted until the end of the war when she was allowed to complete the process of becoming an American citizen.[26]

As during the 1930s, instead of the flower industry being badly hit by the World War II, it thrived. Flowers were seen as a small but cheering, affordable luxury – or a necessity, in cases of mourning. By 1949, Mellano could buy Marquardt Ranch, a larger 10-acre (four-hectare) flower farm in the area and broaden their range of crops to include primrose, caspia, sunflowers and eucalyptus. Early in the 1950s, he broke the mould and became the first wholesaler to run route trucks direct to the flower shops along 'Death Valley', as the flower trade called West Washington Boulevard, where florists served Los Angeles Rosedale Cemetery and mortuaries. Giovanni Mellano's eldest son, Johnny, joined the business fulltime in 1956 after college and military service, keen to expand the delivery routes to more communities. It would be a baptism of fire into the business world for Johnny Mellano as he combined working at the flower market and on the flower farm

*Strelitzia reginae* growing in Bermuda, West Indies, Helen Sharp, water colour, 1890s

when needed. He would rise before midnight to work at the market, then load his truck with customers' orders plus an extra allowance of flowers to supply the day's funerals in Long Beach, Wilmington and San Pedro. Returning home about 1pm, he unloaded the truck then slept for an hour or so before picking eucalyptus or preparing what the field hands had picked for market: 'I came in at five o'clock, showered, ate dinner and was asleep by seven.'[27]

Johnny's younger brother, Michael (known as Mike and ultimately Mike snr, given his son and nephew had the same first name), studied horticulture and eventually graduated with a doctorate in plant pathology from the University of California in 1969. By then, the Mellanos were feeling the pressures of urbanisation, including encroaching housing and rising property taxes. The two brothers decided to take on 'eighty acres [32 hectares] of sagebrush and hard-packed dirt inhabited by rodents and rattlesnakes' in the San Luis Rey valley in northern San Diego County. The farm lies about eight miles inland from the coastal town of Oceanside, its rolling hills not far above sea level with an elevation of about 300 feet (about 90 metres). As the guiding principle of the business has always been 'the bottom line', this means 'some of the foliage and cut flower crops have come and gone' in that time. Beginning with four flower crops, they expanded to about 60 different crops on their 'lush, green hillsides punctuated with patches of color in neat, even rows'.[28]

Growing bird of paradise for the cut-flower market was a significant feature of San Diego County in the 1960s, while Robert Boddy, a grower specialising in rhododendrons in the San Fernando valley, was also producing strelitzia seed that was fetching about US$25 a pound.[29] Mike Mellano snr introduced strelitzia, responding also to the flowers' surge in popularity with the revival of the tropical-garden trend in the 1970s, use of the flower in branding Los Angeles when it hosted the Olympics in 1984 and the flower's status as a singular design icon from the 1990s onwards. Today birds of paradise occupy about four acres (1.6 hectares) of the Mellano farm. They generate sales of about 25 000 to 30 000 stems a year, fetching between 35 and 45 cents a stem at the farm gate. Many of the customers are local, but the Mellano farm also distributes nationally throughout the USA and exports to Canada as well. They grow three or four varieties of birds of paradise, partly to improve looks and yields, but mostly to spread production through the season. Even so, just as in South Africa (see chapter 11), 'the heavy flush comes at Christmas which is hard to sell so we end up with a fair amount that is unharvested during December,' notes Mike Mellano jr.

## FROM FIELD TO VASE

For the Mellano operation, top-grade bird of paradise stems should be at least 40 inches [102 centimetres] long, though usually on average eight or 10 inches [20 to 25 centimetres] longer than that. They are harvested by pulling them off the clump with a slight twist, a procedure also followed in Hawaii. In South Africa, though, flower stalks are cut with secateurs or a large blade.

Stems are packaged in bunches of five with the heads even and sent off in bulk boxes, not as finished arrangements because the Mellano operation is a wholesale shipper. Flowers and stems must be clean of diseases and insects, as well as clear of damage such as lesions from previous infections, according to Mike Mellano jr.

Preventing damage to the tip of the flower sheath is particularly challenging as it easily dries out and, without care, can be 'crimped'. This is another reason why stems are harvested as the flowers prepare to open. Before packaging, each flower is shielded with a paper cap, though for special orders, the older formula of placing these paper caps on the flowers in the field is used. But, says Mike Mellano jr: 'It doesn't seem to make a lot of difference and it is hard to control insects and disease issues.'[30]

Top: Strelitzia dynasty – Giovanni and Maria Mellano with son Johnny, 1930s; bottom: strelitzias buds protected by waxed paper, Mellano fields

## FINDING NEW OPPORTUNITIES

Waandert vander Bruggen had become a stalwart of the California bird-of-paradise industry, showing his flair by winning 'many trophies and prizes' for displays of the flowers he bred and grew. He also served the flower industry more broadly, as a long-time member of Hollywood Park's International Flower Show Committee, the California State Floral Association and the Society of American Florists. Close to home, his passion for birds of paradise inspired two generations of his own family, with both his son and his grandson, Nick vander Bruggen 'hybridizing and selling more colorful and longer-lasting varieties of the stately blooms, until imported Strelitzia stems from South America began appearing at the Market.'[31]

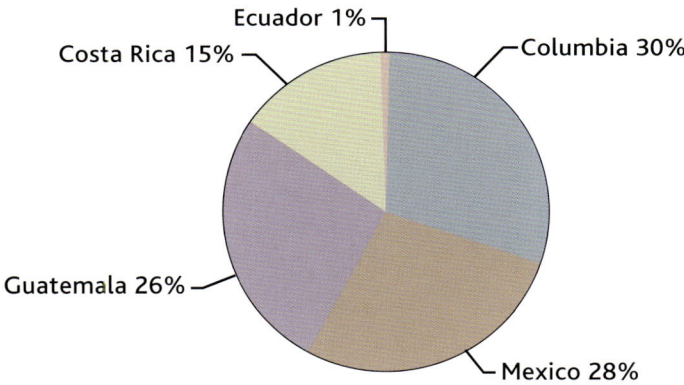

### BIRD OF PARADISE IMPORTS TO USA

Total weekly incoming shipments between April 2013 and the end of March 2014 in order of quantity supplied:

- Columbia 30%
- Mexico 28%
- Guatemala 26%
- Costa Rica 15%
- Ecuador 1%

But the 'Montebello birds' would be under more immediate threat than gradually increasing competition from imports or the third-generation 'reluctant farmer' syndrome experienced by Nick. In fact, having agreed to oversee the farm by himself while his father took a holiday, Nick found the family flower-growing bug bit and bought him out. He served in industry associations, becoming president of the Southern California Floral Association as his grandfather had once been. He grew waxflowers, agapanthus and greens as well as birds of paradise. But it was not this interest in diversification that sounded a death knell for his grandfather's fields of birds.

First, encroaching urbanisation claimed the original Montebello farm. Then Nick moved his operation to land that seemed safe from such threats – under the Southern California Edison's powerlines. Though the powerlines meant that the land could not be used for housing development, Southern California Edison significantly raised the leasing costs in the early 1980s. Nick was one of the area's farmers who could not balance this additional leasing cost when he was also fighting against imports, especially from South America, 'that sold for one-fifth the price of his'. By joining Mas Yoshida's then newly formed GM Floral Company, Nick vander Bruggen was able to continue growing his own flower crops as Regina Farms, combined with handling GM's greens business until he retired in 2003. In the meantime, in 1995, Paradise Hill Wholesaler had also gone out of business in the area. This had belonged to Ken Foltz, husband of Groen's daughter Sylvia, who had taken over Groen's market business as well as farming birds of paradise and other plants in Montebello.[33]

A lingering loss was the vander Bruggen family's carefully bred and nurtured clumps of birds of paradise, which were sold off to landscapers and nurseries. This was also regretted by the likes of bird grower Mike Mellano jr. He, like the vander Bruggens, had inherited a love for and flair for working with birds of paradise from his own father:

> Unfortunately all that hard work is scattered all over the Los Angeles basin. It is too bad as there were several interesting selections that he had developed. We would have gladly maintained a selection had we known. Today we have only one or two of these varieties on our farm.[34]

Two significant bird-of-paradise growers, Johnny Roth and Mark Larson, have managed to survive the imported competitive onslaught in the San Diego area. Mike Mellano jnr partly puts that down to size – they 'are larger than we are'. The fact that they 'focus solely on birds' may also have ensured intense concentration on strategies to combat competition. Meanwhile, at the Mellano family flower ranch, east of Oceanside in northern San Diego County's San Luis Rey valley, production of birds has become 'pretty static' in the face of imports. Yields also declined somewhat from about 2010 because the bird clumps were rather outgrown. The large clumps needed thinning out and resetting:

> We are currently evaluating the market to determine if we want to increase our plantings as we set about the task . . . [The] market seems to be a bit limited but [birds are] still a staple item. Most floral production these days though needs to be targeted at bouquet makers and supplying supermarkets and birds seem to have limited acceptance in that channel so far.[35]

Despite their caution, others still believe, 'If you want to make some easy money, grow the bird of paradise.' The efforts of Meyberg and vander Bruggen have had a lasting impact. To this day individual blooms of *Strelitzia reginae* are in demand in flower shops in Manhattan and Chicago, where they are priced at US$5 each upwards.[36]

## HOW SCIENCE BOOSTS STRELITZIA PRODUCTION

For centuries, top gardeners have relied on their own insights and instincts to help them grow better examples of their favourite blooms or bring to flower an exciting species that they have not grown before. Botanists and plant collectors have sometimes been able to offer advice and guidance based on where the plant was collected or the family to which it belongs. From the 20th century onwards, science's deepening understanding of plant genetics, physiology and reproduction was assisted by increasingly acute observations with electron microscopes and other ever more detailed diagnostic tools. Floricultural scientific research has both helped and hindered commercial South African strelitzia growing, developing techniques to make strelitzia production and sales more attractive around the world.

Micropropagation has made strelitzia growing more commercially viable in areas such as Campania, Italy. Controlling the banana borer (*Opogona sacchari*) became imperative as it infests a range of ornamental crops from strelitzia and heliconia to begonia and pineapple. This moth originated in humid tropical and subtropical regions of Africa, particularly West Africa and some African islands. It now also occurs on the Canary Islands and Madeira, in Central and South America, in Florida, USA, and in Europe. It is a threat to exporters, with exporting countries expected to guarantee their crops are banana-borer free. In the European Union, it is classified as a quarantine pest so if caterpillars or moths are found, plants are quarantined. In 2009, a sex pheromone trap was developed to detect banana-borer infestation.[37]

The opportunities created by new diagnostic tools accelerated the race to develop floricultural experiments in the nursery and research in the laboratories that would improve germination, growth and the quality of blooms. This has had a wide range of spinoffs for strelitzia growers, as shown by this selection:[38]

- *How strelitzia divide:* Propagating new plants is a vital element of most horticultural businesses, making it important for horticulturists and flower-farmers to know how a plant propagates itself and use the most appropriate method to divide the stem. In 1972, R.A. Dyer of South Africa's Botanical Research Institute exploded the general assumption about how strelitzia plants multiply. He showed that strelitzia multiply through a series of branching at the stem axis, known as dichotomous branching, not by developing suckers from the base of the plant, as had long been assumed.[39] Dyer's dual affiliation to the Department of Agricultural Technical Services indicated the anticipated commercial applications of his work.

Figures 8a and 8b compare flowers of *Strelitzia reginae* to other plants, *Annales des Sciences Naturelle: Botanique*, 1837

- **Controlling strelitzia growth with light:** Researchers monitored 30 *Strelitzia reginae* plants from two groups growing a few kilometres apart in Port Elizabeth in the Eastern Cape, South Africa. One group of plants was growing mainly in shade, particularly in winter, the other mainly in sun. The sunny group produced flowers all year round and produced more than twice as many flowers compared to leaves as the shaded group. Winter shading may be particularly damaging for flower development, the authors concluded, noting, 'virtually all the *Strelitzia reginae* plants growing in the wild are found on north-facing slopes'.[40]

- **Boosting strelitzia production:** For 12 months, researchers in Bangalore, India, monitored feeding 80, 100 and 120 per cent of the recommended level of fertiliser to strelitzia plants through an irrigation drip, along with eight to 12 litres of water per plant per day (fertigation). They found that best results were achieved from the highest fertigation combination of 120 per cent of the fertiliser dose with 12 litres a day of water for 12 months. This produced the maximum number of flowerheads early and the maximum number of florets each month, as well as more leaves and larger plants.[41]

- **Where does strelitzia's colour come from?:** Food technologists found orange sepals of Strelitzia reginae contained at least 19 known natural orange colourants (carotenoids) – and about 800 times the carotenoid content of white sepals. Greenish-yellow and white strelitzias sepals contained high percentages of 5,6 epoxides, which help the plant use light in its chemical processes, while yellow and orange sepals contained 5,8 epoxides instead. They suggested that hexagonal-pattern protein crystals could contribute to colour.[42]

- **Detecting strelitzia dormancy:** Strelitzia growers would prefer quick seed germination but researchers found it often took at least eight weeks for just 30 per cent of seeds of shrublike strelitzias to germinate. Results suggested the seeds needed at least 16 weeks' after-ripening – but at least half were innately dormant to a lesser or greater degree.[43]

- **Improving strelitzia germination:** To attempt to speed up germination, Strelitzia reginae seeds were treated with various combinations of sulphuric and gibberellic acids and ethrel, followed by time in an incubator at 25C or 30C. The best combination was five minutes of pre-treatment with sulphuric acid to penetrate the hard seedcoat, followed by gibberellic acid at 50ppm for 24 hours, plus 24 hours in an incubator at 25C.[44]

- **Defeating strelitzia dormancy:** A wide range of treatments to break seed dormancy was tested: 'Low temperatures had a decidedly harmful effect on both dry and imbibed seeds.' Germination was improved by using various chemical treatments such as thiourea, mercaptans, ascorbic acid and ethrel, as well as other methods such as incubation at increased oxygen levels and exposing the embryo within the seed.[45]

- **Looking for an in vitro answer:** Strelitzia growers had not been able to follow other ornamental plant growers into the modern floricultural world of micropropagating by controlled and speedy cloning because strelitzia cuttings oxidise, turning them brown. Citric acid at 200ppm for 24 hours was found the most effective antioxidant, backed up by 1 per cent activated charcoal in the growing medium.[46]

- **Controlling strelitzia pollination:** The sticky substance that typically oozes from strelitzia flowers (exudate) was thought to be sugary to attract some bird and insect pollinators to the bloom – but this assumption of cause and effect has been dismissed as 'a myth of modern botany'. The exudate also plays a role in germination as it contains fats (lipids) that control how much water is available to pollen and a 'surprising' amount of free fatty acids (phenolic) that may affect the speed of germination through their impact on enzymes and metabolic processes.[47]

- **Understanding pollen-grain development:** Despite its robust flowers and seeds, *Strelitzia reginae*'s pollen grains are delicate. These need to be protected and transported for germination by developing speedily into pollen tubes, secured and possibly fuelled by the flower's sticky secretion.[48]

- **The strelitzia pollination syndrome:** In a 'highly specialised pollination syndrome', strelitzia pollen is controlled along its pollination journey by threads formed by cells on the flower's anther. The threads are comparatively 'long and robust', often gathering the pollen together in a mass[49] – and making pollination within a clump of shrubby strelitzia all the more likely.

- **Discovering the strelitzia seed factory:** Splitting microspores creates generative cells and an apparently short-lived vegetative nucleus in the pollen grain. When mature, the pollen grain has a thick outer coating (exine) and a thin inner layer (intine).[50]

# HOME-GROWN HOPES
## *South Africa's strelitzia growers*

*CHAPTER ELEVEN*

# HOME-GROWN HOPES: *South Africa's strelitzia growers*

Since Joseph Banks's splashy announcement of *Strelitzia reginae* in 1790, the revelation of each new relative has been celebrated, particularly in Europe and in North America. European visitors such as natural-history collector William Burchell (see chapter 9) remarked on these striking plants in their travel books. Strelitzias were exotic drawcards in European botanical magazines. Back in southern Africa, though, barriers of local tastes and geography kept strelitzias on the fringes of their home region's botanical and horticultural world for a couple of centuries before their iconic status as charismatic megaflora became established in the late 20th century.

Strelitzias are both more elusive in the wild than most staples of the Cape wildflower trade and more challenging to cultivate. These factors contributed to proteas dominating the historically small but persistent local market for indigenous South African flowers. Yet despite this, strelitzias have gained a foothold in South Africa's national and international flower trade. They are as popular massed in public and corporate landscaping as displayed in dramatic singularity, just one bloom or even just one leaf to a vase. At first, strelitzia were excluded from South Africa's flower markets by geography, costs and logistics (see chapter 9). They then faced very similar problems again in trying to capture a small slice of the world flower market — especially as California's flower growers already had almost a century's start.

When flower farming became California's new gold rush, strelitizias were key exotics, complementing the fields of roses and geraniums, popular as cut flowers and often familiar to immigrants (see chapter 10). By contrast, for many decades in the strelitzias' home country of South Africa, the flower trade was low key and small scale, supplied by wildflowers harvested direct from the veld. Even in the Western Cape, where flower farmers were close to major markets, they were elbowed out of much of the land by wheat and wine farmers so larger-scale flower farming was a much later development.

## CAPE TOWN'S PRIDE

Cape Town's trade in wild and, later, cultivated flowers may have originated early in the 1800s as a way for slaves to earn money but it was the mid-1880s before flower sellers hawking local indigenous flowers became a common sight. These were particularly arum lilies, heaths and proteas simply gathered from the veld on the outskirts of town. By 1886, a local newspaper, the *Cape Argus*, was pushing for a flower market in Cape Town itself. Local businessman William Thorne (1839-1917), a partner with Samson R. Stuttaford in a thriving, new-style department store, noticed the increasing popularity of local flowers and arranged for a local flower-seller to supply the store with two bunches of wild flowers a week. Stuttaford's 20-year-old son joined Thorne in the enterprise and the clientele gradually expanded, with the flower-seller being allocated a sales pitch outside the store. Thorne later served as mayor of Cape Town from 1901 to 1904, before becoming a member of the Legislative Assembly, and was renowned for doing 'much to beautify gardens and open spaces in the Cape Peninsula'.[1]

By 1890, Cape Town's various informal flower-selling initiatives coalesced into the country's first flower market. Flower-selling became more formalised and commercial as the city's trade and institutions consolidated. Birds of paradise, however, one of South Africa's most feted flowers internationally in both the late 18th and the 19th centuries, did not feature in Stuttaford's store venture nor in what became the Adderley Street Flower Market a century later. They were simply not generally available in the city for sale. Their natural habitat was far east of the Cape Town area and they were not yet a major feature of Western Cape gardens except as a proud curiosity in the gardens of the wealthy, as Burchell had noticed (see also chapter 9).[2]

Geopolitical factors can contribute to why some indigenous flowers become fashionable, while other equally striking or interesting species do not. Trains carried Cape flowers up to the Highveld for sale in Johannesburg via the new railway line which had reached the City of Gold from the Cape in 1892. Patches of birds of paradise, though, were distributed sparsely and found inconveniently far away in the eastern reaches of the Cape Colony, about 500 kilometres across challenging terrain from the colonial centre of Cape Town. The prospect of transporting birds of paradise from the Eastern Cape was a challenge that few, if any, would then have believed worthwhile undertaking.

Access to wild-growing strelitzia had been a particular challenge even for professional botanists or enthusiastic amateurs. The habitats where both *Strelitzia reginae*, the bird of paradise, and *Strelitzia juncea*, the rush-leaved strelitzia, flourished were in the area over which nine frontier wars were fought between 1779 and 1879.[3] Even in intervals of peace, though, the area was hardly accessible with the existing routes crossing formidable obstacles through the mountains. When the Swede Anders Sparrman travelled in the Cape in the 1770s (see also chapter 1), he noted: 'It took a hundred hours of heavy driving in a heavy wagon over bad roads to bring timber from Mossel Bay to Cape Town . . . Mountains literally blocked up a large portion of the interior.' He was referring to the Cape fold mountains, which sprawled eastwards across the Cape and were notoriously difficult to navigate while keeping a wagon in one piece. Several dangerous

Overleaf: George French Angas, *Near Wynberg, Cape,* 1849

river gorges towards the east of the Cape region made transport even more hazardous – and expensive. This, combined with the Dutch East India Company's tight control of trade until the British occupation in 1795 meant that it was worth transporting only higher-value products – butter, soap and timber – to the Cape Town market. Even staples such as grain 'could not be marketed at a profit if they were produced beyond a radius of about 100 kilometres from the Cape.'[4] The British administration and immigrant entrepreneurs tried to encourage coastal shipping trade as a cheaper alternative but even after ports were established at Mossel Bay, Knysna and Plettenberg Bay, these vessels carried mainly grain, timber and troops. Shipping comparatively delicate bird of paradise blooms along the notoriously stormy Cape south-east coast was not a priority.

Roads were built enthusiastically in the region from 1806 onwards. In the Western Cape, they encouraged trade. They also provided easier access to the Eastern Cape for British military responses during the frontier wars. Great new passes were at last engineered but at first, these still linked only Cape Town's hinterland with the metropolis. It was Sir Lowry's Pass (1829) that first conquered the Hottentots Holland mountains and opened up the Overberg route. The Bainskloof Pass (1853) and the Hex River Pass (1860s) gave the Cape peninsula easier access to the Boland, while Van Ryneveld's Pass (1832), Michell's Pass (1851) and the Katberg Pass (1860) gave Graaff-Reinet and Grahamstown swifter access to their hinterland. The breakthrough for the Eastern Cape was the Montagu Pass (1847). By 1849, the year Bolus arrived in Grahamstown, this pass had more than halved the time spent on the dreaded journey of nearly 1 000 kilometres from Port Elizabeth to Cape Town into a more acceptable trip of three days by post cart.[5] But perishables such as flowers were not immediate priorities for attempts to sell Eastern Cape goods in Cape Town. Nor would the railway help.

Railways were an urban fashion from 1860 onwards, first in Durban, then in Cape Town. Later, the gold and diamond rushes made a long-distance line into the interior strategically necessary. This line reached Kimberley in 1885, stretched on to Bloemfontein and finally arrived in Johannesburg in 1892. Also in 1892, a line from Port Elizabeth met the Cape Town-Johannesburg line in the interior at De Aar.[6] But this would mean the route to transport birds of paradise between Port Elizabeth and Cape Town would have followed two sides of a triangle, an unheavenly transit for the flowers. Even today the only direct rail link from Port Elizabeth is to Johannesburg. Essentially, however the strelitzia travelled, what little profit might have been made would be spent many times over on transporting the flower crop.

## FLOWER FEVER

In Cape Town, locally harvested flowers were sold by hawkers in the streets and, as in Europe, these flower-sellers were often depicted in romanticised and picturesque images. Over time, the Adderley Street Flower Market (now relocated to Trafalgar Place adjoining Adderley Street) became an internationally known tourist attraction, even though tourists tend to pass through, taking photographs rather than buying flowers. This colourful, created image eclipsed the reality of flower-sellers as shrewd small-time traders, often living hand to mouth.[7] Their livelihoods also became increasingly threatened by impassioned pleas from leading botanists against stripping out indigenous flowers, especially for sale. Pride in indigenous flowers ultimately led to control over what had been a pleasure for all – and with the new regulations including more remote plants such as strelitzias in their net.

The first permits for picking flowers on Table Mountain, issued by the Forestry Department, were introduced in 1893. After indignant public and parliamentary debate, in 1897 flower-picking was banned on Table Mountain altogether. Unfortunately, this simply shifted pressure to more outlying regions as railways expanded and motor transport later became available. In Cape Town's hinterland of the Boland, Caledon developed a thriving international flower trade. Sacks of everlastings (*Helichrysum* species) were gathered in the mountains and exported mostly to Germany and Japan, earning the town up to £15 000 a year. By

Cape fold mountains near Sir Lowry's Pass

contrast to more perishable blooms such as strelitzia, everlastings could be easily dried, which also reduced their shipping weight from about 50 pounds (about 22 kilograms) per collected bag to about 15 pounds (about six kilograms). It has been calculated that 'Some 70 tonnes of fresh flowers and 1,000 tonnes of everlastings were exported overseas during the 1890s', sent north and east by steamship.[8]

Leading figures such as Lincolnshire-born and Cambridge-educated English immigrant Henry Harold Welch Pearson (1870–1916) saw the mission to protect Cape wildflowers much more broadly. He used his speech at the founding meeting of the South African National Society in 1905 as a rallying cry for wildflower conservation. This prompted the formation of a committee where the attorney general led botanists in drafting the 1905 Wild Flowers Protection Bill, which introduced a £3 licence fee for flower-sellers and an approved list of wild flowers for trading. A 1917 amendment tightening regulations further made an explicit appeal to national pride 'to promote . . . the salvation of the beauties of this country.' Even so, confusion reigned over which species were protected and which not, what constituted a legal sale of wildflowers and what made a sale illegal. This loophole of ignorance meant that in reality, the regulations were observed in the breach in the Western Cape. The Eastern Cape's poor transport infrastructure protected strelitzias from the ravages of this commerce but Cape Town's main flower-market days were supplied on Wednesdays and Saturdays for many more decades by 'Lorries laden with stacks of heath [that] came down Sir Lowry's Pass from the Caledon, Elgin and Hermanus districts'. In the 1920s, for example, flowersellers in the Western Cape were policed quite gently by Inspector Farquharson, accepting his suggestion of a five-shilling admission of guilt fine 'philosophically'.[9]

Although strelitizias would become one of the 20th century's most iconic vase blooms, ironically, not being Western Cape flowers, they did not feature in South Africa's early wildflower shows that sprang up in this region. As early as 1887, the Tulbagh wildflower show was so well received that the following year two special trains transported visitors north from Cape Town. One carried a Government House delegation, the other visitors from the general public. To the east of Cape Town, Caledon first held a wildflower show in 1892. Darling's first wildflower show in 1917 was a community initiative launched by the local predikant's wife, Mrs D.J. Malan.[10]

## CONSPICUOUS BY THEIR ABSENCE

Elsewhere, early wildflower shows were often linked with schools and teachers. Margaret Raubenheimer, for instance, had learned about wild flowers in the George area on the way back from the Dutch Reformed Church Nagmaal gatherings and had been taught by her mother to arrange them. Raubenheimer enlisted the school principal's help to launch a wildflower show in Piketberg in 1917. The rarities in Raubenheimer's own displays attracted the interest of renowned German-born, Cape-based botanist Dr Rudolf Marloth (1855-1931), who enlisted her as one of his contributing collectors. In Cape Town itself, wildflower shows succeeded for a while in being doubly desirable – in themselves and, through their sales, as a charity effort to raise funds for troops fighting in the World War I. When the effects of the American and European financial crash of the 1930s echoed round the world, support for entertainments such as wildflower shows faded in the Cape.[11]

At the Cape Town flower market, though, many of the family flower stallholders continued to grow their own violets, marigolds and snapdragons for sale until at least the mid-20th century when forced removals under the 1950 Group Areas Act disrupted their lives and businesses. Tastes in flowers also changed, with sunflowers, irises and freesias becoming more popular. Local blooms lost favour until, by the early 21st century, only a few Cape Town flower stalls still had indigenous wildflowers for sale. From the period after the World War II to the early 21st century, the number of licensed flower sellers at the Adderley Street Market dropped by about two-thirds to probably fewer than 20.[12]

In the meantime, pride in Cape flowers extended further, including sporting emblems. One of the earliest was the popular Pride of Table Mountain, the crimson-pink indigenous orchid, *Disa uniflora*. Embodying a strong sense of the Cape's burgeoning pride in its floral and other natural riches, the flower was adopted as the emblem of both the Mountain Club of South Africa, founded in 1891, and the Western Province Rugby Union, founded in 1883. It was retained as the provincial sporting symbol after the new provincial constitution was signed into law in 1998.[13]

Patricia McCracken

Flower picking was banned on Table Mountain in 1897

## THE DARK SIDE

Cape Flora SA focuses mainly on veld collecting rather than on flower growers or sellers. It also uses the legal definition of fynbos and its 2013 survey found that the Western Cape's areas of highest veld harvesting of wildflowers were Riversdale on the Agulhas coastal plain, followed by Gansbaai and Napier, both in the Overberg area. *Strelitzia reginae* and particularly *Strelitzia juncea* are sparsely scattered through the veld in their natural strongholds in South Africa's more eastern areas, effectively making it harder to wild harvest a profitable number of stems. Only the more widespread and prolific *Strelitzia nicolai* have been thought to be a target and then usually for its seeds.

Cape Flora SA has expressed particular concern over the ethics of veld collectors broadcast-sowing wildflower seed in the veld because this risks changing the natural pattern of plant diversity. The 2013 survey found that a fifth of the wildflower harvesting area around Bredasdorp in the southern Overberg area was broadcast sown.[14] Bird of paradise plants, however, are so challenging to germinate that the risk of anyone attempting this is also rare. Instead, there is a greater risk of wild-growing plants being dug out for gardens and formal landscapes. Removal of *Strelitzia nicolai* was witnessed in July 2014 on the slopes of Inanda mountain, KwaZulu-Natal, by Patricia McCracken and Gerry Bruton, for example. This is the dark side of strelitzias typically being supplied to growers as young, established plants or as divisions of older plants. This is usually explained as ensuring flower cultivars grow true to the original but it clearly speeds up results for those who do not have the patience to wait for slower-growing strelitzias.

## A NEW NICHE FOR THE BIRD OF PARADISE

Climate combined with established farming practices meant that bird of paradise flowers surprisingly found a commercial farming niche in the north-east of South Africa, particularly the Mpumalanga Lowveld, beyond the flowers' natural range. However, the area is closest to the major Johannesburg markets and transport hub and so suited climatically and commercially to farming *Strelitzia reginae*. Export and marketing costs are usually 'the single most expensive' element of being a successful cut-flower exporter, often accounting for more than half of all flower production costs in South Africa. Mpumalanga is close enough to national markets and transport links in and around Johannesburg to help contain freight costs as much as possible. It is also relatively closer to direct flights from O.R. Tambo airport outside Johannesburg to international markets than the Northern Province and KwaZulu-Natal, where few flower farmers operate. In interviews by Himansu Baijnath for a survey of methods and challenges in South Africa today, where most participants preferred to remain anonymous, Farmer A from Mpumalanga remarked, 'Strelitzia production is reasonably popular but not really commercialised in South Africa.'[15]

In South Africa's north-eastern areas, strelitzias can grow well in the open without needing to be cossetted in costly shade structures, tunnels or greenhouses. Larger-scale farmers generally grow strelitzias as a supplementary crop, although they may sometimes be a single crop for a small-scale farmer. In an area of about five hectares, Farmer A grows just over 40 000 plants with 95 per cent of them standard orange *Strelitzia reginae*, which were planted in 2000. The remainder are the yellow Mandela's Gold cultivar, with the first specimens planted about 2008 but the bulk in 2012. Increasing plant holdings and acreage would not be commercially viable, believes Farmer A, although expanding in a small way into Kirstenbosch's newer Centenary Gold variety of *Strelitzia juncea* is a possibility. Farmer A commented:

Growing strelitzias is more a sideline type of hobby – but

*Strelitzia nicolai* poaching, Inanda Mountain outside Durban, 2014

there is very little room for higher production because the market could not cope with greater supply, particularly of *Strelitzia reginae*, as it is already saturated at peak production season.[16]

Both practical and aesthetic reasons contribute to the demand for strelitzias in South Africa:

Strelitzia are popular for display in vases at home, though, because they appear joyous and are long-lasting as cut flowers – especially if the airlock created in the stem by harvesting is cut off by removing the last centimetre of the stem before arranging the flowers in a vase.[17]

In the past decade or so, however, like many other South African flower-farmers, strelitzia growers have found it increasingly difficult to export effectively. Although the exchange rate of the South African currency has fallen sharply against major currencies, this has still not helped growers outcompete their rivals closer to the key international markets, either on production costs or particularly on flower-delivery costs. This has made strelitzias less appealing to flower farmers. No strelitzia farmers were members of the South African Flower Growers Association in 2016, for instance, probably because low levels of production did not make this worthwhile. Despite this, the association, founded in 1959, still retains the strelitzia as its emblem and presents an annual Strelitzia Award for excellence in the flower industry.[18]

## HOME DISADVANTAGE

As cut flowers, strelitzias are dramatic and tend to demand the spotlight. Ironically, these are key reasons why although strelitzias rank among the top five most beautiful flowers in the world, they are not found among the world's top 20 commercial flowers, which account for more than 90 per cent of the total value of flowers sold globally. The top 20 flowers include the rose and the daffodil and are generally more versatile, able either to carry the centrepiece of an arrangement or to act as supporting flowers. Top 20 blooms are sold in such high volumes that they are known in the flower trade as commodity flowers. The demand for one commodity flower alone, the carnation, is about 1 billion plants a year. The narcissus and strelitzia, on the other hand, are among blooms classified as niche flowers.[19]

Even in strelitzias' country of origin, where they have the advantage of being relatively easily cultivated, the bird of paradise remains a niche flower, its sales lagging behind the indigenous gerbera and freesias. Like other flower-producing countries from Malaysia to Colombia, the staples of South Africa's flower trade have become international flowers that are widely in demand across the world, such as roses (30 per cent of South African flower production), chrysanthemum (25 per cent) and carnations (13 per cent). Chrysanthemum are South Africa's best export

opportunity because they cannot be grown successfully in its floristic rival Kenya. It is the fourth largest cut-flower exporter globally compared to South Africa in 21st place. Proteas continue to dominate South Africa's indigenous flower exports. All South African flower growers, however, have been particularly hard hit over the generations by increasing costs of long-haul transporting of blooms and the influx of competition in international markets.[20]

Both in South Africa and internationally, strelitzias became more widely popular in the 1950s as gardeners wanted to create an exotically lush, tropical look. The commercial prospects for strelitzia were helped by improvements in infrastructure and technology over the following decade or so, with both transport links and air-freight logistics developing dramatically. This made it more feasible to ship the more delicate strelitzia blooms across

Eileen Bass, *Strelitzia reginae*, watercolour, 2017

the world, particularly as new cold-chain storage techniques helped prolong the flowers' life.

Strelitzias began slowly playing a supporting role in South African flower exports only from the 1960s although, to this day, proteas remain the national flower so still tend to embody the image of South African indigenous flowers. Proteas, with their hardy stems and closely packed petals, travel well and led South African flower exports from the start – almost as if predicted by that first protea woodcut by Clusius in 1605 depicting the so-called 'Madagascan thistle' (see chapter 1). Proteas also led the development of South African floriculture, with A.C. Buller founding the country's first large-scale growing of protea in Stellenbosch's Banghoek Valley in 1910. A decade later, in 1920, Kate Stanford founded the first protea nursery in the same area. Developments in South Africa's flower-growing industry, or floriculture, tend to go hand in hand with war, peace and changes in government. Dutch immigrants to South Africa were particularly keen founders of floriculture projects during the 1920s and 1930s. By 1945, the industry was large enough to support a central flower auction, Multiflora, which today distributes more than half of South Africa's flower production. About 600 flower growers now supply Multiflora's approximately 400 buyers, creating a turnover of about R300 million a year.[21]

Rising affluence in Europe and Japan from the 1960s onwards saw consumers demanding more, and more exotic, cut flowers. As the formal, structural beauty of strelitzias became internationally admired, the flowers became at least as well known globally as proteas. When the international cut-flower trade boomed from the 1960s onwards, prospects seemed promising enough for the South African flower industry to begin considering a wider range of flower crops for mass cultivation. This was the moment when South African flower growers could attempt marketing and distributing internationally – and profitably. However, the fact that strelitzias usually take two to three years after planting or transplanting to settle and then flower contributes to them remaining so far niche flower products. This, though, is just one reason why strelitzia growing remains an interesting curiosity even in the flower's home country. A series of geopolitical events also prevented South African growers from establishing a strong foothold in export markets so far.[22]

## SHADOWS GATHER

South African growers might have hoped or even expected to benefit along with other countries when the flower trade expanded globally in the 1970s. This was largely prompted by lower-cost flower producers in southern Europe expanding to supply north European markets through Dutch flower auction houses, particularly Aalsmeer, about 15 kilometres south-west of Amsterdam, and Westland near Rotterdam, in the southern Netherlands. Their flowers could be bought and sent on to other European countries or even bought back by dealers and retailers in their countries of origin. They might also be airfreighted into the American market via New York. Any such opportunity for South Africa to expand into the international flower trade proved relatively short lived, though. International flower trading shrank during the global energy crisis of 1973 and 1974 when oil prices doubled to US$4 a barrel and transport costs rocketed. Oil prices continued increasing dramatically. By the 1979 crisis, oil had once again more than doubled in price over that year – but from US$50 to US$101 a barrel.[23]

Lower production costs for South African growers might have helped offset these higher transport costs but they were outmanoeuvred by other international competitors. In contrast to north European growers, who had to cover rising costs of maintaining adequate temperatures in greenhouses and polytunnels, the climate buffered southern European producers, allowing them to offer competively priced blooms. Israeli growers also entered the market, boosted both by government-subsidised transport for export and by a climate which allows them to grow flowers outside all year round.

South African growers were hit by more than the much higher cost of long-haul transport, though. Stricter anti-pest regulations also threatened to curb their exports, particularly to the USA and Japan. The South African floriculture industry might have attempted to overcome these challenges but it was overshadowed by the increasingly widespread international politico-economic boycott of apartheid South Africa in the late 1970s and 1980s. Although strelitzias did not, like proteas, carry the potentially greater stigma of being the national flower of a pariah country, they also suffered with other South African produce as sanctions gradually closed the country's export markets. As a result, South African flower farmers had to concentrate on supplying local markets.

## NEW GROWTH

In 1995, the year after the country's first democratic election, South Africa's flower and horticultural exports were worth just R95 million, contributing less than 0.2 per cent to the year's national gross domestic product of R548.5 billion. By 2008, though, when the country's flower and horticultural exports were welcome back in international markets and the post-apartheid economic resurgence had taken effect, exports had grown more than five times over to more than R530 million. This simply maintained the proportion of their contribution to South Africa's larger gross domestic product of R2 255.8 billion. South Africa remained a small part of the value of the global trade in floricultural products, which amounted to US$14 billion that same year.[24]

Alida Alberts/Angyati Farms

five stems a plant. Just over three-quarters of this production was marketed through Multiflora in Johannesburg and Cape Town. The Johannesburg Multiflora complex in City Deep, operates seven days a week, except for religious holidays, auctioning flowers from 7am, Monday to Friday, to wholesalers and large-scale agents. This farmer has seen auction prices vary from 50 cents a stem to R7, depending on market forces, including supply and demand of both strelitzias and other flowers. Farmer A's remaining strelitzia blooms are sold at a fixed price to other smaller outlets.[26]

Farmer A also sold nearly 15 000 strelitzia leaves through Multiflora in 2012. Leaf material is sometimes harvested to order for clients, too, but this farmer does not generally encourage this as harvesting both blooms and leaves becomes much more time consuming overall. Although dried leaves can usefully supplement income for some flower farms, Farmer A prefers not to add this product.[27]

After fresh pressures since the 2008 financial crisis, particularly due to strong competition from South American growers, South African flower growers today find themselves taking careful decisions on product lines. South African growers have attempted to capitalise on legacy links, such as with Britain and the Netherlands, as a bridgehead into European markets. South Africa still holds 1 per cent of the British flower market, supplying proteas, roses and other exotics as well as strelitzias. Yet as much as 80 per cent of South African cut-flower exports are destined for western Europe, where prices remain poor after Colombia rapidly captured about 6 per cent of the European flower trade. To offset the costs of supplying this long-haul market, Colombian growers maximised their returns by concentrating on high-demand, high-volume carnations and roses. Ecuador, which also supplies the European market, follows this strategy too. South American supplies particularly continue to force down prices. At best, flower prices have been 'stagnant for years' but even this means they have, in fact, been 'declining in real terms'. In the first 16 weeks of 2014, for example, high volumes of strelitzia were sold at Flora Holland when counted by stem – but at lower prices. Proportions of exports reflect how much South African flower growers have concentrated on legacy European markets and made little headway into newer markets. Just 12 per cent of South African flower production goes to the USA, for example, and 8 per cent to Japan, with Dubai a newer but growing market for South African cut flowers.[28]

South Africa's local cut-flower market is considered relatively large in world terms, with the domestic market absorbing about half of all locally produced blooms. In 2008, flowers worth R418 million were grown and sold within South Africa. Most – R314 million worth – were sold through local markets and R105 million were sold directly to consumers. This places South Africa midway between countries such as Australia, China and Japan, where flower production is almost completely for local markets and other more recently established producers, such as Israel, Colombia and Kenya, where flower production is almost totally exported. Despite these substantial overall flower-production figures, South African strelitzia production remains generally relatively small scale, in keeping with indigenous flower farming being often another sideline for South African farmers. In the Western Cape, for example, fynbos can be a secondary crop for farmers who specialise in potatoes, citrus or other products.[25]

In Mpumalanga, Farmer A's existing 40 000 or so plants produced more than 200 000 stems in 2012, an average of about

## GLOBAL EVENTS, LOCAL IMPACT

The range and quantity of cut flowers available on international markets may also be affected by distant factors such as currency fluctuations in importing countries. When the British pound is weak, for example, wholesalers and supermarkets import fewer cut flowers to the United Kingdom. The fallout of the 2008 international financial crisis

South African strelitzia-growing often supplements other crops such as macadamia nuts (in background)

was an extreme example of this. Demand fell in the global north for exotic luxuries, including imported strelitzia blooms, while export costs for growers increased.[29]

As the European and North American markets were still struggling to recover from the financial crisis, producers were hit by a further, unexpected blow. Air freighting flowers had become the norm so the 2010 eruption of the Icelandic volcano Eyjafjallajökull had a global impact on the floriculture industry. In Europe, shortly before the normally lucrative Mother's Day season, more than 60 000 flights were grounded, which in turn affected many thousands more flights across the globe. The South African flower industry was among those suffering severely as a result, with its flower exports to Europe grounded. Protea industry historian Maryke Middelman noted: 'Huge losses were reported with supermarkets in Europe having to cancel their orders. It was estimated that the crisis cost the Kenya flower farmers around US$15.5 million per day.'[30]

## IN SOUTH AFRICA'S STRELITZIA FIELDS

South African flower farms are generally small, averaging about 4.5 hectares. That gives on average employment to about 16 full-time and three part-time labourers.[31] Strelitzia production occupies 10 staff on Farmer A's land in Mpumalanga:

> Both men and women are included in our team. They have all mastered the growing and packing techniques, from seedlings and transplanting to fertilising, watering and harvesting.[32]

The size of strelitzia plants means artificial growing mediums are not feasible, although preparing soil and caring for the plants must be included in labour costs. In Mpumalanga, summer temperatures can reach 38 degrees Celsius and average 26 degrees Celsius so regular watering is essential to prevent heat killing flowers, using micro-irrigation (seven metres by three):[33]

> At high temperatures flowers respire much faster than at lower temperatures. Respiration and transpiration lead to a significant loss of water. Temperature is a critical factor in the whole [cut-flower] quality chain. For every 10-degree temperature decrease, the speed of the physiological processes [of blooming or decay] in plants is on average reduced by two to three times.[34]

Harvesting is labour-intensive work. Volumes harvested depend on the point in the flower production cycle as well as the season's production. Harvest yields rise from February, with Easter providing a seasonal spike in demand. Peak harvest is timed for July and August because demand is highest in the South African winter (May to August), although fewer blooms can then be produced in South Africa's colder areas. As in California (see chapter 10), the Christmas–New Year period remains a neglected trough with 'hardly any' blooms harvested in December and January:

> Unfortunately, there is no production over Christmas. It would be a good time to have flowers.[35]

Choosing the perfect moment to harvest is one of the challenges of flower farming:

> It is difficult to routinely harvest flowers at exactly the right stage of maturity ... stages of optimum harvesting ... vary among cultivars of the same flower type. Furthermore, cutting stages usually differ during winter and summer. The main aim is trying to ensure that the flowers arrive at the customer at just the right opening stage as demanded by them ...
>
> ... different countries or consumers often want their flowers at different opening stages. For example, the United States typically requires export flowers to be in partial bloom, while they need to be tightly closed buds for the Japanese market and wide open for the Russian market.[36]

Protecting strelitzia flowers in the field by covering them in wax bags might help prevent damage but no more. It does not slow down the flowers' opening and would add to costs. Bagging flowers would also slow down harvesting instead of allowing speedy work essential to transfer the cut blooms into the cool of the packhouse. Most markets expect the blooms to be sold as closed to help prevent damage in transit, especially as strelitzias

Alida Alberts/Angiyati Farms

Harvesting strelitzias: picking, transporting fresh to packshed, trimming and boxing (overleaf)

are easily persuaded to open by using a floral preservative in the water.[37] During mid-year peak production, Farmer A's staff harvest about 2 500 stems a day for four days a week. Research findings have recommended harvesting flowers in the late afternoon 'when the carbohydrate reserves are at their highest' but this is considered impractical on most farms.

## PICKING THE CROP

Many farmers prefer to pick early in the morning when flowers have the highest water reserves to protect them during the harvesting and packing process.[38] In Mpumalanga's subtropical summer heat, this certainly means starting to work early:

> Flowers should be harvested in the cool of the day . . . Early in the morning flowers are still . . . full of water and firm. During the heat of the day, cut flowers lose a lot of water through transpiration. Often more water is lost than can be taken up and thus temporary wilting takes place. Even if these flowers are picked and immediately placed in water, the results are not as good and quality is lost.

> The main cause of transpiration is the vapour pressure deficit . . . the combined effect of temperature and relative humidity. When [this] is high, the transpiration rate of flowers is also high. A low transpiration rate, which is ideal, can be obtained by a combination of low temperature and high humidity.[39]

Accessible flowers on the edge of the clump can be harvested easily by pulling on the stalk with a swift tug. It is often difficult to grip flower stalks inside the clump at the right angle to achieve this successfully so in South Africa they are usually cut using pruning shears. Harvesting a flower stalk creates an open wound of less than 10 millimetres on the main stem of the plant, small enough not to be a significant risk for introducing disease to the plants, especially as it is shielded by many layers of leaves at the base of the plant.[40]

Stems 60, 80 and 100 centimetres long are chosen for harvesting. Spindly stems are avoided:

> A healthy cut stem typically weighs more than a poor quality one . . . simply because the healthy stem will be thicker and have larger flowers . . . and more leaves . . . The stem of a flower must be able to support the flower head.[41]

Longer-stemmed blooms fetch a better price because they are more versatile in flower-arranging, noted Farmer A. When harvesting leaves, the staff are trained to choose broader ones with maximum leaf stalk: 'Apart from this, leaf sizes and shapes are not particularly selected as the consumer prefers different sizes.'[42]

## PROTECTING THE STRELITZIA INVESTMENT

Cut flowers are considered one of the internationally traded products most affected by any delay in the time taken from

Alida Alberts/Angiyati Fa

harvest to market. Delicate handling and secure packaging is essential for strelitzias to travel well to markets overseas and then on to consumers' homes, although this ultimately contributes to regularly rising costs. High transport costs must also be included, pulled upwards again as the oil price peaked once more on 4 July 2008, this time at US$145.28. In 2014, a South African flower industry grouping, the Fynbos Marketing Forum, suggested flagging this price element for buyers using a label indicating whether blooms had been flown in or shipped by sea.[43]

Cool rooms became a must for serious flower farmers from the 1990s onwards, with growers and shippers investing heavily to protect blooms in transit. By 2005, Californian researchers had pinpointed the ideal temperature needed within the cold chain after harvesting, as well as looking at ways in which to maintain the cold chain, for instance with cooled delivery vehicles – as far as the customers' door if possible.[44]

After harvesting, strelitzias should be stored in packhouses to be chilled to no higher than 10 degrees Celsius, ideally at 2 to 4 degrees Celsius. Either side of this range, the flowers are liable to suffer heat or chill injuries.[45]

Field heat needs to be removed from flowers as soon as possible after harvesting and a cold chain maintained... A primary factor determining the consumer quality of cut flowers is temperature management during handling and transportation... The main emphasis in maintaining the cold chain is to ensure flower quality, freshness and vase life...

Even in storage, transpiration continues and this can lead to unacceptable moisture loss that cannot be effectively replaced, leading to premature wilting and death of the flower. If the humidity is set too high (over 95 per cent) other problems arise as the condensation drips onto the flowers and also creates an ideal environment for the outbreak of diseases such as botrytis.[46]

For this reason, too, wet flowers must be allowed to dry out before being stored in a coolroom. Similarly, condensation from the coolroom ceiling should not be allowed to drip on cut flowers or leaves as this spoils them, also possibly encouraging mould to develop.[47]

Tools, buckets, clothing, work surfaces, cool-rooms, pack-houses and vehicles must all be regularly cleaned and disinfected to prevent any possible infection of flower crops. If microbes infest flower stems and block the vessels inside them, this prevents their water uptake. To keep cut flowers in the best possible condition, preservative, holding and hydrating solutions are often used to prevent this. Making sure that storage water remains clear is vital because cloudy water indicates the growth of harmful bacteria that shortens vase life. Fortunately, unlike many other flowers, strelitzias are not aged, shrunk or wilted by the gas ethylene, which occurs naturally and is also produced by many mechanical processes.[48]

## ON THE MARKET

Both at auction and in the wholesale trade, overall quality of strelitzia blooms is judged by their intense, sharp colour and uniform shape and size. Strelitzias are sold as 'standards', meaning a single flower on each stem. So in Farmer A's pack-house, when staff check picked strelitzia blooms to ensure that only top-quality heads have been harvested, they also remove double heads.[49] The blooms are then graded into different lengths and grouped five to a bunch of the same length:

Proper grading according to market standards and demands is vital... Flowers in a bunch should be uniform in all aspects. That is in their opening stage, length, colour and grade. Few things irritate a buyer or end-consumer more than a poorly graded bunch of flowers. They often regard this as poor quality or dishonesty on the part of the supplier. Presentation forms a vital part of perceived and real quality of flowers.[50]

The standard 1.2-metre box used for supplying to Multiflora easily accommodates stems from 60 to 100 centimetres long. Grower information such as cultivar, variety, grade, colour and length of the flowers supplied must be noted on the box endpieces. Farmer A also marks stalk length on the box lids. The weight of a box of blooms can vary significantly depending on whether the flowers are open or in bud: 100 closed buds can be

packed into a box, bringing the weight to 15.15 kilograms (including 85 grams for the box). When flowers have opened, only 60 flowers can be packed into the same box, reducing the weight by a third to 10.10 kilograms (including box). Most flowers can often survive for only three or four days in a cool-room before starting to lose quality so it is important to transport boxed flowers to market or direct to the buyer as quickly as possible:[51]

> There is no post-harvest procedure that can improve the quality of cut flowers or potted flowers. After production, quality may, at best, be maintained as the flowers move through the marketing chain to the flower vase. The quicker the flowers are sold and leave the florist shop the better.[52]

## LUSCIOUS LEAVES

In 2008, about half of world sales of floricultural products were made up of cut flowers and about a tenth of cut foliage. Internationally, the bird of paradise is a staple of the horticultural, the cut-flower and the foliage-plant industries. In total, the foliage plant industry supplies more than 100 genera and probably about 1 000 species in various shapes, colours, textures and styles. South Africa and Israel dominate this market, forcing growers elsewhere, such as in Italy, to experiment with a wider variety of niche plants, such as monstera and viburnum, if they want to compete effectively.[53]

At its 2002 peak, the total foliage industry was believed to have had a wholesale value of US$663 million. The industry's global operations include: 'rooted and unrooted cuttings, seedlings, and tissue cultured liners from the Caribbean, Central America, China, India, Korea, Thailand, the Netherlands and elsewhere are grown in the United States and sold nationally and internationally, primarily in Canada and Europe.'[54]

## INTERNATIONAL HURDLES

Long-standing issues of trade protection barriers, whether explicit in import quotas and levies or implicit in hygiene certificates, may delay blooms on the way to market. Before flowers grown in South Africa are exported, the exporter must also often pay for them to be fumigated so they comply with phytosanitary regulations in Europe and Japan, for example. The European Union includes various families of cut flowers as well as some vegetables and fruits imported from countries outside its union in its phytosanitary inspections. Extra requirements may be imposed by importing supermarket chains, usually through the Euro Retailer Produce Working Group Framework for Good Agricultural Practice, and the standards of the Federation of Dutch Flower Auctions. In practice, more than 70 per cent of the cut flowers imported from outside the European Union to Holland were inspected, although the Dutch phytopathological service charged an import tariff whether or not the flowers were in fact inspected. This creates 'a substantial extra burden on the importer', according to the World Bank, and compounding the disincentive to import from beyond the European Union.[55] Finally, importers themselves:

> have their own unwritten quality standards (with European consumers being particularly demanding), and might often be biased against products originating from developing countries, assuming inferior or unprofessional production processes.[56]

Elsewhere, particularly where demand for South African flowers is low, officials of an importing country may enforce non-tariff trade barriers rigorously: 'certain export markets (notably Japan) demand the complete absence of any living insect or mite in imported flowers.' In reality, that can mean protecting the importing country's own flower producers by banning (and often burning) a whole import consignment or fumigating it and sending the original exporter the bill. In future, compensating for carbon footprint may become an additional disincentive to exporting to Europe and the United States. This might give South Africa some competitive edge as it is somewhat closer to Europe, for example, than are China and Indonesia.[57] The introduction of online ordering has already helped cement imports from Ecuador to Russia, for example, and may in future change the way South African flower growers market their produce.

## REGULATING STRELITZIAS

Strelitzia marketing does not fall under newer South African organisations such as the Fynbos Marketing Forum that generally use the South African legal fynbos definition.[58] This harks back to the early 20th century when those drafting the country's laws on collecting and dealing in wild flowers focused primarily on proteas, heaths and watsonia. The 1997 Standards and Requirements regarding Control of the Export of Fresh Cut Flowers and Fresh Ornamental Foliage defined Cape flora, for example, as:

> the sexual reproductive and vegetative parts of all Ericaceae as well as Proteaceae and "fynbos" (of which some of the genera not endemic to South Africa including Telopea and Banksia consist of Bruniaceae, Leucadendron, Leucospermum, Mimetes, Protea, Serruria).[59]

To this day, although South African law does not explicitly recognise strelitzias as forming part of fynbos or Cape flora – despite strelitzia species occurring naturally in the broader fynbos region – although development of commercial strelitzia growing and marketing is controlled within the same legal framework as fynbos.

Mary Jones, *Strelitzia nicolai*, pen and ink, 2017

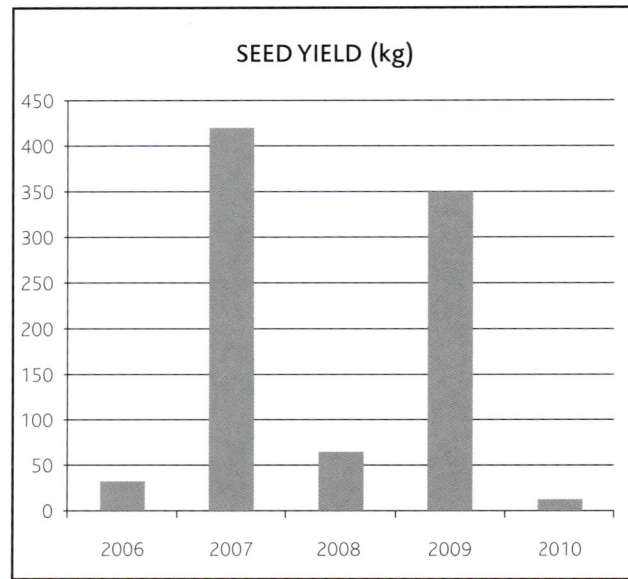

## THE SEED OF AN INDUSTRY

In recent years, a related market for strelitzia seeds has emerged. Although other strelitzia farmers find that selling seeds may maximise the return from strelitzia cultivation in months where there is a glut of blooms, Farmer A currently simply discards surplus blooms, not allowing them to go to seed. There is growing international demand for seeds as experiments in pharmaceutical and general use are developed commercially (see chapter 15). Pragmatically, properly gathered and stored seeds are also much easier and cheaper to transport than cut-flower blooms, making seeds a logistical and logical choice when the flower trade is squeezed.

As a result, some South African strelitzia growers have gradually swung from flower to seed production, supplying both pharmaceutical businesses and floriculture, usually through wholesalers and intermediaries. There is already some demand for seeds, with one report claiming that a single Western Cape supplier was dealing in orders of 100 kilograms of *Strelitzia nicolai* seed, although the source of such quantities of seed was not clear. Knowing such demand exists also encourages some South African strelitzia growers to attempt to supply seeds from their own fields but they often find they encounter three key problems – production, price and logistics. *Strelitzia reginae* seed yields seesawed dramatically for Farmer B over a recent five-year period.[60]

As well as weather affecting seed yields, raids by buck and monkeys are a problem. This prompted Farmer B to instal electric fences around the seed sites, which significantly improved production. Farmer C was horrified when a veld fire swept through his strelitzia plants – but was later impressed that they quickly recovered and none died. He noted, however, that flowering did not increase.[61]

The modest South African demand for seeds tends to mean that just one farmer producing seed at peak levels can saturate the market. Breaking into the export market is arduous, though, as it has tended to be dominated by three or four key seed merchants in South Africa.

The weakened rand has tilted the exchange rate in favour of overseas buyers but even some intermediaries seemed to be struggling, extending payments to suppliers beyond 60 days. Exporting direct is very difficult, something that Farmer B had 'virtually never' attempted, for example. As a result, farmers may find themselves trying to clear a glut by looking for less traditional outlets and offering much-reduced prices in an attempt to interest flower wholesalers in taking on strelitzia seed stock.[62]

Selling strelitzia seed has not cushioned the flower-pricing shock following the global financial crisis of 2008 and subsequent recession years in South Africa. Farmer B, who generally receives a price of between R350 and R450 per kilogram, depending on the seed quality and the amount of sorting done, commented:

> This recession has hit both demand and pricing hard with everyone complaining that they find it hard to make anything out of this seed. It means we are actually selling our seed for lower prices now than we were in 2005/6, but our higher volumes are helping balance this. Even so, one guy we deal with is always complaining about the price and assures us there are people selling in the Eastern Cape for R200 per kg.[63]

Sibonelo Chiliza, *Strelitzia nicolai*, graphite, 2017

## SAMPLE PRICES OF STRELITZIA SEEDS IN SOUTH AFRICA, 2010-2012

| SPECIES/VARIETY | DETAILS | PRICE[64] |
|---|---|---|
| *Strelitzia reginae* | Per kg | R770 |
| *S. reginae* 'Mandela's Gold' | Per kg | R450-R950 |
| *S. reginae* 'Mandela's Gold' | Per kg, excluding shipping & export phytosanitary permit | US$60 |
| *S. reginae* 'Mandela's Gold' | Per 1 000 seeds | US$135.50 |
| *Strelitzia juncea* | Per 1 000 seeds in South Africa | R1 000-R1 750 |
| *Strelitzia juncea* | Per 1 000 seeds for export, excluding shipping & export phytosanitary permit | US$145-210 |
| *Strelitzia nicolai* | Per kg | R300 |
| *Strelitzia nicolai* | Per kg for orders over 100kg | US$25 |

## NEW PROSPECTS

For more than a century, international development economists have dreamed of selecting certain indigenous flowers and transforming them into cash crops that could extend the range of existing agricultural products and generate income for small and medium-sized businesses in rural areas. Over the past half-century, this dream has become increasingly common in developing countries in Asia, Africa and South America. South African floricultural enthusiasts and microeconomic planners are no exception.[65]

Despite the challenges and frustrations faced by existing strelitzia farmers, this small industry has not been forgotten among the various economic policies developed by new South African governments since 1994 aimed at driving transformation and development across various industries. In the flower-producing industry, Timbali Technology Incubator was created in Nelspruit, Mpumalanga, for example, offering skills in growing a range of flowers including gerbera, celosia and strelitzia.[66]

## 'HOW WORMS COSSET OUR STRELITZIAS'

Growing strelitzias commercially means taking special care to produce the best possible blooms every time. At Angiyati Flowers in South Africa's Mpumalanga, Alida Alberts and her team moved to organic, biological growing methods in 2013, having previously used chemical fertiliser and pesticide sprays. Now they make their own vermicompost tea for overleaf irrigation, both for fertilising and spraying, particularly against the leaf disease botrytis.

They dilute vermicompost tea 1:1 and regularly pump it through the main irrigation system, using 250 litres to the hectare. While Alberts knows it is difficult to prove statistically that vermicompost tea has a direct effect on the tea, they have definitely observed healthier, stronger plants with a greater resistance to all predators, botrytis included. She notes that it is important to leave as much organic material as possible on the ground between the plants so that the micronutrients released in the tea can feed on it and transform it into plant nutrients.[67]

While harvesting is in progress, part of the workforce continues to care for the strelitzia plants in the Angiyati fields. The plant rows are crowded since most of the plants have been established there for nearly 15 years. This means production areas must be weeded by hand, pests controlled by regular, targeted herbicide spraying and sick or diseased plants treated or removed. 'Poor cultivation practices and disease are quick to show in the leaves of a plant. . . . The quality of leaves will suffer first at the expense of the actual flower.'[68]

Lack of certain nutrients quickly shows as yellowing along leaf veins, tips or edges. Excessive heat results in burnt tips, while pests and disease can cause leaves to curl, discolour or die. Angiyati has found proper cultivation practice essential. Regular irrigation, proper fertilisation, adequate light and pest management all help ensure that stems are strong, straight and of good length. Flower colour is often adversely affected by cultivation conditions such as excessive heat.[69]

A really thick clump could make some flowers difficult to reach without damaging others, so thinning and dividing strelitzia clumps is important so timing has to be well judged. Slumping production can be a clear indicator that this should happen. On the other hand, productivity can also be lost when dividing because most flowers are produced on the divisions at the edges of clumps. Finally, whether dividing and moving clumps for hundreds of kilometres or from one field to the next, they usually take two years or so to resettle and begin producing again.[70]

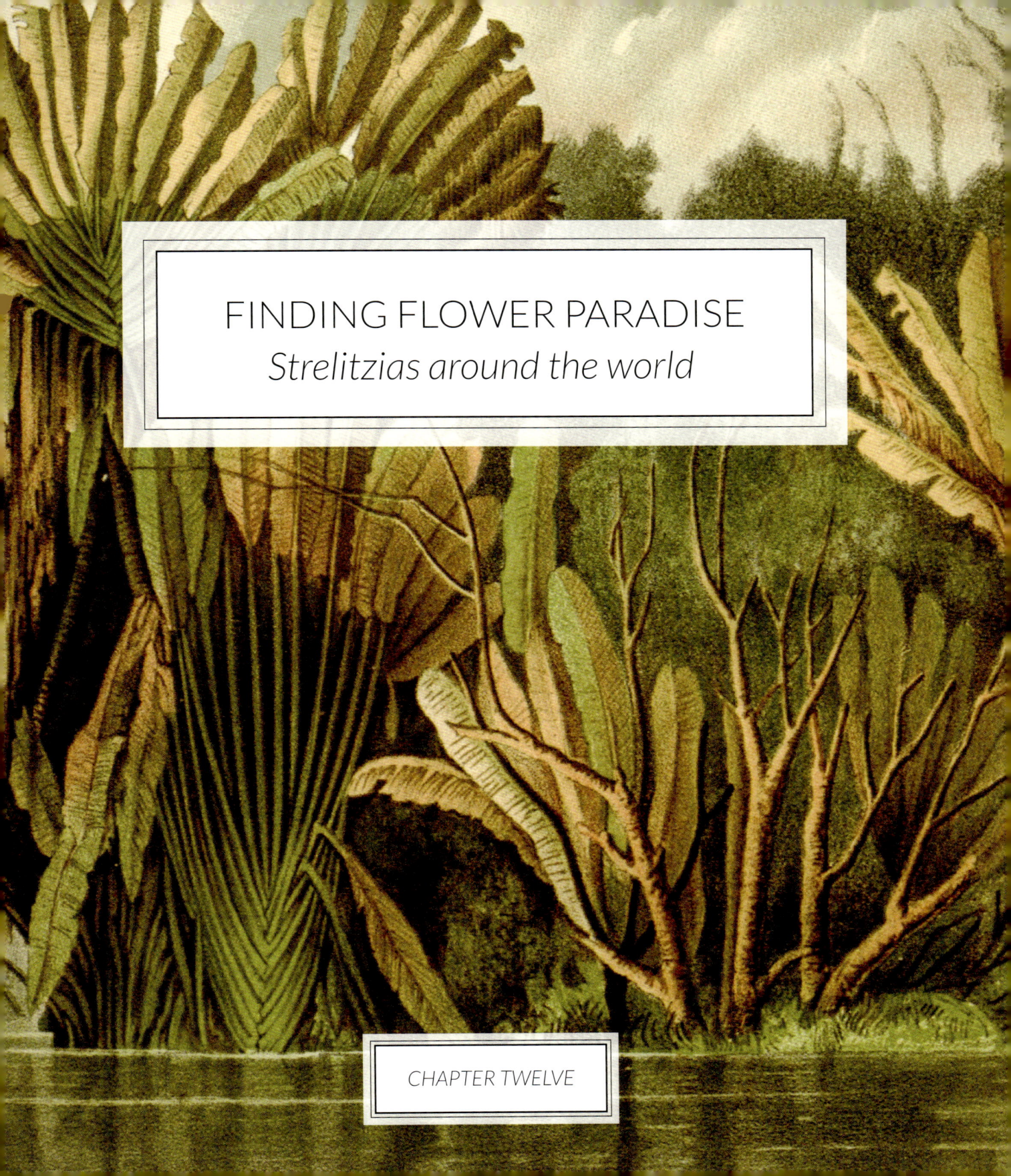

# FINDING FLOWER PARADISE
## *Strelitzias around the world*

CHAPTER TWELVE

# 12

# FINDING FLOWER PARADISE: *Strelitzias of the world*

There is plenty of scope among the strelitzia family for garden enthusiasts who like to grow and display rarities. They were usually launched to the broader public through specialist gardening magazines produced in cooperation with horticulturists who could supply specimens – often at a handsome price.

The earliest dispersal of *Ravenala madagascariensis* from its island home seems to have been when Pierre Poivre transplanted it to Mauritius around 1768 (see chapter 5). The alluring illustration from *L'Illustration Horticole* is a coloured version of the black-and-white illustration first published in the British *Gardeners' Chronicle*. It depicts a scene that is still familiar to present-day visitors to Pamplemousses Botanic Gardens (now Sir Seewoosagur Ramgoolam Botanic Gardens).

Generally, ravenala was a conservatory specimen in much of Europe and, at full growth, could be accommodated only in the more majestic conservatories. There was a ravenala at the imperial summer palace of Schönbrunn, Vienna (see chapter 5). Specimens of the tree travelled from Mauritius to the Jardin du Roi in Paris and in 1871, it was growing in the Jardin des Plantes, as this garden was renamed.

Ravenala was growing at the Royal Botanic Gardens, Kew, in 1858. According to *The Garden* magazine, it was flourishing in the Palm House there in 1874, adding it 'forms a noble ornament to the conservatory at Chatsworth', the duke of Devonshire's estate.[1]

Ravenala was also considered a drawcard in the second hothouse at Liège in Belgium:

*Cet arbre d'un aspect tout-à-fait insolite et bizarre vient d'une contrée où toute la nature, plantes et animaux, présente des caractères extraordinaires. Il croît dans les marécages de Madagascar, et comme tout ce que l'on a rapporté de cette grande île, il s'éloigne des formes que l'on est habitué à rencontrer.*[2]

[This tree looks extremely unusual and bizarre and it comes from a country where all of nature, plants and animals, has extraordinary characteristics. It grows in the swamps of Madagascar, and like everything that is brought back from this great island, it has distanced itself from the forms that we are used to encountering.]

Van Houtte in Ghent was particularly successful in raising ravenala and Leiden Botanic Gardens imported seed from Ceylon in 1859 in the hope of raising their own. Ravenala was considered one of the most interesting species on display in the hothouse at Grenoble Botanic Gardens in Switzerland in 1860 and featured in the central hothouse at Breslau Botanic Gardens in Germany at about the same time. It was listed as occurring in Spain in 1871.[3]

Patricia McCracken

Overleaf: Traveller's tree at Pamplemousses Botanic Gardens, Mauritius, in *Flore des Serres et des Jardins de l'Europe* magazine, 1875
(Left) *Ravenala madagascariensis* at National Botanic Gardens of Ireland, Glasnevin, Dublin, and (centre) at Royal Botanic Gardens, Kew, London; (right) *Strelitzia nicolai* at Schönbrunn Imperial Summer Palace, Vienna

## CULTIVATING COLLECTABLE STRELITZIAS

Ravenala are described as providing an overarching, tropical feel in the greenhouse of Renée Rougon-Macquart, a character in the 1871 novel *La Curée* by French writer Emile Zola (1840-1902). This mention is intended to convey both voluptuous horticultural decor and the power of wealth, as an admiring description of the tree published in Bonn made clear. Indeed, 25 years later, the talk of the European gardening world was that Miss Helen Gould had paid £7 000 (about £650 000 in current value) for a 10-metre-high *Ravenala madagascarensis*. Her purchase was first reported by *Modern Society* and then picked up by the British *Gardeners' Chronicle*, the Belgian *L'Illustration Horticole* – which called it a '*jolie somme* [pretty penny]' to spend on '*Un arbre de prix*', meaning both a pricy and a prized tree – and by *Teysmannia* in faraway Batavia, which remarked that this '*kostbare plant . . . is de hier wel bekende pisang kipas* [costly plant . . . is well known here as the leaning banana]'.[4] It is not surprising then that when Mr D.S. Brown presented a ravenala of similar size to the horticultural department in St Louis, Missouri, in June 1892 that the *American Florist* featured the event as a news item. By contrast, ravenala seedlings of about half a metre high were selling for just a dollar each in Florida in 1894, while in Trinidad, seedlings sold for 50 cents each in 1922. In Singapore, though, the tree was reported to 'grow luxuriantly' on a roadside, as *The Garden* magazine showed in its 1874 illustration.[5]

Roadside traveller's tree, Singapore, *The Garden* magazine, 1874

*Ravenala madagascariensis* had been an early horticultural immigrant to the South American region from its Madagascar home about 10 000 kilometres away, featuring in Rio de Janeiro botanic gardens by the 1870s. Ravenala also appeared as an ornamental in streets and gardens in Paraguay, Nicaragua, Mexico and Belize. In the Caribbean, it was '*très bien acclimatée*' in the French Antilles and also found in San José in Costa Rica, where 'a handsome specimen' had been planted in front of the Ministerio de Relaciones Exteriores (Ministry of Foreign Affairs). One was reported as fruiting in St Kitt's in the West Indies in 1905 and a specimen collected in Tobago in 1932 is part of London's Natural History Museum herbarium, for example.

## TRAVELLING WITH THE TRAVELLER'S TREE

The tree was generally well enough established in the region by the 1950s to give the title of the first and award-winning book by Patrick Leigh Fermor (1915-2011) on his travels in the Caribbean, *The Traveller's Tree*. In turn, this book was quoted frequently by Ian Fleming (1908-1964) in his second James Bond novel, *Live and let die*, published in 1954. In Africa, Belgian botanist Paul van Oye had found ravenala plentiful at Eala in the then Belgian Congo in the 1920s (see chapter 6). Nearly a century later, it was flourishing as an accent plant flanking the entrance to the Bujumbura International Airport terminal in Burundi, central Africa. Botanic gardens also played an important role in making unusual plants such as the strelitzia family better known.[6]

In the 19th century and most of the 20th century, botanic gardens were central points in their region where visitors could see rare plants from all over the world and often buy seeds or seedlings to experiment in their own gardens. British naturalist and painter Marianne North (1830-1890) depicted ravenala in the Rio de Janeiro Botanic Gardens in 1872 and again in 1876 beside the Sarawak River in Borneo. In 1892, North Borneo issued a two-cent stamp bearing the image of the ravenala, an adopted icon. Durban Botanic Gardens was growing the traveller's tree by 1883. A second *Ravenala madagascariensis* was planted in 1894, as was its close relative *Phenakospermum guyannense* (listed as *Ravenala guianensis*). The botanic gardens may well have supplied seedlings to create the ravenala avenue fanning round the amphitheatre at nearby Jameson Park. Durban Botanic Gardens also grew four southern African strelitzias: *Strelitzia reginae*, *Strelitzia juncea*, '*Strelitzia farinosa*' from the Cape and '*Strelitzia augusta*' from KwaZulu-Natal (in reality, *Strelitzia nicolai*).[7]

To many Victorian and Edwardian travellers and landscape gardeners, ravenala had the glamorous and exotic looks which immediately evoked the tropics – and it was a horticultural challenge. La Mortola on the French (now Italian) Riviera succeeded in growing it in the open, a major achievement as a ravenala tree in the Hamma *jardin d'essai* (experimental garden) in Algeria had been killed by the severe frosts of the winter of

1877/1878. These experiences make the success of London's Regent's Park in growing a ravenala as a showpiece all the more striking. In a general description of the park, a familiar portrait of ravenala is repeated: 'the Mound . . . accommodates a good specimen of the Traveller's Tree, *Ravenala madagascariensis*, whose leaves when pierced at the base are said to furnish a supply of clear liquid to the weary traveller.'[8]

Across the Atlantic in Toronto's Reservoir Park, 'the favourite picknicking ground of all the city parks', it was noted in 1892: 'A large plant of *Ravenala Madagascariensis*, the travellers' tree, was an object of interest as well as of beauty.' The tree was featured in the US Department of Agriculture's 1891 *Catalogue of economic plants and seeds,* recorded as being imported into the United States in April 1902 from seedsmen Haage & Schmidt in Erfurt, Germany. A dramatic mural featuring the traveller's tree in a swampy Madagascan setting was added to the Hall of Plant Life in the Chicago Field Museum of Natural History in 1936, where the tree was described as 'one of those striking exotic plants that never fail to attract attention wherever planted, whether in botanical gardens or elsewhere for ornament.' Today, when there are ever greater demands for botanic gardens to focus on their own region's endangered flora, strelitzia may well become rarities once more as the stately specimen trees gradually reach the end of their lives. *Strelitzia reginae*, on the other hand, is now considered more or less naturalised in places as widespread as California, parts of Australia, and the Canary Islands.[9]

## SURPRISED IN JAPAN

Despite *Strelitzia reginae* becoming an internationally established horticultural standard, finding strelitzia in bloom in a Japanese garden in November 2013 was a surprise and delight for South African landscape designer Leon Kluge. He was visiting Japan to take part in the Gardening World Cup and was 'very particular' about what he needed to complete his design:

> Searching for material for the display requires some cunning and resourcefulness. Contestants are not allowed to bring anything with them but have to rely on their begging and borrowing skills. [Leon] arrived two days before the competition to start to scout for material and plants. While walking around a little village, he spotted a strelitzia on a veranda, knocked on the door, and asked the little 90-year-old lady who answered . . . to lend it to him.[10]

That strelitzia helped Kluge win one of his several horticultural-show gold medals but the fact that he could stumble across it in Japan at all was largely due to the passionate enthusiasm of one man, Yutaro Suzuki. The intricate but sharp-edged shape of the bird of paradise flower makes it look as if it were created to appeal to the nation that developed the art of origami, folding paper into intriguing angular shapes, and today origami versions of strelitzia and instructions to make them can easily be found online. Like some of the myths and symbols that have been tracked from one culture to another round the world,[11] the appeal of this diva flower seems to be universal. Suzuki remained devoted to the plant for decades, even travelling repeatedly to South Africa to observe the plants in their natural habitats.

The iconic bird of paradise is by far Japan's favourite member of the strelitzia family, according to Suzuki. He dates its appearance in Japan to the late 19th century but notes that it did not become 'widely popular' until after World War II. It was only in a nursery in the mid-1950s that he himself first saw the plant in flower while still a young teacher. This seems to have been a gardening moment of love at first sight. Suzuki immediately decided to buy the whole stock of 450 plants imported from Germany. He recalled how he had no knowledge about the species and that the plants cost the equivalent of five months' salary: 'Those days *Strelitzia* was rarely grown by commercial growers and its prospect in the market was uncertain. A big difference comparing to now that it is seen in every flowershop almost all year round'. He believes being an early strelitzia enthusiast in Japan helped contribute to his own horticultural success.[12]

Japan is now one of the top 10 world importers of strelitzia blooms, supporting Suzuki's observations. It is estimated that Japan imports about 4 per cent of the world's strelitzia trade. Suzuki himself led the way in creating new bird of paradise cultivars, bringing out two varieties in 1980, Gold Crest and African Gold, and two more varieties the following year, Orange Prince and Orange Princess. It is not clear how consistent these

Gail de Smidt, *Strelitzia reginae*, watercolour, 2017

new cultivars were nor how long they survived in horticulture but Suzuki's passion for South African strelitzias was deep rooted and he made four trips to the country in the late 1970s and early 1980s, eager to familiarise himself with the conditions in which strelitzias live in the wild as part of his research for his short monograph on his beloved flower.

## IN LOVE WITH THE ORIGAMI FLOWER

In June 1981, for example, Suzuki saw *Strelitzia alba*, 'the queen of Tsitsikamma Forest', growing alongside huge yellowwood, often on deeply shaded slopes. At Uitenhage, Suzuki found *Strelitzia juncea* and observed that the plants produce fewer flowers than *Strelitzia reginae*. He also saw *Strelitzia juncea* north of Port Elizabeth, where he noted that the annual rainfall in the area was just 400mm – and that the dawn temperature was only about 5 degrees Celsius when he photographed their sunrise silhouette, with the afternoon temperature also falling sharply as the sun dropped. He noted that *Strelitzia juncea* prefers gentle slopes, while *Strelitzia reginae* tended to appear near the hilltops and 'never at the bottom of valleys'.[13]

In the Ecca Pass, about 20 kilometres out of Grahamstown, he noticed that the *Strelitzia reginae* were more 'compact', only about 50 to 60 centimetres high. If anything, he found strelitzias somewhat reticent in their growth. At Gonubie Nature Reserve, he observed, 'The soil [was] moderately moist with enough space to increase around the clumb, but the *Strelitzia* are not so thriving as one may expect.'[14]

Accompanied by J.R. Odell, Suzuki visited Pluto's Vale, about 30 kilometres northeast of East London, where he was fascinated that the apparently showy bird of paradise flowers could merge into the hillside thickets, appearing as just scattered, small silvery dots among the tall euphorbia, rather in the way that *Strelitzia juncea* had grown side by side with aloes in the Uitenhage area.[15]

Only where there was a little more soil and on the boulder-strewn cliffside of the Kwelegha River, north of East London, did Suzuki find birds of paradise growing 'plentifully'. Here, where the 'serene, tranquil' atmosphere made him 'feel like standing in Paradise', he was also particularly interested to see many *Strelitzia nicolai* growing beside the river, especially on the west bank. Nearby at Fuller's Bay, he suggested that the warmer climate might account for the many strelitzia already setting seed in June, while they were still flowering in Port Elizabeth. He was delighted to find 'a small seedling of *Strelitzia*', which he believed to be very seldom [seen] in the wild' both here and at the nearby Nahoon Dam as he had not managed to find seedlings on his previous three visits.[16]

Suzuki even went as far as the Drakensberg and Mpumalanga, looking for *Strelitzia caudata*, which he found preferred eastern-facing slopes. On a field trip with Dr H. Grobler and Mr Joubert, they scanned the Graskop landscape with binoculars in the hope of picking out the species against the surprisingly effective camouflage of the inaccessible terrain, while it was all but invisible at Piesang Kop even with Dr Lindsey Milne's expert guidance. He did manage to find two *Strelitzia caudata* seedlings in the Piesang Kop forest. Suzuki observed that in the

Yutaro Suzuki published his short, photographic tribute to strelitzias in 1982

wild, *Strelitzia caudata*'s flower was about 30 centimetres wide but about 40 centimetres wide when cultivated in a garden. He predicted a bright future for the plant: 'It should become popular in other countries too when it is commercially produced.'[17]

## BIRDS OF PARADISE ON THE MEDITERRANEAN RIVIERAS

Producing strelitzias for horticulture or for the cut-flower trade remains a niche market but one often born from the initial zeal of enthusiasts such as Suzuki in Japan and Lowell Swisher and Mike Mellano in California (see chapter 10). Suzuki's fervour for birds of paradise echoed in several ways the introduction of strelitzias to the Riviera coastline, which embraces the coasts of Provence in southern France and Liguria in north-western Italy. Sweeping alongside the Mediterranean through Cannes and Antibes, through Nice and Monte Carlo to Menton and Ventimiglia, Portofino and La Spezia, this area became popular with wealthy, sometimes eccentric British visitors from the early 19th century. The invasion began at Nice in the 1820s, was followed by Cannes in the 1830s and then Menton. Many adopted the area, undeterred by geopolitical changes, such as Nice and Menton being transferred from the kingdom of Piedmont-Sardinia to France in the 1860s. They took on the climate with similar nonchalance. Henry Brougham (1778-1868), lord chancellor of Great Britain from 1830 to 1834, shipped in loads of turf to create sweeping lawns around his villa. The newcomers often created new hybrid gardens which combined the English woodland-style garden with the exciting opportunities to grow exotic treasures from the subtropics. One of the best examples, known both as La Mortola or Villa Hanbury, was created by the Hanbury family as a private botanic gardens, occupying what was then the most easterly promontory on the French Riviera but now within Italy. By the early 20th century, Villa Hanbury was considered 'the largest and best subtropical garden in Europe'.[18]

The gardens' founder, Thomas Hanbury (1832-1907), had started his career at a London tea brokerage, then sailed for China in 1853, aged 21, where he made his own fortune in the silk trade. The newly fashionable Riviera became a favourite place to visit during his leave. His father, a partner in the pharmaceutical company of Allen, Hanbury & Barry, had insisted that horticulture and the science of plants should be an important part of his sons' education, contributing to Thomas's elder brother Daniel becoming a botanist and pharmacologist. Daniel Hanbury was the treasurer of the Linnean Society and a friend of Sir Joseph Hooker, whom he joined on a collecting trip to Lebanon and Syria in 1860, 'when civil war was raging.' In 1867, a sailing trip decided Thomas Hanbury to make a holiday home along this scenic Mediterranean coast. He had been 'much impressed by a house which was situated halfway up the limestone cliffs' between Menton and Ventimiglia, buying it and some neighbouring land to create a new estate of more than 40 hectares. He called this La Mortola, a local name believed to derive from the widespread wild myrtle. Inspired by luxuriant gardens he had seen along the French Riviera, Hanbury had planned to create a garden that showcased 'a collection of indigenous Mediterranean plants' and featured subtropical and some tropical species. This would be making a longstanding dream come true – 'to make a garden in a southern climate, and to share its pleasures and botanical interests with his . . . brother.'[19]

Daniel Hanbury joined Thomas at La Mortola for the first time in July 1867, just two months after Thomas had confirmed the property's purchase. Much of the land fell steeply towards the sea, had 'thin, poor and dry' soil, and had been deforested for firewood and overgrazed by goats, an environmental problem that Daniel had studied in Lebanon. To help it recover, the brothers encouraged areas of wild garden by sowing seeds of indigenous shrubs and trees. They also mixed in more exotic favourites of the period, such as cedars of Lebanon, roses and geraniums, passion flower and peonies. Thomas was able to add rare cultivars from the far east to the collection of more than 20 citrus varieties, while Daniel also used his botanical network to locate rare plants for the new garden. On a journey in May 1868 from England to La Mortola, Daniel collected an interesting contribution of plants, particularly Australian and Chilean palms, from the Jardin des Plantes in Paris.[20]

Moving south, Daniel was also the guest of Jules-Emile Planchon, professor of botany at Montpellier University (see chapter 4), who gave him several interesting plants. Planchon had already been working the outbreak of phylloxera that had been ravaging French vineyards for the previous two years. Two months after Hanbury's visit, Planchon became part of a commission appointed by Société Centrale d'Agriculture de l'Hérault to investigate the costly disease and eventually became internationally renowned as part of the international team of three botanical detectives who pinpointed the cause.[21]

The Hanbury gardens took shape largely thanks to Daniel's efforts, said German botanist Alwin Berger, an agave specialist who was curator of La Mortola from 1897 to 1914 and later director of the Stuttgart Museum of Natural History and an agave specialist:

> Through [Daniel Hanbury's] proficiency in botany, and his numerous botanical acquaintances at home and abroad, he was able to obtain rare and valuable plants from all parts. One of his chief endeavours was to procure those of economic and, especially, pharmaceutical importance . . . The collection of Australian, South African, and American plants must have been a notable one already in the very first years, though no catalogue of that period exists, except of *succulents*.[22]

By 1872, just five years after Hanbury had bought the property, the gardens were opened to the public on Mondays and Fridays: 'many of them remember with appreciation how courteously the owner himself acted as guide'. Opening the gardens did not mean the Hanburys thought they had completed their garden project. They kept adding new treasures. In February 1873, for example, they bought *Strelitzia augusta* and *Strelitzia reginae* from Algeria's Hamma *jardin d'essai*. Nearly a quarter-century later, on 16 June 1897, another *Strelitzia augusta* was planted on the south side of the restored house. It had reached 3.3 metres high by 1912. *Strelitzia parvifolia*, as *Strelitzia juncea* was then known, and a cross of *Strelitzia reginae* and *Strelitzia parvifolia* came from the Genoa Botanic Gardens in 1901. The gardens had become an attraction included in the classic late 19th-century Baedeker guidebooks and celebrated by visitors such as Paris-based A-R. Proschavsky as the only garden on the Mediterranean coast where he knew of *Ravenala madagascariensis* being successfully grown in the open. Several members of European royal families visited, including Queen Victoria.[23]

The work of creating the gardens as a partnership between the brothers was cut short, though. Daniel died suddenly of typhoid in March 1875, soon after finishing his major work, *Pharmacographia*, jointly written with Swiss-born Professor Friedrich Flückiger (1828-1894) and considered probably 'among the few classic productions of modern pharmaceutical literature'.[24] But for the rest of his own life, Thomas Hanbury, remained deeply attached to the gardens he and his brother had begun to create together:

> [He] was never happier than when there, surrounded by his family and by friends who shared his love of Nature. He knew almost every individual plant in his garden, and the most precious to him were those which reminded him of his beloved brother.[25]

Thomas Hanbury became known as a philanthropist, in keeping with his Quaker upbringing and practice 'using his wealth for public benefit'. In 1892, he sponsored the Istituto Hanbury, a building to house the herbarium, library and laboratories of the University of Genoa botanical department. The following year, he presented Kew with 'a collection of 30 valuable volumes on botanical subjects.' A volume of *Curtis's Botanical Magazine* was dedicated to him in 1893 by Sir Joseph Hooker 'in token of his contributions to horticulture.' Thomas Hanbury was knighted in 1901. In 1903, he bought an estate of about 25 hectares near Weybridge in Surrey for the Royal Horticultural Society, which became the core of their Wisley gardens, replacing their earlier Chiswick garden. When Thomas Hanbury died at La Mortola in March 1906, 'according to his wishes, his ashes were interred in his garden, amidst the old cypress trees' and until the 1940s, his descendants tried to maintain his garden.[26]

Being situated in Liguria meant that La Mortola had a classic Mediterranean climate of warm, dry summers and relatively mild winters, similar to the Western Cape. Strelitzias had been relatively early exotics to arrive at La Mortola and the Hanburys' interest in strelitzias combined with Thomas Hanbury's own well-maintained and widespread botanical correspondence gave him two reasons to celebrate the excitement of his brother Daniel's friend, Professor Jules-Emile Planchon of Montpellier University, at describing his great find of a yellow-flowering strelitzia (see chapter 4). The Hanbury garden was admired for experimenting with interesting and unusual plants from across the world and maintained a wide international network of sources. In 1912, Berger named 71 collaborators from 26 countries, islands

Hanbury brothers Thomas and Daniel featured strelitzias when creating a private botanic gardens near Ventimiglia, Italy, late 1800s

and regions, as well as noting donations from 40 international botanic gardens. Enthusiasts sent South African plants both from the continent and from Europe:

> For plants from *South Africa* we are indebted chiefly to the late Prof. MacOwan, who was one of the oldest and most generous correspondents of Daniel and Thomas Hanbury; to Mr Arderne and the late Mr. Harry Bolus of Cape Town, Dr. Brunnthaler of Vienna, Prof. Burtt-Davy of Pretoria, Mr. Hislop of Pietermaritzburg, to Mr. Hutchins, formerly of Cape Town, to the late Max Leichtlin of Baden-Baden, to Dr. R. Marloth of Cape Town, Mr Medley Wood of Durban, and to Dr. S. Schönland of Grahamstown . . . From Delagoa Bay [now Maputo], seeds have been sent by the Bishop of Lebombo.[27]

In return, seeds and plants were 'distributed to almost every botanical establishment in the world, and to many private gardens, in increasing numbers'. In 1908, for example, 13 085 packets of seeds were sent out.[28]

Fortunately, though, Thomas Hanbury did not live to see the 'devastating' effect of World War I on the gardens nor the 'second disaster' when Allied troops shelled the area and looted the house during World War II. The Botanical Society of Italy campaigned from 1957 for the gardens to be saved and in 1960, the Italian government bought it to be managed by the University of Genoa. By 1985, its traditional New Year list of plants in flower was still less than half of what it had been a century earlier. Restoration and renewal have continued, though, and in 2000, it became a 'regional protected area' under a new Italian law, applying for World Heritage Site status. A 'thick group' of *Strelitzia nicolai* still marks the final stretch of the ascent up the gardens' *Grande Route*.[29]

## *OISEAUX DE PARADIS* IN FRANCE

The first strelitzias for La Mortola had been brought from Algeria across the Mediterranean rather than introduced from local sources, such as the Jardin Botanique de la Villa Thuret in Antibes. This private botanic gardens was founded in 1857 by French botanist Gustave Thuret (1817-1875) in counterpoint to his professional speciality in algae. Thuret focused on 'rare and tender' plants that could not be grown easily in mainland Europe, bringing together curiosities from South Africa, New Zealand and the Canary Islands. In *Lettres d'un voyageur* (A traveller's letters) French novelist George Sand referred to it as an 'Eden', calling it the most beautiful garden she had ever seen. His five-hectare garden was bought and donated to the French nation after he died.[30]

Thuret supplied the Hanburys with 'so many interesting and valuable plants', although these did not include strelitzias.[31] This suggests the Hanburys' Algerian-sourced strelitzias may well have been the first strelitzias to be introduced to the Riviera. Today the plants now thrive so well in France and Italy that they are competition for South African strelitzias imported to the Netherlands and have the advantage of being grown closer to the major Dutch flower auctions.

Since La Mortola's establishment and Planchon's later delight in the yellow-flowering Lemoinier strelitzia, the bird of paradise flower has become established in France as a niche plant, known as *oiseau de paradis*, that extends the range of the local floriculture market. France is Europe's fourth-largest producer of cut flowers and pot plants. French authorities offered up to 50 per cent subsidies from 2004 to growers able to diversify into 'new varieties' of cut flowers, including strelitzia. The 2008 theme of the *Bouquet d'Aujourd'hui* (Today's Bouquet) competition at Salon du Végétal in Angers, France's key horticultural exhibition, was promoting the growing and use of strelitzia, carnations and mimosa in flower-arranging.[32]

Like the Villa Thuret, Charles Huber's 'nursery-garden' at Hyères in France's Provence-Alpes-Cote d'Azur region was one of 'several interesting gardens and horticultural establishments' already thriving on the French Riviera when Thomas Hanbury bought La Mortola. Together with Nabonnand at Golfe Juan, 'it was a source for the substantial amounts of plants that were bought for the gardens, as well as later helping supply the gardens' recreation. Today the Hyères flower market mostly sells very locally grown flowers – including strelitzias. The Marché aux Fleurs d'Hyères is supplied by about 600 growers, the majority within a radius of about 20 kilometres. It attracts about 350 buyers, raising a turnover of €45 million in 2006. The bulk of this trade, 85 per cent, was in cut flowers, with the remaining 15 per cent foliage. About one in 50 of the flowers sold were strelitzias, contrasting with roses, by far the most popular at about one in four flowers sold.[33]

## *UCCELLI DEL PARADISO* IN ITALY

In the early 1880s, painters Pierre-Auguste Renoir (1841-1919) and Claude Monet (1840-1926) escaped the Paris winter at Bordighera, about 15 kilometres east of Ventimiglia. The tropical luxuriance of the gardens is said to have given a fresh lushness to Monet's style, culminating eventually in his acclaimed paintings of his own garden at Giverny. Monet particularly favoured the gardens of citrus merchant Francesco Moreno. It was at Bordighera that a memorial was erected to Prussian botanist and nurseryman Ludwig Winter (1846-1912), who had been a gardener at both La Mortola and to Empress Eugénie (1826-1920) at Cap Martin. Winter later ran a nursery in Bordighera and was a leading figure among the German horticulturists 'who were the motor force of early floriculture in Liguria', particularly cultivating new varieties of asparagus, gerbera and strelitizia.[35]

## COMPARING FLOWER SALES AT FRANCE'S MARCHÉ D'HYÈRES[34]

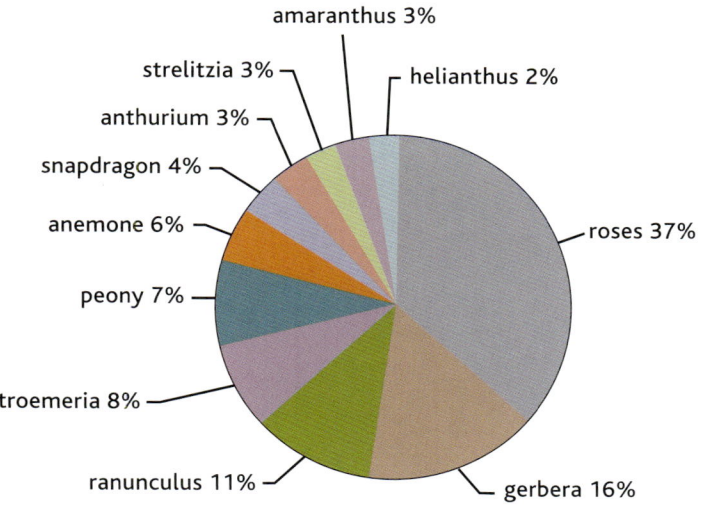

- amaranthus 3%
- helianthus 2%
- strelitzia 3%
- anthurium 3%
- snapdragon 4%
- anemone 6%
- peony 7%
- stroemeria 8%
- ranunculus 11%
- gerbera 16%
- roses 37%

It was the Hanburys, though, who had become 'the most famous British gardeners on the Italian riviera'. They attracted kindred souls and migrant English gardening enthusiasts spread southerly along the Ligurian coast, where the temperate climate happily accommodates palms and exotic plants. The Hanburys' friend Ellen Willmott, 'famous for bankrupting herself for her love of gardening' and for supporting plant collectors' journeys through the USA's Arnold Arboretum, often visited La Mortola. In 1905, she bought land near Ventimiglia to establish her third garden, La Boccanegra. In Alassio, about 85 kilometres south east of La Mortola, Scotsman General William Montagu Scott McMurdo had already followed the Hanburys' lead in creating a garden in 1875. He was able to call it 'one of the wonders of the riviera' by 1908.[36] It is now part of the Grandi Gardini Italiani network.

Strelitzias had become familiar to connoisseur gardeners such as the Hanburys over the previous century and a half. Today bird of paradise flowers – *uccelli del paradiso* – have also become part of Italian floriculture. They are grown in areas such as the south-eastern Apulia region of Italy, a dry area although with 'abundant' groundwater.[37] In Liguria, the thousands of family-run nurseries had shrunk to just a few hundred even before the 2008 financial crisis, so niche crops such as strelitzia have a special role here today. They are part of a survival strategy of the few remaining younger growers, who focus on improving quality and combining modern techniques of micropropagation with family labour as they strive to recreate the good financial returns of floriculture in the 1980s.[38]

Strelitzias in Durban Botanic Gardens' gold-winning stand at Chelsea Flower Show, 2014; Leon Kluge using strelitzias in one of his show gardens

In Campania, flower production has fragmented, with about 2 000 nurseries competing in an area of only about 1 800ha. Production in 2008 was estimated at between €210 million and €400 million. However, with 70 per cent of the trade concentrated between just 200 wholesalers who generally sold on to exporters and traders based in Liguria, the Campania growers took action to give their produce a local identity, labelling it: 'Guaranteed standard – Flowers from Campania'. They also took part in the Costiera dei Fiori (Flower Coast) promotion, extending from Vesuvius to Capri and Ischia. Priorities have been defined as lower production costs, longer shelf life, lower energy use and less environmental impact. Strelitzia production is small – forming part of the 'other' floriculture on 277 hectares out of 1 000 hectares under glass and in the open. But along with aspidistra, strelitzia is seen as a particularly 'promising' candidate for in-vitro micropropagation, which avoids often 'genetically unstable' plant sexual reproduction and also usually produces better-quality and larger flowers that fetch higher prices.[39]

## IMPORTS FLY IN

The strelitzia's relatively long post-harvest life had been exploited in railing it from California to the newly enthusiastic markets on the US east coast in the 1950s, followed later by shipments from Hawaiian growers (see chapter 10). This robustness also made it a good candidate as a floriculture exports developed globally. In South America, for example, the Colombian flower industry was founded in 1966. It developed quickly thanks to the combination of marketing understanding from Colombian Edgar Wells, followed by agricultural research insights from an American master's student, David Cheever, and sheer excitement over new opportunities from Colombian cattle-rancher John Vaughan. To this day, the combination of reliable sunshine and comparatively cheap labour make Colombian flower-growing a leading force in world floriculture.[40] Colombia's main growing region on the Bogotá plateau has an ideal climate for horticulture, that is situated almost on the equator and combines high-altitude at 4 260 metres high with both mild temperatures and strong light all year round. Strelitzia, heliconia and other exotics are grown on the coast, however, where the climate is much hotter and tropical.

Colombia's young flower-growing industry was also helped by two unexpected boosts. As an oil-producing country, it benefitted instead of suffering when an oil-production embargo by the Organisation of Arab Petroleum Export Countries sent the price of a barrel of oil quadrupling from US$3 in October 1973 to US$12 in March 1974. Oil costs soared even higher after the Iranian revolution in early 1979, reaching US$39.50 per barrel over the next year. A different kind of revolution was felt in the international flower-production industry. Transport costs for growers supplying the USA from Africa and Europe were severely hit. Basic production costs shot up for European growers particularly (see chapter 11). Even those in more marginal climates, such as parts of California and Florida, found the cost of heating shade-houses in cooler spells became prohibitive. Many US growers either changed flower crop or left production altogether and turned to wholesaling. European growers, especially those in the Netherlands, turned to producing seedlings and cuttings for their erstwhile cut-flower rivals, such as Colombia.[41]

Colombian flower-growers received a further boost nearly two decades later. As part of its attempt to cut back drug trafficking into America, the USA's 1991 Andean Trade Promotion and Drug Eradication Act included suspending US import duties on Colombian flowers. The aim was to make coca farming comparatively less attractive and profitable and to create more legitimate job opportunities: 'The results were dramatic, though disastrous for US growers' over the next four decades as the trade balance tilted and eventually was reversed:

> In 1971, the United States produced 1.2 billion blooms of the major flowers . . . and imported only 100 million. By 2003 . . . the United States imported two billion major blooms and grew only 200 million.[42]

Colombia's main flower cash crops remained roses and carnations, followed by orchids from its central region. Miami International Airport usually receives 10 to 12 flights a day from Colombia, often loaded with flowers, and during the peak Valentine season this can quadruple to about 40 flights a day. In the year 2013/14, Colombia exported 310 million carnation stems, 230.5 million rose stems and 1.7 million strelitzia stems to the USA, making it the top import supplier in all categories. Only as a supplier of cymbidium orchids was it seriously challenged, its 36 000 stems coming a very poor third to the 1 million stems from the Netherlands and 566 000 from New Zealand.[43]

Since the heyday of US-Colombian aid in the 1990s, steady increases in US customs duties on imported flowers gave US growers greater anti-dumping protection. But since South American suppliers are closer geographically, with only a two and a half-hour flight to Miami, they have weathered this challenge better than those in Africa. Among other South American strelitzia suppliers, from 2007 Costa Rica chose to concentrate more on lilies than birds of paradise and other tropical flowers. Even so, generally South American suppliers of bird of paradise now comprehensively eclipse US imports from South African growers. In the year between April 2013 and the end of March 2014, only 3 000 stems – or 0.05 per cent of weekly incoming shipments – were shipped into the USA from 'other sections'. The remaining 99.5 per cent of bird of paradise imports came from leading supplier Colombia, as well as Costa Rica, Ecuador, Guatemala and Mexico.[44]

## MEASURING THE GLOBAL FLOWER TRADE

The world's largest flower-farming areas are found in China, especially in Yunnan, and India, followed by the USA, Japan and the Netherlands. But all these countries except the Netherlands produce mainly for their own domestic flower demand. Like Colombia and other South American countries, most African countries produce flowers mainly for export and have only a small domestic market. South Africa ranked as the 21st largest exporter of cut flowers in 2003. Among other African countries, Kenya was fourth largest (with about a quarter of imports into Europe) and Zimbabwe eighth.[45]

## *AVES DEL PARAÍSO* IN THE CANARY ISLANDS

In the past three decades or so, flower growers closer to the more lucrative European markets have been inspired by the success of and competition from countries such as Colombia, Guatemala and Costa Rica to seize the initiative in introducing the strelitzia. In Europe, production of the flower now stretches as far west as the Canary Islands, where the flower is considered the emblem of Las Palmas, the capital of Gran Canaria. The botanical gardens in Las Palmas grows *Strelitzia reginae*, *Strelitzia juncea*, *Strelitzia nicolai* and *Ravenala madagascariensis*.

A significant number of fields were already flourishing on the island by the 1980s and flowers are regularly seen for sale on the streets. Techniques such as withholding water and calibrating irrigation are used both to prompt stalk elongation and to time flowering to coincide with good prices in the market. A floriculture research station on Tenerife has also investigated finetuning these practices.[46]

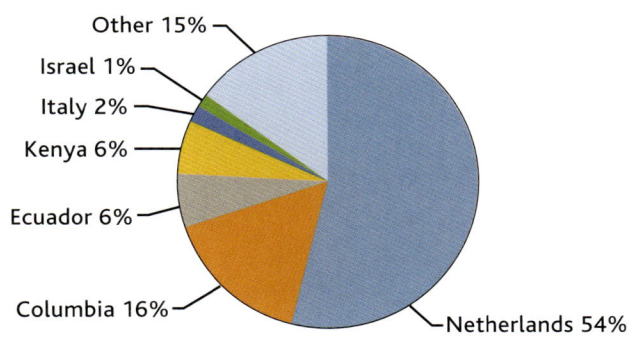

### PERCENTAGE OF WORLD FLOWER TRADE: EXPORTERS

- Other 15%
- Israel 1%
- Italy 2%
- Kenya 6%
- Ecuador 6%
- Columbia 16%
- Netherlands 54%

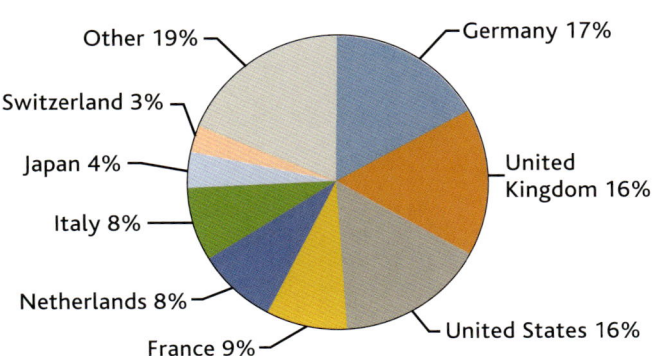

### PERCENTAGE OF WORLD FLOWER TRADE: IMPORTERS

- Other 19%
- Switzerland 3%
- Japan 4%
- Italy 8%
- Netherlands 8%
- France 9%
- Germany 17%
- United Kingdom 16%
- United States 16%

Sue Cochrane, *Strelitzia reginae*, watercolour, 2015

## THE FLOWER QUEEN IN INDIA

Floriculture has also become an important aspect of agribusiness across the Indian subcontinent and Asia. In India, it is grown in subtropical and warmer temperate areas. These include Himachal Pradesh, Kalempong, Darjeeling, West Bengal, the Nilgiri Hills, Western Ghats and Karnataka. In Palampur in Himachal Pradesh, strelitzias flower for eight or nine months of the year, providing they are protected from the sun during summer and from frost during winter.[47]

Strelitzias flower in Palampur in two phases, from February to May and August to November. Peak flowering for the year usually occurs in March and April, with a complete break in flowering both in June and July and December and January. Where strelitzias are grown in a polyhouse in the midhill area of Solan Nauni, though, they flower steadily from August to May, peaking between September and December. Indian researchers have concluded that the best temperatures to promote early flowering are between 10 and 13 degrees Celsius at night and 20 and 22 degrees Celsius during the day. They also noted that although strelitzias grow happily in daytime temperatures of between 17 and 27 degrees Celsius, anything higher than this tends to 'promote leaf production and inhibit flowering'.[48]

Healthy strelitzia clumps produce 10 to 15 flowers a year but, warned the researchers, the length of the flowers' stems significantly affects their post-harvest quality and survival. To boost survival of harvested strelitzia blooms, they recommend pulsing the cut flowers with a solution of 10 per cent sucrose, 250 ppm 8HQC and 150ppm citric acid for 48 hours.[49]

## 大鹤望兰 IN TAIWAN

Further east, in Taiwan, strelitzias relatively recently became one of the island's top 15 commercial flower crops, known in Chineses as 大鹤望兰 Using brightly coloured flowers is a distinctive feature of the region's flower-arranging style. The main orange or yellow colour of the blooms helped drive popularity as locally red, orange and yellow flowers are considered auspicious or lucky, making them especially popular for festivities such as Chinese New Year. Like other speciality flowers 'with a strong geographic image', strelitzia flowers are available and bought all year round, with a peak production period from July to September.[50]

Flower farming became organised in Taiwan in 1971, when only 234 hectares of flowers were grown on the island. A development plan was drawn up to concentrate on core flower farmers, establish flower markets and promote flower buying. With Taiwan's economy booming, by 1978 flower growing had expanded dramatically to 1 241 hectares. By 2012, floriculture stretched over 12 486 hectares. The value of flowers and flower products also soared, from NTD5.3 million in 1971 to NTD300 million in 1978 – and NTD12.5 billion in 2012 (about US$422.9 million or SAR3.5 billion).[51]

Most commercial Taiwanese flower farming is concentrated in the island's west central lowlands, near Taichung, Changhua and Nantou, where winters are usually sunny and dry. Though the Pul-li area is within the hotter central part of the island, nearby mountains keep it cooler and it has become a centre for temperate cut flowers. Generally, floriculture is a small family business with farms averaging about 1.1 hectares in the Changhua region. Strelitzias are well suited to this because they are not considered 'demanding', as long as the soil is well drained. They can survive for some time without water and are straightforward to harvest by hand. As a result, they have become more and more widely cultivated over the years.[52]

Strelitzia are grown in the open air. By 1991, they were planted in about 50 hectares. By 1996, strelitzia worth A$112 995 (US$88 485 or SAR 380 512) were sold on the Taipei flower market. Staples such as oriental lilies and carnations sold much greater volumes, A$6 440 830 and A$4 133 908 respectively. Following Japanese preferences, stem length is the main standard. Profit margins within the Taiwanese flower market depend on the distribution level: 10 to 20 per cent for importers; an average of 30 per cent for wholesalers; and for retailers, a wide range up to as high as 500 per cent for especially painstaking arrangements.[53]

Flowers have been auctioned in Dutch-style against the computerised clock since 1989, with about four-fifths of cut flowers in Taiwan sold through wholesale markets. The number of bunches (usually with 12 blooms per bunch) peaked at 79.7 million in 2002. The global impact of financial crisis meant this dropped back to 67.7 million bunches in 2009 but recovered to 75 million bunches by 2011. The value of cut flowers traded was NTD3.2 billion (US$94 million or SAR814 million) in 2001 and NTD3.9 billion (US$132.7 million or SAR962.5 million) in 2011, having also recovered from the global recession.[54]

Floriculture in Taiwan originally supplied a largely local market. In 1994, floriculture generated about A$367 million (US$268.5 million or SAR953.8 million), with 90 per cent of the products sold in Taiwan and 10 per cent exported. By 2001, floriculture exports peaked at 47 499 tonnes, though most of this consisted of live plants and seedlings, with cut flowers forming only 4 per cent. In 2011, exports were 42 354 tonnes but cut flowers by then made up 10 per cent. Chrysanthemums and gladiolus originally dominated Taiwanese flower exports but newer products are becoming 'key floriculture merchandise'. In 2011, Taiwan's most valuable export markets were Japan, the USA, Korea and Holland. Japan spent US$19 million (SAR137.8 million) on Taiwan's cut flowers, taking 86 per cent of its exports – though the major Japanese earthquake that year reverberated through the whole country's economy including its flower imports.[55]

Julie Wilson, *Strelitzia reginae*, watercolour, 2017

## BIRD OF PARADISE IN MALAYSIA

In Malaysia, strelitzias thrive in the main flower-growing area in the cooler Cameron Highlands. In 1995, strelitzias ranked 13th with 0.02 per cent of Malaysian flower production, the same as gypsophila and more than alstromeria, dahlia and helichrysum. Roses, chrysanthemum and carnations dominated production, however, with 33.5 per cent, 22.6 per cent and 9.0 per cent respectively. High rainfall in the Cameron highlands means that flowers are often protected, even temporarily, by plastic sheeting for temperate flowers or shade cloth for hardier flowers, which is laid across three-metre-high poles. Surface and drip irrigation is still needed outside the rainy seasons of April and May and October to December.[56]

Flower-growing in Malaysia remained principally a 'hobby industry' until it began to boom in the 1980s. Agriculture, particularly commodity plantations, had been the mainstay of the country's economy up to and beyond independence in 1957. While palm oil joined rubber as a key plantation commodity for the country, its traditions of smallholder farming were also expanded to include flower production. Between the mid-1980s and the mid-1990s, Malaysian flower growing expanded tenfold and its exports twelve-fold. By 1995, it was estimated that more than 1 000 farmers were involved in floriculture in Malaysia, generating more than RM370 million a year and with an annual growth rate of 24 per cent. Small growers acted through export agents and larger growers deliver directly to importers. Markets included Belgium, Finland, Ireland and other countries in northern Europe, the Middle East and eastern Asia. More than 30 per cent of vegetable-growing areas converted to flower production between the early 1990s and 1995 because there appeared to be less risk of flower crops suffering from pests and disease and because flower prices were considered more stable.[57]

Cut flowers were named as a priority crop in Malaysia's National Agriculture Policy for 1992 to 2010. Floriculture exports rose steeply from 1985's RM6.5 million to 1994's RM54 million. Instead of mirroring this eightfold increase, market prices for cut flowers fell, making their export value fall in turn by more than half from 1992's RM40.6 million to 1995's RM18.3 million. This and the 1997 Asian financial crisis highlighted a fundamental unsustainability in the Malaysian flower industry: relying too heavily on imported seedlings cut into any competitive advantage it was trying to create. Fortunately, the domestic market for cut flowers had expanded during the Asian tiger boom years – fivefold between 1990 and 1997, for example – as flowers became regular decorative and gift items. As the domestic market matured and export production developed, demand for 'Exotic and newer varieties' beyond roses and carnations, chrysanthemum and jasmine, grew to embrace the colourful and dramatic strelitzias. Domestic Malaysian demand for flowers peaks during Chinese New Year and Deepavali but, overall, local flower buying has still been dismissed as 'pathetically low compared to per capita consumption in other countries'.[58]

Ironically, though, different economic imperatives could encroach on flower-growing land, which shrank from about 2 000 hectares in 1996 to 1 787 hectares in 2006 – far short of the target of 7 800 hectares in the country's third National Agricultural Plan. The number of producers dropped by 84 to 458 in 2006, partly due to consolidation in the industry as company-owned farms by then already owned 86.6 per cent of the flower-growing area. Company-owned farms had the clout of size, too, averaging 5.3 hectares compared to the individual smallholder's average of 0.94 hectares. Malaysian flower growing increased tenfold between 1995 and 2005 to export RM164.4 million but short-term land tenure particularly and environmental concerns generally could prevent it expanding further. Overall, this sub-sector is now considered high risk, given its rising input and marketing costs, reliance on migrant foreign labour and the perceptions of 'unfair pricing practices' and 'complete lack of local product innovation'.[59]

Polish handbook on strelitzia-growing, 1990

## STRELITZIAS GO ONLINE

Expanding enthusiasm for strelitzias means that seeds of *Strelitzia reginae*, *Strelitzia juncea* and *Strelitzia nicolai* are easily found among the many plant species on sale online (see chapter 11). One leading online sales site, e-bay, offers a wide range of sources of strelitzia seed. In a snapshot survey, the first 50 listings alone of e-bay's more than 400 strelitzia seed listings offered seed from the United States, United Kingdom, Australia, China, Hong Kong, Cyprus, Spain, Portugal, Germany, Romania, Hungary and Poland. Strelitzias are popular garden plants in Brazil, where they have been proposed as a potentially successful crop in the country's tropical floriculture. In the Netherlands and eastern Europe, they are prized for flower arranging and as conservatory plants. Apart from Suzuki's tribute to strelitzias, another rarity is the pocket-sized floricultural handbook *Strelicja* by Professor Jerzy Hetman and Dr Elzbieta Pogroszewska, published in Polish in 1990 as part of a series on cultivating exotic flowers.[60]

'Birds' produce flowers that are both bulky and delicate, so packaging and transport have always been a challenge, making logistics and cost drive their sourcing in international markets. Today strelitzias have been adopted widely around the world. As well as the major growers in California and Hawaii, Colombia and Costa Rica, France and Italy, Israel and Malaysia, strelitzias also add interestingly different visual elements to the gardens and flower crops of countries such as Australia, New Zealand, Guatemala and Venezuela.[61]

Fashions for flowers and cycles in colour trends, often set by the worlds of fashion and interior design and popularised by the media, can mean that the range of cut flowers demanded on markets can also vary over time. Anticipating and working with changing fashion trends has created another major marketing challenge for South African strelitzia suppliers. They need to try to ensure that their bird of paradise flowers remain available yet affordable for enthusiastic customers so they do not move their affections to a different exotic. Competition in global cut-flower sales is so strong that developing difficulties in one major market usually means redoubling marketing efforts rather than moving focus to another country.

## THE STRELITZIA SALESMEN

If there were a prize for top strelitzia salesman, the final contest would probably be between two 19th-century Belgian magazines, *Flore des Serres* and *L'Illustration Horticole*. Editor of *L'Illustration Horticole* Charles Lemaire enticed collectors to invest when featuring *Phenakospermum guyannense* in 1860, pointing out its luxuriant leaves – '*d'un beau vert luisant*' [of a beautiful shining green] – and that:

> *Il représente absolument dans l'Amérique méridionale les Strelitziae du Cap, même port, même feuillage, mais avec une inflorescence tout-à-fait différente.*

> [It absolutely represents in South America the Cape Strelitziae, same stature, same leaves, but with a totally different flowerhead.]

He concluded:

> *Le Phénicosperme de la Guiane, même en l'absence de ses fleurs, si belles et si amples, de ses fruits si richement vélus, serait, par son port seul, éminemment ornamental et d'un grand effet parmi les autres plantes d'une serre chaude . . .*[62]

> [The Phenakospermum from Guyana, even in the absence of its very beautiful and full flowers, or of its richly enveloped fruits, by its stature alone is eminently ornamental and has a grand effect among the other plants in a hot house . . .]

Where Lemaire had inferred the exclusivity of the phenakospermum for a few fortunate collectors, Louis van Houtte on *Flore des Serres* was more direct. Regel and Körn's announcement of the new species, *Strelitzia nicolai* (see chapter 4), clearly represented a marketing opportunity for van Houtte.

In 1858, he urged collectors to ensure that they had both the treelike southern African strelitzias then known as *Strelitzia augusta* (now *Strelitzia alba*) and *Strelitzia nicolai*. He bracketed the account of the new species with reminders that his nursery had a few specimens of the newly recognised species for those who wanted to invest. He concluded by openly soliciting trade:

> *Nous avons la bonne fortune de posséder quelques exemplaires de cette magnifique plante, ces exemplaires sont déjà d'une bonne force, 0m,45 – 0m,75 – 2m50; on en trouvera le prix dans notre catalogue No.82 qui paraîtra vers le 14 août prochain.*[63]

> [We have the good fortune of stocking some specimens of this magnificent plant, these specimens are already sturdily grown, 0.45m – 0.75m – 2.50m; the price will be available in our catalogue no.82 which will appear around 14 August next.]

Van Houtte delicately does not state whether the price of the specimens that he has identified already existing in his nursery would attract a new rarity premium.

Detail from Angela Beaumont, *Strelitzia reginae* 'Mandela's Gold', 2003

# NEW STRELITZIA TRIBUTES
'Mandela's Gold' & 'Centenary Gold'

CHAPTER THIRTEEN

# NEW STRELITZIA TRIBUTES: *'Mandela's Gold' & 'Centenary Gold'*

The bird of paradise flower's popularity in public, corporate and domestic landscaping both in its home country and internationally grew initially from its renewed status, this time as a cut flower in the second half of the 20th century. It was also boosted by the stamp of approval from cities such as Los Angeles – which had even adopted the bird of paradise flower as its emblem and planted it far and wide. By the beginning of the 21st century, though, the bird of paradise had to some extent become the victim of its own success, too omnipresent – 'familiarity breeds contempt' even. Some gardeners in southern California had begun to think of them as 'unruly clumps' of 'flowers that spring up with weedy abandon'. A single *Strelitzia reginae* bloom still costs several dollars and upwards (about R100) in California flower shops but more fashionable Californian gardens have begun to showcase instead the rarer and more expensive *Strelitzia juncea*. The lines of its straight spiky leaves complement the architectural appeal of the flower in the most minimal way, especially as the recently developed cultivars are usually mid to pale yellow, in contrast to its usual, more vivid orange form.[1]

In the southern African region, the bird of paradise began to fulfil the role of a statement plant for the indigenous gardening movement, in line with the dream of Kirstenbosch founder Professor Harold Pearson for indigenous horticultural independence (see chapter 9). Since then, the strelitzia's dramatic looks have been appreciated as an excellent foil for the minimalist lines of modern buildings and it has increasingly been planted en masse. In the right situations, it is robust and hardy, making it desirably waterwise. These factors combined with showy looks have helped establish it high in the planting lists of top contemporary South African landscapers, such as Keith Kirsten, Elsa Pooley, Jan Blok and Lindsay Gray.

## QUEST FOR NOVELTY

Cultivar development can also help boost public interest in a plant. Victorian hybridists and plant breeders such as Lemoinier had already known (see chapter 4) that 'consumers constantly look for novel plants and new cultivars with unique characteristics'.[2] The impact of cultivars could be more like shooting stars, as Yutaro Suzuki's strelitzia cultivars seem to have been in Japan in the early 1980s (see chapter 12). But in 1978 John Winter (1936–2014), aged 43, had been appointed curator at South Africa's Kirstenbosch National Botanic Gardens and he would lead the team creating two new strelitzia cultivars – and succeed in matching the international excitement unleashed by the first strelitzia to reach the outside world two centuries earlier.

Winter was born a Capetonian, but educated in Bloemfontein (in the then Orange Free State) and in Bulawayo (in then Southern Rhodesia). He trained as a horticulturist at Pretoria Technical College and went on to three years of further training at the Royal Botanic Gardens, Kew, which first awakened 'his lifelong interest in botanical gardens . . . and this rather specialized aspect of horticulture'. A shipboard romance while horticulturist on Cunard's liner *Queen Elizabeth* led to marriage to his wife Meg and brought him back to Cape Town, making his energy and enthusiasm for botanic gardens available to Kirstenbosch. Before eventually being appointed curator, Winter served as assistant curator for 12 years. This began the process which would eventually result in 'a hugely rejuvenated Kirstenbosch' by the time he moved on from the post of curator in 1998 to become deputy director of the then new National Botanical Institute (NBI), becoming responsible for all national botanical gardens in the country.[3]

Winter began by redeveloping Kirstenbosch's layout, including creating the upper garden, installing irrigation, modernising and upgrading nursery and propagation facilities and rebuilding specialist collections. He became particularly interested in the families of erica, asparagus, leucadendron and clivia but was also generally renowned as 'a dedicated propagator and selector of superior forms. He had a plantsman's eye for what was attractive and different.'[4] He recalled how he was intrigued by some strelitzias in the corner of the nursery at the gardens:

> In the early 1970s I decided that something needed to be done with the 5 Strelitzia plants that were growing in containers in the nursery. I realised that they were special as they had flowered producing yellow flowers . . .
>
> A special area was prepared in the nursery where the 5 plants were planted out providing them with ample space to grow and develop into mature plants. Denis Jacobs, a senior nurseryman who was responsible for producing all of the Proteaceae and Ericaceae for the garden was given the added responsibility of looking after these special Strelitzia. This involved regular watering and feeding to speed up their growth. The prime objective was to increase the number of plants, firstly for display in Kirstenbosch and the regional gardens and secondly to promote this new introduction to the horticultural industry.[5]

Winter, assisted by specialist nurserymen Denis Jacobs and Dickie Peterson, spent years labouring to breed these curiosities true and so create a Kirstenbosch speciality that might capture the interest of the horticultural trade and generate some income for the gardens: 'Painstakingly, by selection and line breeding, [Winter] built up stocks of a chance yellow mutant of *Strelitzia reginae*'.[6] In the process, South Africa's growing wave of botanical nationalism became harnessed to the celebration of its Rainbow Nation.

## THE HOME-GROWN GARDEN REVOLUTION

Most of Cape Town's early European inhabitants were not instinctively excited by the region's indigenous plants. In the second decade of the 19th century, explorer and naturalist William Burchell had castigated them for this indifference. A century or so later, this began changing to a wave of enthusiasm, acquiring nationalistic undertones as the 20th century progressed. As the city had expanded, new householders moved into newly built homes in the recently developed collar of suburbs around the city bowl and Table Mountain. Increasing numbers of householders had space for a decorative garden so were able to progress from admiring exotic blooms in Cape Town's public gardens and parks to competing to maintain their gardens. The latest gardening styles became status symbols and this new generation of enthusiastic suburban gardeners was supplied by at least 90 nurseries around the city. Significantly, but not surprisingly, this was more than half of all nurseries in the Cape Colony.[7]

Despite Burchell's dismay, Cape florophiles mirrored the inclinations of booming new settlements from Europe to America to Australia and did not initially celebrate the indigenous plants around them. The gardeners wanted it to be obvious that they had invested work and money in their showpiece gardens. Like the Californians who embraced the exotic novelty of the strelitzia in public and private gardens soon after its introduction in 1853, Cape Town gardeners were similarly more enthusiastic about exotic flowers and trees than about those found on their doorstep. Many exotics grew easily, sometimes too easily, in the Cape's modified Mediterranean climate. Some indigenous showpieces such as strelitzias also required considerable patience and stamina from gardeners if they were to be coaxed to thrive. This meant they did not make easy candidates for commercial nurseries where the large part of sales would be to beginner gardeners looking for undemanding plants and easy results.

In the Cape as the 19th century drew to a close, however, Professor Peter MacOwan (1830–1909), the Colonial Botanist, was already leading voices of concern over how maintaining many exotic plants in the Cape climate would strain natural resources, particularly water. It would be another 30 years or so, though, before indigenous gardening was promoted significantly. In the meantime, Cape Town was becoming the heart of a new regional imperialism in the name of botany. As well as being the first director of the Kirstenbosch Botanic Gardens, Pearson had established himself as a botanical politician. The crux of his new plant empire was that the new Kirstenbosch botanic gardens would head a network of national botanic gardens (see also chapter 9).[8]

One of Pearson's driving philosophies was that the new Kirstenbosch botanic gardens would innovate by boldly celebrating indigenous flora. This early botanical nationalism was also supported academically in the Transvaal, with Pole Evans at the Pretoria National Herbarium founding in 1921 the journal series *Flowering plants of South Africa* (*Flowering plants of Africa* from 1940). The Botanical Society of South Africa had been founded in the Cape in 1913, the same year as Kirstenbosch, to promote the protection and growing of indigenous flora by the broader public. Within 15 years, the society's membership had more than quadrupled to above 1 000, with the Cape peninsula home to more than half of this membership. Here they enthusiastically tended their indigenous gardens, judging by reports of them besieging Kirstenbosch staff for free indigenous seed and gardening advice.[9]

## THE FIRST SHOOTS

Pearson's enthusiasm for featuring indigenous flowers in botanic gardens was carried forward by Kirstenbosch's first curator, Kew-trained Joseph (Jimmy) W. Mathews (1871–1949). Born in Bunbury in the English county of Cheshire, Mathews had come to Cape Town in 1895 to work at the Cape Town Public Gardens. He later learned about marketing flowers to public taste by spending nearly a decade running his own floristry and nursery business from Roeland Street. Mathews was also credited by Kirstenbosch's Professor Robert H. Compton (1886–1979) as being one of the founders of floriculture in South Africa. But more importantly for Pearson, as Kirstenbosch's first curator, Mathews was also a pioneer in both encouraging non-specialist gardeners to grow indigenous South African plants and bringing indigenous plants into cultivation.[10]

At a time when understanding the art of growing indigenous South African flowers was just dawning in the 1920s, Edith Struben's Luncarty, close to Kirstenbosch, was another showpiece of Cape gardening. Luncarty's setting was also dramatic: it shared the Table Mountain backdrop with Kirstenbosch and also had a view to the Hottentots Hollands mountains nearly 50 kilometres away. Around the house, the garden was more formal, with trimmed hedges, terraced lawns, hydrangeas, poplars and other exotics. This was surrounded by an indigenous wild garden, gently landscaped both for aesthetics and to remove unwanted exotics

'Mandela's Gold' strelitzias in flower near Nelson Mandela bust, Kirstenbosch National Botanical Gardens

such as pines. Struben saw her garden as a pioneering step in the direction of using indigenous plants to their full potential in local gardening. More broadly, it seemed to Struben and her contemporaries a clever, modern Cape reinterpretation of the classic design popularised in the British Isles in the 18th century — the formal villa surrounded by parkland which gradually merges into the surrounding landscape.

Struben's approach was also an extension of the British vogue for wild and cottage-style gardening — although this in fact promoted informal placing of common and colourful garden flowers rather than reflecting the wild flowers in nearby woodland, fields or mountains. In 1870, when internationally acclaimed Irish gardener William Robinson (1838–1935) had proclaimed the benefits of using hardier indigenous plants in his now classic book *The wild garden*, he was seen as both controversial and pioneering. Key to its influence was the appeal of compromise — mixing indigenous wild plants with prized exotics — combined with the deceptive ease of maintaining naturalistic rather than formal design and presentation. Robinson saw gardens as a place where nature would be, 'Embellished by the choice and delightful representatives of the flora of other temperate parts of the world — from China to South America.'[11] The push of nationalism did not become a significant influence on gardening in Britain until comparatively late, in the 1990s. This has been reinforced more recently by trends to make gardening more easy-care and reduce water usage and has been supported by a greater variety of indigenous plants coming into horticulture.

In the Cape during the 1920s and 1930s, Mathews built up support for the use of indigenous planting in both showpiece and more modest gardens through his gardening columns in the *Cape Argus*. His contributions to the Botanical Society's journal, *Veld & Flora*, drew attention to indigenous gardening around South Africa. It would be another half century, though, in the 1980s before indigenous gardening truly began to flourish more widely in South Africa in both private and corporate landscapes, thanks to the international environmental movement. Until then, most South African gardeners and landscapers continued to give pride of place to exotic flowers, from roses and tulips to hibiscus and frangipani. The local flower-production industry similarly focused on exotic flowers, such as roses.

## PROTECTING SOUTH AFRICA'S WILDFLOWERS

Strelitzias have attracted international attention from kings and queens, artists and designers over the two and a half centuries since they were introduced to the world. But the legacy of geographical circumstance, growing far from the prime areas of veld-collecting and being generally more difficult for collectors to access, made strelitzias latecomers to the Cape flower trade and insulated them from early exploitation. By the 1950s, though, strelitzias were as much at risk as other indigenous plants from the latest and greater than ever threat, agriculture. By then, Ellen Vlok

Cecilia Pienaar, *Strelitzia reginae*, watercolour, 2017

in Piketberg was seen to be holding the bar for 'local veld species . . . against the rapacious mechanical tractor.' She maintained a renowned wildflower garden from the late 1910s for about 25 years and supported the call by Japie Krige, the Stellenbosch horticulturist who had produced internationally successful sparaxis hybrids in 1932, that each farmer should 'put aside a small portion of veld as a wild-flower reserve.' Enthusiasm like theirs led to wildflower reserves being founded across the Western and Eastern Cape. Early environmental campaigners came to see agriculture as the greatest enemy of wildflower survival and excluded any option of partnership with indigenous floriculture.[12]

Even the South African Publicity Association had supported this movement for patriotic, environmental and business reasons. As early as the association's 1932 conference, it pointed out that public nature reserves could help overseas visitors 'more easily obtain a first-hand knowledge of our unique flora.' The call to prevent destruction of South Africa's valuable floral heritage – then estimated at about 16 000 species, more than six times that of Australia – was renewed at its 1956 conference by the Wild Life Protection Society's F.R. Long.[13]

Cape Town's flower sellers had offered a colourful image of the city for well over a century. Their images have been reproduced repeatedly in both privately created and officially produced depictions of the city aimed at tourists. Yet tourism associations were facing a contradiction as increased legal restrictions on picking from the wild and reselling indigenous flowers endangered this picturesque trade. The economic heft of tourism helped push forward the flower-reserve project in the 1930s. Tools for enforcing South African wildflower protection were created by the likes of Harriet Bolus (1877-1970), known to her associates as Louisa and the great-niece of Harry Bolus (see chapter 9). She produced, for example, a volume illustrating local protected flower species for the magistrate and public prosecutor of Caledon and Grabouw in 1944.[14] In the long term, in many areas this social control distanced the bulk of South Africa's population from an everyday familiarity with an understanding of indigenous plants.

Much of the movement to protect wild flowers centred in the Cape so Kirstenbosch and its indigenous programme were seen as a flagship in the 'fellowship of the veld':

> When Kirstenbosch itself has been more fully developed for its appointed task, it is the intention of the Government to start similar botanical gardens in each of the other three provinces. They will be subsections of Kirstenbosch under the control of Kirstenbosch. Thus the time is envisaged when the flora of the entire country will be the special care and inheritance of the Cape Floral Kingdom.[15]

Nationalism had been boosted in 1910 by the unification of former colonies in the Union of South Africa and later the Republic of South Africa in 1960. Pearson had proclaimed protecting and growing indigenous flowers as a national duty in South Africa. This idea was widely promoted and government support confirmed when the South African president, C.R. Swart, addressed the 1963 celebration of the 50th anniversaries of the founding of Kirstenbosch and of the Botanical Society.[16] Kirstenbosch was also the base for the Botanical Society's seed distribution system so plants from the Cape Floral Kingdom came to symbolise indigenous gardening for the whole country: the regional had become a proxy for the new vision of the national. The growth and development of professional botany, though, would be key to new waves of understanding and interest in South African flowers.

Botanical adventuring in the field combined with botanical taxonomy in the herbarium extended the strelitzia family, unexpectedly revealing a new species in the 1940s and a new subspecies in 2006 (see chapters 9 and 14). Strelitzias are a key, well-known South African group, surrounded by colourful drama that has not dulled over the centuries. Despite being more elusive in the wild than most staples of the Cape wildflower trade and more challenging to cultivate, they have gained a foothold in South Africa's national and international flower trade. They would also yield a horticultural pot of gold for the birth of the Rainbow Nation in 1994. This new wave of botanical nationalism, culminating in the announcement of the 'Mandela's Gold' strelitzia form, reflected the transforming state and reverberated through the corridors of international botanical power.

## THE FORGOTTEN FLOWER

In December 1989, a Kew horticultural taxonomist was amazed by what she saw when visiting Kirstenbosch's nursery area. In front of her was 'a small colony of a yellow-flowered *Strelitzia reginae*, something I had never seen or heard of before.' Stumbling on these latest successes from Winter's Kirstenbosch breeding programme, she recalled, 'I was open-mouthed with astonishment when I saw it. It is [as rare as] a black tulip or a blue rose.'[17]

The original yellow-flowering strelitzias, she would discover, were a rare sport largely forgotten by the worlds of botany and horticulture. Although Susyn Andrews had been born and educated in Dublin, Ireland, there were several reasons why she made it her mission to find out more about the surprising sight in the corner of Kirstenbosch's nursery. From 1971 to 1973, she had been an awardwinning horticultural student at the Irish National Botanic Gardens in Glasnevin, where South African plants, including strelitzias and their Madagascan relative ravenala, are features of the historic conservatories. In early 1976, Andrews moved to the Royal Botanic Gardens, Kew, as a horticultural taxonomist and in December 1988, she had married Brian Schrire, a former South African Botanical Liaison Officer at Kew. Andrews was visiting South Africa a year later when she explored behind the scenes at Kirstenbosch.[18]

Andrews was 'determined' that some plants of this golden-flowered strelitzia should be displayed in Kew's Temperate

House, where she had been working on the relandscaping. By then, Kirstenbosch curator Winter and his team had by then been working for about a decade to breed a yellow-flowered form of the bird of paradise. Andrews succeeded in persuading Winter to allow three plants to be transported from Kirstenbosch to Kew by a Kew student. By 1991, the plants were settling in at Kew, due to be planted out in a public area in 1993. But then the yellow strelitzias sprang a surprise by flowering unexpectedly in December 1992. They were put on display and became a media sensation. By late January, Andrews told Winter, 'the public are ooh-aah-ing the yellow Strelitzia all over the place!!'[19]

Kew had not exactly scooped Kirstenbosch, however. Early versions of the yellow strelitzia cultivar had been highlights of Kirstenbosch's gold-medal winning display at the Chelsea Flower Show in the 1980s, keen gardener John Ashmore of Berkshire, England, later recalled. 'At that time,' he said, 'I was told it was a rarity – not easy to produce'. Kirstenbosch's efforts to breed a new cultivar had become something of an open secret. In May 1992, Winter had already told Andrews that the horticultural trade had also 'shown a great deal of interest in "Kirstenbosch Gold",' as this new colouration was already being called. He had added that he 'hoped to release this Strelitzia to the trade in the near future' but needed 'to do some selecting as there is some variation in the inflorescence.' Kew's luck in flowering their plants soon after arrival helped maintain and even boost pre-launch excitement in the international horticultural market. It took Kirstenbosch some time to overcome the problem of inconsistency in the flowers and it was March 1994 before the cultivar was launched commercially, with plants going mostly to local growers and seeds to California.[20]

Excitement at Kew over the yellow-flowering strelitzias stirred institutional memory: had there not been other yellow-flowering strelitzias at some time in the past? Fired by the horticultural mystery, Andrews checked the herbarium specimens and library. The yellow strelitzia had indeed been a rare novelty but it was not completely unknown. The thread meandered from Lemoinier's plant, described with such delight by Planchon in 1880, through three specimens appearing in Kew's Mexican House (now the south block of the Temperate House, where Winter's plants from Kirstenbosch would eventually flower) during the late 19th and early 20th centuries to the yellow cultivars that Suzuki had succeeded in breeding in Japan (see chapters 4 and 12).

Female southern double-collared sunbird on 'Centenary Gold' strelitzia at Kirstenbosch National Botanic Gardens

## HORTICULTURAL ALCHEMY

In horticultural terms, developing an unexpected colour for a well-known bloom is the equivalent of finding the alchemist's stone that would turn base metal to gold. As rare as 'a black tulip or a blue rose', as Andrews had called the yellow strelitzia – and a certain bet for commercial success. The quest was not just an intellectual exercise for botanists and horticulturists. It had the potential to bring income and renown to the Kirstenbosch National Botanic Gardens.

This was the culmination of a painstaking task for staff Kirstenbosch and elsewhere, including Bruce Bayer at the Karoo Botanic Gardens, which had supplied another two yellow-flowering plants to help Winter's efforts. Winter recalled:

> For the past fifteen years at Kirstenbosch, we have been involved in selecting and hand-pollinating our original stock of seven plants with the purpose of eventually introducing this yellow 'sport' into horticulture . . . Left to their own devices the seeds from these yellow forms will not breed true.[21]

Tissue culture was not successful so Jacobs was taught how to cross pollinate the flowers, which proved 'very successful'. Strelitzia germination techniques developed at the University of Port Elizabeth using ethral and heat helped the team make the most of increasing amount of seed produced (see chapter 12) – and safe from Kirstenbosch's grey squirrels which had relished its discovery.[22] The team was teaching themselves as they went along how best to make the seedlings flourish, using hand pollination to avoid orange progeny:

> The young seedlings were planted out into 2 pint black plastic bags. Within a year the young plants were bursting out of the bags and we discovered from experience and observation that by restricting root development the development of the plant was also restricted. We needed to flower these young plants as soon as possible to discover whether we were successful in producing a yellow form of Strelitzia reginae. To encourage rapid growth we planted the young plants out into the open ground. Three years from sowing some of the young plants flowered yellow and I must say that there was never an occasion where the young plants produced flowered orange. We were very particular about the pollination and for this reason I think we were successful in producing the yellow form of Strelitzia reginae.[23]

## SOURCE OF GOLD

Were the yellow-flowered strelitzias really an occasional sport? Or did they point to having been sourced from a localised form of *Strelitzia reginae*, Andrews wondered. Winter could offer little help, replying regretfully in May 1992: 'It has always bothered me that I have no habitat information on the yellow-flowered Strelitzia reginae.' The yellow-flowering strelitzias which caught Winter's eye in the Kirstenbosch nursery when he took over as curator had been sitting there for some time and he had accepted them as the sport that had 'been known for a number of years . . . but never in any quantity, always just an isolated plant.' Andrews prodded memories trying to trace Kirstenbosch's original yellow-flowered strelitzias back to their source, chasing correspondents from across South Africa. Winter, known as 'a man of few words but many deeds', tried to clarify:[24]

> I have discussed [this] with John Rourke [curator of the Compton Herbarium, Kirstenbosch, from 1972 to 2005] and we feel it is a mutant which crops up now and again and is well distributed in horticulture, although not in any great quantity. John has seen a flowering plant in Kings Park and another in a suburban garden in Perth. I first saw this yellow form at the Karoo [botanic] garden and Kirstenbosch. I have also seen a specimen in California.[25]

Sue Cochrane, *Strelitzia reginae,* watercolour, 2015

Once Winter's quest for the yellow-flowering strelitzia cultivar was under way, Bruce Bayer had supplied another two yellow-flowering plants from the Karoo Botanic Gardens. Winter speculated that as, 'The previous curator of the Karoo Garden was Frank Stayner [(1907-1981), curator of the gardens from 1959 to 1973] who came from Port Elizabeth . . . I have always wondered whether he did not bring this yellow form of Strelitzia reginae to the western Cape.'[26]

Pauline Perry, who had worked at the Karoo Botanic Gardens in Worcester from 1976 to 1989, recalled having seen yellow strelitzias there in the late 1980s. She echoed Winter in pointing the way to former curator Bruce Bayer, who had been curator of those gardens from 1969 to 1987 and also mentioned Stayner's role. Bayer recalled that Stayner, 'brought a yellow-flowered form with him to the garden' when he arrived, which he had said 'was from the Uitenhage district and I gather that someone must have brought it to him as he didn't take particular interest in it.' Stayner had trained at Kew from 1933 to 1934 and Winter recalled him as 'a keen plantsman spending a great deal of his time in the field collecting'. This suggested to Winter that it was 'quite possible' that the yellow form had been collected in the Uitenhage area.[27]

Bayer confirmed that he had split the yellow-flowering strelitzia plant into clumps and given two 'to John Winter at Kirstenbosch'. Even at that relatively early date, 'Kirstenbosch had announced their intention of multiplying them' so staff at the Karoo gardens felt, 'They weren't in our brief'. The Karoo gardens' yellow strelitzia clumps did not seem to thrive, said Bayer: 'they never showed any vigour at all and seldom flowered.' Winter added that the two clumps received at Kirstenbosch 'from the Karoo Garden took a long time to recover and [this] confirms the fact that they do not like being divided.'[28]

While Winter and his team were working on improving the horticultural prospects and consistent colouring of their yellow strelitzia cultivar, Bayer was nurturing his own specimens after moving to the Worcester Veld Reserve in April 1987. He took two 'struggling' clumps with him, though he could no longer be sure if these were further subdivisions from his three original subdivisions nearly 20 years earlier. The plants' fortunes had improved and 'they are doing well with me now and I have about 20 seedlings from self-pollination – eventually the plants will go back to the Karoo Garden'.[29]

## LOOKING FOR YELLOW STRELITZIAS

Bayer suggested to Andrews that there could have been a second source for the Kirstenbosch plants: 'If I had to commit myself I would say that the plants came into the collection of the Port Elizabeth Parks Dept. under F.R. Long prior to 1950.' These different sources could have contributed to the flower variations which Winter had noticed at Kirstenbosch, Andrews suggested. But in May 1993, H.L. Reyneke, then director of parks in Port

Elizabeth, could tell her no more than that no yellow-flowering form still existed 'in our present collections'. However, Reyneke did add, 'I have no doubt that it did exist in the days of F.R. Long'. East London, about 280 kilometres further north east along the Indian Ocean coast, was home to 'The only yellow-flowered form, to my knowledge, which exists at present in the Eastern Cape', according to Reyneke. This was grown by Eric Dodds of Gonubie's Pioneer Nurseries, however, there was no response to Andrews' correspondence sent to the nurseryman.[30]

Winter confirmed Andrews' reconstruction of events, admitting: 'I had forgotten that we obtained two divisions from Karoo Garden . . . in 1974.' This meant that all seven of his plants which Andrews had seen in the Kirstenbosch nursery were technically defined as 'ex hort', having come from the garden or horticultural sources, rather than direct from the wild.[31]

Andrews next tried to determine whether the yellow sport was indeed rare or more widespread internationally than had been realised. She found that that 'a citron-yellow form was known in East African gardening circles' in the 1940s. The Kew herbarium's collection contains a 'Strelitzia reginae (yellow)' accessioned on 30 October 1954, donated by W. and C. Gowie of Oatlands Park Nursery in Grahamstown, Eastern Cape. It was planted in the Temperate House and verified as Strelitzia parvifolia on 19 April 1977 by W. Marais. Although the popular 1971 Reader's Digest Complete Guide to Gardening in South Africa described the 'crane flower' as having only orange and blue flowers, a decade later South African gardening specialist Kristo Pienaar mentioned a Strelitzia reginae var. lutea [yellow] in his bestselling local gardening handbook What flower is that?[32]

Dickie Peterson (left) and John Winter (right) with their successful 'Mandela's Gold' strelitzias, 2010

Angela Beaumont, *Strelitzia reginae* 'Mandela's Gold', 2003

Elizabeth McClintock, an associate of the herbarium at the University of California, Berkeley, told Andrews that she had not seen a yellow-flowered strelitzia in California and found the only reference to such a sport in standard horticultural texts stated that it was cultivated 'abroad'. But on the opposite side of the world in Australia, Dr Eleanor Bennett, principal display botanist of Perth's Kings Park and Botanic Gardens, researched strelitzia flower colours carefully. Using the Royal Horticultural Society colour chart, she identified the "pale yellowish-orange sepals" of local yellow-flowering strelitzia as 23B in the yellow-orange group and the "tongue" as 86C in the violet group. These were not quite the same as those at Kew, which Andrews identified as yellow-orange group 21A and violet group 93B. At Kirstenbosch, Winter identified the sepals as yellow-orange group 14A, although the corolla there was also violet group 93B. Andrews felt that these results were close enough to remain inclusive, especially given the more intense light in Australia and South Africa compared to Britain.[33]

Bennett commented that all strelitzias flowering in Perth's King's Park in July 1992 were orange, although by October the more common flower colour was yellow. She also sent Andrews photographs of *Strelitzia reginae* in flower, appropriately beside the South African War Memorial in the city. John Ashmore later added that he had seen yellow-flowered strelitzias in 1990 'at Charco Azul, San Andres on the island of La Palma' in the Canaries.[34]

## THE STRELITZIA SUPERSTAR

Gratifyingly for Kirstenbosch, the release of its new Kirstenbosch Gold cultivar was greeted with a flurry of interest and then supported by a steady flow of sales. But out of the public eye, a new confusion arose between Kirstenbosch and South African pride on the one hand and the International Code of Nomenclature for Cultivated Plants on the other. More than two years after Kirstenbosch had launched its 'Kirstenbosch Gold' strelitzia and nearly three years after the early plants had flowered at Kew, it was relaunched under the name of Mandela's Gold, marking the visit of President Nelson Mandela to Kirstenbosch in August 1996. Photographs of Mandela with the plant would make headlines round the world.

It was explained that 'following the inauguration of President Mandela on 31 May 1994, NBI management proposed that a very special plant be named after the first President of the new democracy'. One suggestion from 'an outside organisation', according to Professor Brian Huntley, NBI chief executive officer, was discarded because the oxalis in question had an inappropriately weedy growth. Other wild species in the queue for scientific naming at the NBI were considered 'too insignificant' and renaming a new hybrid protea which unusually grew along the ground was also dismissed.[35]

Coincidentally, the 'Kirstenbosch Gold' strelitzia had been released by Kirstenbosch just a few weeks before South Africa's first democratic elections which overwhelmingly voted Mandela and the African National Congress into power. Huntley was a keen advocate of giving it the new name, 'Mandela's Gold', writing:

> With the establishment of the National Botanical Institute in 1990, a new logo was sought that would reflect the unique beauty of South Africa's botanical treasures. The bird-of-paradise *Strelitzia reginae* – an endemic of the east coast – was chosen . . .
>
> Following the inauguration of President Mandela on 31 May 1994, NBI management proposed that a very special plant be named after the first President of the new democracy. Many recommendations were received, but the choice of a rare yellow colour form of *Strelitzia reginae* won the day.[36]

This suggests that Huntley felt it particularly appropriate for the NBI to commemorate Mandela with a striking new colour form of the plant which formed the organisation's own logo. Ultimately, Kirstenbosch's golden bird of paradise was also considered more apt than other suggestions given 'the stature of the honoree'. Despite Huntley's enthusiasm, he found that it took until 1996 for the National Botanical Institute to be 'granted permission to re-name it in honour of Nelson Mandela', usually given by the Nelson Mandela Foundation which regulates use of the Mandela name.[37]

President Nelson Mandela at Kirstenbosch National Botanic Gardens with the 'Mandela's Gold' strelitzia, August 1996

# PAINTING FOR MADIBA

As the SANBI staff hurried to prepare the National Botanic Gardens at Kirstenbosch for President Nelson Mandela's visit on 21 August 1996, the only botanical artist employed by SANBI at that time, Gillian Condy, was busy with one of the most urgent and most important commissions of her career.

Condy had won awards for her botanical art and for her stamp designs but the enormity of her newest task was not lost on her. As she was based at the National Herbarium in the Pretoria National Botanic Gardens, which did not yet have any Mandela's Gold plants, a plant in a very large flower pot was sent up to her office from Cape Town. Condy combines scientific accuracy with artistic principles in her work and ensured that her depiction of this individual plant also reflected its scientific description.

Using watercolour, the medium preferred by SANBI for its illustrations, meant she could work fast to produce the large portrait of the full plant: "I had only this one plant to work with but it had a couple of beautiful flowers. Rather than including its pot, I chose to depict the plant as if it were growing out of the ground."[38]

SANBI botanical artist Gillian Condy painting 'Mandela's Gold'; Gillian Condy (left) and John Winter (centre) presenting the painting to President Nelson Mandela (right) at Kirstenbosch National Botanic Gardens

Cape Town's winter rain disappeared for the day of Mandela's visit, 21 August 1996, with the sun shining golden on Kirstenbosch, President Mandela and the glowing Mandela's Gold strelitzia. Condy's framed painting of the flower was presented to Mandela as a keepsake of the occasion. Condy was flown to Cape Town for the day so that Mandela could meet the artist and she was introduced to him along with Winter. ''It was,'' she recalled, "one of the most memorable moments of my career. President Mandela really did have an extraordinary aura about him and I felt so in awe of him." Dazzling as the occasion might have seemed, Condy remembers best Mandela's obvious determination to talk naturally with members of the crowd and his warmly humorous reassurance to her as she nervously posed for photographs with him: "It's all right – you can smile!"[39]

Reporting on the historic announcement of the new name downplayed the earlier 1994 publication:

> although several have been released to other botanical gardens and nurseries, the National Botanical Institute had been waiting for a special occasion to release this plant to the world. And what better person to honour than President Mandela![40]

Halfway round the world, the plant was already listed as 'Kirstenbosch Gold' in the Royal Horticultural Society's standard reference *RHS Plantfinder*. As a result, this name still prevails there, with Kirstenbosch's new preference for 'Mandela's Gold' being better known in Africa.

The last words rest with the late President Mandela himself. Displayed now beside the Kirstenbosch clump of Mandela's Gold is his comment:

> I am happiest when I am in the wild because I can listen . . . I always feel the force of that sentiment when in this environment. I am very happy that you have done me the honour of being associated with this remarkable place, Kirstenbosch.[41]

## ANOTHER GOLDEN MOMENT

South Africa and South African National Biodiversity Institute ultimately triumphed in naming its new strelitzia cultivar 'Mandela's Gold' after the country's iconic liberation hero. But a conundrum arose in 2013 when Kirstenbosch wanted to launch a companion cultivar, a yellow-flowering *Strelitzia juncea*, to celebrate the centenary of the gardens' founding. The epithet Kirstenbosch Gold would have been ideal in the circumstances, yet reviving it would rekindle any earlier confusion about the 'Mandela's Gold' renaming. It might have been fitting for the world's gardens to have a Winter's Gold strelitzia, honouring the previous Kirstenbosch curator who had persevered so long in leading the team breeding the new yellow-flowering cultivars,

others wistfully observed. In the event, the new cultivar was named quite simply 'Centenary Gold'.

This latest cultivar 'symbolizes partnership and mutual respect across social and cultural barriers', said Phakamani M'Afrika Xaba, the Kirstenbosch horticulturist who took over the project leadership from John Winter in 2007. He was helped by specialist nurseryman Mluleki Mbutse; Thandile Mzukwa, who assisted with pollination; and De Wet Bösenberg, who helped with field work and pollen analysis in the laboratory. This time the Centenary Gold portrait was painted by Ebraime Hull, then a specialist nurseryman at Harold Porter National Botanical Gardens, as well as a wildflower artist and mentored by botanical artist Vicki Thomas.[42]

This new project had been prompted when Winter received a wild, yellow-flowering specimen of *Strelitzia juncea* from a Port Elizabeth nurseryman. Winter recalled: 'I managed to obtain a yellow form of *Strelitzia juncea* which we selfed and managed to produce seed after a number of attempts and some of the seedlings flowered earlier this year [2010] which are yellow so in time that will be another new introduction.' Winter retired from his final NBI post as deputy director in 2001 but continued to curate and maintain the Kirstenbosch cycad and clivia collections. He also still worked at plant breeding with nurseryman Dickie Peterson, producing 'the remarkable Apple Blossom strain of *Clivia miniata*', as well as trying to produce this second and different true yellow strelitzia cultivar.[43]

The plant which Winter had received seemed to be a clone as Winter and Peterson struggled to self-pollinate it. The sport itself was so rare that Xaba himself never saw it in the field: 'Although I visited a number of wild populations, I never came across a yellow flowering form'. He concluded that this could mean the sport results from 'ecological and not genetic factors', such as soil composition, for example, although he acknowledged this needed to be tested. Despite this conundrum, other genetic factors would still be one of the keys to developing the cultivar.[44]

The spiky, rush-like leaves of *Strelitzia juncea* are one of the obvious physical differences between it and *Strelitzia reginae*. Although the two species have different habitats, these overlap in the Patensie area of the Eastern Cape, where hybrids of the two can be found due to the fact that both have the same number of somatic chromosomes ($2n=14$). Knowing this, Winter and Peterson hoped they could make progress by cross-pollinating their yellow *Strelitzia juncea* clone with *Strelitzia reginae* Mandela's Gold. Unfortunately, the young plants produced orange flowers. The breakthrough came when Peterson persevered with self-pollinating the original *Strelitzia juncea* clone. Although this produced only 'a handful of seeds', it was progress at last. The team was able to improve seed yield by enhancing the techniques of self-pollination they used but they were still disappointed when nearly half of the seedlings eventually produced flowers that were 'orange or an undesirable shade of yellow'.[45]

## SOLVING THE JUNCEA SEED PUZZLE

The signal that seeds of the new yellow *Strelitzia juncea* cultivar have set and are mature enough to harvest is when the seed pods dry up and split open. This takes about eight months after pollination, Phakamani M'Afrika Xaba noted. For the breeding programme seeds were stored in a cool, dry place until spring or early summer (September and October) when they would be prepared for sowing using techniques investigated by researchers such as Albertus Van de Venter.[46]

First the orange fluffy arils were removed from the seeds, then they were soaked in sulphuric acid to combat dormancy. After three to five minutes in the acid bath, the seeds were placed under running water for a few more minutes to prevent damage and rinse off acid residue. Next they were soaked in a growth regulator (the team used Ethephon) before being sown 'at a depth one-and-a-half times the size of the seed' and then kept at between 25 and 28 degrees Celsius to germinate. During this process, they were kept moist but not wet and transplanted in autumn or winter when they had developed a secondary leaf.[47]

Several years of further development saw them working intensively to create a consistent, deep-yellow cultivar with 'a homozygous and dominant bold yellow gene', due to consistent pairs of identical, dominant yellow genes. To achieve this, all orange-flowering plants were culled. Then they used not one but three breeding techniques to produce more reliable seed faster: self-pollination, sibling mating and back crossing. The result of this painstaking work from two generations of Kirstenbosch horticultural experts was a stunning new cultivar with a long flowering spathe. It is straightforward to grow, thriving in well-drained ground in full sun.[48]

Recently, mature plants of *Strelitzia juncea* Centenary Gold were selling for as much as R2 400 each. Developing the 'Mandela's Gold' and 'Centenary Gold' colour forms put strelitzias in the headlines round the world and also re-emphasised strelitzias as South African intellectual property. Kirstenbosch's strong horticultural tradition had hit gold again.

## HELPING THE YELLOW JUNCEA FLOURISH

To produce plants of the new yellow-flowering *Strelitzia juncea* cultivar as quickly as possible required regimes that were much more intensive than general cultivation of the plants in Kirstenbosch National Botanic Gardens. Although plants in the gardens and the greenhouse are watered just once a week, the young greenhouse plants were given plenty of space so their roots were not restricted and optimum temperature, light and feeding, Phakamani M'Afrika Xaba explained. By growing the young plants at 28 degrees Celsius in the Kirstenbosch greenhouse, the flower stalk appeared within three years, at least twice as quickly as when planted outside. The disadvantage was that they became 'vulnerable to Mealy Bug while . . . still young', instead of being 'generally hardy' outside.[49]

Young seedlings transplanted in autumn (March and April) into 20-centimetre pots containing a well-draining custom-made potting mix develop secondary leaves by early winter. The speciality potting mix was made up of 'two parts loam, two parts sand, three parts bark, three parts compost, 3.1.5, half part bone meal and one part Bounce Back'. Instead of ordinary maintenance feeding, the young plants were fed weekly in winter (April to October) with a mixture of 'one part super phosphates, one part 3.1.5, two parts Bounce Back, one part bone meal and half part dolomitic lime'. Plants in the Kirstenbosch gardens are 'only fertilized once a year in April' but to boost production and growth among the greenhouse cultivars, they were fed seaweed-based fertiliser twice a month in summer (September to March), with a generous helping of compost added to the roots twice a year in January and June.[50]

'Centenary Gold' in flower at Kirstenbosch National Botanic Gardens, June 2018

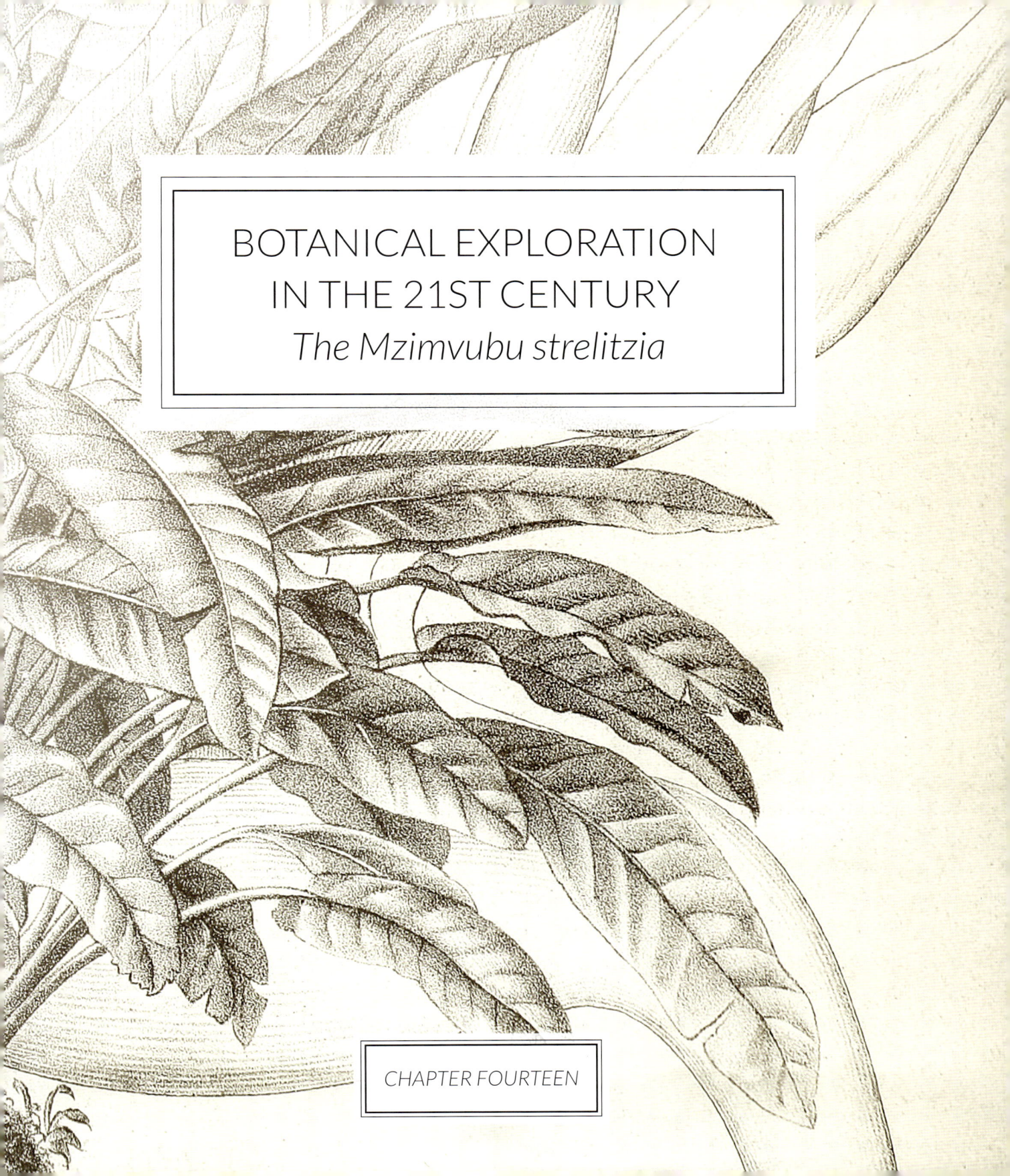

# BOTANICAL EXPLORATION IN THE 21ST CENTURY
## *The Mzimvubu strelitzia*

CHAPTER FOURTEEN

# BOTANICAL EXPLORATION IN THE 21ST CENTURY: *The Mzimvubu strelitzia*

Swimming across a river in flood and clambering up precipitous gorges in full climbing gear do not seem like job requirements for botanists, trained in peaceful lecture rooms and laboratories. In centuries gone by, the likes of Sparrman, Löfling and Gerrard showed great spirit of adventure and this is probably even more, not less, a job requirement for plant collectors today. Around the world, professional and amateur botanists have probably already collected and identified most plants within range of access roads. In almost inaccessible corners of our planet, though, botanical surprises still await the daring. An enthusiastic promoter of ecological gardening, Western Cape horticulturist Ernst van Jaarsveld collects plants both for their scientific interest and as possible new ornamentals for indigenous gardens. A new subspecies of *Strelitzia reginae*, among a feast of other botanical discoveries, was his reward when he led three expeditions – in April and August 2002 and in August 2006 – in the north east of South Africa's Eastern Cape along the Mzimvubu River gorge.[1]

The Mzimvubu River's name sounds its warning – in the local isiXhosa language, it means 'home of the hippo'. The river itself bucked and snorted, capsizing van Jaarsveld and fellow paddler Godfrey Zwide and their rubber canoe twice in just one series of rapids. Back in the 1790s, according to Sir John Barrow (see chapter 9), the area was renowned for 'the hippopotamus that abounds in all the great rivers.' In the 19th century, hippo at the eponymous Sea Cow Lake in Durban, about 350 kilometres north along the Indian Ocean coast, had combined with other dangerous water animals to be the downfall of plant collectors. The aptly named Robert Plant (d.1858), then curator of Durban Botanic Gardens, had 'stood across [an] alligator at Sea Cow Lake'. His successor Mark McKen (1823-72) was chased up into a tree there by a hippopotamus. For four hours, McKen had to cling to his perch, waiting for the hippopotamus to move on.[2] Fortunately, though sadly, for van Jaarsveld and Zwide, the iconic hippos, renowned as Africa's most dangerous animal, were no longer to be found at the river named after them.

Van Jaarsveld had become renowned within the botanical community and specifically at Kirstenbosch National Botanic Gardens, where he worked from 1976 to 2015, for the daring and stamina required by anyone joining one of his expeditions. He has collected plants across all South Africa's nine provinces and in neighbouring countries such as Namibia, tackling expeditions as both a professional and a physical challenge: 'Some of these expeditions were done on foot over long distances – a 55km trek to Slangpoort [Snake Pass] in the Slangkloof [Snake Gorge area of the Baynes Mountains in northern Namibia's Kaokoveld] . . .; others have involved climbing mountain peaks and cliffs . . . or long canoe trips.' The Slangkloof lived up to its name and, unusually, van Jaarsveld found 'Snakes were common everywhere' including a rare, 1.5-metre Angola dwarf python, the western keeled snake and a black mamba.[3]

Patricia McCracken

Overleaf: Detail from Jeanette Loedolff, *Strelitzia reginae* var. *mzimvubuensis*; *Strelitzia reginae* in *Gartenflora* magazine, 1875
Above: *Strelitzia nicolai* at a lagoon on South Africa's Transkei Wild Coast

For his doctorate, van Jaarsveld chose to research the adaptations of succulent plants that live on cliffs. This habitat also increased the hazard level of his expeditions. At 'a hotspot of plants not previously recorded for Namibia' near Kaokoland's Epupa Falls, van Jaarsveld targetted an aloe that was 'tantalizingly out of reach'. On went the climbing harness and he brought back the type specimen of a plant that would be named *Aloe amavandae*. Van Jaarsveld returned from one expedition to Kudikela, along the Mzimvubu River, with a broken leg. He had tumbled down a rockface after the piece of rock on which he was leaning broke loose. On another occasion, van Jaarsveld was attacked and pursued by a swarm of bees after he abseiled down a cliff face near Barberton, Mpumalanga, landing unsuspectingly at the entrance to a bee hive.[4]

## A NEW TREASURE HOUSE

Van Jaarsveld's first Kaokoland expedition confirmed his sense that cliffs are the least-explored territories in the world.[5] The following year he resolved to continue his quest at the promising cliffs along the Mzimvubu River, which often reach 800 metres above sea level. Canoeing the Mzimvubu, the Eastern Cape's largest river, would be the most direct way to reach the gorge's remote plant treasures. The river's source in the Matatiele area of South Africa's Eastern Cape, near the Lesotho border, draws water from a large part of the southern Drakensberg, flowing about 250 kilometres south east to its mouth at Port St Johns on the Indian Ocean coast. Its upper stretches run through 'a fairly narrow gorge' with thickets of semi-arid vegetation and savanna, 'as well as subtropical forest in protected kloofs'. Although the river becomes wider later, huge boulders, looming river cliffs and sometimes precipitous rocky banks make it difficult to navigate. Rainfall steadily increases as the river makes its way to the sea to an annual average of about 1 000 millimetres, falling mainly from spring to autumn but sometimes in winter.[6]

During the first Mzimvubu expedition, the four-man team of van Jaarsveld and Zwide and Kirstenbosch's Adam Harrower and Phakamani Xaba was able to cover the section from the Welsh Bridge, on the Thabankulu Road outside Mount Frere, to Majola, a distance of about 65 kilometres along the river, passing also junctions with the river's tributaries, the Tina and Tsitsa. Humans navigating the Mzimvubu for any distance are a rare sight as almost impenetrable vegetation and steep cliffsides make it difficult to reach the river and a particularly daunting series of rapids is strung along the river's length.[7] As van Jaarsveld noted:

> The terrain was very mountainous with grassland, shrublands and thickets on the slopes . . . The river meanders past steep banks and cliffs with many S-bends and fairly long, straight sections followed by short, steep, rocky rapids.[8]

The expedition's aims were 'to collect and document . . . succulent and bulbous cliff-dwelling plants . . . [and] other plants showing horticultural promise.' This was a vital scientific task as the many river systems of the Eastern Cape have fragmented plant territories. Such isolation means different species have developed locally, usually with narrow endemic ranges. The Mzimvubu is a key river within the Pondoland centre of plant endemism, which stretches north from the Kei River to the Umtamvuna River and includes the spectacular Oribi Gorge. This small area of about 180 000 hectares is considered 'critically important', containing flowering plants which do not grow naturally anywhere else in the world. Some of its plants appear endemic to tiny localities. It has been estimated that only between 200 and 500 Pondo bushman's tea tree (*Lydenburgia abbottii*) exist, for instance, all within a range 'of no more than 10 km from end to end'.[9]

## THE ADVENTURE BONUS

The challenges encountered in 2002 by the first expedition were fairly typical of what also awaited later expeditions, noted van Jaarsveld. That first expedition was scheduled for the beginning of the drier winter season, which meant coping with low temperatures at night while camping on the riverbank but also a day of rain while rowing. Although the team had not anticipated that the steeply dropping altitude would create such a series of rapids along the river, they had taken precautions against being buffetted by the waters. Cameras, for example, were secure in a water-tight drum and spare clothing in dry bags. These, fortunately, proved effective but three years later an expedition along the Mzinyati River, a tributary to the Tugela, in November and December 2003 was not so lucky. The Mzinyati team had to swim for long distances where the gorge was too narrow to walk along and 'In spite of placing our rucksacks in huge survival plastic bags, water managed to infiltrate into everything (except our cameras which were in a special waterproof container).'[10]

There were other challenges to negotiate on the Mzimvubu, though:

> We first had to get acquainted with the canoes' paddling and steering, and soon realized getting wet on the rapids was normal. Fortunately it was sunny most of the time . . .
>
> Because of the drop towards the coast the river flows relatively fast with frequent rapids in the rockier parts. The first few rapids went well with the normal adrenalin rush at each encounter, but our main task was to focus on the plants – the adventure was a bonus.[11]

Xaba climbed a rock face to collect an unusual cliff ox-tongue (*Gasteria croucheri*), for example. But the team also developed another collecting technique that was less physically

challenging but required good hand-eye coordination. They used catapults 'to shoot down succulent plant cuttings from the cliff face – a simple, fairly safe and effective method – saving us the trouble and risk of climbing it.' Any and all methods of collecting plants were considered as all the team needed to take back to Kirstenbosch was a branch or even a leaf that they could root.[12]

Local people live on the plateaus above the Mzimvubu River but tend to avoid fighting their way down precipitous and thickly vegetated gorges to the river. This has largely isolated the river from expanding human settlement, small-scale farming and modern infrastructure, naturally protecting its dense flanking vegetation.[13] While the team paddled along, as well as watching carefully for any novelties, they kept a checklist of familiar plants. Beside the river grew plenty of small-leaved willow (*Salix mucronata* subsp. *capensis*), sweet thorn (*Vachellia* [formerly *Acacia*] *karroo*), river euphorbia (*Euphorbia triangularis*), Kei apple (*Dovyalis caffra*) and agony bush (*Smodingium argutum*). On steep, dry slopes spekboom (*Portulacaria afra*) and Cape bitter aloe (*Aloe ferox*) were seen. Invasive plants were seen on the often degenerated plateau grasslands and were beginning to appear along the river as it carried down more and more of their seeds. In that sense, the Mzimvubu River expeditions were part of a race against time by plant collectors to map the detailed variation and even discover plants unknown to science in these specialised microhabitats: "We knew it was vital to take notes of everything we saw and where we collected specimens as we could not be sure what we would find if we returned," said van Jaarsveld.[14]

Where the river was wider and shallower, its boulders were more obvious hazards, 'like a huge obstacle field'. Exercising caution, the team hauled their canoes through instead of paddling

The Mzimvubu expeditions aimed to collect and record plants from inaccessible places

them.[15] Even so, on the third day, they were faced with 'a series of violent rapids':

> The rapids dictate split second decisions, keeping to the main stream and avoiding rock obstacles. After the first capsizing [Godfrey and I] recovered all our gear but after the second I lost an old iron framed rucksack. It was fixed with an empty [plastic] bottle to keep it afloat, but the bottle came loose and shot out. Fortunately the plant bag came loose as well and we recovered it, but our rope and other collecting gear sank. I tried several dives but because of the strong current could not recover anything. I was very surprised when, later in the day, a *Gasteria* leaf with a collecting number [attached] came floating past!
>
> . . . Rapid-wise this was the worst day and we were glad when we reached our destination, a cliff face south-east of Isilindini with plenty of firewood to dry out . . . our tents and . . . clothes.[16]

The team sometimes met the inhabitants of nearby villages, discussing with them any hazards that lay ahead. They took local warnings of 'a big waterfall' seriously and accepted help where necessary and possible.[17]

## FINDING NEW TREASURE

The excitingly diverse plant life along the river compensated for the journey's hazards:

> There was a new surprise round every corner. I remember finding a plectranthus dangling in front of my eyes and immediately knowing that it was new to science.[18]

Despite the team's attempts at caution, Harrower and Xaba overturned on another day when 'confront[ing] many rapids'. In the process, they lost a paddle which forced them 'to take turns paddling'. As the team was already running late for their meeting with the backup crew, this new blow made progress more difficult. Van Jaarsveld called a final halt to the first expedition at Majola, just 25 km short of their goal at Port St Johns. It turned out to be a sounder decision than they realised as while van Jaarsveld dozed in the grass waiting for the pickup, he suffered a spider bite on his leg that 'burned intensely'. The site of the bite swelled to the size of a cricket ball and turned black.[19]

The dense collars of vegetation make it difficult to penetrate gorges like the Mzimvubu although in practice, human settlement is quite close. Xaba and Harrower could comfortably climb out of the gorge to walk a few kilometres to find cellphone reception, allowing them to check in with the backup crew. When a participant in a later expedition had to pull out, he was able to walk out of the gorge and soon hail a minibus taxi to take him to the nearest town.[20]

Follow-up expeditions completed the route, varying collecting spots explored. It was three months later, on the second expedition along the Mzimvubu's lower reaches in August 2002, that a strelitzia was first collected. The terrain of the river valley changes regularly from an open expanse to a narrower space squeezing between 'steep embankments', although the central section has less shale and more forest. Van Jaarsveld, accompanied again by Xaba and Harrower, was also joined by David Styles, a KwaZulu-Natal based environmental consultant and amateur botanist. Van Jaarsveld soon realised that the river had, in fact, been quite low during their first expedition. By August, early spring rains had been good and the river was in flood: "The water was flowing strongly. We were flung out of the boats and got quite a fright," he recalled.[21]

In the river's lower reaches, up to 1 750 milimetres of rain falls annually, encouraging growth of 'dense subtropical forests' along the riverbanks. On the fringes of riverine forest, the team spotted 'a colony of *Strelitzia* (clearly related to *S. reginae*)' that seemed to fall into their brief to bring back plants 'showing horticultural promise'. None of the strelitzias was in flower. The team collected 'a single young plant . . . to be grown at Kirstenbosch National Botanical Garden.' It was the 50th plant that they had collected on that trip and would become one of eight plant species so far identified as new to science that were collected on the three 2002 expeditions. None of the team suspected the surprise that the plant had in store for them.[22]

Riparian forest along the Mzimvubu River

## UNEXPECTED SURPRISE

When van Jaarsveld saw the strelitzia that the team had gathered from the banks of the Mzimvubu River blooming at last in Kirstenbosch in February 2006, three and a half years later, he was taken aback. The crest of the plant's flower consisted of the three, typical orange sepals he expected. But the inner petals were not the typical *Strelitzia reginae* dark purple. Instead, they were white.

The difference had been quite unsuspected as the team had previously assumed that they had collected simply another example of *Strelitzia reginae*, which are quite common across the Eastern Cape.[23] When studied closely, it became clear that this plant differed from its relative *Strelitzia reginae* subsp *reginae* in other ways. It flowers from February to June. It also became clear that stigma size, overall size and preferred habitat were different, too:

> We realised that the plant represented a new subspecies of the widespread *S. reginae* . . . It is hypothesised that the white colour of the petals . . is a response to its low light-intensity forest habitat, making the flowers more visible to pollinators . . . the possibility of fruit bats as a possible pollinating agent must be considered.[24]

The following month van Jaarsveld, teaming this time with Gerrit Visser, was back in the area of the Mzimvubu River where the original plant had been collected, near Lutengele, about 40 kilometres downstream of the end of the first 2002 expedition at Majola and about 70 kilometres inland from Port St Johns. Van Jaarsveld and Visser had two aims. The first was to check flowering strelitzia specimens to ensure that the individual plant which had been collected 'did not represent a freak form'. This time they did not have a boat so van Jaarsveld swam across the river to collect plant specimens. Unlike Swedish traveller Anders Sparrman in April 1772, who stripped off (except for his hat) to go botanising on an islet in the Berg River near Paarl, van Jaarsveld regards swimming as necessarily functional:

> I love swimming and the water wasn't ice cold. I had to be careful because the current was very strong in places and coming back, I used a log to help me float across.[25]

To ensure that the plant was not a rare sport, a fleeting genetic mutation:

> The population was thoroughly investigated and all plants seen had flowers with white petals. Plants were again observed growing mainly in the forest but also exposed on the river banks.[26]

Gerrit Visser at Mzimvubu River with blooming *Strelitzia reginae* subsp. *mzimvubuensis*, 2006

## ALL IN THE DETAILS

It was an unexpected petal colour which first alerted Ernst van Jaarsveld to the revelation that the strelitzia he had growing in the Kirstenbosch greenhouse might be a new subspecies. The Mzimvubu flower combines two white lower petals instead of the 'dark purple-blue' of *Strelitzia reginae* subspecies *reginae*. Overall, the Mzimvubu plant and its flowers are slightly smaller than this cousin and its two orange upper sepals turn white towards the base, instead of remaining completely orange. The stigma of the Mzimvubu plant is also short – only about 10 to 15 millimetres long. The stigma in *Strelitzia reginae* subspecies *reginae* may be up to three times as long, 'at least 30mm long.' Van Jaarsveld also recorded a series of subtle differences on the Mzimvubu's leaves compared to its strelitzia relatives:

> especially in younger plants, [the leaf blade] differs in surface texture, often being minutely corrugated and showing a clear reticulate pattern when held up against the light. In subsp. *reginae* the leaves are smooth, not minutely corrugated.[28]

This time the team gathered several wild, flowering specimens to confirm their observations and 'to serve as the type of the new name', which commemorates the remote location where it was first found. This expedition also helped van Jaarsveld to pinpoint the new strelitzia's distribution, which 'appear[ed] to be restricted to the lower Mzimvubu River, occurring mainly on rocky banks . . . consisting of Dwyka Tillite . . . rock in the shade of subtropical forest and below 500m. Occasionally it is found in exposed sites along the river.' The plant was found together with typical forest trees such as umzimbeet, Zulu cabbage tree and wild plum (*Millettia grandis*, *Cussonia zuluensis* and *Harpephyllum caffrum*) and forest plants such as fairy crassula, large-leaved dragon tree and Eastern Cape giant cycad (*Crassula multicava*, *Dracaena aletriformis* and *Encephalartos altensteinii*).[27]

From its usual habitat to its white inner petals, the new Mzimvubu strelitzia spelt shade-loving plant. This should make it useful in horticulture, particularly as it flowers earlier in the year than its *reginae* subspecies cousin, as well as flourishing in shady situations while the *reginae* subspecies prefers semi-shade to full sun. Van Jaarsveld sees a good horticultural future for the new subspecies, from the shrubbery to an indoor container, where the

*Strelitzia reginae* subsp. *mzimvubuensis* has rounded, webbed leaves and white inner petals

Mzimvubu strelitzia's compact size would be a benefit. It flowers well and 'is also easily grown' by a patient gardener, making it potentially a popular garden plant.[29] Like the *Strelitzia reginae* subspecies *reginae*, the Mzimvubu subspecies grows slowly, taking 'about four to five years' to flower from seed. From late summer to midwinter (February to June in South Africa), it produces from three to six blooms per flower head. Each 'is presented for four days, replaced by a new flower in succession.' Like other strelitzias, it can be propagated by division or grown from seed, which is best sown in spring or summer in areas that offer its perfect germination temperatures between 23 and 26 degrees Celsius.[30]

Since the plant's discovery was announced in 2007, seeds have been donated to the Royal Botanic Gardens, Kew, and requests for seeds have also come from across South Africa and around the world, including the strelitzia-lovers of Japan. "There is a definite market for the plant and seed if we can get enough material," Van Jaarsveld believes. However, Kirstenbosch's plant and seed propagation programme still needs to source more wild seeds and produce enough stock before it can be distributed to the public.[31]

## THE NEW PLANT HUNTERS

Both early professional plant collectors and their enthusiastic amateur counterparts were intoxicated by the abundant variety of the new habitats they explored. The next wave of plant collectors typified the 19th-century colonial appreciation of cataloguing plants as well as exploiting them for their economic potential. As the 20th century passed, collectors' labels show how their records expanded, including detail from grid references on maps to precise habitat accounts. Digital cameras make it easier to take photographs in areas where film might be damaged by extreme heat or humidity. They also allow plant collectors to take many more images, both documenting plants collected and their habitat and incorporating additional layers of information including GPS coordinates. The photographs can also, for example, be used to form a visual diary as part of the overall expedition record, illuminating the numbered sequence of plants and seeds and the jottings of a field notebook.[32]

Today's plant collectors usually try more than ever to understand the concerns, rights and culture of local peoples living in collecting regions. Van Jaarsveld, for instance, recorded how the team would discuss uses of trees and plants with local people whom they met on their Mzimvubu River expeditions. This is similar to the approach of Amazonian researchers, who often draw together botany, anthropology and indigenous knowledge systems such as in the conservation-focused projects researching natural habitats and sustainable use of plants of William Milliken, head of the Tropical America Programme at the Royal Botanic Gardens, Kew (see chapter 15).[33]

Whether within southern Africa or far beyond, all the strelitzia family members grow naturally in areas where environmental protection is becoming increasingly precarious. Their natural habitats are all also marginal areas which, because soils tend to be poor, have not been primary targets for farming. But the pressure of increasing human population means human activities have begun to eat into these habitats. Within South Africa, in the Eastern Cape's Albany centre of plant endemism, farming, housing and infrastructure needs have all extended further into grasslands. Concerns over loss of the natural and tourist value of the local flora have re-echoed once more following proposals for eucalyptus plantations and mining – and already violent, even deadly confrontation has occurred.

Patricia McCracken

Pondoland's coastal plateau sourveld more or less covers the same area as the Pondoland centre of plant endemism. This type of sourveld is 'the smallest of the 70 South African veld types' in Acock's 1970 classification, for example, but has the highest grassland plant density and about one in 16 of its 1 800 vascular plants is endemic or near-endemic. Yet this grassland type is so 'seriously threatened by overgrazing, agriculture and excessive burning' that it is 'rare outside conservation areas'. Much of the area's grassland occurs on a bed of quartzite sandstone, which initially made it less useful for growing staple crops and so it was 'sparsely settled until recently'. This grassland's naturally poor fertility may ironically contribute to the large numbers of diverse species which have developed there. So far there has not yet been any significant response to calls dating back to the 1970s for a Pondoland national park to be declared.[34]

Pondoland's seriously threatened grassland plateau

Sally Townshend, *Strelitzia reginae*, watercolour, 2017

Pondoland's own geography, criss-crossed by river gorges, may have reinforced its biodiversity. Cliffs can be more than a barrier to humans and animals. In these remote situations such terrain encourages 'local in-breeding and little gene exchange . . . from neighbouring [plant] populations . . . which leads to local speciation', believes van Jaarsveld.[35] Often remote gorges shelter and protect palaeo-endemics, evolution's survivors, that probably 'previously occupied a wider range spanning the Eastern and Western Cape during a moister climatic regime.' This has enabled the wild strelitzias we know and encounter today to have succeeded in surviving as continents shifted and evolution moved on. Southern Africa, for example, is left only with remnants of the warm, damp forest and thick woodlands that once covered much of it about 60 million years ago.[36]

Strelitzias are not alone in having been hard hit by changes in climate and vegetation that in southern Africa made them retreat towards the region's eastern coast. Occasional members of other plant families were left stranded in a habitat that has undergone dramatic change. The much-admired clivia family is even more of a forest dweller than strelitzias yet there was a major upset in botanical thinking in 2001 when a new clivia was discovered in the unlikely setting of the arid Northern Cape. This is about 1 400 kilometres west of the family's stronghold near to the Indian Ocean coast. Conservation officer Wessel Pretorius from the Oorlogskloof Nature Reserve near Nieuwoudtville, Namaqualand, included it with a batch of herbarium specimens sent to the Compton Herbarium in Cape Town for identification. Within days, the keeper of the herbarium, John Rourke was at Oorlogskloof to see for himself this extraordinary find growing in its natural habitat. Rourke also wondered whether it might not eventually be found 'further down the Oorlogskloof Canyon . . . as numerous suitable habitats occur there.'[37] Compared to the clivia's usual damp, shady forests, its adaptation to the dusty, dry Northern Cape was so marvellous that Rourke named it *Clivia mirabilis*:

> Rarely can such an extravagant epithet as *mirabilis* be confidently applied, yet in the case of this extraordinary *Clivia*, so unusual in its distribution and characters, its usage seems entirely appropriate.[38]

## STRELITZIA LOSES GROUND

Many millions of years ago, at a time when the whole world's climate was mainly 'warm and humid', strelitzias thrived in a very different kind of moist forest and thick woodland stretching far west and north across southern Africa. Then a long and important period of ancient climate change began. The South Atlantic high-pressure cell became established about 30 million years ago, making the subcontinent's climate drier and causing seasonal patterns to begin developing. During the miocene period, from 24 million years ago to five million years ago, the vegetation of southern Africa altered dramatically as the region's climate changed substantially. It was influenced also by a huge plateau developing in central southern Africa and the Benguela current becoming established along the west coast. Large swathes of the region then became much drier, even arid, as winter rainfall patterns became established in the subcontinent's west. The forest and woodland retreated ahead of advancing fynbos, which was better able to survive and thrive in the new conditions. Up until about 15 000 years ago, phases of climate change were still forcing more subtropical flowers such as strelitzia and clivia to retreat back towards the eastern coast where the climate remained warmer and more humid. The climatic changes left them unable to withstand the fierce competition from fynbos species and they lost large extents of their previous ranges:[39]

> adaptations to drought and fire gave a selective edge to the Cape floral clades [groups] over the tropical flora, and as summer drought increased during the Miocene, as indicated by fluctuations in strength of the Benguela upwelling, the area occupied by fynbos increased. This could have led to the competitive displacement and eventual exclusion of Palmae (Arecaceae), Casuarinaceae and Winteraceae, which were present in the Cape Floral Kingdom during the Miocene . . . fynbos did not invade into a vacant area but competitively displaced the existing vegetation.[40]

Jeanette Loedolff, *Strelitzia reginae* var. *mzimvubuensis*, colour pencil, 2007

Aeons of climate change have meant that strelitzias have often been left with only a small number of related species nearby – so far as we currently know. Instead their other closest genetic relatives are half a world away or on farflung islands, with the family's strongholds bridging South America, southern Africa and Madagascar.[41]

It is not yet clear whether *Strelitzia reginae* subsp. *mzimvubuensis* is a relic, ancient palaeo-endemic or, in fact, a relatively younger neo-endemic. This type of endemic may have developed from plants isolated in remote areas such as inaccessible river gorges. Or they may have migrated north east during climate change over the past five million years or so during the plio-pleistocene era, contributing significantly to the links between Pondoland's plants and those in a string of other specialised centres of endemism scattered up South Africa's east coast, from Albany through Pondoland to the Drakensberg in the west and Maputaland in the north:

> There is . . . a clear path of migration between the Pondoland Centre and the KwaZulu-Natal Drakensberg (Drakensberg Alpine Centre) via the Ngeli range. Plants may have migrated from Pondoland northwards via the Natal Group sandstone outcrops to the Afromontane regions of the north-eastern Drakensberg Escarpment, including the Barberton, Wolkberg and Sekhukhuneland Centres.[42]

It is probably no coincidence that two of the new strelitzias recognised in the past century in southern Africa, *Strelitzia caudata* and *Strelitzia reginae* subsp *mzimvubuensis*, were found in underexplored areas of this migration path.

Botanists working in Pondoland believe the area still hides plenty of unusual novelties waiting to be discovered, especially as a detailed study of the neighbouring Albany centre of plant endemism estimated its species population as between 5 000 and 6 000 species, up to 50 percent higher than earlier estimates and also with a higher proportion of endemics.[43] Understanding these centres of endemism better might also explain current mysteries such as distribution of *Strelitzia caudata* and gaps in *Strelitzia reginae*'s distribution between extreme southern KwaZulu-Natal and the Hluhluwe area of Zululand. Most of all, the process might yet reveal more strelitzia novelties – or perhaps change the way we understand the strelitzia family.

## THE CONSERVATION DILEMMA

The Mzimvubu strelitzia is so far known only from the small area in the lower reaches of the river gorge where Ernst van Jaarsveld first found it. The United Nations Development Programme (UNDP) has recognised that a major challenge which South Africa faces in conserving biodiversity is that 'many of its globally important habitats are home to poor communities who rely on natural resources to sustain their livelihoods and further action is needed to integrate biodiversity conservation efforts into sustainable development initiatives designed to abate poverty.'[44]

Much of the hinterland of the Wild Coast, where the Mzimvubu River meets the Indian Ocean, is 'exceptionally rich floristically' but many of the forested areas 'are heavily utilised for firewood, medicinal barks and poles for domestic use.'

The rural population has already needed to use grasslands for subsistence farming, including grazing, leaving most of them 'degraded . . . [with] loss of floristic diversity and increase in unpalatable grass *Aristida junciformis*.' The United Nations Development Programme has concluded that even the 'business-as-usual scenario' means that:

> the biodiversity of global and local significance will continue to degrade due to increasing pressures from the subsistence needs of local communities who have few other livelihood alternatives other than the use of natural resources. In the absence of an effective conservation and sustainable use planning and implementation structure, increasing commercial development pressures from the tourism industry, mining and construction could lead local authorities to make land use decisions that are contrary to sustainable development principles.[45]

Within the 3 260 hectares of just one small nature reserve, Umtamvuna, researchers recorded more than 1 300 vascular plants. By comparison, this is only slightly less than the whole number of plant species in the Kruger National Park – or the whole of Great Britain:[46]

> The Pondoland centre is the only local centre of endemism in southern Africa outside the Cape Floristic Region to harbour an entire family of flowering plants, . . . the Rhynchocalycaceae . . . The Pondoland Centre is characterised by rugged plateaux deeply dissected by narrow river gorges.[47]

On van Jaarsveld's Mzimvubu expedition, the high riverside cliffs of Beaufort shale repaid their attention as they were home to plants such as a blue scilla or slangkop (*Merwilla plumbea*), blue-leaved cliff daisy (*Senecio talinoides*) and klipbloom (*Crassula perfoliata* var *perfoliata*). At the nearby campsite, trees found underlined the expedition's importance in establishing and broadening the range of both known and unknown plants as they noted the southernmost record for a tambotie (*Spirostachys africana*), as well as a northern record for the Kei bauhinia (*Bauhinia bowkeri*), 'formerly only known from the Bashee (Mbashe) River and further south.'[48]

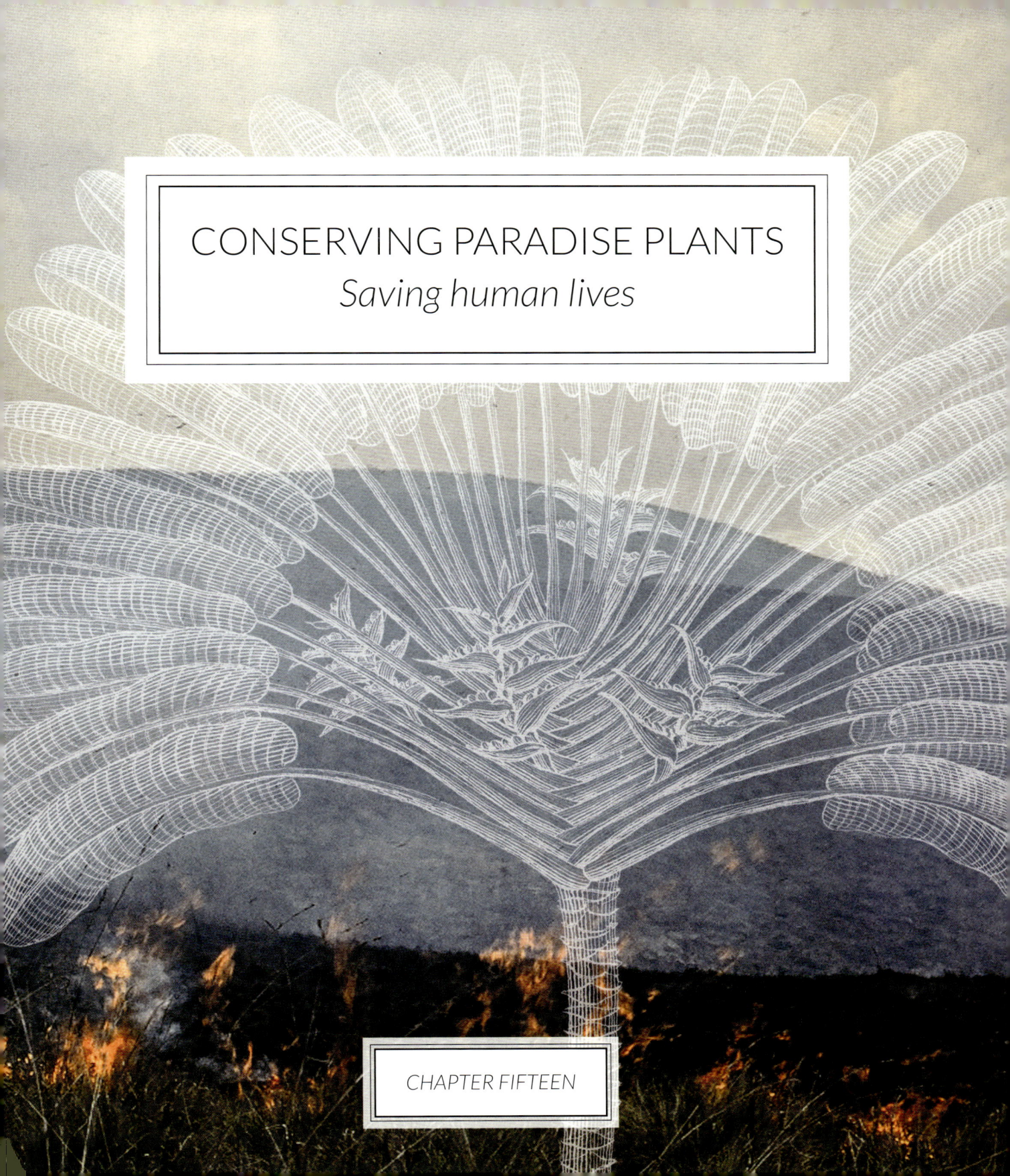

# CONSERVING PARADISE PLANTS
*Saving human lives*

CHAPTER FIFTEEN

# 15  CONSERVING PARADISE PLANTS: Saving human lives

All members of the strelitzia family – from Madagascar's traveller's tree (*Ravenala madagascariensis*) and South America's big palulu (*Phenakospermum guyannense*) to the various treelike and shrubby birds of paradise of southern Africa – favour remote natural regions renowned for their diversity of plants and animal species. Some strelitzias occur in thick forests or almost impenetrable swamps. Others are found in semi-barren, rocky grassland. Until relatively recently, all these areas have been less exploited by humans, often because they are remote and conditions there can be inhospitable. Amazonia's dense forests, for example, are also poor in nutrients, which initially encouraged any humans living there to forage far and wide as hunter gatherers (see chapter 8). Now, however, people instead tend to convert land for farming, housing and industry, with roads and other infrastructure developed to serve expanding human populations. This has intensified the tension between transforming natural habitats and investigating the threatened species that live there. Researchers are still attempting to identify scientifically all the species in such areas, with 1 730 vascular plant species identified in 2016. Meanwhile bioprospectors are searching to improve the treatments for old diseases in humans and to discover cures for new ones. In the past quarter-century, astonishing discoveries at minute biochemical levels have meant members of the strelitzia family have become one of the groups of plants that could promise effective cures for expanding diseases such as diabetes.[1]

In some strelitzia territories, the slash and burn that often marks humanity's encroachment has been happening for centuries. However, instead of light exploitation by nomadic hunter-gatherer societies or occasional visits from hunters, such regions now struggle. They are part of the 340 million hectares 'of the earth's vegetated surface' that burns each year.[2]

Guyana, which has been a 'fragile frontier' since the 16th century when Dutch and Spanish traders founded the logging trade there, includes part of the heartland of the big palulu, for example. Within half a century of coastal land being cleared to grow crops for export, the forest and the indigenous peoples who lived there had retreated into near-invisibility for about 200 years. But as sales of Brazil's hardwood grew sharply in the 1980s and timber began to be harvested increasingly from the Amazonian forests, international investors began eyeing the areas that had seemed too remote or too challenging to attract commercial interests. By the late 20th century, Guyana's forests once more seemed to promise 'limitless wealth' to multinational lending institutions and, this time, to 'companies based in southern, rather than northern, nations' such as Malaysia. By the 21st century, Guyana had once again become a hub for logging, mining, plantations and intensive farming, as well as establishment of significant hydropower and industrialisation.[3]

Ecuador similarly lost half its surviving forest, 'mainly due to illegal logging', between 1975 and 2005. It is not clear yet what the impact of this human exploitation will be on the environment in general on the range and quantity of species that can survive and thrive in particular and on the forces of climate change. But in the once-remote rainforests of the Amazon region and northwards across the equator to the Guyanas and Venezuela, the scattered communities of the big palulu are as threatened as their relatives in Madagascar and southern Africa.[4]

© The Trustees of the Natural History Museum, London

Overleaf: Bush fire, Madagascar; detail from Pierre Sonnerat, *Ravenala madagascariensis*, 1772
Above: Franz Bauer, *Strelitzia reginae* from *Strelitzia depicta*, 1818

## FLAGSHIPS AND UMBRELLAS

The traveller's tree and the bird of paradise flower are so instantly recognisable that they have helped make the strelitzia family a recognisable conservation flagship species in their regions. Like several areas of South America, including Amazonia, Madagascar was recognised by a 2013 study as one of 10 priority areas globally where biodiversity hotspots overlap with potential impact of climate change on small-scale farmers.

Four out of every five Madagascan plants occur nowhere else on earth, making the island home to more than 11 000 endemic species of vascular plants. In Madagascar's north-eastern Masaola peninsula, increasingly heavy human settlement in the four decades between 1950 and 1990 caused most of the deforestation, with the traveller's tree also at high risk because it is widely used by local inhabitants.

One study found, 'Local inhabitants already use more than 100 species of rain forest trees destructively', killing plants during harvesting for firewood, building houses, making dug-out canoes, handicrafts or for food. Ravenala was the most important species in three of the four house construction categories (floors, walls and roof) and one of three food categories (palm heart), with 97 per cent of local inhabitants surveyed using it in one or more ways. Most local inhabitants were surviving within a subsistence economy, with generally only about one in 20 inhabitants of each of the seven villages surveyed having 'truly entered into a cash economy'.[5]

Previous attempts to conserve umbrella species such as lemurs do not seem to have succeeded in conserving the biodiversity of their environment as well as originally hoped. In one experimental conservation approach, a 310 000-hectare section of rain forest in the region was split into two levels of protection.

About two-thirds (210 000 hectares) would be declared a national park with 'absolute protection', while the forest in the 100 000 hectares surrounding this core would be managed to prevent agriculture but allow multiple sustainable uses, such as harvesting species already used by local villagers. New markets for forest products that were not previously exploited, such as harvesting palm seeds to supply the ornamental palm market, would also be created:[6]

> the two prongs of the . . . strategy . . . will have important influences on the way that local people use plant resources, with subsequent influences on rain forest ecology and the local economy. By expanding access to existing domestic markets and developing new international markets, they will begin to have access to cash, which may encourage them to substitute purchased items for products they formerly got from the forest (e.g., tin roofs instead of thatch).[7]

Patricia McCracken/courtesy: Museum of Economic Botany, Royal Botanic Gardens, Kew

## 'THE MORE THEY LOOK, THE MORE THEY FIND'

Amazonia combines the world's largest tropical rainforest – estimated as about the size of western Europe – and the world's largest river network in its ecosystem. Its vast size of 17.4 million square kilometres makes it a major focus of international concern to preserve as much of earth's biodiversity as possible. The region's half a billion hectares of forests contain almost 400 billion trees, made up of about 16 000 different species. About a quarter of these have not yet even been scientifically collected as a first step to identifying them correctly. This extremely rich diversity may have resulted from plants adapting in different ways to survive in this difficult environment.[8]

Over the centuries, Amazonia's forests have lost as much as a fifth of their estimated original area of 3.5 million square kilometres. In just four months in 1997, satellite images showed 'almost 45,000 separate fires' in the Amazon, the majority running disastrously out of control after being started by landowners and causing hundreds of square kilometres of forest to disappear. The following year, fires raging during an El Niño drought 'destroyed over 1,1 million hectares' of forest. Amazonian rainforests are especially vulnerable to fire because it is rare in their generally damp conditions. Most tree species there 'possess thin bark and can be killed by even low-intensity surface fires.'[9]

Rice pan, bottle holder and basket collected in Madagascar, made using *Ravenala madagascariensis* fibres

About 80 per cent of the Amazonian rainforest survived up until 2005, with two thirds lying within Brazil. However, one of Brazil's two main rainforest areas, the Mata Atlântica that 'once covered huge areas along the country's Atlantic coastline', had already been reduced to 'only slivers' by then. The country's construction industry turned its attention to the Amazonian forest at the same time as soya-bean farming expanded dramatically in the area. At the peak of this boom, about 30 000 square kilometres – an area larger than the American state of New Jersey' – was being lost every year. Brazil slowed deforestation by 80 per cent between 2005 and 2013 but 6 000 square kilometres of forest were still turned into new farmland every year. There was a 'meteoric rise' in cattle ranching, with the Brazilian national herd almost doubling to 35.5 million cattle between 1985 and 1996. Soya-bean farming 'increased dramatically from 33 [hectares] in 1975 . . . to 1.66 million [hectares] in 1996' and was the country's leading agricultural export by 2005. It has been estimated that in the three and a half decades between 2015 and 2050, between 9 per cent and 28 per cent more of the Amazonian forest could be lost. This would affect soil erosion, availability of water and other natural goods, generally impoverishing habitats. It is feared that deforestation and change of land use could further intensify changing climate patterns, already affected by the huge amounts of carbon dioxide released over the past century and a half through burning fossil fuels such as coal, oil and natural gas.[10]

## FIRE AND RAIN IN AMAZONIA

The Amazon basin provides about a sixth of the world's freshwater and is home to about two-fifths of all the rainforest that remains in the world, including up to 30 per cent of the world's plant species thanks to its many different habitats allowing a huge diversity of species to thrive. But '[b]ased on current trends in agricultural expansion, 40 per cent of the Amazon forests are predicted to be eliminated by 2050'.[11] Already satellite photographs helped researchers estimate that the Amazonian cleared forest land area expanded nearly fourfold between January 1978 and August 1998 from 155 200 square kilometres to 551 782 square kilometres.

More than 10 per cent of the earth's vegetated surface demonstrates high sensitivity to climatic variability. Forests tend to act as so-called carbon sinks, absorbing many times more carbon dioxide in their photosynthesis than agricultural crops. Losing Amazonia's trees could speed up global warming by releasing up to 140 billion metric tonnes of carbon. Fewer trees produce less water vapour, which would cause lower humidity and also lower rainfall – possibly intensifying conditions where drought wildfires can take hold. In a vicious circle, rainfall can then lessen even more as water vapour is trapped in the ceilings of smoke lowering over the forests.[12]

*Phenakospermum guyannense* habitat destroyed by Amazonian slash and burn

While politicians and environmentalists negotiate boundaries to development within wild habitats, balancing 'local values' against 'global norms', natural scientists are in a race against time to catalogue unexplored and underexplored areas. Their results point to the huge numbers of species on the planet about which we know little or nothing. Over the four years from 2010 to 2013, hundreds of scientific expeditions in the Amazon rainforest found on average more than 100 new species a year, or two new species a week. The total of 441 species newly described in scientific terms for the first time during this period included 258 plants, 84 fish, 58 amphibians, 22 reptiles, 18 birds and one mammal. Insects and a further 90 invertebrates were not included in this tally. The mammal was a Caqueta titi monkey (*Callicebus caquetensis*) from Colombian Amazonia whose young 'purr like cats'. A thimble frog (*Allobates amissibilis*) and a vegetarian piranha (*Tometes camunani*) were other fascinating discoveries.[13]

Like the Mzimvubu strelitzia and other new plants from southern Africa's crescent of endemism (see chapter 14), the new Amazonian species are thought to be particularly at risk because they are distributed over very small ranges. The river habitat in Brazil's Pará state where the vegetarian piranha lives is threatened by dam projects and mining. The thimble frog may be lost again almost as soon as it was found because the small area in Guyana which it inhabits may shortly become a tourist destination. A fish species from Peru's Loreto region lives in just one small lake and has, intriguingly, adapted to low-oxygen water.[14] These and many more finds confirm that:

the extraordinary Amazon remains one of the most important centres of global biodiversity . . . The richness of the Amazon's forests and freshwater habitats continues to amaze the world. But these same habitats are also under growing threat. The discovery of these new species reaffirms the importance of stepping up commitments to conserve and sustainably manage the unique biodiversity and also the goods and services provided by the rainforests to the people and businesses of the region.[15]

While governments, environmentalists and industrialists attempt to find a balance between the often competing demands of people and the rich environments that members of the strelitzia family favour, some scientists are proposing that taxonomic data should be used to assess and decide conservation priorities. '[T]here remain large tracts of the basin where botanists have yet to set foot and substantial differences of opinion even over the number of plant species that occur in the Amazon,' commented Kew's William Milliken, a British ethnobotanist who specialises in biodiversity issues and has been working in Amazonia since the 1970s to map and discover their plant wealth.[16]

(Clockwise from top) Tapir; caiman; seeds of *Phenakospermum guyannense* with their characteristic red arils

## THE 'INDIANA JONES OF THE PLANT WORLD'

Typical of the adventures that earned Milliken the nickname of 'the Indiana Jones of the plant world' was his exploration of the heart of phenakospermum territory in mid-2015. He was airlifted by the French Foreign Legion into the border area of French Guyana and Brazil to accompany their 'Raid of the Seven Stones'. He joined them for the first half of their exploration of a region from the Tumuc-Humac mountains towards the headwaters of the Oyapock River, most of which had never before even been properly mapped geographically, let alone botanically, with French botanist Guillaume Odonne taking over for the second half.[17]

All around him, Milliken found "many plants new to me and probably to science." He and Odonne had set themselves four main goals: to describe changes of vegetation along the border, where possible quantifying this; to collect specimens to help document the area's flora; to record any plants that suggested indigenous people had previously lived in the area; and to map the 'important pioneer tree genus *Cecropia*' in the area.

Labouring under the weight of a 30kg pack, his feet blistering and festering in the tropical heat, Milliken gathered small branches, flowers and fruit. He made a visual diary of panoramic photographs with georeferencing by using the Google Camera app on his smartphone, which he also used to note changes in vegetation type, complete with GPS points. Whenever the team struck camp, he had to begin the process of preparing herbarium specimens, enfolding each one in a double-page of newsprint and then sealing all the day's samples in a thick plastic bag to protect them from any tropical rain or unexpected duckings in a river or stream. On rest days from the 180-kilometre route march, Milliken made a modified survey of a plot of vegetation, hacking off a bark sample with a machete to allow for DNA analysis, also saving the time and risk of climbing 30 metres or so to the canopy to collect a plant specimen.[18] This is the kind of painstaking research that may reveal, as van Jaarsveld did with the Mzimvubu strelitzia, a new species or subspecies of phenakospermum flourishing in an isolated corner of Amazonia.

Such expeditions, and all the detailed identification work back in herbaria and laboratories afterwards, combine the thrill of discovering new species with the awe of understanding more fully our planet's plant diversity. Examples of this wealth of diversity have been calculated using a botanical version of a stock check, a floristic inventory, which records all plants found in one small plot. During a 1989 expedition to the Rio Camanaú in Brazilian Amazonia to study Wimire Atroari Indians' traditional uses of plants, Milliken made an inventory of a one-hectare plot. This section of Amazonian *terra firme* forest yielded 645 individual trees, representing 201 species. This small area was so diverse in trees alone that each of those 201 species was represented by only an average of 3.2 trees each. The trees were split between 95 genera and, beyond that, between 34 families, meaning about 19 trees per plant family in that single small plot.[19]

## WHEN THE TRAVELLER TAKES ROOT . . .

Nearly 13 000 vascular plants are 'naturalised outside their native region and nearly 5 000 species have now documented as invasive. Natural habitats of members of the strelitzia family are under threat but because their striking looks have made them welcome in many other parts of the world for more than three centuries, they have also proved invasive in a few areas, too. Ravenala, for example, is increasingly under threat in its native Madagascar although the showy good looks of the statuesque traveller's palm have won it many admirers. Landscapers particularly prize it for use as an accent plant and it has been widely planted throughout the tropics, particularly in the Pacific and Indian Ocean islands.

By 1912, Professor Hubert Winkler of the University of Breslau in Germany could comment that *Ravenala madagascariensis* was: 'Als ornamentale Zierpflanze überall in den Tropen verbreitet [an ornamental plant spreading everywhere in the tropics]'. It was a feature of Perediniya Royal Botanic Gardens in Ceylon, as much as it was of Mary Foster's garden in Honolulu, Hawaii. In the Caribbean, it was '*très bien acclimatée*' in the French Antilles and also featured near the main gate and driveway at the Dominica Botanic Gardens. Paul C. Standley (1884-1963) noted 'a handsome specimen' had been planted in front of the Ministerio de Relaciones Exteriores (Ministry of Foreign Affairs) in San José, Costa Rica. A decade and a half later, he commented that ravenala occurred horticulturally in Guatemala, 'planted occasionally for ornament in gardens of Guatemala City, in

Rare Mauritian kestrel uses invasive alien *Ravenala madagascariensis* in Mauritius for lookout perch

the bocacosta of Quezaltenango and San Marcos, and doubtless in other places.'

In a sense, Winkler was prophetic as ravenala has become naturalised in the three Mascarene islands, Reunion, Rodrigues and, particularly, Mauritius where it is no longer a welcome visitor but an alien invader. In its native Madagascar, ravenala plays a natural role in forest succession. But in the different ecology of Mauritius, this behaviour causes it to 'vigorously colonize the highly disturbed montane forests', creating a 'jungle'. It is particularly at home 'on mid-altitude (3 000 to 6 000 metres) south and west facing slopes in and around the Black River Gorges National Park.' There it spreads quickly, creating 'dense stands that shade out all other plant species with the large leaves'. These sizeable areas become virtual ravenala 'monocultures, with little or no native vegetation left.' In the Vallée de l'Est in Mauritius, ravenala and other aliens have been cleared and the land replanted with indigenous plants.

Elsewhere ravenala might be less aggressive and is 'widely cultivated in suburban gardens throughout the tropics' but it shows intriguing signs that it might be adapting itself to its adopted environment. In New Guinea, for example, a variant with a flowerhead at the apex of the stem was recorded. In the north-eastern Australian city of Darwin, the tree seems to be assisted to set seeds by visits from bats (*Pteropus alecto gouldii* and *Macroglossus lagochilus*) and birds, particularly the honeyeaters (Meliphagidae, especially the banana-bird or blue-faced honeyeater, *Entomyzon cyanotis*) although these seem to prefer paperbarks (Melaleucaceae, members of the myrtle family), if available, to ravenala seeds. The ravenalas' blue-jacketed seeds are sometimes being eaten by orioles (*Oriolus sagitattus*). In Darwin, ravenala appeared to propagate both through local pollinators and through self-fertilisation but these tend to reduce seed quality and quantity, which might point to ravenala making itself at home but not turning invasive in that setting.[20]

| OTHER PLACES WHERE THE RAVENALA HAS BEEN INTRODUCED INCLUDE:[21] | | | |
|---|---|---|---|
| REGION | COUNTRY | PLACE | STATUS |
| Pacific | Cook Islands | Aitutaki Atoll | Introduced; cultivated |
| Pacific | Cook Islands | Rarotonga Island | Introduced; cultivated |
| Pacific | Micronesia | Pohnpei Island | Introduced; cultivated |
| Pacific | Fiji | Viti Levu Island | Introduced; cultivated |
| Pacific | French Polynesia | Moorea Island, Society Islands | Introduced; cultivated |
| Pacific | French Polynesia | Raiatea (Havai) Island, Society Islands | Introduced; cultivated |
| Pacific | French Polynesia | Tahiti, Society Islands | Introduced; cultivated |
| Pacific | Guam | Guam Island | Introduced |
| Pacific | State of Hawaii | Hawaiian Islands | Introduced; cultivated |
| Pacific | Japan | Bonin (Ogasawara) Islands | Introduced |
| Pacific | Marshall Islands | Kwajalein (Kuwaljleen) Atoll, Ralik Chain | Introduced |
| Pacific | New Caledonia | Ile Grande Terre, New Caledonia Archipelago | Introduced; cultivated |
| Pacific | Niue | Niue Island | Introduced; cultivated |
| Pacific | Palau (Belau) | Babeldaob Island | Introduced |
| Pacific | Philippines | Philippine Islands | Introduced; cultivated |
| Pacific | Samoa | Upolu Island, Western Samoa Islands | Introduced; cultivated |
| Pacific | Solomon Islands | Solomon Islands | Introduced; cultivated |
| Pacific | Wallis and Futuna (Horne) Islands | Wallis (Uvea) Island | Introduced; cultivated |
| Pacific | Ecuador | Santa Cruz Island, Galapagos | Introduced; cultivated |
| Pacific rim | China | Hong Kong | Introduced; cultivated |
| Pacific rim | Singapore | Singapore | Introduced; cultivated |
| Indian Ocean | Australia | Christmas Island, Christmas Island Group | Introduced; cultivated |
| Indian Ocean | La Réunion | La Réunion Island | Introduced; cultivated; invasive |
| Indian Ocean | Mauritius Islands | Mauritius | Introduced; cultivated; invasive |
| Indian Ocean | Mauritius Islands | Rodrigues | Introduced; invasive |
| Indian Ocean | Mayotte Islands | Mayotte Island | Introduced; cultivated |

Vikash Tatayah of the Mauritian Wildlife Foundation inspects invasive ravenala in Mauritius

## USING PLANT TREASURES

Sustainable conservation depends on carefully calculated sustainable use of diminishing natural resources. It also ideally requires finding new, preferably rewarding reasons to protect areas at risk. One hope is discovering miracle foods to nourish the world's ever-increasing population, especially as westernised diets revolve around only about 30 of the world's 80 000 or so plant species that have edible parts or products, among them the fruits and even heart of members of the strelitzia family. As well as food security, the hope has also long been cherished that plants hidden deep in the rain forest or in some other almost inaccessible region might contain an unknown ingredient that could save humanity from deadly diseases such as cancer, haemorrhagic fevers, TB and malaria, the effects of unexpected new viruses such as AIDS or zika and other scourges which might appear in the future, or replace current treatments where drug resistance has developed.

This hope is nourished by the high proportion of pharmaceuticals developed so far from plants, although interest has waxed and waned over the past couple of centuries. A quarter of medicines prescribed in the USA between 1959 and 1980 were developed using active plant ingredients, for example, despite the pharmaceutical companies having tended to shun plant-based medicines in favour of drugs developed in the laboratory.

Conversely, more than 28 000 plant species 'are currently recorded as being of medicinal use'. For centuries, traditional healers had already used the healing properties of plants by harnessing, for example, the willow bark's natural painkilling salicylates, which were developed into aspirin about 150 years ago. Similarly, for more than 1 500 years, Chinese herbalists used leaves of sweet wormwood (*Artemisia annua*) to treat malaria. It was not until the 1960s, though, that the active ingredient, the chemical artemisinin, was identified and isolated from the plant. Extensive testing and pharmaceutical development over the following 50 years led to nearly 80 countries adopting artemisinin combination therapies for treating malaria. The World Health Organisation currently recommends these therapies for treating malaria parasite infections caused by both *Plasmodium falciparum* and *Plasmodium vivax*, whether uncomplicated or severe, but warns against using artemisinin by itself in case of encouraging drug resistance.[22]

Above: *Ravenala madagascariensis* shown as a picturesque exotic in French Antilles, *L'Illustration Horticole* magazine, 1871; Opposite: *Strelitzia reginae* in *Gartenmagazin*, 1810

Researchers also succeeded in creating the first plant-based vaccine ever to be licensed by the US Food and Drug Administration (FDA) in 2006. Three years later, Pfizer and Protalix agreed to produce a treatment for the hereditary Gaucher's disease, a metabolic disorder, that would be based on therapeutic plant proteins although these are generally both 'complex and hard to make'.[23]

Fewer than 5 000 (about 15 percent) of all the plants so far recorded as having medicinal uses 'are cited in a medicinal regulatory publication'. Reports of therapeutic proteins produced using plants include: anticoagulants; thrombin inhibitors; growth hormones; blood substitutes; collagen replacement; and antimicrobial agents. Treatment for and/or prevention of anaemia; hepatitis; liver cirrhosis, burns, cystic fibrosis, haemorrhage, hypertension, HIV, diabetes and organophosphate poisoning have also been produced from plants. The World Health Organization has also pointed out that about four out of five people in less developed countries rely on or regularly use traditional medicines derived from plants.[24]

## A NEW WORLD OF BIOPROSPECTING

By the 1980s, the pharmaceutical industry was discarding its earlier antipathy to plant-based medicines, soon to be called phytoceuticals. Within a decade, perhaps as many as 250 companies were experimenting with phytoceuticals in the hope of cutting the costs of developing a new medicine – as much as US$500 million by the end of the 20th century – and also shortening the development time of up to 15 years.[25]

Thanks partly to its eyecatching blue aril, ravenala was one of the first members of the strelitzia family to attract interest from phytoceutical developers when researching the benefits of medicinal plants expanded dramatically. The search engine Pub Med maintained by the US National Library of Medicine at the National Institutes of Health featured 115 articles on the subject of plants' antimicrobial activity in almost three decades between 1966 and 1994 but almost twice as many (307) in a single decade from 1995 to 2004.[26] Investigating the potential of traditional herbal remedies seemed increasingly attractive given the likelihood of potentially harnessing possibly centuries of indigenous knowledge, reducing research and development costs.

By 1998, harnessing ravenala's traditional medicinal use became one of '21 breakthroughs that could change your life in the 21st century' according to a cover story in *Life* magazine's autumn special issue, 'Medical miracles'. This research was a project of young biotechnology company Shaman Pharmaceuticals, founded by biochemist and entrepreneur Lisa Conte in San Francisco, California, in 1989 to create prescription drugs from plants.[27] It raised more than US$250 million in funding on stock exchanges from New York to London, as well as from wealthy entrepreneurs and philanthropists. Both Conte and Steven King, who later rose to senior vice president at Napo Pharmaceuticals, the successor to Shaman Pharmaceuticals, experienced personal revelations in the wild about the power of medicinal plants. Conte recalled how during a holiday in Tanzania

in 1989, while working as a venture capitalist analysing biotech companies, she climbed Mount Kilimanjaro:

> I noticed that one of the other climbers, who was suffering from altitude sickness, was ingesting some sort of plant-based conction he had bought from a trailside pharmacy, which actually seemed to help him. There was something about the simplicity of it . . . and just how natural it all felt that inspired me.[28]

King, who took charge of Napo Pharmaceuticals' sustainable supply, ethnobotanical research and intellectual property issues, spent his graduate fellowship studying medicinal plants for a project in which the New York Botanic Gardens partnered with pharmaceutical giant Merck. King's personal epiphany had come when he was studying food and medicine plants in Peruvian Amazonia in the late 1970s. *Sangre de drago* (dragon's blood), the bright red sap of a croton, was the best treatment he tried for the burning fungal infections and blisters on his feet. The sap turned out to be highly valued across South America, used to treat upset stomachs, colds, machete cuts and sore gums and even haemorrhages after childbirth: 'because it contains latex, it's also a liquid bandage, forming a second skin over open wounds'.[29] Such experiences as King's and Conte's own convinced her to send Dr Tom Carlson, an ethnobotanist, academic at the University of California and Shaman Pharmaceuticals, senior director of ethnobotanical field research, bioprospecting in Guinea in West Africa.

## LOOKING FOR A RAVENALA MIRACLE

One strelitzia family member, the Madagascan ravenala, was so well naturalised in both West Africa and India that traditional healers were already using its leaves to treat diabetes and kidney stones.[30] In West Africa, traditional healers also brew a decoction from the tree's leaves and stalks to treat diarrhoea and anorexia. This made ravenala a strong candidate for Carlson to investigate on his bioprospecting trip to Guinea. If Napo Pharmaceuticals could isolate an effective active ingredient from ravenala leaves, it could perhaps become a contender to be the next metformin, an anti-diabetes medication marketed as glucophage and based on the traditional remedy goat's rue (*Galega officinalis*) that made US$579 million in sales in 1997 alone for manufacturers Bristol Myers Squibb.[31]

In Guinea, Carlson and his team of 27 local doctors, botanists, traditional healers and translators[32] found the incidence of type 2 diabetes was rising fast. Traditional healers in the country, regardless of ethnic or linguistic group, were familiar with both what type 2 diabetes symptoms to look for and also how to stabilise the condition. They described treating patients with various typical symptoms including:

> fatigue, increased urination, urine that tastes sweet, and when [the patients] urinate on the ground the urine attracts ants. [Patients] may also have foot sores that heal very slowly. The healers report that when they treat the patients with botanical medicines their foot sores heal, their urination returns to normal, and when they urinate on the ground, the ants do not go to the urine.[33]

Both traditional healers and patients assured Carlson that the ravenala brew worked faster than the most modern treatments for the increasingly widespread type 2 diabetes. They demonstrated how it normalised blood sugar overnight and explained how it had corrected diabetes side-effects such as excessive, sweet-smelling urination and blurred vision. Carlson ordered several kilograms of ravenala leaf samples to be sent back to Shaman Pharmaceuticals' development lab.[34]

Only a small proportion of would-be pharmaceuticals manage to complete all the development, testing and regulatory process to be launched commercially, however. Reproducing results seen out in the field consistently and safely for the majority of patients is the challenge that faces any company attempting to develop a new pharmaceutical.[35] For the Guinean supporters of ravenala's potential, the hope of useful long-term results was slimmer than they knew.

Ravenala was one of nearly two dozen plant species reputed to have anti-diabetic properties submitted from Guinea alone to Carlson's lab at Shaman Pharmaceuticals in San Francisco. The team returned with ethnomedical information on a further 145 Guinean plant species from 118 genera, all used to treat symptoms of type 2 diabetes. Of these, it had already shortlisted 21 as showing potential for development and sent back about 40 kilograms of specimens of these plants. In fact, of 2 300 medicinal plants collected across the globe by Shaman Pharmaceuticals, the company had already tested more than 300 plants for anti-diabetic properties. It eventually filed patents on nine.[36]

Independent Indian scientific studies on traditional medicinal use of ravenala later found that the plant's leaf extracts significantly reduced blood glucose 'in a dose dependent manner', as well as also having a significant impact on heart and kidney side-effects in diabetes by improving blood fat ratios. A Danish study of 40 Madagascan plant species traditionally used to treat diabetes, including ravenala, however, found none held 'promise as specific inhibitors of [two key] carbohydrate-hydrolyzing enzymes'. Similarly, although the seeds had been recorded as having antiseptic properties, a Bangladesh study of plant antimicrobial properties found that extracts of ravenala leaves showed 'very weak antimicrobial activity'.[37] Such conclusions reflect the more cautious approach to plant-based medications since the bioprospecting boom in the 1980s and 1990s.

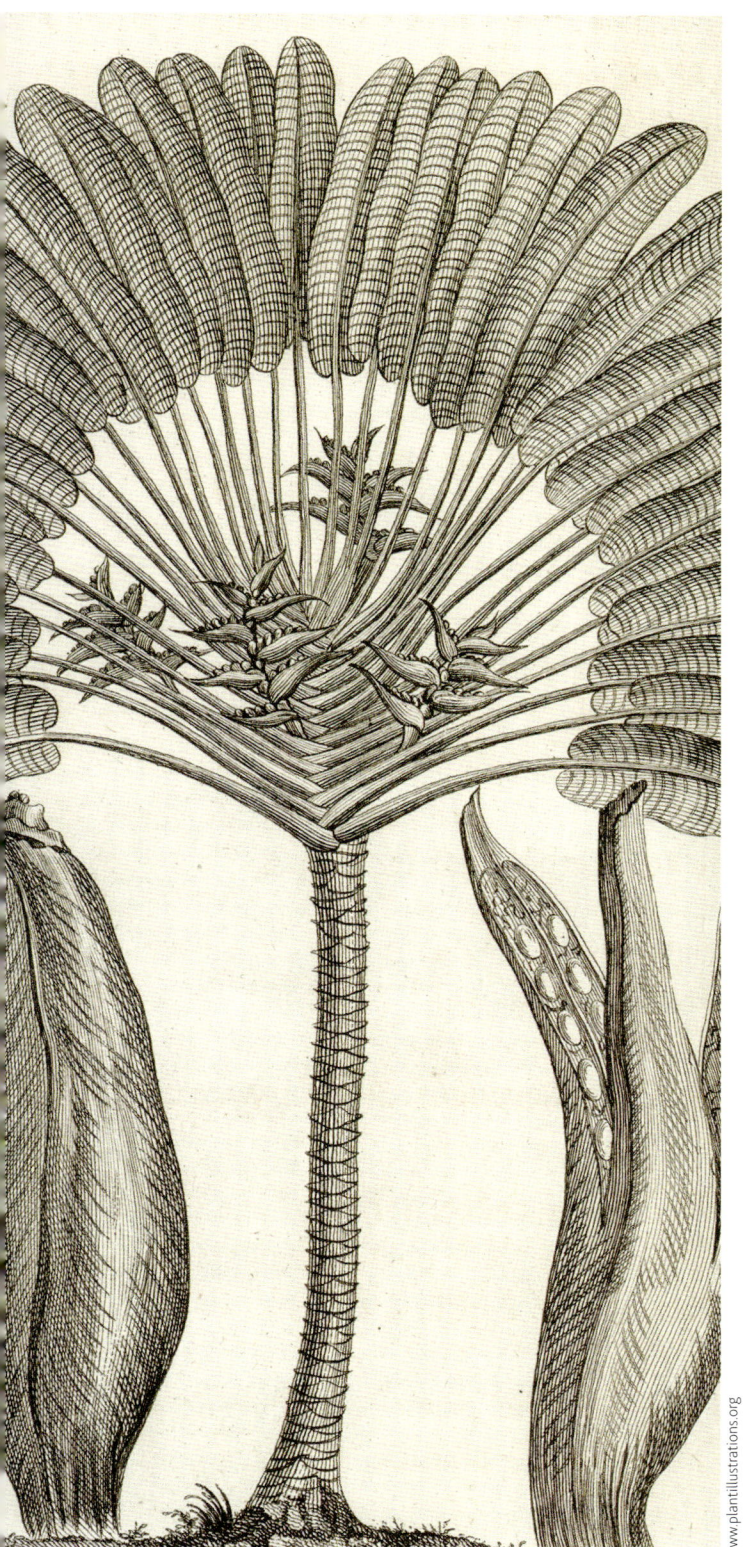

*Ravenala madagascariensis* in Pierre Sonnerat's 1772 *Voyage aux Indes et à la Chine*

## SHARING THE SPOILS

The other question clouding bioprospecting was how to balance and fairly reward the different contributions of traditional healers and pharmaceutical developers. Part of the problem was imbalances of power at several levels, especially as much greater biodiversity could be found in tropical countries of the global south, while 'technology-rich countries, with resources to sustainably develop biodiversity, are primarily in the temperate North.'[38] The climate in tropical countries has helped develop plants that offer rich chemical as well as biological diversity:

> In temperate climates, winter kills many plant predators, and temperate plants flourish in the spring before predator populations increase. But since tropical species have minimal seasonal respite from predators, many have evolved chemical protection from countless predators. The plant chemicals that have evolved to increase plant resistance against bacteria and other infectious organisms of tropical plants may also provide protection and be therapeutically useful for human health.[39]

Conte had envisaged the original Shaman Pharmaceuticals as an ethical pharmaceutical business that would focus extending traditional healers' most effective treatments to benefit the rest of the world's population, while allowing a share of any profits to healers and local communities where the remedies were sourced. This did not mean that Conte saw the business as a non-profit organisation: "You have to think that you're going to make a boatload of money – this isn't charity." Later, she and Carlson incorporated the principles of the Convention on Biological Diversity and Pharmaceutical Research agreed at 1992's Rio de Janeiro Earth Summit into the Shaman Pharmaceuticals' code of practice. They paid particular attention to applying the principle of 'cooperation between government authorities and private sector in developing methods for sustainable use of biological resources' and to establishing 'fair and equitable benefit sharing in contemporary pharmaceutical research on tropical plants.'[40]

It was a fraught area because within a year of the Rio Earth Summit, bioprospecting had become increasingly identified with biopiracy. This pejorative term suggested pharmaceutical companies snatching indigenous knowledge and rights but giving little or nothing in return and was coined in 1993 by Pat Mooney of the Rural Advancement Foundation International (renamed ETC Group in 2001). Establishing and rewarding the intellectual property of indigenous knowledge became an increasingly controversial subject, raised in courtroom battles and lobby groups. Katy Moran, executive director of Shaman Pharmaceuticals' Healing Forest Conservancy, writing on how the company's code had been put into practice in Nigeria, broadly acknowledged that 'Seventy-four per cent of the main 121 plant derived drugs have the same, or similar, use by native cultures.'[41]

## THE HUNT FOR PLANT TREASURE

By the time Carlson went to Guinea, Shaman Pharmaceuticals had already drawn up a system designed to reward local collectors for their work and expertise. Carlson recorded that Shaman Pharmaceuticals paid Guinean stakeholders US$82 000 for collecting the 40 kilograms of specimens sent back to its California lab.[42] A further US$62 920 had been paid, more than half as 'immediate compensation', including daily wages to the Guinean team, the rest in 'reciprocity' initiatives from workshops to improving water supplies or assisting schools. Planned long-term benefits would follow Shaman Pharmaceuticals' bioprospecting, 'If a marketable product is developed from research on plants collected from Guinea or any other collaborating country', with half the benefits going to government conservation programmes and half to communities to support 'conservation of cultural and biological diversity.'[43] But the Shaman Pharmaceuticals researchers did not succeed in making a breakthrough drug from ravenala and the Guinean collaborators on the ravenala project missed their long-term payday.

Shaman Pharmaceuticals became one of many casualties in the biotech sector in the closing years of the 20th century. Conte slashed staffing from about 90 to a dozen, concentrating on subcontracting and outsourcing its product development. This was mainly due to the long, arduous and expensive process of isolating active plant ingredients and formulating a drug that would win approval from the US Food and Drug Administration. The company was finally liquidated in 2005 but Conte had already reappeared as CEO of Napo Pharmaceuticals in 2001.

After discovering that *sangre de drago*'s sap had binding properties that also made it an effective treatment for diarrhoea, Shaman and Napo focused on this application of the new treatment. Napo partnered with Salix Pharmaceuticals to steer crofelemer, trademarked as Fulyzaq, the drug which finally emerged from the *sangre de drago* research, to FDA approval. But this would be a lengthy saga of repeated human trials, bankruptcy and lawsuits before the FDA finally approved the medication at the very end of 2012. Crofelemer was only the FDA's second approval of a botanical drug, which it defines as 'consist[ing] of vegetable materials'.[44] It draws a line between 'Botanical drug products [that] often have unique features, for example, complex mixtures, lack of a distinct ingredient, and substantial prior human use' and 'Fermentation products and highly purified or chemically modified botanical substances', noting that: 'A botanical drug's special features require consideration and adjustment during the FDA review process'.[45] All the development, testing and negotiation had succeeded — but it had taken a quarter of a century to achieve approval for just one rainforest pharmaceutical.

The long journey from plant to pharmaceutical often put off better-resourced organisations than Napo Pharmaceuticals. Eli Lilly's anti-cancer drug derived from Madagascar periwinkle (*Catharanthus rosea*) is said to make the company US$100 million a year.[46] But, pointed out King from Shaman and Napo, the US National Cancer Institute 'which had screened some 35,000 plants and animals since 1956 in its quest to find anti-cancer compounds, abandoned the program in 1981 after largely failing to identify new treatments.'[47]

The lengthy and fraught process of achieving FDA approval for crofelemer caused a schism between Napo and Salix. Conte re-emerged in 2014, though, raising US$70 million in funding as CEO of Jaguar Animal Health, whose main product would treat diarrhoea in dogs. Such persistence could bring substantial rewards as in 2015, the US market for dietary supplements was valued at about US$30 billion in 2015. It was also estimated to be growing at about 7 per cent a year overall — and double that in the online sales sector.[48]

## ARIL DETECTIVES

Those with hopes for the pharmaceutical future of ravenala leaves are still waiting for that spotlight to return to them. In the meantime, strelitzias still moved to the forefront of biochemical research following a revelation which ensured textbooks on human, animal and plant evolution had to be updated. While Conte, Carlson and King looked for new pharmaceutical treasure in jungles and forests, laboratories and boardrooms, on the opposite coast from Napo's California headquarters, biology postgraduate student Cary L. Pirone was working on the structure of strelitzia arils in Miami, Florida. She was trying to identify the pigments that so dramatically attract attention to strelitzia seeds, from the sky-blue jackets on ravenala seeds to the deep red tufts on phenakospermum and the flaming orange arils on giant white bird of paradise seeds.[49]

Ravenala seeds are cloaked in a fibrous sheath so blue that its colour has 'an intensity and depth quite unusual in nature', two Australian researchers observed, prompting them to find out why.[50] Jeffrey A. Crowther and Graeme R. Hanson were particularly intrigued as 'throughout the plant kingdom in general, the colour blue tends to occur less frequently and usually as a pale blue or purple-blue' that is produced by a flavonoid plant colourant, anthocyanin. But the ravenala blue was so intense that it strongly reminded Crowther and Hanson of hydrated copper sulphate or nitrate. After failing to extract samples from arils using solvents, they tried spectrometry and concluded that the resonances they observed could be caused by a combination of copper, manganese, an organic radical and iron. A 'novel blue protein pigment complex' was later identified by Pirone and her colleagues, David Lee and Leung Kim at the Florida International University and W. John Kress of Washington's Smithsonian National Museum of Natural History, an expert in the broader Zingiberales group that includes bananas, gingers, heliconias and strelitzias.[51]

*Strelitzia reginae* in *Revue Horticole* magazine, 1909

Plant colours can be greatly influenced by their ecology, physiology and development and, like the Australian aril detectives, from 2006 onwards Pirone and the Florida team were trying to fill in some of the major gaps in our understanding of how and even why plant species have different colours from each other. Plant colours are created by colour compounds from one of four classes, the chlorophylls, carotenoids, flavonoids or betalains. In 1958, for instance, American chemist Harold Strain had noted a 'trace' of α-carotene in *Strelitzia reginae* and 'some' in *Ravenala madagascariensis* compared to 'much' in heliconia. Pirone and the team noted in 2006, 'Numerous lesser pigments also exist, but little information is available regarding their chemistry and biology.' Carefully examining arils from ravenala, phenakospermum and the giant white bird of paradise showed that the arils of all three had similarly sized cells – but the areas where pigment occurred varied from one species to another, as did the shape and size of the pigment cells. In ravenala arils, they discovered, colour occurs in many tiny centrally located air pockets (vesicles), each about one micron wide. In phenakospermum, pigment forms long crystals. But colour crystals are shaped more like cubes in the giant white bird of paradise.[52]

The mystery still facing the researchers was about more than shape, size or position of the pigment cells, though. They were perplexed to find that they could not at first identify 'several' of the pigment compounds they found in the arils – because they did not fit anywhere in the generally accepted four classes of plant pigments.[53] It was a botanical riddle that demanded to be solved.

The researchers decided to focus on the giant white bird of paradise, *Strelitzia nicolai*. Finding arils to work on would be easy as Florida is one of the tropical and subtropical areas where this tree of South African origin has become widespread. But what neither Pirone nor her colleagues immediately suspected, though, was the enormous surprise waiting for them at the end of a trail strewn with negatives. In fact, it had seemed so unlikely that only 'Pirone's persistence and curiosity' convinced colleagues 'that she was on the right track'.[54]

## THE BILIRUBIN SURPRISE

If questioned, most biologists would probably have assumed that the strelitzia's arils were coloured orange by carotenoids, a group of plant chemicals that put the orange in carrots, the yellow in pawpaw (papaya) and the red in tomatoes and red peppers. Or perhaps it could be, as Crowther and Hanson had suspected, flavonoids including anthocyanins, which create the red, purple and blue shades in fruit and vegetables such as brinjals (aubergines), raspberries and cherries. Both carotenoids and flavonoids are 'pervasive in the plant kingdom', whereas the third possible metabolic pathway to creating colour in flowers and fruit, betalain, occurs only in plants of the Caryophyllales order, from beetroot to bougainvillea.

The chemicals in the strelitzia 'demonstrated unusual properties' in preliminary tests. In fact, at first the orange compound in the strelitzia arils proved what it was not as they eliminated one by one likely identities for the increasingly mysterious pigment colouring the arils blazing orange. It did not belong to the orange-red of the carotenoid family of plant colourings because it did not absorb light wavelengths under UV-Vis spectroscopy in the pattern that a carotenoid would. Testing the pigment's reaction to polar or nonpolar solvents showed that it did not fit into the flavonoid family of plant colourings, found in red grapes. Nor did it even belong to the phenalenones, 'a rare group of pigments' earlier discovered in several strelitzias and their relatives but ruled out in this case as not 'play[ing] a significant role in colour production'.[55]

Pirone moved on to calculate the mass of a molecule of this mysterious pigment using mass spectrometry. This at least created a shortlist of possible identities. For the first time ever, the astonishing possibility arose that the pigment causing the aril's fiery colour was bilirubin. Until then, bilirubin was believed to occur only in humans and some animals. Bilirubin is created in

Karl Schumann, *Strelitzia augusta* in *Vegetation der Erde*, 1910

animals during the process of breaking down the blood's haemoglobin. The yellowish shades produced in a wide range of physical functions in humans and animals, from bruises, jaundice-sufferers to faeces, are all caused by bilirubin. It was even suggested that bilirubin-dominant species tend to be carnivores or omnivores with substantial amounts of protein in their diets, as opposed to biliverdin-dominant species which are usually herbivores.[56]

The Florida researchers had made an astounding discovery by documenting the crossover of bilirubin into the vascular systems of the higher levels of plants, suggesting it was also a byproduct of chlorophyll in plants breaking down: 'The pigment bilirubin, long known as a leftover in the breakdown of animal blood, [had] turned up in the blood-free world of plants'.[57]

When Pirone and the Florida International University team presented preliminary results at the Botanical Society of America's 2008 conference, she was awarded a student travel prize in the phytochemical category.[58]

## MAPPING A NEW FRONTIER

Bilirubin, called one of the 'pigments of life', is produced along a pathway of tetrapyrroles, chemicals structured in four-ring structures consisting of pyrroles with a carbon unit. But the team's findings suddenly revealed that bilirubin was the first tetrapyrrole to act as 'the primary pigment' and so create 'conspicuous colour in a plant reproductive structure'. In the process, this discovery reversed the previous assumption that 'no tetrapyrroles were known to generate display colour in plants.' Both human and animal haemoglobin and the plant pathways that allow chlorophyll to use light for photosynthesis are built on similar tetrapyrrole structures. But there is a key difference. In plants, the metabolic pathway for plants to produce chlorophyll is found in the chloroplasts. In animals, the enzymes needed to break down bilirubin are split between the cell's cytoplasm and mitochondria.[60]

Succeeding in identifying the arils' mystery pigment had required painstaking preparation and tenacity. Pirone needed to isolate and gather enough of the purified compound to allow her

## THE BILIRUBIN ALERT

Humans produce between 250 and 400 milligrams of bilirubin every day, about 80 per cent from breaking down haemoglobin. This process is 'the only natural progression in humans' which can be tracked by colour. It is the familiar pattern seen when bruises heal. Bruises appear when an impact damages small blood vessels under the skin. A bruise's early blackish purple colour is caused when the red haemoglobin in blood has broken down into haem ('heme' in the USA), together with globin and iron. When the bruise turns greenish, the haem has oxidised into biliverdin. The final yellow stage of a bruise shows that bilirubin reductase has broken down the biliverdin into bilirubin. Bruises are the most visible evidence of this sequence which occurs repeatedly in the human liver, spleen and bone marrow as red blood cells are replaced about every 120 days.

Bilirubin is usually processed by the human liver and disposed of through bile and in faeces. If this process does not work efficiently enough, high levels of bilirubin (hyperbilirubinaemia) can build up in the body, signalled by the yellow skin and yellowed eyes of jaundice. This might happen because the liver has become infected or inflamed in one of the forms of hepatitis. When levels of bilirubin rise too high, this can become toxic and may cause brain damage (kernicterus). This is especially monitored in newborn and especially premature babies. They are at greater risk for two reasons: they are normally born with high bilirubin levels and their livers also often take a few days to begin working efficiently.[59]

*Strelitzia nicolai* aril using light photomicroscopy

to put it through nuclear magnetic resonance spectrometry. Although she needed only a milligram, the purification process she had developed did not scale up effectively. Collecting just that minimal quantity meant she had to run the purification process a couple of dozen times. Performing a battery of tests on the strelitzia aril's bilirubin confirmed that this was the same as bilirubin produced from degradation of haem in the human body, rather than one of the other half dozen or so compounds made up of the same chemical formula as bilirubin but differently arranged and having different properties (isomers). The tests included mass spectrometry tests, where the compound produced spectrum 'match[ing] those of authentic bilirubin standard and previous published data'; and high-performance liquid chromatography to check the compound was the same as human bilirubin-IX α.[61]

Their discovery was 'the first example' of the bilirubin molecule occurring in a higher plant, as opposed to algae and cyanobacteria which 'synthesize molecules similar to bilirubin'. It was such 'a wonderful new discovery', said Kress, that it might mean revising our contemporary understanding of how major plant chemicals such as chlorophyll, essential for photosynthesis, develop and operate, as well as reconsidering the path of plant evolution.

Plant biochemist Dean DellaPenna of Michigan State University remarked, "The most interesting thing is that this suggests the first couple of steps of [bilirubin] degradation are identical in plants and animals." By the time that the discovery of the 'first example of bilirubin in a plant' was published in 2009, the investigation had been broadened. The researchers found bilirubin was not unique to *Strelitzia nicolai* arils but is also present in the arils of *Phenakospermum guyannense* and of *Strelitzia reginae*. While bilirubin appeared as 'granular bodies irregularly distributed throughout the cell' in arils, it was later found in 'elongated structures' in the bird of paradise's sepals where the different plant colour chemicals carotenoids had previously been thought to occur.[62]

Whether the veld is green and studded with many different flowers or parched and monotone, bilirubin had been revealed as the chemical colour which is the signal making bird of paradise flowers stand out like a flag to attract sunbirds and other potential pollinators – or in the thick, dim forests of South America, its vivid arils attract birds to eat phenakospermum's seeds.

## THE NEXT HORIZON

As bilirubin was found in more plants, the question became whether it occurs only in strelitzias or more broadly. Although the testing particularly concentrated on members of the strelitzia family and their broader Zingiberales relatives because they would be more likely to have similar metabolic pathways, there were more surprises in store.

Flowers and fruit from the chosen plants were both tested, except for *Costus lucanusianus*, or spiral ginger, a medicinal plant from Nigeria, as fruit was not available. At least one member of each family in the banana group (Musaceae) was tested except for Lowiaceae because its species are so rare that it was impossible to find enough plant material to analyse. Even so, this was enough to show just how widely bilirubin might be found – in 80 per cent of the plant species tested. This included more distant strelitzia relatives such as *Gastrococos [Acrocomia] crispa*, the Cuban swollen-belly palm from the Arecales, and *Eugenia luschnathiana*, a Brazilian flowering plant from the order Myrtales, which showed 'bilirubin is not restricted to the Zingiberales and may be broadly distributed throughout the plant kingdom.' But given that bilirubin was not found in avocado, for example, the results also showed bilirubin is 'not universal in plants at levels detectable by mass spectrometry.'

Levels of bilirubin are not constant from one family to another, nor even within one species, researchers found. In *Phenakospermum guyannense*, for instance, bilirubin was almost twice as concentrated in one sample compared to another (3.041 mg g$^{-1}$ in the first and 5.787 mg g$^{-1}$ in the second).[63] This may reflect researchers' earlier observation that bilirubin levels in *Strelitzia nicolai* appear to vary according to the plant's development as they were highest 'in mature tissue'.[64]

Pirone and her colleagues have acknowledged that their surprising discovery raises many questions, including whether bilirubin produces colour in plants other than strelitzias, exactly how plants' biochemical pathways produce the compound and whether it is more than 'a mere . . . waste product' of that metabolic process.

Green wood hoopoe with *Strelitzia nicolai* seed

## STRELITZIAS AND THE 'PIGMENTS OF LIFE'

It had been a mystery why the human body apparently squanders energy on converting biliverdin to the potentially dangerous bilirubin, which must then be transported through the body's waste processes and excreted, using yet more energy. Then research began to suggest that far from being only a potentially toxic waste product in humans, bilirubin had more positive properties. In 1954, bilirubin in humans was shown to help prevent oxidation (and potential toxicity) in fatty acids (lipids) such as linoleic acid and vitamin A. Half a century later, it was found that the antioxidant properties of human bilirubin can help protect the body's cells, acting as a cytoprotectant.[65]

These are just two of the various studies that have extended our understanding of how bilirubin works in the human body. It appears to reduce the risk of cancer and heart disease, as well as atopic dermatitis and brain-cell deterioration, including the neurological disease, Lou Gehrig's disease (amylotrophic lateral sclerosis). Appropriate levels of bilirubin appear to counterbalance the harmful effects of free radicals and reduce the oxidative stress that contributes to ageing and age-related diseases. Higher levels of bilirubin in the blood have been linked to a lower risk of diabetes by improving control of blood-sugar levels. They also appear to help prevent colorectal and lung cancers and to protect the heart and blood vessels. A 1994 study linked 50 per cent decrease in total bilirubin to 'a 47 per cent increase in more severe coronary artery diseases'. Doctors in Australia have also shown that infusing the heart with bilirubin after a heart attack reduced damage and improved heart function while recovering.[66]

Although strelitzia arils can still appear vivid orange after being stored in herbaria for more than a century, it has not yet been shown whether bilirubin works in plants, as it appears to do in humans, acting as powerful antioxidant to help prevent ageing and destruction of plant cells.[67] More detailed investigations could once more return the strelitzia family to the frontiers of pharmaceutical research and development – and even suggest a possible reason why ravenala leaves seemed to contribute to controlling type 2 diabetes in West Africa.

## SHEDDING LIGHT WITH STRELITZIAS

Like pharmaceuticals, for centuries skincare products have been developed and formulated using plants. The potential use of strelitzia-sourced bilirubin was not lost on this multibillion-dollar global industry – and is believed to have created a fast-growing market for giant white bird of paradise seeds. From the USA to the Far East, the ingredient is being developed to help skin creams deliver the glowing, clear and young-looking skin that they so often promise. Plants containing bilirubin must also have a means of controlling and breaking down this chemical, researcher Philip Ludwig suggested. In humans, the skin around the eyes is very fine and blood vessels are particularly close to the surface, contributing to the effect of dark circles that are partly caused by bilirubin breaking down.

Together with haemoglobin and melanin, bilirubin and collagen are four key substances that affect skin colour and luminosity (chromophores). Ageing skin generally tends to look more sallow because its collagen fibres have been damaged to some extent by free radicals and by UV light: 'decreased and damaged collagen is less effective at reflecting light back, resulting in a dull, less luminous reflection.'[68] But cosmetics developers hoped that if an extract of *Strelitzia nicolai* could degrade bilirubin, it might reduce dark circles round the eyes and improve sallow skin, for instance.

## WHY COLLAGEN MATTERS

The 'most abundant protein' that mammals produce is collagen. It is found throughout the skin of humans, making up between 70 and 80 per cent of the skin's dry weight. Three types of collagen have been identified, with types I and III making up at least 90 per cent of our skin. The smooth elasticity and healthy glow of the skin of children and young people is a result of the fixed ratio between these two collagen types. But as people age, this ratio is skewed because collagen III levels fall twice as fast as collagen I. As a result, skin treatments which can help prevent the collapse of collagen III are the elixir of the beauty business.[69] The arils of the giant white bird of paradise tree might just be an answer, bringing profits to beauty companies and hope to consumers.

Although success in degrading bilirubin concentrations depended on the concentration of aril extract, levels of collagen III, a marker for ageing skin, improved at three different concentrations during in vitro tests in the laboratory. Tests for 60 days on 50 humans produced better skin tone and radiance, as well as reducing under-eye circles and puffiness and lightening age spots: 'The natural, plant-based Strelitzia aril extract was found to be a skin revitalizer that improves skin tone to yield an illuminated complexion.'[70]

Strelitzias, the plant family that helps epitomise the Gondwana continental drift, is poised to bring new hope to the pharmaceuticals and skin treatment industries. It could also be a key to understanding lingering questions over evolutionary links between plants and animals. Strelitzias were golden horticultural rarities in the 18th century. Today they are as prized in the gardens of the world as they are in the powerhouses of pharmaceutical gold – perhaps ultimately helping ensure their lasting protection, the survival of their farflung natural habitats and possibly discovery of new members of the strelitzia family.

# REFERENCES

*Abbreviations: BM(NH): Natural History Museum, London; K: Herbarium, Royal Botanic Gardens, Kew; PRE: SANBI National Herbarium, Pretoria; SBT: Bergius Foundation Herbarium, Stockholm; UPS: Museum of Evolution, Uppsala; n.d.: no date*

## CHAPTER 1

[1] Yutaro Suzuki, *The strelitzia world*, (Japan, 1982), p.31.
[2] Mary Gunn & Leslie Codd/Hugh Glen & Gerrit Germishuizen, *Botanical exploration of southern Africa*, (Pretoria, 2010), 2nd edition as Strelitzia 26, 'Masson, Francis', p.287-290; Ray Desmond, *Dictionary of British and Irish botanists and horticulturists*, (London, 1977), 'Banks, Joseph', pp.35-36.
[3] William Aiton, *Hortus kewensis*, (London, 1789), vol.1, p.285; Miles Hadfield, Robert Harling & Leonie Highton, *British gardeners: A biographical dictionary*, (London, 1980) 'Aiton, William', p.12; BM(NH), *Strelitzia reginae*, BM000810157, Gascoyne, Masson & Kew, all probably ex hort [online]: http://data.nhm.ac.uk/object/d1e7b807-bbbd-4844-be69-0e60d7303fc7, accessed 29 July 2016, examined 6 October 2016.
[4] Harold Carter, *Sir Joseph Banks*, (London, 1988), p.304; *Curtis's Botanical Magazine*, vol.IV, 1790, plate 119; see Miranda Mollendorf, 'The world in a book: Robert John Thornton's Temple of Flora (1797-1812)', doctoral dissertation, Harvard University, [online]: https://dash.harvard.edu/handle/1/11169773, pp.11-12, accessed 5 July 2015; Abraham Rees, *The cyclopaedia; Or, Universal dictionary of arts, science and literature*, (London, 1819 edition), vol.XXXVI, 'Strelitzia'; Abraham Rees, *The cyclopaedia; Or, Universal dictionary of arts, science and literature*, (London, 1819 edition), vol.XXXIV, 'Strelitzia'; Wilfrid Blunt, *In for a penny*, (London, 1978), pp.61-65.
[5] Anna Pavord, *The naming of names*, (London, 2005), p.67; Bill Laws, *Fifty plants that changed the course of history*, (London, 2012 ed.), pp.6-7 & 198-199; Miles Hadfield, *A history of British gardening*, (London, 1969), pp.226-7.
[6] Richard Gorer, *The growth of gardens*, (London, 1978), p.38.
[7] Desmond, *Dictionary*, 'Stuart, John', p.591; Hadfield et al, *Dictionary*, 'Augusta, Princess of Saxe Gotha', p.18; Blunt, *Penny*, p.19 & 22; see Ray Desmond, *Kew*, (London, 1995), chapter 2, pp.30-40; Hadfield, *History*, pp.228 & 230.
[8] Hadfield et al, *Dictionary*, p.66; Desmond, *Kew*, p.64; Anon., 'Queen Charlotte, a loyal wife', *Daily Telegraph*, 14 April 2009.
[9] Desmond, *Kew*, p.64; Blunt, *Penny*, p.57.
[10] Blunt, *Penny*, p.73, quoting Olwen Hedley, *Queen Charlotte*, (London, 1975), p.300.
[11] Hadfield, *History*, pp.96-98, 143 & 145; Hadfield et al, *Dictionary*, 'Miller, Philip', pp.204-205.
[12] David Mabberley, 'A note on some adulatory botanical plates distributed by Sir Joseph Banks', *Kew Bulletin*, vol.66, 2011, pp.475-477; *Strelitzia reginae*, BM000810157.
[13] Rees, 'Strelitzia'; ed. L. Nanier & J. Brooke, 'Gascoyne, Bamber', [online]: http://www.historyofparliamentonline.org/volume/1754-1790/member/gascoyne-bamber-1725-91, accessed 29 July 2016; Desmond, *Dictionary*, 'Dryander, Jonas', p.197; Staffan Müller-Wille, 'Carolus Linnaeus', [online]: https://global.britannica.com/biography/Carolus-Linnaeus, accessed 15 August 2016.
[14] Carter, *Banks*, p.251.
[15] James Britten, 'John Frederick Miller and his "Icones"', p.256, *Journal of Botany, British and Foreign*, 1913, vol.51, p.256; Desmond, *Dictionary*, 'Miller, John Frederick', p.438; T.A. Sprague, 'Bibliographical notes: Cl. John Sebastian Miller's 'Icones Novae', *Journal of Botany, British and Foreign*, vol.74, 1936, pp.208-209; Mabberley, p.475; *Curtis's Botanical Magazine*, vol.IV, 1790, plate 119.
[16] Mary Watson-Wentworth, dowager marchioness of Rockingham, to Banks, 21 December 1785, summarised in ed. Warren R. Dawson, *The Banks Letters*, (London, 1958), p.859.
[17] R.E.W. Maddison & Raymond E. Maddison, 'Spring Grove, the country house of Sir Joseph Banks, Bart, PRS', *Notes and Records of the Royal Society of London*, vol.11, no.1, (Jan. 1954), p.93.
[18] Hadfield et al, *Dictionary*, 'Sowerby, James', pp.268-269, 'Aiton, William', p.12.
[19] Car. O. Linné, *Mantissa plantarum*, (Holm, 1771), p.211; World Checklist of Selected Plant Families [online]: http://apps.kew.org/wcsp/namedetail.do?name_id=248319, accessed 30 July 2016.
[20] *Curtis's Botanical Magazine*, plate 119; Sprague, pp.208-209; Global Plants, 'Heliconia bihai', [online]: http://plants.jstor.org/compilation/Heliconia.bihai, accessed 27 February 2017.
[21] Linné Online, 'Family life', [online]: http://www.linnaeus.uu.se/online/life/8_2.html, accessed 15 August 2016; Carolo a Linné [Linnaeus], *Supplementum plantarum*, (Braunschweig, 1781), pp.157-158.
[22] *Hortus kewensis* (1789), p.284.
[23] Sprague, p.209; *Curtis's Botanical Magazine*, plate 119; Joseph Banks to Antoine Laurent de Jussieu, 29 June 1788, D.T.C. 6.40, summarised in Dawson, p.482, 8; ed. Lady [Pleasance] Smith, *Memoir and correspondence of the late Sir James Edward Smith, M.D.*, (London, 1832), vol.II, pp.82-83; Mabberley, p.475-476.
[24] Endpapers, *Hortus kewensis* (1789); Marchioness of Rockingham to J.E. Smith, 3 March 1793, in ed. Lady Smith, *Memoir and correspondence*, vol.II, pp.59-60; Hadfield et al, *Dictionary*, 'Johnes, Thomas', p.165; NHM Archive, Correspondence of the Honourable Sir Joseph Banks Bart, vol.X, 1796-1797, Thomas Johnes to Banks, 26 March 1796, also summarised in Dawson, p.475; Desmond, *Dictionary*, 'Salisbury, Richard', pp.537-538; Richard A. Salisbury, *Prodromus stirpium in horto ad Chapel Allerton vigentium*, (London, 1796), pp.144-145.
[25] Ed. William Townsend Aiton, *Hortus kewensis*, (London, 1811, 2nd edition), vol.2, pp.54-56; James Britten, 'John Frederick Miller and his "Icones"', *Journal of Botany*, (1913), p.256; Britten later published a correction noting that the artist had been J.F. Miller's father, John Miller (Johann Sebastian Müller): James Britten, '"John Frederick Miller and his Icones"', *Journal of Botany*, (1919), p.353; Sprague, p.208; Mabberley, p.1; *Curtis's Botanical Magazine*, plate 119.
[26] *Curtis's Botanical Magazine*, plate 119; and Desmond, *Dictionary*, 'Curtis, William', p.167.
[27] Banks to Dryander, 22 October 1791, summarised in Dawson, p.278.
[28] Henry Andrews, 'Strelitzia reginae', *Botanical Register*, (1797), vol.6, plate CCCCXXXII.
[29] Anon., 'Groningen', December 1793, *Nieuwe Nederlandsche Jaarboeken*, (Leiden & Amsterdam, 1793; 28th edition), pp.1973-1974; link given in Colin Roberts, 'What's in a name', [online]: http://www.thoughtandawe.net/biology/botany/whats-in-a-name-2/, published 1 October 2014, accessed 19 May 2015.
[30] Susyn Andrews, 'Golden Bird of Paradise', *The Garden*, May 1994, p.229; Anna C. Saltmarsh, 'Francis Masson: Collecting plants for king and country', *Curtis's Botanical Magazine*, vol.20, no.4, (November 2003), p.238, Ray Desmond, *A celebration of flowers*, (London, 1987), p.30.
[31] Desmond, *Dictionary*, 'Aiton, William Townsend', p.5, 'Thiselton-Dyer, William', p.605; W.T. Thiselton-Dyer, 'Historical account of Kew to 1841', *Bulletin of Miscellaneous Information*, no.60, December 1891, p.1; Gunn & Codd/Glen & Germishuizen, 'Masson', p.289; Desmond, *Kew*, p.95.
[32] Carter, *Banks*, pp.24-5; Desmond, *Dictionary*, 'Lyons, Israel', p.401; Carter, *Banks*, pp.26-27.
[33] Anon., 'The Endeavour botanical illustrations', [online]: http://www.nhm.ac.uk/our-science/departments-and-staff/library-and-archives/collections/cook-voyages-collection/endeavour-botanical-illustrations/about.dsml, accessed 13 August 2016; Gunn & Codd/ Glen & Germishuizen, 'Solander, Daniel', p.399 Desmond, *Dictionary*, 'Solander, Daniel', pp.572-573, 'Parkinson, Sydney', p.480; Carter, *Banks*, p.30, 35-39, 55-57, 60 & 69; see 'Buchan, Alexander', in *South Seas Companion* [online]: http://southseas.nla.gov.au/biogs/P000328b.htm, accessed 13 August 2016.
[34] Leonard Thompson, *A history of South Africa*, (Sandton, 1990), pp.32-33; T.R.H. Davenport, *South Africa: A modern history*, (Cape Town, 1988 edition), pp.24-25.
[35] Anon., 'Dias, Bartolomeu', [online]: https://global.britannica.com/biography/Bartolomeu-Dias, accessed 13 August 2016; Andrew Duminy, *Mapping South Africa*, (Johannesburg, 2011), pp.8-13; Gunn & Codd/Glen & Germhuizen, p.5.
[36] 'Drake, Francis', [online]: https://global.britannica.com/biography/Francis-Drake, accessed 13 August 2016; Thompson, pp.31-33.
[37] T.R.H. Davenport, *South Africa: A modern history*, (Cape Town, 1988 edition), pp.22-3; Anon. 'Who discovered how to prevent scurvy?', [online]: https://www.pharmaceutical-journal.com/opinion/blogs/who-discovered-how-to-prevent-scurvy/11072182.blog, accessed 13 August 2016; Gunn & Codd/Glen & Germhuizen, pp.12 & 14.
[38] Anon., 'Carolus Clusius', [online]: https://www.hortusleiden.nl/en/the-hortus/history/carolus-clusius, accessed 12 March 2013.
[39] Gunn & Codd/Glen & Germishuizen, pp.26-27.
[40] K ex Herbarium Benthamiensis, ex Herbarium of late Bishop Goodenough, *Strelitzia reginae*, no location, 1825; K ex Herbarium Hookerianum, Strelitzia reginae, Dr. [W.] Gill, no date [before 1867], Keiskamma area, Eastern Cape; Gunn & Codd/Glen & Germishuizen, 'Gill, William', p.191.
[41] K, *Strelitzia reginae*, T. Cooper, Cape of Good Hope, purchased March 1873, 'Flora Capensis'; K, T. Cooper, *Strelitzia reginae*, 1225, n.d., no location, Kew Negative No.13047; K ex Herbarium Hookerianum, *Strelitzia reginae*, W.J. Burchell, 3670, marked 'Between Blue Krantz and Kowie Poort (Albany Div.)', probably by R.A. Dyer; BM(NH), *Strelitzia reginae*, Bowie, C.B.S. in the District of Albany near the Cowie River, n.d.; BM(NH), *Strelitzia reginae*, J. Bowie, C.B.S. on the Rocky Heights of Uitenhage, n.d; K ex Herbarium Hookerianum, *Strelitzia reginae*, Drège, 1867; BM(NH) ex Herbarium of C.E. Moss, *Strelitzia reginae*, 2562, 18 July 1919, Banks of Buffalo River, East London, Cape; Gunn & Codd/Glen & Germishuizen, 'Moss, Charles Edward' and 'Moss, Margaret', pp.304-305; K, *Strelitzia reginae*, M.C. Gillett 2018, 22 March 1937, Port Alfred, Albany district; Gunn & Codd/ Glen & Germishuizen, 'Gillett, Margaret (née Clark)', p.191; 'Smuts, Jan', pp.397-8; 'Hutchinson, John', pp.223-226; 'Gillett, Jan', p.191.
[42] Gunn & Codd/ Glen & Germishuizen, pp.11-14, 26, 46-47; F. de Nave & D. Imhof, *Botany in the Low Countries*, (Antwerp, 1993); Desmond, *Dictionary*, 'Petiver, James', pp.491-492.
[43] Desmond, *Dictionary*, 'Sherard, William', p.555; Gunn & Codd/ Glen & Germishuizen, pp.26-27; Hadfield et al, *Dictionary*, 'Sloane, Hans', p.264.
[44] Carter, *Banks*, p.93.
[45] Gunn & Codd/ Glen & Germishuizen, 'Oldenland, Heinrich', pp.319-321; Joseph Banks, 'Some account of the Cape of Good Hope', April 1771 *The Endeavour journal of Sir Joseph Banks* [online]: http://gutenberg.net.au/ebooks05/0501141h.html#apr1771, accessed 12 March 2013; Lawrence G. Green, *Grow lovely, growing old*, (Cape Town, 1952 ed.), p.98; Donal P. McCracken & Eileen M. McCracken, *The way to Kirstenbosch*,

(Cape Town, 1988), Annals of Kirstenbosch Botanic Gardens, vol.18, p.14.
[46] Gunn & Codd/Glen & Germishuizen, 'Banks, Sir Joseph', p.87.
[47] James Lee to J.E. Smith, *Memoir and correspondence*, vol.2, pp.117, 118n; W.T. Aiton, *Hortus kewensis* (1811), pp.54-56.
[48] Francis Masson, 'An account of three journeys from the Cape Town into the Southern Parts of Africa undertaken for the Discovery of New Plants, towards the Improvement of the Royal Botanic Gardens at Kew. By Mr Francis Masson. One of His Majesty's Gardeners. Addressed to Sir John Pringle, Bart. P.R.S.' *Philosophical Transactions of the Royal Society of London*, volume 66 (1776), pp.268-317.
[49] William J. Burchell, *Travels in the interior of South Africa*, (Cape Town, 1967), vol.1, p.388, quoted in Elizabeth Green Musselman, 'Plant knowledge at the Cape: A study in African and European collaboration', *International Journal of African Historical Studies*, vol.36, no.2, (2003), p.382.
[50] Mia C. Karsten, 'Francis Masson, A gardener-botanist who collected at the Cape: Masson's journeys at the Cape', *Journal of South African Botany*, vol.XXV, pp.169-170.
[51] Siegfried Huigen, *Knowledge and colonialism: Eighteenth-century travellers in South Africa*, (Leiden, 2009), pp.36-42; Anon., ' Thunberg, Carl Peter', [online]: https://plants.jstor.org/stable/10.5555/al.ap.person.bm000008462, accessed 13 August 2016; Karsten, 'Masson', pp.170-171; Gunn & Codd/Glen & Germishuizen, 'Sparrman, Anders', pp.401-402; Anders Sparrman, *A voyage to the Cape of Good Hope*, ed. V.S. Forbes, (Cape Town, 1975), Van Riebeeck Society, 2nd series, no.6, vol.1, p.2; Carl Thunberg, *Travels at the Cape of Good Hope, 1772-5*, ed. V.S. Forbes, Van Riebeeck Society 11-17, 2nd series, no.17, (Cape Town, 1986), 'Preface to vol.I English edition 1793', pp.xxviii-xxxix, 'Preface to vol.II English edition 1793', p.xliv.
[52] Karsten, 'Masson', part 2, pp.177-178, 199; Graham Duncan, Barbara Jeppe & Leigh Voigt, *The Amaryllidaceae of southern Africa*, (Pretoria, 2016), p.54.
[53] Toby Musgrave, Chris Gardner & Will Musgrave, *The plant collectors*, (London, 1998), p.94.
[54] Gunn & Codd/Glen & Germishuizen, 'Thunberg, Carl', p.422.
[55] Gunn & Codd/Glen & Germishuizen, 'Auge, Johann Andreas', p.82; Thunberg, *Travels*, ed. Forbes, pp.87-88 & 90-1.
[56] Thunberg, *Travels*, ed. Forbes, p.92; Peter MacOwan, 'President's annual address, July 28, 1886: Personalia of botanical collectors', *Transactions of the South African Philosophical Society*, vol.IV, (1888), p.xxxvi.
[57] Gunn & Codd/Glen & Germishuizen, p.38; Thunberg, *Travels*, ed. Forbes, p.xii.
[58] Carl Thunberg, *Resa uti Europa, Africa, Asia, forrättad åren 1770-1779*, (Uppsala, 1788), vol.1, p.216; Thunberg, *Travels*, ed. Forbes, p.92.
[59] Donal P. McCracken, *Gardens of empire*, (Leicester, 1997), pp.100-3.
[60] UPS: *Strelitzia reginae*, C.P. Thunberg, n.d., Cape of Good Hope: UPS:BOT:V-005894 & UPS:BOT:V-005895; Strelitzia reginae, C.P. Thunberg, n.d.,Cape, SBT11174, sheet nrs. 1.3.12.7 & 1.3.12.8;
Anon., 'Bergius, Benedictus', [online]: http://kiki.huh.harvard.edu/databases/botanist_search.php?mode=details&id=65822, accessed 13 August 2016; Anon., 'Bergius, Peter Jonas', [online]: http://plants.jstor.org/stable/10.5555/al.ap.person.bm000000635, accessed 13 August 2016; http://www.bergianska.se/english/collections/the-bergius-herbarium, accessed 30 July 2016. For more on the relationship between Thunberg, Linnaeus and the Bergius brothers, see chapter 2.
[61] Liesl van der Walt, 'Strelitzia reginae', [online]: http://plantzafrica.com/frames/plantsfram.htm, posted July 2001; updated by Phakamani Xaba, August 2011; accessed 15 August 2016.
[62] Gunn & Codd/Glen & Germishuizen, 'Auge', p.82.
[63] Carl Thunberg, *Travels*, vol. 1, (London, 1794), pp.105-6; ed. Forbes, Thunberg, *Travels*, p.28; [online]: 'Michael Grubb and the Swedish East India Company': http://www.bergianska.se/english/collections/the-bergius-herbarium/michael-grubb-and-the-swedish-east-india-company, accessed 26 February 2017.
[64] Gunn & Codd/Glen & Germishuizen, 'Oldenland, Heinrich', pp.319-321; ed. E.E. Mossop, *Journals of the expeditions of the Honourable Ensign Olof Bergh (1682 and 1683) and the Ensign Isaq Schriver (1689)*, (Cape Town, 1931), Van Rieebeck Society, series 1, no.12.
[65] Gunn & Codd/Glen & Germishuizen, 'Oldenland', p.320; Desmond, *Dictionary*, 'Smith, James Edward', p.568.
[66] Gunn & Codd/Glen & Germishuizen, 'Tulbagh, Rijk', p.428; ed. Forbes, Thunberg, *Travels*, pp.53 & 86.
[67] Gunn & Codd/Glen & Germishuizen, p.143; Manonmani Filliozat, 'D'Après de Manneville, Captain and hydrographer to the French East India Company (1707-1780)', *Indian Journal of History of Science*, vol.29, no.2, (1994), p.333; Gunn & Codd/Glen & Germishuizen, 'Bladh, Pehr Johann', p.98; 'Bergius, Peter Jonas', [online]: http://plants.jstor.org/stable/10.5555/al.ap.person.bm000000635, accessed 13 August 2016; Gunn & Codd/Glen & Germishuizen, 'Sonnerat, Pierre', pp.400-401; Gunn & Codd/Glen & Germishuizen, 'König (Koenig), Johann Gerhard', pp.244-245; Conrad Lighton, *Cape floral kingdom*, (Cape Town, 1960), p.14.
[68] BM(NH), *Strelitzia reginae*, BM000810157.
[69] BM(NH), *Strelitzia reginae*, BM000410152, Prom. b. spei, Nelson; Gunn & Codd/Glen & Germhuizen, 'Nelson, William', pp.312-313.
[70] Gunn & Codd/Glen & Germhuizen, 'Nelson, David', p.312.
[71] K ex Herbarium of W.H. Gower, Strelitzia reginae, no collector, no date, Cape of Good Hope (possibly cultivated at Kew or elsewhere).
[72] Desmond, *Dictionary*, 'Gower, William Hugh', p.261.
[73] Eds. Hazel Crampton, Jeff Peires & Carl Vernon, *Into the hitherto unknown*, (Cape Town, 2013), Van Riebeeck 2nd series, no.44, (Cape Town, 2013), p.xix & appendix 1: 'Deposition of Hendrik de Vries de Jonge and Hendrik Scheffer de Jonge, 10 July 1737', pp.178-181; V.S. Forbes, *Pioneer travellers in South Africa*, (Cape Town, 1965), 'August Frederik Beutler', pp.13 & 18.
[74] Anon., 'Hendrik Beenke', [online]: https://www.openarch.nl/show.php?archive=ghn&identifier=d7a67fac-f37f-4d3d-825f-a32e61d2212e&lang=en, accessed 26 February 2017; Forbes, *Pioneer travellers*, p.7 & 9-10.
[75] Filliozat, pp.329 & 332-334; Forbes, *Pioneer travellers*, pp.8 & 13-19.
[76] Hadfield, *History*, p.260.
[77] MacOwan, 'President's annual address', p.xxxiv; Anon., 'Royen, David van', [online]: http://plants.jstor.org/stable/10.5555/al.ap.person.bm000007222, accessed 25 April 2016; Gunn & Codd/Glen & Germishuizen, p.57, after T.T. Barnard, 'The story of *Nerine undulata* (L.) Herb.', *Nerine Society Bulletin*, 3, 1968, pp.8-10.
[78] Hadfield, *History*, p.227; McCracken, *Empire*, pp.100-3.
[79] http://www.nhm.ac.uk/research-curation/scientific-resources/collections/botanical-collections/clifford-herbarium/about-clifford/index.html, accessed 25 April 2016; Gunn & Codd/Glen & Germishuizen, pp.63-64; and Hadfield et al, *Dictionary*, 'Repton, Humphry', p.237-241.
[80] Hadfield et al, Dictionary, 'Compton, (Hon. Rev.) Henry', p.77; Hadfield, *History*, pp.136-137; Gunn & Codd/Glen & Germishuizen, 'The Claudius illustrations', p.35; Phyllis I. Edwards, 'Sloane Manuscript 5286: An important source for Decade 9, 'Herbarium Capense' of Petiver's Gazophylacium Naturae and Artis', *Journal of South African Botany*, vol.xxxiv, part iv, July 1968, pp.243-253; John Claudius Loudon, *Arboretum et fruticetum Britannicum*, (London, 1838), pp.40-41.
[81] Carter, *Banks*, pp.115-119; Anon., 'Jean Nicolas Sébastien Allamand', [online]: http://data.bnf.fr/12462542/jean-nicolas-sebastien_allamand/, accessed 15 August 2016.
[82] Gunn & Codd/Glen & Germishuizen, 'Masson', pp.289-290; Desmond, *Dictionary*, 'McNab, William', p.412; E. C. Nelson, 'William Ramsay McNab's herbarium in the National Botanic Gardens, Glasnevin', Glasna, new ser., 1 (1990), pp.1-7.
[83] 'Strelitziaceae', World Checklist of Selected Plant Families [online]: http://wcsp.science.kew.org/qsearch.do, accessed 9 May 2018.

## CHAPTER 2

[1] Gunn & Codd/Glen & Germishuizen, 'Lichtenstein, Martin H(e)inrich Carl', pp.267-270; Cuthbert J. Skead, *Historical plant incidence in South Africa*, eds. John C. Manning & Nicola C. Anthony, (Pretoria, 2009), as *Strelitzia* 24, p.141, quoting Hinrich Lichtenstein, *Travels in southern Africa*, translated by Anne Plumptre (London, 1812-1815; 1930 reprint), vol.2, p.251; and Thunberg, *Travels*, ed. Forbes, p.90 and fn.253.
[2] Mia Karsten, *The old Company's garden at the Cape & its superintendents*, (Cape Town, 1951), p.150; Burchell, *Travels*, vol.1, chapter 1, p.24; Jenni Evans, pers. comm., 20 May 2018.
[3] Quoted in Gunn & Codd/Glen & Germishuizen , 'Lichtenstein', from Lichtenstein, *Travels*, vol.2, p.134.
[4] Gunn & Codd/Glen & Germishuizen , 'Swellengrebel, Hendrik', pp.411-412; and Thunberg, *Travels*, ed. Forbes, p.53.
[5] Karsten, *Old Company's garden*, p.146; Carl Peter Thunberg & Nicolaus Gustavus Bodin, *Genera nova plantarum*, part 9, (Uppsala, 1798), pp.132-3; Carl P. Thunberg, *Prodromus plantarum capensis* (Uppsala, 1794), p.viii ; Karsten, *Old Company's garden*, p.148; and Gunn & Codd/Glen & Germishuizen, 'Auge', p.82.
[6] C.C.de Villiers, 'Anthonij Alexander Faure', *Genealogies of old Cape families*, ed. C.Pama (Cape Town, 1966) p.227; Karsten, *Old Company's garden*, pp.147-8, quoting Lichtenstein, pp.168-9.
[7] Karsten, *Old Company's garden*, p.147.
[8] Karsten, *Old Company's garden*, p.147; Lichtenstein, vol.2, pp.169-71.
[9] Karsten, *Old Company's garden*, p.148; Lichtenstein, vol.2, pp.169-71.
[10] Karsten, *Old Company's garden*, p.148; Lichtenstein, vol.2, pp.169-71.
[11] Karsten, *Old Company's garden*, p.147; Lichtenstein, vol.2, pp.169-71.
[12] Karsten, *Old Company's garden*, p.147; Lichtenstein, vol.2, pp.169-71.
[13] R.A. Lubke & Y. van Wijk, 'Terrestrial plants and coastal vegetation', in ed. Roy Lubke, Fred Gess & Mike Bruton, *A field guide to the Eastern Cape coast*, Grahamstown, 1988; and Karsten, *Old Company's garden*, p.149. See also Vit Grulich, 'Strelitzia alba (L.f.) Skeels – strelície', http://botany.cz/cs/strelitzia-alba/, posted 4 January 2013, accessed 24 April 2016.
[14] Karsten, *Old Company's garden*, pp.131-135 & 148-149; Gunn & Codd/Glen & Germishuizen, 'Auge', p.82; Thunberg, *Travels*, ed. Forbes, p.86.
[15] See C. Plug, 'Oldenburg, Frans', [online]: http://www.s2a3.org.za/bio/Biograph_final.php?serial=2061, accessed 19 January 2017; and BM(NH), Oldenburg, *Heliconia albiflora* no.839; *Strelitzia augusta*, no.1414, Lydenburg, Transvaal Republic; *Strelitzia augusta*, 1225 & 1226, T. Cooper, Natal, 1862.
[16] MacOwan, 'President's annual address', p.xxxiv; Thunberg, *Travels*, ed. Forbes, pp.90 & 92 and fn.253; and Bergius Herbarium, SBT11174, sheet 1.3.12.7 and 1.3.12.8, no date, no location.
[17] John Barrow, *Travels in the interior of southern Africa*, (London, 1806), vol.1, p.302; and Skead, p.139.
[18] SBT11173, sheet 1.3.12.6, no date, no location; Gunn & Codd/Glen & Germishuizen,

'Barrow, Sir John', p.91.
[19] Linnaeus jnr, *Supplementum plantarum*, p.157; *Hortus kewensis* (1789), 'Heliconia Bihai', p.284, and 'Strelitzia Reginae', p.285.
[20] *Curtis's Botanical Magazine*, plate 119.
[21] Masson to Banks, 21 December 1789, State Library of New South Wales, Papers of Sir Joseph Banks, section 5: Gardeners and collectors, series 13.51.
[22] Liesl van der Walt, 'Strelitzia reginae', [online]: http://www.plantzafrica.com/plantqrs/strelitziaalba.htm, posted July 2000, updated by Phakamani Xaba August 2011, accessed 22 August 2016; Phakamani Xaba, 'Strelitzia alba', [online]: http://www.plantzafrica.com/plantqrs/strelitziaalba.htm, posted November 2010, accessed 22 August 2016.
[23] William Townsend Aiton, *Hortus kewensis*, 2nd ed, vol.2, (London, 1811), 'Strelitzia augusta', p.55. See Desmond, *Dictionary*, 'Aiton, William Townsend', p.5; and BM(NH), BM000810155, marked 'Strelitzia augusta, Hort. Kew. 1797', [online]: http://data.nhm.ac.uk/object/0978e00f-23f8-4b2c-aac1-63af8cbcc02e; BM(NH), BM000810153, marked 'Prom. b. spei. Hortus Mr Masson', [online]: http://data.nhm.ac.uk/object/803c9606-17cd-47fb-9133-c67c54c6002f, both accessed 17 May 2016, viewed 6 October 2016.
[24] Masson to Banks, 7 March 1791, State Library of New South Wales, Banks papers, section 5, series 13.41.
[25] See ed. M.D. Nash, *The last voyage of the Guardian, Lieutenant Riou, Commander, 1789-1791*, Van Riebeeck Society, series 2, no.20, (Cape Town, 1990); and Dawson, *Banks letters*, p.704, quoting Banks to Riou, 15 May 1791, BM. Add. MS. 33979.97.
[26] Library and Archives, Natural History Museum (BM), London; Hadfield et al, *Dictionary*, pp.179-180.
[27] F.A. Stafleu, 'Dates of botanical publications 1788-1792', *Taxon*, 12 (2), 1963, p.50.
[28] Henry Andrews, 'Strelitzia reginae', *Botanical Register*, (1806), vol.6, plate CCCCXXXII.
[29] BM00081053, Prom. b. spei (Cape of Good Hope). Hortus. Mr Masson; Gunn & Codd/Glen & Germishuizen, 'Masson', pp.287-9.
[30] Peter Bergius, *Descriptiones plantarum ex Capite Bonae Spei*, (Stockholm, 1767).
[31] Thunberg, *Travels*, ed. Forbes, pp.xii & xv; and Gunn & Codd/Glen & Germishuizen, 'Thunberg', p.424.
[32] Nils Svedelius, 'Carl Peter Thunberg (1743-1828)', *Isis*, vol.35, no.2 (spring, 1944), pp.130; and John P. Rourke, 'Thunberg's botanical achievements at the Cape and their implications', in Thunberg, *Travels*, ed. Forbes, p.xx; and L.C. Rookmaker, *The zoological exploration of southern Africa*, (Rotterdam, 1989), pp.149-50.
[33] Carl Peter Thunberg, *Nova genera plantarum*, part 7 (Eric Carl Trafvenfeldt), 2 June 1792, pp.111-114.
[34] Linné, *Supplementum*, pp.157-158; and 'Strelitzia reginae', (1789), p.285.
[35] Andrews, 'Strelitzia reginae', Botanical Register, vol.6, plate CCCCXXXII; Barrow, *Travels*, vol.1, p.302; William Townsend Aiton, Hortus kewensis, vol.2, (London, 1811), p.285; ed. Carl Ludwig Willdenow, Carl Linnaeus *Species plantarum*, vol.1, part 1, (Berlin, 1798), p.1190; Jerome J. Bylebyl, 'Willdenow, Karl Ludwig', [online]: http://www.encyclopedia.com/doc/1G2-2830904661.html, accessed 24 December 2013; and William Hooker, 'Strelitzia augusta', Curtis's Botanical Magazine, (1845), vol.1, 3rd series, (or vol.LXXI), tab.4167 & 4168.
[36] Adapted from 'Strelitzia alba (L.f.) Skeels', Global Biodiversity Information Facility: http://www.gbif.org/species/273106, accessed 30 June 2016; Salisbury, Chapel Allerton, p.144.
[37] See SBT11173, sheet 1.3.12.6; ed. Niklas Wikström, 'Professor Bergianus from past to present': http://www.bergianska.se/english/about-us/history-of-the-garden/professor-bergianus-from-past-to-present, updated 10 August 2016, accessed 21 August 2016.
[38] Bo Peterson, 'Johan Emanuel Wikström, with historical notes on the genus *Wikstroemia*', *Pacific Science*, vol.50, no.1, (1996), pp.77-83.
[39] Peterson, p.77.
[40] Gunn & Codd/Glen & Germishuizen, 'Drège, Carl', p.153, 'Drège, Johann', pp.154-6.
[41] K, *Strelitzia alba*, ex Hooker Herbarium, Drège, 1840, Cape; K *Strelitzia alba*, ex Bentham Herbarium, Drège, 1840, southern Africa; *Strelitzia alba*, K carpological collection, 170.07, Drège, 1834, southern Africa; *Strelitzia alba* Thunb., K carpological collection, 170.07, K, carpological collection, *Strelitziaceae* 3, *Strelitzia augusta*, 170.07, Feilden, cult. Egypt, 27 October 1914; K, *Strelitzia augusta*, Feilden, cult. Egypt, Oct.1914.
[42] F.L.M. du Mont de Courset, *Le botaniste cultivateur*, vol.VII (Supplement), (Paris, 1814, 2nd ed.), p.88; Pierre-Aimé Lair, 'Description des jardins de Courset, aux environs de Boulogne-sur-mer', *Mémoires de la Société Royal d'Agriculture et de Commerce de Caen*, vol.4, (1836), [online]: http://www.bmlisieux.com/normandie/courset.htm, posted 21 December 2003, accessed 23 August 2016.
[43] *Strelitzia augusta* [*Strelitzia alba*], K, marked 'Hort Liverpool 1838' and 'Herbarium Benthamiensis 1854'; *Strelitzia augusta*, K, ex Hooker Herbarium, marked 'Hort Kew, December 4th 1861'.
[44] Heinrich Göppert, 'Der botanische Garten der Universität Breslau', ed. A.E. Fürnrohr, *Flora*, (1855), new series XIII, pp.377-378; Anon., 'Göppert, Heinrich Robert', [online]: http://www.idref.fr/035080023X, accessed 22 August 2016; Hooker, 'Strelitzia augusta'.
[45] Hooker, 'Strelitzia augusta'; Desmond, *Dictionary*, 'Fitch, Walter Hood', p.224; Anon., 'Walter Hood Fitch', [online]: http://www.avictorian.com/Fitch_Walter_Hood.html, accessed 22 August 2016; Anon, 'Louis van Houtte', [online]: http://www.victoria-adventure.org/water_gardening/biographies/louis_van_houtte.html, accessed 22 August 2016.
[46] Charles Lemaire, 'Strelitzia augusta', *Flore des Serres et des Jardins*, (1846), vol.2, p.27; Anon, 'Charles Antoine Lemaire', [online]: http://www.illustratedgarden.org/mobot/rarebooks/author.asp?creator=Lemaire,+Charles+Antoine&creatorID=3, accessed 23 August 2016.
[47] Louis van Houtte, 'Strelitzia augusta: Culture', *Flore des Serres et des Jardins*, (1846), vol.2, p.28.
[48] *Seeds and plants imported during the period from July 1 to September 30, 1911, Inventory no.28; Nos.31371 to 31938*, (Washington, 1912), p.57; for Skeels, see Anon., 'Skeels, Homer Collar' [online]: http://kiki.huh.harvard.edu/databases/botanist_search.php?botanistid=7665, accessed 23 August 2016; for Fairchild, see Anon., 'David Grandison Fairchild', [online]: http://everglades.fiu.edu/reclaim/bios/fairchild.htm, accessed 28 June 2016.
[49] *Seeds and plants, July to September 1911*, p.57.
[50] Lemaire, 'Strelitzia augusta'.
[51] John Phillips to Midlands conservator of forests (Cape), 6 June 1926, Phillips' ref: 90/B.5500, *Strelitzia alba* file, SANBI Herbarium, Pretoria.
[52] Philips to Midlands conservator.
[53] R.A. Dyer, 'Plate 995. Strelitzia alba', 'Plate 996. Strelitzia nicolai', 'Plate 997. Strelitzia caudata', all *Flowering plants of South Africa*, 1946; and J.H. Keet to R.A. Dyer, R.701, 5 November 1943.
[54] Keet to Dyer, November 1943.
[55] For Edith Burges, see 'Concise dictionary of South African botanical art', ed. Marion Arnold, South African botanical art, (Cape Town, 2001), p.189; SANBI Archives, Artists' file; Dyer, 'Plate 995. Strelitzia alba'; 'Strelitzia alba', K & PRE, Keet, 27476, November 1943, Knysna Division, Groot River Pass, Cape Province, sheets 1-5.
[56] J.H. Keet to R.A. Dyer, R.701, 5 November 1943.
[57] Keet to Dyer, 5 November 1943.
[58] Keet to Dyer, 5 November 1943; see H.G. Fourcade, *Checklist of the flowering plants of the divisions of George, Knysna, Humansdorp, and Uniondale*, Memoirs of the Botanical Society of South Africa no.20, (Pretoria, 1941); Clare D. Storrar, *The four faces of Fourcade*, (Cape Town, 1990), pp.104, 109-112.
[59] See Sandra Knapp, Gerardo Lamas, Eimear Nic Lughadha & Gianfranco Novarino, 'Stability or stasis in the names of organisms: The evolving codes of nomenclature', *Philosophical Transactions of the Royal Society London* B, vol.359, (2004), pp.611-622.
[60] Knapp et al, 'Evolving codes', p.612; for a discussion of the origin of the terms 'lumper' and 'splitter', see Glenn Branch, 'Whence lumpers and splitters', [online]: http://ncse.com/blog/2014/11/whence-lumpers-splitters-0016004, accessed 5 January 2016.
[61] Knapp et al, 'Evolving codes', p.613.
[62] http://www.theplantlist.org/browse/A/Strelitziaceae/Strelitzia/, accessed 19 April 2014.
[63] Cron et al, p.614.

## CHAPTER 3

[1] Catherine II to Voltaire, 25 June 1772, quoted in Louis Réau, *L'Europe française au siècle des lumières*, (Paris, 1938).
[2] KA, Kew record book, 1793-1809, pp.38-42.
[3] H.B. Carter, 'Sir Joseph Banks and the plant collection from Kew sent to the Empress Catherine II of Russia 1795', *Bulletin of the British Museum (Natural History)*, historical series, vol.4, no.5, London, 1974, pp.281-385.
[4] Viktor Zakharov, 'Foreign merchant communities in 18th-century Russia', in eds. Viktor N. Zakharov, Gelina Harlaftis & Olga Katsiardi-Hering, *Merchant colonies in the early modern period*, (Cambridge, Pickering & Chatto, 2012), pp.103-126; Lindsey Hughes, 'Peter I', [online]: http://www.encyclopedia.com/topic/Peter_I_(Russia).aspx, created 2004, accessed 25 August 2016; Orlando Figes, *Natasha's dream*, (London, Allen Lane, 2002), p.8; Anon, 'The botanical garden', [online]: http://www.saint-petersburg.com/parks/botanical-garden/, accessed 9 July 2016; Emil Bretschneider, 'Dr Fischer and the Imperial Botanical Gardens, St Petersburg', *History of European botanical discoveries in China* (London, 1898), pp.317-318; Peter Hayden, 'The Russian Stowe', *Garden History*, vol.19, no.1, (spring, 1991), pp.21-27; Blunt, *Penny*, p.47.
[5] Ed. Neil Chambers, *Letters of Sir Joseph Banks: A selection*, 1768-1820, (London, 2000), p.168, n.3; John Gorton, 'Whitworth (Charles, earl)', *General biographical dictionary*; Peter Hayden, 'Gardens of the tsars', *Garden History*, vol.25, no.2, (winter, 1997), p.239.
[6] Grenville to King George III, 15 November 1793, quoted in Carter, 'Russia', p.329.
[7] King George III to Grenville, 16 November 1793, quoted in Carter, 'Russia', p.329.
[8] Banks to Burges, 6 May 1795 quoted in Carter, 'Russia', pp.330-331; Lionel Kochan, *The making of modern Russia*, (London, 1962), pp.127-128; Anthony Cross, 'Russian gardens, British gardeners', *Garden History*, vol.19, no.1, (spring, 1991), pp.12-20.
[9] Carter, 'Russia', pp.288-291, 295-296 & 302.
[10] Carter, 'Russia', p.288; John Brooke, 'Burges, James Bland', [online]: http://www.historyofparliamentonline.org/volume/1754-1790/member/burges-james-bland-1752-1824, accessed 28 November 2016; Burges to Banks, 4 May 1795, quoted in Carter, 'Russia', p.329.
[11] Banks to Burges, 5 May 1795, quoted in Carter, 'Russia', pp.329-330.
[12] Banks to Burges, 5 May 1795, quoted in Carter, 'Russia', pp.330-331.
[13] Burges to Banks, 5 May 1795, quoted in Carter, 'Russia', p.330.
[14] Banks to Burges, 5 May 1795, quoted in Carter, 'Russia', pp.330-331.
[15] Banks to Burges, 'Memorandums', 6 May 1795, quoted in Carter, 'Russia', p.332; Banks to Burges, 6 May 1795 & 15 June 1795, quoted in Carter, 'Russia', pp.330-334; Carter, *Banks*, pp.253-254; ed. Nash, *Last voyage of Guardian*, p.xxiii; Banks diary note,

15 'Articles of charge', 6 May 1795, quoted in Carter, 'Russia', p.332.
16 Banks diary note, 'Articles of charge'.
17 Desmond, *Dictionary*, 'Forster, Edward', p.230; *Essex Naturalist*, [online]: http://www.essexfieldclub.org.uk/archivetext/s/020/o/0074, accessed 20 August 2016; Carter, 'Russia', p.336, n.4.
18 Forster to Banks, 8 May 1795, quoted in Carter, 'Russia', p.332.
19 Banks diary notes, 8 May 1795, quoted in Carter, 'Russia', p.343; Anon, 'Thomas Raikes', [online]: http://www.clement-jones.com/ps28/ps28_044.htm, accessed 28 November 2016.
20 Banks diary notes, 8 May 1795.
21 Banks diary notes, 14 June 1795, quoted in Carter, 'Russia', p.344; Banks, 'Sketch plan by Sir Joseph Banks of the dimensions of the hold in the *Venus*', 14 June 1795, reproduced by Carter, 'Russia', p.333; Banks to Burges, 15 June 1795, quoted in Carter, 'Russia', pp.334-334; Greg Dening, 'William Bligh', [online]: https://www.britannica.com/biography/William-Bligh, n.d., accessed 23 August 2016.
22 Banks to Burges, 15 June 1795.
23 Banks to Burges, 15 June 1795; Banks to Robertson, 17 June 1795, quoted by Carter, 'Russia', p.336.
24 Banks to Burges, 15 June 1795; Banks to Forster, 22 June 1795, quoted in Carter, 'Russia', pp.334 & 336; Desmond, *Dictionary*, 'Aiton, William Townsend', p.5.
25 See Banks diary notes, 24 June 1795, quoted in Carter, p.344.
26 Banks diary notes, 14 June 1795, 25 June 1795 & 4 July 1795, quoted in Carter, 'Russia', pp.344-345; Banks, 'Sketch plan'.
27 Banks to Forster, 22 June 1795, Marsden to Banks, 26 June 1795, Banks to Vickerman, 28 June 1795, Vickerman to Banks, 30 June 1795, all quoted in Carter, 'Russia', pp.336-338; Banks note of 1 July 1795 on copy of letter to Vickerman, 28 June 1795.
28 Aiton to Banks, 2 July 1795; quoted in Carter, 'Russia', pp.338-343.
29 Hadfield, *History*, p.232; Gorer, pp.63-64; W.T. Aiton to Banks, 2 July 1795, quoted in Carter, 'Russia', p.338; Brian Halliwell, 'Some Australian plants in cultivation in England by 1800', [online]: http://anpsa.org.au/APOL29/mar03-4.html, posted March 2003, accessed 23 August 2016.
30 Desmond, Kew, p.98; Banks to Burges, 4 July 1795, quoted in Carter, 'Russia', p.347; Chambers, p.168, n.5.
31 Banks diary note, probably February 1796, quoted in Carter, 'Russia', p.362.
32 Banks, diary note, February 1796; Banks to Noe, 4 July 1795, quoted in Carter, 'Russia', pp.349-350.
33 Banks to Noe, 4 July 1795, quoted in Carter, 'Russia', p.350.
34 Noe to Banks, 13 July 1795, quoted in Carter, 'Russia', pp.351-352.
35 Noe to Banks, 25 July 1795.
36 Anon, 'Paul I', [online]: http://www.rusartnet.com/biographies/russian-rulers/romanov/tsar/paul-i, accessed 25 August 2016; Anon, 'Empress Maria Fyodorovna', [online]: http://www.rusartnet.com/biographies/russian-rulers/romanov/family-of-paul-i/wives/empress-maria-fyodorovna, accessed 25 August 2016; Noe to Banks, 8 August, St Petersburg, quoted in Carter, 'Russia', p.352; Anon, 'Pavlovsk', [online]: http://www.gardenvisit.com/gardens/pavlovsk, accessed 25 August 2016; Banks diary note, 29 December 1795, Noe to Banks, 8 August 1795, quoted in Carter, 'Russia', p.352 & 359.
37 Carter, *Banks*, p.303.
38 John T. Alexander, 'Orlov, Grigory Grigorievich', [online]: http://www.encyclopedia.com/doc/1G2-3404100963.html, created 2004, accessed 25 August 2016; Anon, 'Gatchina Palace Garden', http://www.gardenvisit.com/gardens/gatchina_palace_garden, accessed 25 August 2016; Banks, diary note, 29 December 1795, quoted in Carter, 'Russia', p.356.
39 Adapted from Noe to Banks, 'Extraordinary expences', [1796 n.d.], quoted in Carter, p.361.
40 Banks to Noe, 4 July 1795, Banks diary note, 29 December 1795, Noe to Banks, 12 November, 1795, quoted in Carter, 'Russia', pp.350, 355 & 359; Carter, *Banks*, p.304.
41 Banks diary note, 4 July 1795 & 29 December 1795, Noe to Banks, 12 November 1795, quoted in Carter, pp.345, 355-359; Hadfield et al, *Dictionary*, p.127; Anon, 'Eduard von Regel', *Nature*, vol.46, pp.60-61, (19 May 1892), [online]: http://www.nature.com/nature/journal/v46/n1177/abs/046060a0.html, accessed 25 August 2016; Anon, 'Nikolai T. Aksakov', [online]: http://impereur.blogspot.com/2015/08/1797-1882.html&prev=search, posted 29 August 2015, accessed 20 August 2016; E. Regel, *Catalogus plantarum quae in horto Aksakoviano coluntur*, (St Petersburg, 1860), p.135; Anon., *Strelitzia humilis* & *Strelitzia ovata*, [online]: http://www.theplantlist.org/tpl1.1/record/kew-267262, accessed 27 November 2016.
42 Global Plants, 'Fischer von Waldheim, Aleksandr Alexandrovich', [online]: http://plants.jstor.org/stable/history/10.5555/al.ap.person.bm000383458, accessed 19 September 2016; Anon, 'Russian horticultural exhibition', col.4, & notification, col.1, *Chicago Tribune*, 12 September 1898.
43 'Eduard von Regel', *Nature*; Regel & Körnicke, 'Strelitzia nicolai', fig.253, *Gartenflora*, (1858), vol.7, pp.265-267.
44 S.L. Naidu, S. Ramdhani & H. Baijnath, 'The floral biology of Strelitzia nicolai Regal & Koern. and the role of honey bees as opportunists', poster presented at South African Association of Botany conference, 1992.
45 *Curtis's Botanical Magazine*, tab. 4167 & 4168.
46 *Flores des Serres*, tab. 173 & 174.
47 Regel & Körnicke, 'Strelitzia nicolai', p.265; translation by Dr Hugh Glen.
48 Regel & Körnicke, 'Strelitzia nicolai', p.266.
49 'Kornicke, Friedrich August', [online]: http://kiki.huh.harvard.edu/databases/botanist_search.php?mode=details&id=811, accessed 9 August 2016.
50 Regel & Körnicke, 'Strelitzia nicolai', pp.266-267.
51 Regel & Körnicke, 'Strelitzia nicolai', p.266.
52 Desmond, *Dictionary*, 'Hooker, Sir Joseph', p.318; Joseph D. Hooker, 'Strelitzia nicolai', *Curtis's Botanical Magazine*, vol.115, (1889), plate 7038.
53 Gunn & Codd/Glen & Germishuizen, p.289; Dawson, Masson to Banks, 4 February, 27 May & 12 December 1777, p.590.
54 Regel & Körnicke, 'Strelitzia nicolai', p.267.
55 Global Plants, 'Fischer, Sebastian', [online]: http://plants.jstor.org/stable/10.5555/al.ap.person.bm000043219, accessed 29 May 2016; Renate Matzke-Karasz & David Damkaer, 'Sebastian Fischer (1806-1871), physician and naturalist in Munich, Cairo and St Petersburg', *Joannea Geologie und Paläontologie*, vol.11, (2011), pp.119-121; Bretschneider, 'Dr Fischer', pp.316-319; 'Letter from F.E.L. Fischer to Sir William Jackson Hooker; from St Petersburg; 12 July 1825', [online]: http://plants.jstor.org/stable/10.5555/al.ap.visual.kmdc1744, accessed 11 June 2017.
56 Desmond, *Dictionary*, 'Darwin, Charles Robert', p.173; Charles Darwin, *The effects of cross and self fertilisation in the vegetable kingdom*, (London, 1876), chapter X, p.371n; Stefano Mancuso, 'Federico Delpino and the foundation of plant biology', *Plant Signaling & Behavior*, vol.5, no.9, (September, 2010), pp.1067-1071; F. Delpino, 'Rapporti tra insetti e nettari extranuziali nelle piante', *Bollettino della Società Entomologica italiana*, vol.6, (1874), pp.234-239; F. Delpino, 'Funzione mirmecofila nel regno vegetale', *Memorie della Reale Accademia delle Scienza di Bologna*, vol.7, (1886), pp.215-323; Nickolas M. Waser, Jeff Ollerton & Andreas Erhardt, 'Typology in pollination biology: Lessons from an historical critique', *Journal of Pollination Ecology*, vol.3, no.1, (2011), pp.1-7.
57 Gunn & Codd/Glen & Germishuizen, 'Scott-Elliot, George Francis', pp.388-389; Desmond, *Dictionary*, 'Scott-Elliot, George Francis', p.548; G.F. Scott-Elliot, 'Ornithophilous flowers in South Africa', *Annals of Botany*, vol.IV, no.XIV, (May 1890), pp.265-280.
58 G.F. Scott-Elliot, 'Note on the fertilisation of Musa, Strelitzia reginae, and Ravenala madagascariensis', *Annals of Botany*, vol.IV, no.XIV, (May 1890), pp.260-261.
59 Gunn & Codd/Glen & Germishuizen, ''MacOwan, Peter', p.281.
60 Scott-Elliot, 'Note on fertilisation', p.261.
61 Quoted in C.J. Skead, 'Sunbirds and strelitzias', *Bokmakierie*, vol.14, no.23, (1963), p.24; see also: Emil Werth, 'Blütenbiologische Fragmente aus Ostafrika. Ostafrikanische Nectarinienblumen und ihre Kreuzungsvermittler. Ein Beitrag zur Erkenniniss der Wechselbeziehungen zwischen Blumen- und Vogelwelt', *Voranisches Centralblatt*, January 1900, LXXXIV, pp.188-192; Raju J.S. Aluri & C. Subba Reddi, 'Explosive pollen release and pollination in flowering plants', *Proceedings of the National Academy of Sciences, India*, (1995), B61, no.4, pp.323-332; Gunn & Codd/Glen & Germishuizen, 'Marloth, Hermann Wilhelm Rudolf', pp.285-286; Universität für Bodenkulture, 'Porsch, Otto', [online]: http://www.boku.ac.at/universitaetsleitung/rektorat/stabsstellen/oeffentlichkeitsarbeit/themen/geschichte/rektorinnen/porsch/ accessed 17 September 2016.
62 Skead, p.24.
63 Skead, pp.24-25.
64 Skead, pp.25-26.
65 Skead, p.25.
66 Skead, p.26.
67 M.K. Rowan, 'Bird pollination of *Strelitzia*', *Ostrich*, vol.45, (1974), p.40.
68 R. Cowgill, S.B. Davis & D. Harebottle, 'Firefinches and sunbirds on the move', *Bird numbers*, vol.12, no.1, (July 2003), p.39.
69 S.K. Frost & P.G.H. Frost, 'Sunbird pollination of *Strelitzia nicolai*', *Oecologica*, vol.49, no.3, (1981), p.383.
70 Frost & Frost, p.381.
71 Frost & Frost, pp.379-380 & 383.
72 Frost & Frost, p.384.
73 Skead, p.25; Frost & Frost, p.384.
74 Quentin Cronk & Isidro Ojeda, 'Bird-pollinated flowers in an evolutionary and molecular context', *Journal of Experimental Botany*, vol.59, no.4, (2008), pp.715-719; G. Coombs & C.I. Peter, 'Do floral traits of Strelitzia reginae limit nectar theft by sunbirds?', *South African Journal of Botany*, vol.75, (2009), pp.751-752 & 755-756.
75 *Blumengarten*, 1896; J. Medley Wood, *List of trees, shrubs and a selection of herbaceous plants, growing in the Durban Municipal Botanic Gardens*, (Durban, 1915), pp.2 & 7; R.A. Dyer, 'Strelitzia nicolai', *Flowering plants of Africa*, volume 25 (1946), plate 996.
76 Eve Palmer, *A field guide to the trees of southern Africa*, (London, 1977), p.80.
77 A.W. Bayer, 'Discovering the Natal flora', *Natalia*, vol.4, (1974), p.42; Gunn & Codd/Glen & Germishuizen, 'Drège, Carl', 'Drège, Isaac' & 'Krauss, Christian', pp.153-156 & 245-248; C. Plug, 'Krauss, Dr Christian F.F.', [online]: http://www.s2a3.org.za/bio/Biograph_final.php?serial=1574, accessed 17 September 2016; BM(NH), *Strelitzia augusta*, Krauss, S. Africa, Natal, s.n., n.d.
78 C. Plug, 'Delegorgue, Mr Louis Adulphe Joseph', [online]: http://www.s2a3.org.za/bio/Biograph_final.php?serial=689, & 'Wahlberg, Mr Johann August', [online]: http://www.s2a3.org.za/bio/Biograph_final.php?serial=3018, both accessed 17 September 2016; Adulphe Delegorgue, *Travels in southern Africa*, translated Fleur Webb, (Pietermaritzburg, 1990), pp.34 & 51; Bayer, p.43.

[79] Eds. Adrian Craig & Chris Hummel, Johan August Wahlberg, *Travel journals and some letters, 1836-1856*, Van Riebeeck Society series 2, no.23, (Cape Town, 1994 for 1992), '7 May', p.47.
[80] Elsa Pooley, *The complete guide to the trees of Natal, Zululand & Transkei*, (Durban, 1993), p.60.
[81] Ed. R.J. Mann, Henry Brooks, *Natal: A history and description of the colony*, (London, 1876), p.166.
[82] Hooker, 'Strelitzia nicolai'.
[83] Gunn & Codd/Glen & Germishuizen, 'Wood, John Medley', pp.464-466; J. Medley Wood, *A guide to the Natal Botanic Gardens containing plan of the gardens, byelaws etc., & catalogue of plants*, (Durban, 1883), p.45; J. Medley Wood, *List of trees, shrubs and a selection of herbaceous plants, growing in the Durban Municipal Botanic Gardens*, (Durban, 1897 and 1915 editions), nos.6, 18 & 41; Donal P. McCracken, *A new history of Durban Botanic Gardens*, (Durban, 1996), pp.6-9. Donal P. McCracken, 'The old Durban forest', in ed. Mark Mattson, *The Durban forest*, (Durban, 2015), pp.45-89.
[84] Medley Wood, *List*, (1915), p.2, no.6.
[85] Pooley, p.60.
[86] Y. Singh & H. Baijnath, 'Some interesting observations in the family Strelitziaceae', poster presented at South African Association of Botany conference, 1992.
[87] Naidu et al.
[88] Naidu et al.
[89] Frost & Frost, pp.381-382.
[90] Frost & Frost, p.382.
[91] S.H. Foord, R.J. van Aarde & S.M. Ferreira, 'Seed dispersal by vervet monkeys in rehabilitating coastal dune forests at Richards Bay', *South African Journal of Wildlife Research*, vol.24, no.3, (1994), pp.57 & 58, table 2.
[92] Foord et al, p.59.
[93] Mac van der Merwe & Rebecca L. Stirnemann, 'Group composition and social events of the banana bat, *Neoromicia nanus*, in Mpumalanga, South Africa', *South African Journal of Wildlife Research*, vol.39, no.1, (April 2009), pp.48-56; Hans J. Baagøe, 'Observations on the biology of the banana bat, *Pipistrellus nanus*', *Proceedings of the Fourth International Bat Conference*, eds. R.J. Olembo, J.B. Castelino & F.A. Mutere, (Nairobi, 1978), pp.275-282.
[94] PRE: ex Union of South Africa Department of Agriculture Government Herbarium, *Sterelitza augusta* [sic], J. Burtt-Davy, 2404, June 1904; Gunn & Codd/Glen & Germishuizen, 'Burtt Davy, Joseph', pp.121-123; Figures based on holdings at K, September 2012, and PRE, February 2016.
[95] K: *Strelitzia nicolai* Regel & Körn, St Petersburg Botanic Gardens, Nov. 1867; Desmond, *Dictionary*, 'Smith, Matilda', p.569; PRE, *Strelitzia nicolai*, H.M.L. Forbes, 27163, Sept. 1943; Pooley, p.60.
[96] W. John Kress, 'The phylogeny and classification of the Zingiberales', *Annals of the Missouri Botanic Gardens*, 1990, pp.678-721.
[97] T.R.E. Southwood 'Foreword', in William T. Stearn, *The Natural History Museum at South Kensington*, London, 1981, p.xx .
[98] Kress, 'Phylogeny and classification of Zingiberales', p.699; Otto Georgius Petersen, 'Musaceae, Zingiberaceae, Cannaceae, Marantaceae' in eds. A. Engler & K. Prantl, *Die natürlichen pflantzenfamilien*, Leipzig, 1889, vol.II, section 6, pp.1-43; K. Schumann, 'Musaceae', in A. Engler, *Das pflanzenreich*, IV.45, pp.28-33; Nakai, 'Essential results obtained from my observation on tropical plants in Java,' in Bulletin of Tokyo Science Museum, no.22, (1948), pp.39-43; J. Hutchinson, 'Zingiberales' in *The families of flowering plants*, 1959, vol.II: *Monocotyledons*, pp.581-90; R. Dahlgren, H.T. Clifford & P.F. Yeo, 'Strelitziaceae' in *The families of Monocotyledons*, (Berlin, 1985), pp.358-9.
[99] Kress, 'Phylogeny and classification of Zingiberales', p.701.

## CHAPTER 4

[1] Mandy Kirkby, *The language of flowers: A miscellany*, (London, 2001), p.183; Desmond, *Kew*, pp.64-65.
[2] Desmond, *Kew*, pp.73-74 & 78-79; W.J. Bean, *Royal Botanic Gardens, Kew*, (London, 1908), p.17; Mollendorf, pp.11-12; Banks to de Jussieu, 29 June 1788, DTC 6.40, summarised in Dawson, p.482, 8, also quoted in Desmond, *Kew*, p.79; Banks to Hamilton, 4 July 1794, BM Egerton Ms.2641. 151-152, summarised in Dawson, p.390, 52.
[3] Desmond, *Dictionary*, pp.385-386; James Britten, 'The herbarium of John Lightfoot', *Journal of Botany, British & Foreign*, vol.53, (1915), pp.269-271; Desmond, *Kew*, p.79; Johan Anders Murray to Joseph Banks, 5 April 1788, BM, Add Ms., 8097.102-103, summarised in Dawson, p.627, 13.
[4] Ruth Hayden, *Mrs Delany: Her life and her flowers*, (London, 1980; 2005 edition), pp.161-169; and Mrs Delany to Mrs Port, pp.319-322, & Mrs Delany to the Viscountess Andover, pp.323-325, in ed. Sarah Chauncey Woolsey, *The autobiography and correspondence of Mrs. Delany*, vol.2, (Boston, 1879). For a list of Mrs Delany's known surviving works, see R. Hayden, pp.172-188; & ed. Woolsey, vol.2, 'Appendix', pp.484-492.
[5] Blunt, *Penny*, p.56; Desmond, *Dictionary*, 'Delany, Mary', p.181, 'Meen, Margaret', p.433; Desmond, *Kew*, pp.79 & 109; Kelly M. McDonald, 'Margaret Meen: A life in four letters', pp.2 & 4, [online]: https://www.academia.edu/11307303/Margaret_Meen_A_Life_in_Four_Letters, accessed 18 August 2016; W. Botting Hemsley, 'Margaret Meen', *Gardeners' Chronicle*, 17 February 1894, pp.197-198; Andrews, 'Golden bird'; Wilfrid Blunt & William T. Stearn, *The art of botanical illustration*, (London, 1950), p.194; McDonald, p.2, quoting *Bulletin of Miscellaneous Information*.
[6] *Curtis's Botanical Magazine*, plate 119; see also plate 432, *Botanist's Repository*, vol.4; Desmond, *Dictionary*, 'Sowerby, James', p.575; 'Edwards, Sydenham', p.206; 'Fitch, Walter', p.224.
[7] *Curtis's Botanical Magazine*, plate 119.
[8] Desmond, *Celebration*, pp.30, 32 & 42.
[9] Desmond, *Dictionary*, 'Bauer, Ferdinand', p.47, 'Sibthorp, John', p.558; Anon, 'Flinders voyage', http://www.slsa.sa.gov.au/encounter/flindersvoyage.htm, accessed 29 November 2016.
[10] Bauer memorial in St Anne's Church, Kew, viewed 25 September 2012; quoted in full in W.T. Thiselton-Dyer, 'Historical account of Kew to 1841', *Bulletin of Miscellaneous Information*, no.60, December 1891, p.303.
[11] Alice M. Coats, *The book of flowers*, (London, 1984 ed.), pp.18-21; Desmond, *Kew*, pp.79-80.
[12] Coats, text to plate 65 & pp.14-15, 20-21; H. Walter Lack & Victoria Ibáñez, 'Recording colour in late eighteenth century botanical drawings', *Botanical Magazine*, vol.14, no.2, May 1997, p.91; Hu Walsh, 'Ferdinand Lukas Bauer, 1760-1826', [online]: http://www.illustratedgarden.org/mobot/rarebooks/author.asp?creator=Bauer,%20Ferdinand%20Lukas&creatorID=96 , accessed 6 August 2016; 'Art themes: Drawing conclusions', [online]: http://www.nhm.ac.uk/nature-online/art-nature-imaging/collections/art-themes/drawingconclusions/more/strelitzia_mo, accessed 6 August 2016; Bauer memorial, St Anne's Church, Kew; Anon, 'Illustration and representation' [NHM online].
[13] Julia Buckley, 'A royal flower – Bauer's Strelitzias', [online]: http://www.kew.org/discover/blogs/library-art-and-archives/royal-flower-%E2%80%93-bauer%E2%80%99s-strelitzias, posted 19 November 2014, accessed 4 May 2016; Marian Boddy-Evans, 'Camera lucida: An optical illusion for artists', [online]: http://painting.about.com/od/oldmastertechniques/ss/camera_lucida.htm, accessed 16 September 2016; Neolucida, 'The camera lucida and camera obscura', [online]: http://neolucida.com/history/obscura-and-lucida/, accessed 16 September 2016; Blunt & Stearn, *Botanical illustration*, (London, 1950), p.3, quoted by John Rourke, 'Beauty in truth', in ed. Marion Arnold, *South African botanical art*, p.64.
[14] Stephan A. Welz, 'Preface', in ed. Arnold, *South African botanical art*, p.7; Joseph Banks, *Delineations of exotic plants*, (London, 1796), preface, p.7; Anon, 'Illustration and representation', *The botany of empire in the long eighteenth century*, [online]: http://www.doaks.org/library-archives/library/library-exhibitions/botany-of-empire/illustration-and-representation, accessed 5 August 2016. See also Coats, pp.13-17.
[15] Peter & Frances Mallary, *A Redouté Treasury*, (New York, 1985), pp.16-19, 36-37 & plates 28 & 29.
[16] Buckley, 'Royal flower'.
[17] Blunt & Stearn, *Botanical illustration*, Appendix A, 'Botanical drawing' by Walter Hood Fitch, p.280.
[18] Anon., 'Kew diamond jubilee', [online]: http://florilegia.info/kew-diamond-jubilee-collection.php, accessed 7 August 2016.
[19] Carter, *Banks*, p.521; Shirley Sherwood & Martyn Rix, *Treasures of botanical art* (London, 2008), p.240; Desmond, *Dictionary*, 'Hooker, William Jackson', p.319; Bauer memorial, St Anne's Church, Kew, viewed 25 September 2012.
[20] Richard Perkins, *St Anne's Church, Kew Green*, (Much Wenlock, 1993), pp.3. & 9
[21] Perkins, p.14; parish history display, St Anne's, viewed 25 September 2012.
[22] Anon, 'Sir Richard Westmacott', [online]: http://www.victorianweb.org/victorian/sculpture/westmacottr/index.html, accessed 29 August 2016.
[23] Desmond, *Dictionary*, 'Thornton, Robert', p.611; Handasyde Buchanan, 'On collecting old flower books', in ed. Miles Hadfield, *The gardener's album*, (London & Wisbech, 1954), p.41
[24] Buchanan, p.38.
[25] Buchanan, p.38.
[26] Robert Thornton, *Temple of Flora*, [online]: http://digicoll.library.wisc.edu/cgi-bin/DLDecArts/DLDecArts-idx?id=DLDecArts.ThornTempFlo, accessed 8 August 2016.
[27] Dr Maurice, 'Verses addressed to Dr Thornton', in Thornton, *Temple of Flora*.
[28] Maurice, 'Verses'.
[29] Buchanan, pp.38 & 41; Coats, p.8; Patricia Fara, *Erasmus Darwin*, (Oxford, 2012), pp.75-92.
[30] Coats, p.9.
[31] Coats, p.9.
[32] Coats, p.9.
[33] Coats, p.9; Albrecht Dürer, [online]: http://www.albrecht-durer.org/, accessed 20 September, 2016; Lack & Ibáñez, p.98; S. Grayer, 'The Royal Horticultural Society's Colour Chart: An everyday tool for use in the herbarium. Its past, present and future', NatSCA News, issue 18, p.20; Grayer, pp.21-22.
[34] Coats, p.9.
[35] Coats, p.9; Mallary & Mallary, plate 28, p.36.
[36] Linné, *Supplementum*, p.157; Thunberg, *Resa*, vol.1, p.216; Charles Thunberg, *Travels in Europe, Africa, and Asia, made between the years 1770 and 1779*, (London, 1795, 2nd edition), vol.1, p.191; ed. Forbes, Thunberg, *Travels*, p.92.
[37] Masson to Banks, 2 June 1790, State Library of New South Wales, Papers of Sir Joseph Banks, section 5: Gardeners and collectors, series 13.46; William T. Stearn, *Botanical Latin*, (Portland, 2004 ed.), p.444; *Supplementum plantarum*, p.157; *Hortus kewensis* (1789)

vol.1, p.285.
[38] Masson to Banks, 6 August 1790, State Library of New South Wales, Banks Papers, section 5, series 13.47.
[39] J.E. Alexander, *An expedition of discovery into the interior of Africa, through the hitherto undiscovered countries of the great Namaquas, Boschmans and Hill Damaras*, (London, 1840), vol.2, p.195, quoted in Skead, p.226.
[40] Scott-Elliot, p.261, including n.2; Charles Morren, 'Strelitzia reginae var. rutilans hort', *Annales de la Société Royale de l'Agriculture de Gand*, (1846), vol.II, pp.53-4 & plate 53; Joseph Paxton, 'Select list of stove perennials', *Paxton's Botanical Magazine*, (1836), vol.2, p.207; Thomas Baines, *Greenhouse and stove plants*, (London, 1894), p.322; Desmond, *Dictionary*, 'Baines, Thomas', p.29; William Fawcett, 'Plants allied to the banana', *The banana: Its cultivation, distribution and commercial uses*, (London, 1921), p.201; eds. John Lindley & Thomas Moore, *Treasury of botany*, (London, 1866), part 2, p.1103.
[41] Maggie Campbell-Culver, *The origin of plants*, (London, 2012), p.363.
[42] Kate Lowe, 'Welcome to a journey of 175 years of horticulture', [online]: http://www.hortweek.com/welcome-journey-175-years-horticulture/article/1398958, posted 16 June 2016, accessed 29 August 2016; Buchanan, pp.43-44.
[43] George W. Taylor, 'Poor men's treasures', ed. Hadfield, *Album*, pp.142-152; quoted by Taylor, p.147; Rowland H. Biffen, *The auricula*, (Cambridge, 2014), p.36.
[44] Morren, 'Strelitzia reginae var. rutilans hort'; Anon, 'Morren, Charles', [online]: http://data.bnf.fr/10682560/charles_francois_antoine_morren/, accessed 29 November 2016; Morren, p.54; William Townsend Aiton, *Hortus kewensis*, (1814), p.55; World Checklist [online]: http://www.theplantlist.org/tpl1.1/record/kew-267254, accessed 29 August 2016.
[45] Morren, 'Strelitzia reginae var. rutilans hort', p.54.
[46] Morren, 'Strelitzia reginae var. rutilans hort', pp.53-54.
[47] Morren, 'Strelitzia reginae var. rutilans hort', p.54.
[48] J.E. Planchon, 'Strelitzia reginae Banks ex Aiton var. lemoinierii', *Flore des Serres et des Jardins de l'Europe*, series 2, vol. 23, (1880), t.2370-1, pp.1-3; K: *Strelitzia reginae* var. *lemoinierii*, K000308004, Lemoinier, 3 May 1870; herbarium sheet also contains *Strelitzia reginae*, Herbarium of J. Gay, presented by Joseph Hooker, February 1868.
[49] Andrews, 'Golden Bird', p.228.
[50] Planchon, 'Strelitzia reginae', p.3.
[51] Planchon, 'Strelitzia reginae', p.3.
[52] K: *Strelitzia reginae var. citrina*, December 1887, Palm House, Hort. Kew.
[53] Edwin Tidmarsh to William Watson, 3 February 1888, note attached to: K: *Strelitzia reginae* var. *citrina*.
[54] Gunn & Codd/Glen & Germishuizen, pp.424-5, 'Tidmarsh, Edward'.
[55] A. Siebert & A. Voss, 'Gattung 1107. Strelitzia', *Vilmorin's Illustriete Blumengärtnerei*, (Berlin, 1896), p.959; Th. Rümpler, *Vilmorin's Illustriete Blumengärtnerei*, Berlin, 1879; Anon, '1868-1931: Siesmayer's Actiengesellschaft', [online]: http://www.palmengarten.de/#/en_GB/about/history, accessed 29 August 2016.
[56] 'A hybrid strelitzia', *Bulletin of Miscellaneous Information*, (1910), p.65; K: *Strelitzia kewensis*, 14 February 1910, 'cult. in hort. bot. reg. Kew'.
[57] 'Hybrid strelitzia'.
[58] C.Q., 'The bird of paradise flower', 'Correspondence', *The Garden*, vol.LXXVIII, no.2200, p.30, 17 January 1914; Mark Sparrow to Susyn Andrews, 5 January 1993, Andrews archive; William Fawcett, 'Plants allied to the banana', *The banana*, (London, 1921 ed.), p.261; Reader's Digest, *Complete guide to gardening in South Africa*, (Cape Town, 1971), vol.2, pp.419-20; Kristo Pienaar, *The South African What Flower Is That?*, (Cape Town), 1984, p.318; Andrews. 'Golden bird', p.228.
[59] Morren, 'Strelitzia reginae var. rutilans hort', p.54.
[60] Extracted from Thomas J. Givnish, Timothy M. Evans, J.C. Pires & K.J. Sytsma, 'Polyphyly and convergent morphological evolution in Commelinales and Commelinidae', Molecular Phylogenetics and Evolution, vol.12, no.23, (Aug. 1999), pp.360-85; W. John Kress, Linda M. Prince, William J. Hahn & Elizabeth A. Zimmer, 'Evolutionary radiation of the Zingiberales', *Systematic Biology*, vol.50, no.6, (Nov.-Dec. 2001), pp.926-44: Table 3, p.939.
[61] Thomas J. Givnish, J. Chris Pires, Sean W. Graham, Marc A. McPherson, Linda M. Prince, Thomas B. Patterson, Hardeep S. Rai, Eric H. Roalson, Timothy M. Evans, William J. Hahn Kendra C. Millam, Alan W. Meerow, Mia Molvray, Paul J. Kores, Heath E. O'Brien, Jocelyn C. Hall, W. John Kress & Kenneth J. Systma, 'Repeated evolution of net venation and fleshy fruits among monocots in shaded habitats confirms *a priori* predictions: Evidence from an ndhF phylogeny', *Proceedings of the Royal Society*, 272, (2005), pp.1481-99, DOI: 10.1098/RSPB.2005.3067
[62] Cron et al, pp.614-6; Givnish et al, 'Repeated evolution', p.1481; W. John Kress, 'Phylogeny of the Zingiberaneae: Morphology and molecules', in eds. P.J. Rudall, P.J. Cribb, D.F. Cutler & C.J. Humphries, *Monocotyledons: Systematics and evolution*, (London, 1995), pp.443-460.
[63] P. Barry Tomlinson, 'An anatomical approach to the classification of Musaceae', *Journal of the Linnean Society*, Botany, vol.55, (1959), pp.779-809; P. Barry Tomlinson, 'Strelitziaceae', *Anatomy of the monocotyledons*, (London, 1969), pp.325-333; P. Barry Tomlinson to W. John Kress, no date (probably early 2002): http://botany.si.edu/cuatrecasas/medal2002.cfm, accessed 17 April 2016.
[64] Irwin E. Lane, 'Genera and generic relationships in Musaceae', *Mitteilungen der Botanischen Staatssammlung München*, vol.13, (1955), pp.114-31.
[65] Lane, pp.125-7; Kress, 'Phylogeny of Zingiberanae', p.699; Givnish et al, 'Polyphyly'.
[66] John Manning & Peter Goldblatt, 'Chromosome number in *Phenakospermum* and *Strelitzia* and the basic chromosome number in Strelitziaceae (Zingiberales)', *Annals of Missouri Botanic Gardens*, vol.76, (1989), pp.932-3.

[67] Manning & Goldblatt, p.933.
[68] Kress, 'Phylogeny and classification of Zingiberales', p.719; W. John Kress, Linda M. Prince, William J. Hahn & Elizabeth A. Zimmer, 'Unraveling the evolutionary radiation of the families of Zingiberales using morphological and molecular evidence', *Systematic Biology*, vol.50, no.6, (Nov-Dec 2001), pp.926-44.
[69] Austin Boyd, 'Musopsis N. Gen.: A banana-like leaf genus from the early tertiary of eastern North Greenland', *American Journal of Botany*, vol.79, issue 12, (1992), pp.1359-1367.
[70] Kress et al, 'Unraveling', p.939 & Table 3; John C. Benedict, 'Zingiberalean fossils from the Late Paleocene of North Dakota, USA, and their significance to the origin and diversification of Zingiberales', Ph.D. dissertation, Arizona State University, 2012, [online]: https://repository.asu.edu/attachments/93705/content/tmp/package-YwGKxo/Benedict_asu_0010E_11865.pdf, accessed 15 June 2016.
[71] Stephanie T. Chen & Selena Y. Smith, 'Phytolith variability in Zingiberales: A tool for the reconstruction of past tropical vegetation', *Palaeogeography, Palaeoclimatology, Palaeoecology*, vol.370, (2013), pp.1-12; Chen & Smith, pp.6 & 8-10; Cron et al, pp.606 & 616.

## CHAPTER 5

[1] Anon, 'Travellers Palm', [online]: http://www.calpalms.com/products.asp, accessed 23 December 2013; L.J. Dorr & P.G. Parkinson, 'Proposal to conserve the spelling 1320 *Ravenala* (Strelitziaceae)', *Taxon*, vol.39, Feb. 1990, pp.131-132.
[2] Central Intelligence Agency, 'Africa: Madagascar: Geography', [online]; Library of Congress Country Studies, 'Madagascar: Geography', [online]: http://www.loc.gov, accessed 11 November 2013; Olive Murray Chapman, 'Primitive tribes in Madagascar', *Geographical Journal*, vol.96, no.1, (June, 1940), pp.14-25; Pier Martin Larson, 'Colonies lost: God, hunger, and conflict in Anosy (Madagascar) to 1674', *Comparative Studies of South Asia, Africa and the Middle East*, vol.27, no.2, (2007), p.345; Central Intelligence Agency, 'Africa – Madagascar', *The world factbook*, [online]: https://www.cia.gov/library/publications/the-world-factbook/geos/ma.html, updated 15 August 2016; accessed 11 November 2013; Anon, 'The history of Madagascar', [online]: http://www.masombahiny.com/History_of_Madagascar.html, accessed 13 November 2013; Hubert de Vries, 'Madagascar: Heraldry', [online]: http://www.hubert-herald.nl/Madagascar.htm, posted 9 February 2009, updated: 23 February 2009, 3 June 2010, 2 July 2014, 22 January 2015; accessed 5 November 2013; Anon, 'Madagascar history: Geological history', [online]: http://www.eurekaencyclopedia.com/index.php/Category:Madagascar_History#Geological_History, accessed 13 November 2013; J. Hutchinson, 'The flora of Madagascar', *Nature*, vol.145, (23 March 1940), pp.448-451; Jo van Ass, Johann du Preez, Leslie Brown & Nico Smit, *The story of life & the environment: An African perspective* (Cape Town, Struik Nature, 2012), pp.331-2; Maarten de Wit, Margaret Jeffrey, Hugh Bergh & Louis Nicolaysen (University of Witwatersrand), 'Gondwana reconstruction and dispersion', http://www.searchanddiscovery.com/documents/97019/index.htm, accessed 21 December 2013; www.calpalms.com, accessed 23 December 2013.
[3] Gregg F. Gunnell, 'Biogeography and the legacy of Alfred Russel Wallace', in *Geologica Belgica*, (2013), 16/4: 211-216; Ulrich Meve & Sigrid Liede, 'Floristic exchange between mainland Africa and Madagascar', in *Journal of Biogeography*, vol.29, no.27, (Jul.2002), pp.865-873; C. Galley & H.P. Linder, 'Geographical affinities of the Cape flora, South Africa', in *Journal of Biogeography*, vol.33, no.2, (Feb. 2006), pp.236-250; Gillian Feeley-Harnik, '*Ravenala madagascariensis* Sonnerat: The historical ecology of a "Flagship Species" in Madagascar', *Ethnohistory*, 48:1-2 (winter-spring 2001), pp.31-86; John Hutchinson, 'Order 94. Zingiberales . . . 367 Strelitziaceae', *The families of flowering plants*, vol II, 'Monocotyledons', (1959), p.582; Alfred Russel Wallace, *Island life*, (London, 1892, 2nd ed.), p.441.
[4] Patrick Blanc, Annette Hladik & Claude Marcel Hladik, 'L'arbre du voyageur dans toute sa diversité', 1er partie, *Hommes & Plantes*, p.47.
[5] Mervyn Brown, *A history of Madagascar*, (Cambridge, 1995); Anon, 'Madagascar-History', http://www.nationsencyclopedia.com/Africa/Madagascar-HISTORY.html, accessed 13 November 2013.
[6] Encyclopedia of the Nations, 'Madagascar – History'; Milijoana Randrianarivelojosia, 'History of malaria treatment in Madagascar', [online]: http://www.mmv.org/sites/default/files/uploads/docs/artemisinin/2010_Madagascar_History_of_Malaria_Treatment.pdf, accessed 25 December 2013.
[7] William Ellis, *History of Madagascar*, (London, [1838]), vol.1, pp.2 & 4; Gwyn Campbell, 'Imperial rivalry in the western Indian Ocean and schemes in colonised Madagascar, 1769-1826', *Revue des mascareignes*, no.1, (1999), p.77; Henry Yule, *The book of Ser Marco Polo*, (London, 1871), Book 3, chapter 33, [online]: https://en.wikisource.org/wiki/The_Travels_of_Marco_Polo/Book_3/Chapter_33, accessed 10 September 2016; Silvio A. Bedini & David Buisseret, *The Christopher Columbus encyclopedia*, 'Cabral, Pedro Alvares', pp.87-89; Larson, p.345.
[8] Quoted in Richard H. Grove, *Green imperialism* (Delhi, 1995), p.82; Science Museum, 'Garcia de Orta', [online]: http://broughttolife.sciencemuseum.org.uk/broughttolife/people/garciadeorta, accessed 10 September 2016.
[9] Larsen, p.346; London merchant Richard Boothby quoted in Ellis, pp.4-5; Campbell, pp.76-78, 84, 93; Ellis, pp.4-5 & 33-34.
[10] Brian Whitenton, 'The difference between pirates, privateers and buccaneers', parts 1 & 2, The Mariners' Museum and Park Blog, [online]:

http://www.marinersmuseum.org/blog/2012/09/the-difference-between-pirates-privateers-and-buccaneers-pt-1/, posted 20 September 2016; http://www.marinersmuseum.org/blog/2012/10/the-difference-between-pirates-privateers-and-buccaneers-pt-2/, posted 4 October 2012; accessed 11 September 2016; Campbell, pp.77 & 87; Library of Congress Country Studies, 'Madagascar: Precolonial era, prior to 1894', [online]: www.loc.gov, posted August 1994, accessed 11 November 2013; Travel Madagascar, 'Sainte Marie (Nosy Boraha)', [online]: http://www.travelmadagascar.org/CITIES/Sainte-Marie.html, posted 2013, accessed 16 November 2013; Maritime Heritage Project, 'Madagascar – Boutres', http://www.maritimeheritage.org/ports/madagascar.html, accessed 13 November 2013.
[11] Piaulian & Viètte, p.509.
[12] Campbell, p.77; Claude Allibert, 'Etienne de Flacourt', Archives de France, [online]: http://www.archivesdefrance.culture.gouv.fr/action-culturelle/celebrations-nationales/2008/litterature-et-sciences-humaines/etienne-de-flacourt-dictionnaire-de-la-langue-de-madagascar-et-histoire-de-la-grande-isle-madagascar, accessed 11 September 2016; Encyclopaedia Britannica 1911, 'Etienne de Flacourt', [online]: http://www.theodora.com/encyclopedia/f/etienne_de_flacourt.html, accessed 11 September 2016; Campbell, p.77.
[13] Ed. Ruth Dorrel, 'Charles T. Price', Biographies of Deceased United Brethren Ministers, based on Adam Byron Condo, *History of the Indiana Conference of the Church of The United Brethren in Christ*, (1926), [online]: http://genealogytrails.com/ind/united-brethren-ministers.html; quoted in Feeley-Harnik, p.57.
[14] New World Encyclopedia, 'Cardinal Richelieu', [online]: http://www.newworldencyclopedia.org/p/index.php?title=Cardinal_Richelieu&oldid=682745, posted 2 April 2008, accessed 8 September 2016; George Padmore, 'Madagascar fights for freedom', *Left*, no.132, October 1947, transcribed by Christian Hogsbjerg for the Marxists' Internet Archive 2007, [online]: https://www.marxists.org/archive/padmore/1947/madagascar.htm, accessed 13 November 2013; Library of Congress Country Studies, 'Madagascar: Precolonial era, prior to 1894'; New World Encyclopedia, 'Napoleon Bonaparte', [online]: http://www.newworldencyclopedia.org/p/index.php?title=Napoleon_Bonaparte&oldid=987574, posted 27 April 2015, accessed 8 September 2016; Campbell, p.82.
[15] Le Sieur de Flacourt, *Histoire de la Grande Isle Madagascar*, (Paris, 1658 ed.), pp.119 & 124; Angela J. Beaumont, H. Baijnath & T.J. Edwards, 'Ravenala madagascariensis', Flowering Plants of Africa, (2005), vol.59, pp.44-50.
[16] De Flacourt, p.124.
[17] De Flacourt, p.124.
[18] Charles Lemaire, 'Ravenala madagascariensis', *L'illustration Horticole*, (February 1860), vol.7, plate 234.
[19] De Flacourt, pp.119, 123-124.
[20] De Flacourt, p.124.
[21] De Flacourt, pp.123-124.
[22] Belle McPherson Campbell, *Madagascar*, Missionary Annals (Chicago, 1889), p.23; E. Caustier, 'La flore et les forêts de Madagascar', *Bulletin de la Société Nationale d'Acclimatation de France*, (1896), 87.
[23] Global Plants, 'Adanson, Michel', [online]: http://plants.jstor.org/stable/10.5555/al.ap.person.bm000000048, accessed 11 September 2016; Michel Adanson, *Familles naturelles des plantes*, (Paris, 1763), pp.61-67, ix: Familles: Les Gingembres: *Zingiberes*; Giovanni Scopoli, *Introductio ad historiam naturalem, sistens genera lapidum, plantarum et animalium hactenus detecta, caracteribus essentialibus donate, in tribus divisa, subinde ad leges naturae* (Prague, 1777), p.96.
[24] Global Plants, 'Adanson, Michel'; Randy Smith, 'Book of the week: Familles des plantes', Biodiversity Heritage Library Blog, [online]: http://blog.biodiversitylibrary.org/2012/06/book-of-week-familles-des-plantes.html, posted 15 June 2012, accessed 16 November 2013.
[25] Adanson, pp.61-67; Steve Jones, *No need for geniuses*, (London, 2016), p.56; Global Plants, 'Jussieu, Antoine Laurent de', [online]: http://plants.jstor.org/stable/10.5555/al.ap.person.bm000004180, accessed 11 September 2016; Encyclopaedia Britannica, 'Bernard de Jussieu', [online]: https://global.britannica.com/biography/Bernard-de-Jussieu, accessed 11 September 2016; Global Plants, 'Adanson'.
[26] E. Decrock, 'Sur l'assise silicifère du tegument seminal des Ravenala', Proceedings of 22 May 1911 meeting, Académie des Sciences, Paris, in *Comptes Rendus Hebdomadaires des Séances de l'Académie des Science*, vol.152, (Jan-Jun 1911), pp.1406-1407; see http://www.sio2.ca/what-is-silicon-dioxide/, accessed 3 July 2016; Friedrich Czapek, 'Die Verhältnisse im reifen Samen', section 2, chapter 25, subsection 1, pp.372-373, *Biochemie der pflanzen* (Jena, 1925).
[27] Global Plants, 'Sonnerat, Pierre', [online]: http://plants.jstor.org/stable/10.5555/al.ap.person.bm000008002, accessed 11 September 2016; Dorr & Parkinson, p.131; Pierre Sonnerat, *Voyage aux Indes orientales & à la Chine*, (Paris, 1782), Tome second, livre 5, 'Le ravenala', p.225.
[28] Bibliomonde, 'Pierre Poivre', [online]: http://www.bibliomonde.com/auteur/pierre-poivre-2268.html, accessed 11 September 2016; Encyclopaedia Britannica, 'Physiocrat', [online]: https://global.britannica.com/topic/physiocrat, accessed 11 September 2016; Concise Encyclopedia of Economics, 'François Quesnay', [online]: http://www.econlib.org/library/Enc/bios/Quesnay.html, accessed 11 September 2016; Murray N. Rothbard, 'Richard Cantillon: The founding father of modern economics', [online]: https://mises.org/library/richard-cantillon-founding-father-modern-economics, posted 12 June 2010, accessed 11 September 2016.
[29] Bibliomonde, 'Pierre Poivre'; and online archive: www.pierre-poivre.fr.
[30] Etienne Taillemite, 'Dumas, Jean-Daniel', in *Dictionary of Canadian Biography*, vol.4, [online]: http://www.biographi.ca/en/bio/dumas_jean_daniel_4E.html, posted 2003, accessed 24 December 2013.
[31] Yves Laissus, 'Commerson, Philibert', [online]: http://www.encyclopedia.com/doc/1G2-2830900963.html, posted 2008, accessed 8 November 2013; Strait Through, 'Louis-Antoine de Bougainville', [online]: http://libweb5.princeton.edu/visual_materials/maps/websites/pacific/bougainville/bougainville.html, accessed 11 September 2016; Philippe Henrat & Marjorie Jung, 'Poissonnier, Pierre-Isaac', [online]: http://cths.fr/an/prosopo.php?id=114732, posted 5 May 2013; updated 10 October 2015; accessed 11 September 2016; Laissus, 'Commerson'.
[32] Global Plants, 'Commerson, Philibert' [online]: https://plants.jstor.org/stable/10.5555/al.ap.person.bm000001607, accessed 11 September 2016; Strait Through, 'Louis-Antoine de Bougainville'; Pierre-Yves Beaurepaire, 'Pèlerinage à l'île de Cythère', [online]: https://www.histoire-image.org/etudes/pelerinage-ile-cythere-dit-embarquement-cythere, posted May 2013, accessed 11 September 2016; Internet Encyclopedia of Philosophy, 'Jean-Jacques Rousseau', [online]: http://www.iep.utm.edu/rousseau/, accessed 11 September 2016; Grove, p.238; Jean-Paul Morel, 'Philibert Commerson à Madagascar et à Bourbon', p.7, downloaded from: www.pierre-poivre.fr, posted December 2012, accessed 13 November 2013.
[33] Jennie Cohen, 'First woman to circle the globe honored at last', [online]: http://www.history.com/first-woman-to-circle-the-globe-honored-at-last, posted 3 January 2012; accessed 12 September 2016; Global Plants, 'Commerson, Philibert'; Grove, pp.241, 253.
[34] Encyclopaedia Britannica, 'Jacques-Henri Bernardin de Saint-Pierre', [online]: https://global.britannica.com/biography/Jacques-Henri-Bernardin-de-Saint-Pierre, accessed 11 September 2016; Dennis Wood, 'Jacques-Henri Bernardin de Saint Pierre', *New Oxford companion to French literature*, (Oxford, 1995); Larousse, 'Henri Bernardin de Saint-Pierre', [online]: http://www.larousse.fr/encyclopedie/personnage/Henri_Bernardin_de_Saint-Pierre/108558, accessed 22 December 2013; Robin Howells, 'Bernardin de Saint-Pierre's founding work: The *Voyage à l'île de France*', *Modern Language Review*, vol.107, issue 3, (July 2012), p.756; Académie Française, 'Jacques-Henri Bernardin de Saint Pierre', [online]: http://www.academie-francaise.fr/les-immortels/jacques-henri-bernardin-de-saint-pierre, accessed 11 September 2016.
[35] Nikolaus Jacquin, *Plantarum rariorum horti caesarei schoenbrunnensis*, (Vienna, 1797), vol.1, p.x & '93 Ravenala madagascariensis', pp.47-49; Santiago Madriñan, 'The illustrious Nikolaus Jacquin', *Botanica Collecta*, March 2014, p.30.
[36] Sonnerat, 'Des Iles de France et de Bourbon', pp.360-361; Wikipedia, 'Jean-Francois Charpentier de Cossigny', [online]: https://fr.wikipedia.org/wiki/Jean-Fran%C3%A7ois_Charpentier_de_Cossigny, accessed 11 September 2016; Blanc, Hladik & Hladik, p.26; Campbell, p.80; William Roxburgh, 'Urania speciosa', *Flora indica*, vol.2, (Serampore, 1832), p.114; BM(NH), *Ravenala madagascarensis*, Dr Roxburgh, Hort. Calcutta, 1808; David Prain, 'Ravenala', *Bengal plants*, vol.2, (Calcutta, 1903), pp.1049-1050; Desmond, *Dictionary*, 'Prain, Sir David', pp.502-503; H.F. Hamilton, *A handbook of tropical gardening and planting*, (Colombo, 1914, 2nd edition), pp.312-314.
[37] Morel, pp.1-2.
[38] Morel, pp.2-3.
[39] Morel, pp.3-4; Madeleine Ly-Tio-Fane, 'Sonnerat, Pierre' in *Complete dictionary of scientific biography*, [online]: http://www.encyclopedia.com/doc/1G2-2830904084.html, posted 2008, accessed 23 December 2013.
[40] Quoted in Morel, p.4; Commerson to Lalande, quoted in Morel, p.5.
[41] Thomas L. Hankins, 'Lalande, Joseph-Jérôme Lefrançais de', [online]: http://www.encyclopedia.com/doc/1G2-2830902430.html, accessed 12 September 2016.
[42] Morel, p.5
[43] Morel, p.5.
[44] Quoted in Morel, p.6; Pierre Poivre online archive, 'Le baron de Clugny, visite aux environs de Foulpointe', Clugny to Dumas, 15 September 1768, [online]: http://www.pierre-poivre.fr/doc-68-9-15b.pdf, accessed 9 September 2016; Wikipedia, 'Marc Antoine Nicolas Gabriel de Clugny', [online]: https://fr.wikipedia.org/wiki/Marc_Antoine_Nicolas_Gabriel_de_Clugny, accessed 23 December 2013.
[45] Morel, p.4; Understanding Evolution, 'Extinctions: Georges Cuvier', [online]: http://evolution.berkeley.edu/evolibrary/article/0_0_0/history_08 & 'Early concepts of evolution: Jean Baptiste Lamarck', [online]: http://evolution.berkeley.edu/evolibrary/article/history_09, both accessed 13 November 2013; Grove, pp.242-7.
[46] Quoted in Feeley-Harnik, p.39.
[47] Quoted in Jeannine Monnier, Anne Lavondès, Jean-Claude Jolinon & Pierre Elouard, *Philibert Commerson, Le découvreur de Bougainvillier*, (Chatillon-sur-Chalaronne, 1993), p.124; and in Feeley-Harnik, p.39.
[48] Morel, pp.2-3, 7-8 & 9-11; Londa Schiebinger & Claudia Swan, *Colonial botany: Science, commerce and politics in the early modern world*, (Philadelphia, 2007), p.197; F.B. de Montessus, *Martyrologie et biographie de Commerson*, (Châlon-sur-Saone, 1889); C. Plug, 'Oldenburg'.
[49] Commerson to Jérôme Lalande, quoted in Morel, p.4; Stephen Ellis, 'A history of

50 William Ellis, 'Madagascar', *International Institute of Asian Studies Newsletter*, [online]: http://iias.asia/iiasn/iiasn7/iswa/ellis.html, accessed 11 November 2013.
51 William Ellis, p.34.
52 Desmond, *Dictionary*, pp.248-249; Ellis, p.36; McCracken, *Gardens of empire*, p.160.
53 Karl Shuker, Shuker Nature [blog], 'Sonnerat's non-existent penguins (and kookaburra) of New Guinea', posted 1 May 2013, accessed 23 December 2013; Ly-Tio-Fane, 'Sonnerat'.
54 Ly-Tio-Fane, 'Sonnerat'.
55 Shuker, 'Sonnerat's non-existent penguins'.
56 Pierre Sonnerat, *Voyage aux Indes*, Tome premier, title page-p.x, (Paris, 1782).
57 Ly-Tio-Fane, 'Sonnerat ; Sonnerat, *Indes*, vol. 2, book 4, chap.3, 'De l'île de Madagascar', pp.315-358.
58 Sonnerat, *Indes*, vol. 2, p.318.
59 Sonnerat, *Indes*, vol. 2, pp.330-331.
60 Sonnerat, *Indes*, vol. 2, pp.327 & 329; Annette Hladik, Patrick Blanc, Nicolas Dumetz, Vololoniaina Jeannoda, Nelson Rabenandrianina & Marcel Hladik, 'Données sur la repartition géographique du genre *Ravenala* et sur son rôle dans la dynamique forestière à Madagascar', *Mémoires de la Société de Biogéographie de Paris*, (2000), p.94.
61 Sonnerat, *Indes*, vol. 2, p.322; Louis Fauchenet, 'Pailles et fibres diverses employees à Madagascar dans la chapellerie et la vannerie', *L'agriculture pratique des pays chauds* 33, no.11, (Mar-Apr 1903), p.619.
62 Sonnerat, *Indes*, vol. 1, book 5, chap.3, 'Plantes', p.223; BM(NH), *Ravenala madagascariensis* type, marked 'Sonnerat'.
63 Sonnerat, *Indes*, vol. 1, pp.224-225.
64 Sonnerat, *Indes*, vol. 1, plates 124-126.
65 Ly-Tio-Fane, 'Sonnerat'.
66 Ly-Tio-Fane, 'Sonnerat'.
67 Johann Schreber, *Genera plantarum*, (Frankfurt, 1789), 'Ravenala madagascariensis', p.212; Tropicos, 'Schreber, Johann Christian Daniel von', [online]: http://www.tropicos.org/Person/200?projectid=%2011, accessed 22 December 2013; Botanischestaatssammlung, 'Johann Christian Daniel von Schreber', [online]: http://www.botanischestaatssammlung.de/herbarium/plants/schreber.html, accessed 12 September 2016; Anon., 'Gmelin, Johann Friedrich', [online]: http://www.societe-belge-de-malacologie.be/s-dictio-Gmelin.html, accessed 29 May 2018; ed. J.F. Gmelin, *Systema naturae*, (Leipzig, 1788-1793), 13th edition, vol.2, p..434.
68 HUH Index of Botanists, 'Willemet, (Pierre) Remi', [online]: http://kiki.huh.harvard.edu/databases/botanist_search.php?id=66227, accessed 24 December 2013; John Gorton, *General biographical dictionary*, (London, 1833), vol.3, 'Willemet, Pierre'; Pierre Rémi Willemet, 'Heliconia ravenala', in Annalen der Botaniek, vol.18 (1796), p.22; Global Plants, 'Raeuschel, Ernst Adolf', [online]: https://plants.jstor.org/stable/10.5555/al.ap.person.k35867, accessed 24 December 2013; Ernst Raueschel, 'Urania madagascariensis', *Nomenclator botanicus omnes plantas*, ed.3, (Liepzig, 1797), p.91; ed. Karl Willdenow, *Species plantarum*, vol.2, (Berlin, 1799), 'Urania speciosa', p.7; Global Plants, 'Richard, Achille', [online]: http://plants.jstor.org/stable/10.5555/al.ap.person.bm000006989, accessed 24 December 2013; Achille Richard, 'Urania ravenala', *Nova Acta Physico-Medica Academiae Caesareae Leopoldino-Carolinae Naturae Curiosorum Exhibentia Ephemerides sive Observationes Historias et Experimenta*, vol.15, (1831), p.15; Richard Pulteney with William George Maton, *A general view of the writings of Linnaeus*, (London, 1805 ed.), p.59.
68 Robert Furlong, 'Those exceptional citizens', *Discover Mauritius*, July-Sept. 2012, p.43; Association La Gens Fi.et, 'La Bigorne', [online]: http://www.filet.org/FilGens/labigorne.htm, posted 10 May 2011, accessed 16 November 2013; Anon, 'Sainte Marie (Nosy Baraha)', [online]: http://www.travelmadagascar.com/CITIES/Sainte-Marie.html, accessed 8 September 2016.
69 Jean-Claude Legros, 'La reine Betty, l'une des plus belles femmes que l'on pût voir', [online]: http://7lameslamer.net/spip.php?page=imprimir_articulo&id_article=1352, posted 17 May 2015, accessed 13 September 2016.
70 Unifrance, 'La Bigorne', [online]: http://www.unifrance.org/film/140/la-bigorne-caporal-de-france, accessed 13 September 2016.
71 Julien Durup, 'The forgotten French commandant of Rodrigues Island in the Seychelles', *Seychelles Weekly*, 14 February 2016, [online]: http://www.seychellesweekly.com/February%2014,%202010/top6_french_commandant.html, accessed 16 November 2013.
72 Richard Via and Vikash Tatayah to Himansu Baijnath, email, 2 January 2013; Tatayah to Baijnath, email, 10 January 2012; Desmond, *Dictionary*, 'Balfour, Sir Isaac', p.32. BM(NH), 'Strelitzia sp.', Dr I.B. Balfour, Rodrigues, Aug-Dec 1874, originally marked 'Canna indica', annotated 'Afr Strelitzia Not a Canna'; Is. B. Balfour, 'Collections from Rodriguez: Botany', in 'An account of the petrological, botanical, and zoological collections made in Kerguelen's Land and Rodiguez during the Transit of Venus Expeditions', *Philosophical Transcactions of the Royal Society of London*, vol.168, 1879, p.307.

## CHAPTER 6

1 Marco Turco, *A visitor's guide to Madagascar*, (Cape Town, 1988), p.43; Alex Shoumatoff, 'Our far-flung correspondents: Madagascar', *New Yorker*, 7 March 1988, [online]: http://blog.dispatchesfromthevanishingworld.com/our-far-flung-correspondents-madagascar/, accessed 24 December 2013.
2 Feeley-Harnik, pp.57-58, 63 & 65; see [online]: http://www.airmadagascar.com/fr, accessed 13 September 2016.
3 Hladik et al, 'Données sur la repartition géographique', pp.99-100.
4 P. Blanc, A. Hladik, N. Rabendrianina, J.S. Robert & C.-M. Hladik, 'Strelitziaceae: The variants of *Ravenala* in natural and anthropogenic habitats', in eds. S.M. Goodman & J. Benstead, *The natural history of Madagascar*, University of Chicago Press, (London, 2003), p.473.
5 Chapman, p.22; Library of Congress Country Studies, 'Madagascar: Precolonial era'; Institute of American History, 'Radama I: Background', [online]: http://www.gilderlehrman.org/collections/a4fe9327-64b1-4fd1-b4e7-422514664945, accessed 13 September 2016.
6 McPherson Campbell, pp.21-22; Oliver, p.306.
7 McPherson Campbell, p.21.
8 Charles W. Forman, 'Sibree, James', [online]: http://www.bu.edu/missiology/missionary-biography/r-s/sibree-james-1836-1929/, accessed 13 September 2016; Sibree quoted in Oliver, pp.307-308; Chapman, pp.17 & 22.
9 Oliver, pp.310-311; Sibree quoted in Feeley-Harnik, p.65; Oliver quoted in Feeley-Harnik, p.65.
10 Oliver, p.313.
11 Blanc et al, 'Strelitziaceae', pp.474-5.
12 R. Decary, 'Les formations littorals dans la region de Mananara', Bulletin de l'Académie Malgache, new series, vol.6, (1922-1923), p.65; Blanc et al, 'Strelitziaceae', pp.474-5.
13 Blanc et al, 'L'arbre du voyageur', 1, pp.42-47; Hladik et al, 'Données sur la repartition géographique', p.98; Feeley-Harnik, p.72 n.7, quoting from Rev. John Richardson's 1885 London Missionary Society dictionary.
14 Oliver, p.314.
15 Hladik, Blanc & Hladik, 'L'arbre du voyageur', 1, pp.44-47; Oliver, p.314.
16 Hladik et al, 'L'arbre du voyageur', 1, pp.44-47.
17 Oliver, p.315.
18 Lemaire, 'Ravenala madagascariensis'.
19 Hladik et al, 'Données sur la repartition géographique', p.98.
20 Quoted in Feeley-Harnik, p.66.
21 Milijoana Randrianarivelojosia, 'History of malaria treatment in Madagascar', http://www.mmv.org/sites/default/files/uploads/docs/artemisinin/2010_Madagascar/History_of_Malaria_Treatment.pdf, accessed 25 December 2013; Feeley-Harnik, p.65.
22 Russell E. Mittermeier, 'Primate diversity and the tropical forest', in ed. E.O. Wilson, *Biodiversity*, (Washington, 1988), pp.145-154.
23 Feeley-Harnik, pp.44-54; Desmond, *Dictionary*, 'Ellis, Rev. William', p.209.
24 Quoted in Feeley-Harnik, p.44.
25 Quoted in Feeley-Harnik, p.44.
26 Karl Eggert, 'Malagasy commentary', in *Ethnohistory*, vol.48, no.1-2, winter-spring 2001, pp.309-318; Feeley-Harnik, p.55.
27 See Eggert.
28 Global Plants, William Ellis to Sir William Jackson Hooker, 26 November 1858, [online]: http://plants.jstor.org/stable/history/10.5555/al.ap.visual.kldc8171, updated 5 March 2013, accessed 13 September 2016; Dictionary of African Christian Biography, 'William Ellis', [online]: http://www.dacb.org/stories/madagascar/ellis_william.html, accessed 13 September 2016.
29 Feeley-Harnik, pp.52-56.
30 Hamilton, pp.313-314.
31 Flanders Marine Institute, 'Paul van Oye', *Liber memorialis*, part 4, pp.164-174; Anno Faubel, 'Memories, Biographies and Bibliographies of Famous Turbellariologists', p.26, [online]:www.iwane-je.ed.jp/af/memories, accessed 28 December 2013.
32 Kibungu Kembelo, 'The botanical gardens of Zaire', [online]: http://www.bgci.org/worldwide/article/0037/, posted December 1996, accessed 28 December 2013; Congo Biodiversity Initiative, 'Botanical gardens of Eala-Mbandaka', [online]: http://www.congobiodiv.org/en/infrastructure/botanical-gardens/eala-mbandaka, accessed 28 December 2013.
33 Nobel Prize, 'Andre Gide', [online]: http://www.nobelprize.org/nobel_prizes/literature/laureates/1947/gide-bio.html, accessed 13 September 2016; André Gide, *Voyage au Congo*, (Paris, 1927), chap.2, 11 September.
34 A-R. Proschavsky, 'Sur le *Ravenala madagascariensis*', *Bulletin de la Société National d'Acclimatation de France*, June 1897, p.273; A.D. Frederickson, *Ad orientem*, (London, 1889), pp.172-173; H.H. Rusby, 'Guide to the Economic Museum of the New York Botanical Garden', *Bulletin of the New York Botanical Garden*, vol.11, no.41, p.96; Anon., 'Miscellaneous notes', *Bulletin of Miscellaneous Information*, no.63, (March 1892), p.73; Karl Afzelius 'Ravenala', p.53, 'De resandes träd', *Fauna och Flora*, (Uppsala, 1914), pp.69-70; ed. George Nicholson, *Illustrated dictionary of gardening*, Division 6, (London, 1887), 'Ravenala', p.279.
35 Paul van Oye, 'Recherches sur la biologie de Ravenala madagascariensis Sonner', *in Revue de la Zoologie Africaine*, XI, 2, Suppl. Bot., 15 July 1923, pp.21 & 23.
36 Van Oye, p.19.
37 Van Oye, p.19; G.M. Ryan, 'The wild plantain', *Journal of the Bombay Natural History Society*, vol.15, part 4, (June 1904), p.589.

[38] William Fawcett, 'Ravenala madagascariensis', Bulletin of the Botanical Department, Jamaica, new series, vol.1, (1894), p.195.
[39] P. Baccarini, 'Intorno ad un singolare accumulo d'acqua nel sistema lacunare delle guaine foliari di una *Musa ensete*', *Bulletino della Società Botanica Italiana*, 1904, pp.276-281; van Oye, p.25; Ryan, p.589.
[40] Van Oye, pp.26-27.
[41] Van Oye, p.27.
[42] Van Oye, p.28.
[43] Van Oye, pp.28-29.
[44] Van Oye, p.31.
[45] Van Oye, pp.32-34.
[46] R. Piaulian & P. Viette, 'An introduction to terrestrial and freshwater invertebrates', in eds. Steven M. Goodman & Jonathan P. Benstead, *Natural history of Madagascar*, (Chicago, 2003), chap.8, 'Invertebrates', p.510.
[47] J-B. DuChemin, O. Ramiljaona Ravoahangimalala & G. Le Goff, 'Culicidae, mosquitoes, Moka Gasy', in eds. Goodman & Benstead, *Natural history of Madagascar*, p.715; Piaulian & Viette, p.510; DuChemin, Ravoahangimalala & Le Goff, p.713.
[48] Duchemin, Ravoahangimalala & Le Goff, p.714; Ralph Harbach, 'Orthopodomyia', [online]: http://mosquito-taxonomic-inventory.info/simpletaxonomy/term/6222, posted 11 February 2008, accessed 8 December 2016.
[49] Duchemin et al, p.714.
[50] Neil Cumberlidge, Danté B. Fenolio, Mark E. Walvoord & Jim Stout, 'Tree-climbing crabs (Potamonautidae and Sesarmidae) from phytotelmic microhabitats in rainforest canopy in Madagascar', *Journal of Crustacean Biology*, vol.25, issue 2, (2005), p.306.
[51] Cumberlidge, Fenolio, Walvoord & Stout, p.306.
[52] A.P. Raselimanana & D. Rakotomalala, 'Systematic accounts: Skinks', in eds. Goodman & Benstead, *Natural history of Madagascar*, pp.989-992.
[53] Markus Fölling, Christoph Knogge & Wolfgang Böhme, 'Geckos are milking honeydew-producing planthoppers in Madagascar', in *Journal of Natural History*, 2001, 35, pp.279-284.
[54] Christopher J. Raxworthy, Colleen M. Ingram, Nirhy Rabibisoa & Richard G. Pearson, 'Applications of ecological niche modelling for species delimitation: A review and empirical evaluation using day geckos (*Phelsuma*) from Madagascar', in *Systematic Biology*, 56:6, December 2007, 907-923.
[55] Steven M. Goodman & Beverley A. Lewis, 'Description of the Réserve Naturelle Intégrale d'Andringitra, Madagascar', *A floral and faunal inventory of the eastern slopes of the Réserve Naturelle Intégrale d'Andringitra, Madagascar*, *Fieldiana* (Zoology), (1996), new series no.85, p.8; H. Schliemann & S.M. Goodman, '*Myzopoda aurita*, Old World sucker-footed bat' in eds. Goodman & Benstead, Natural history of Madagascar, pp.1303-1306; J.L. Eger & L. Mitchell, 'Systematic accounts: Bats', in eds. Goodman & Benstead, *Natural history of Madagascar*, pp.1291-1292.
[56] J. M. Ekstrom, 'Psittaciformes: *Coracopsis* spp., Parrots', in eds. Goodman & Benstead, *Natural history of Madagascar*, pp.1098-1099.
[57] Ekstrom, pp.1099 & 1102.
[58] Ekstrom, pp.1099 & 1101.
[59] Ekstrom, pp.1098-1099; J.M. Hutcheon, 'Frugivory by Malagasy bats', in eds. Goodman & Benstead, *Natural history of Madagascar*, p.1205.
[60] Hutcheon, pp.1205-1207; J.L. MacKinnon, C.E. Hawkins & P.A. Racey, 'Pteropodidae, Fruit Bats, Fanihy, Angavo', in eds. Goodman & Benstead, *Natural history of Madagascar*, p.1302.
[61] MacKinnon et al, pp. 1299-1302; IUCN: 'Pteropus rufus', 'Rousettus madagascariensis', [online]: http://www.iucnredlist.org/details/18756/0, http://www.iucnredlist.org/details/19750/0, both accessed 29 May 2018.
[62] MacKinnon et al, pp.1299-1301.
[63] MacKinnon et al, p.1300.
[64] MacKinnon et al, p.1300.
[65] Hutcheon, pp.1205 & 1207; Eger & Mitchell, p.1292; Hladik et al, 'L'arbre des voyageurs', 1, p.25; W.J. Kress, G.E. Schatz, M. Adrianifahanana & H.S. Morland, 'Pollination of *Ravenala madagascariensis* by lemurs in Madagascar: Evidence for an archaic coevolutionary system?', in *American Journal of Botany*, 81, pp.542-551.
[66] Bibliothèque Nationale de France, 'Henri Jumelle', [online]: http://data.bnf.fr/12388659/henri_jumelle/, accessed 1 January 2014; Martin W. Callmander, Bruno Vila, John Dransfield & Henk Beentje, 'The legacy of Henri Jumelle in Marseille: An overlooked collection of palms from Madagascar', *Candollea*, vol.66, issue 2, (2011), pp.419-421; HUH, 'Perrier de la Bâthie, Joseph Marie Henry Alfred', [online]: http://kiki.huh.harvard.edu/databases/botanist_search.php?id=455, accessed 1 January 2014; Bibliothèque Nationale de France, 'Henri Perrier de la Bâthie', [online]: http://data.bnf.fr/12327409/henri_perrier_de_la_bathie/, accessed 1 January 2014; J.W. White jnr & Landis W. Doner, 'Honey composition and properties', *Beekeeping in the United States*, Agriculture Handbook no.335, (Washington, 1980), pp.82-91, [online]: http://www.beesource.com/resources/usda/honey-composition-and-properties/, accessed 28 December 2013; Janice Wood, 'Suddenly discovering my great-great-grandfather was a writer', [online]: https://blog.britishnewspaperarchive.co.uk/2012/08/07/your-bna-stories-suddenly-discovering-that-my-gg-grandfather-was-a-writer/, posted 7 August 2012, accessed 31 December 2013.
[67] Hladik et al, 'L'arbre du voyageur', part 1, p.20.
[68] Kress, Schatz, Andrianifahanana & Morland, pp.542-551.
[69] Christopher R. Birkinshaw & Ian C. Colquhoun, 'Pollination of *Ravenala madagascariensis* and *Parkia madagascariensis* by *Eulemur macaco* in Madagascar' *Folia Primatologica*, vol. 69, issue 5, (Oct 1998), p.255.
[70] Studies in 1970s listed by Kress, Schatz, Andrianifahanana & Morland, p.543; see also pp.542-549.
[71] Claude Marcel Hladik, Patrick Blanc & Annette Hladik, 'L'arbre du voyageur: Reproduction, usages et diffusion horticole', 2eme partie, in *Hommes & Plantes*, no.41, Apr.-Jun. 2002, p.20.
[72] Hladik et al, 'L'arbre du voyageur', 2, pp.20-21.
[73] Hladik et al, 'L'arbre du voyageur', 2, pp.20-21; Lesley A. Sharp, 'Wayward pastoral ghosts and regional xenophobia in a northern Madagascar town, *Journal of the International African Institute*, vol.71, no.1, pp.38-81.
[74] Hladik et al, 'L'arbre du voyageur', 2, p.21; Claire Kremen, Isaia Raymond, Kate Lance & Andrew Weiss, 'Monitoring natural resource use on the Masaola peninsula, Madagascar: A tool for managing integrated conservation and development projects', in eds. K. Saterson, R. Margoluis & N. Salafsky, *Managing conservation impact*, (Washington, 1999).
[75] John Belling in 'Gleanings', *Agricultural News*, vol.4, (January to December 1905), p.44; BM(NH), *Ravenala madagascariensis*, W.E. Broadway, Botanic Station, Tobago, West Indies, 8 November 1932; Desmond, *Dictionary*, 'Bourne, Sir Alfred Gibb', p.79; A.F. Judd in eds. Sidney Blake & Cary Atwood, *Geographical guide to the floras of the world*, US Department of Agriculture Misc. Publications, Hafner, 1967, p.110; Wikipedia, 'Albert F. Judd jnr', [online]: https://en.wikipedia.org/wiki/Albert_Francis_Judd,_Jr., accessed 3 January 2014; Wikipedia, 'Albert Francis Judd [snr]', [online]: https://en.wikipedia.org/wiki/Albert_Francis_Judd, accessed 3 January 2014; K specimen file: 'Ravenala madagascariensis, cultivated'; Gunn & Codd/Glen & Germishuizen, 'Schweickerdt, Herold', pp.387-388.
[76] Feeley-Harnik, pp.34-36.
[77] Nathalie Messmer, Pierre Jules Rakotomalaza & Laurent Gautier, 'Structure and floristic composition of the vegetation of the Parc National de Marojejy, Madagascar', *Fieldiana* (Zoology), (August 2010), new series no.97, pp.67 & 94; Nick A. Helme & Pierre Jules Rakotomalaza, 'An overview of the botanical communities of the Réserve Naturelle Intégrale d'Andohahela', *Fieldiana* (Zoology), (June 1999), new series no.94, pp.12-13; Goodman & Lewis, 'Andringitra', p.8.
[78] Berthe Raminosoa Rasoanalimanga, 'Queen Ranavalona II', Dictionary of African Christian Biography, [online]: http://www.dacb.org/stories/madagascar/ranavalona2.html, posted 2008, accessed 15 September 2016; Edouard Ralaimihoatra, *Histoire de Madagascar*, (Paris, 1958), p.40; Hubert de Vries, 'Madagascar: Heraldry'.
[79] J.G. Baker, 'Flora of Madagascar', in Samuel Pasfield Oliver, Madagascar, (London, 1886), vol.1, p.547.
[80] Baker, pp.547-548.
[81] Baker, p.549.

## CHAPTER 7

[1] Bureau of Plant Industry, 'Plant material introduced by the Division of Foreign Plant Introduction, Bureau of Plant Industry, April 1 to June 30, 1932', *United States Department of Agriculture Inventory No.111*, (May 1934), p.28. L.H. Bailey, *The standard encyclopedia of horticulture*, (1958), pp.2914-2915.
[2] Tom Hollowell, Lynn J. Gillespie, V.A. Funk & Carol L. Kelloff, 'Smithsonian Plant Collections, Guyana: 1989-1991, Lynn J. Gillespie', *Contributions from the United States National Herbarium*, vol.44, (Washington, 2003), p.97.
[3] W. John Kress & Donald E. Stone, 'Morphology and floral biology of *Phenakospermum* (Strelitziaceae), an arborescent herb of the neotropics', *Biotropica*, 25(3), 1993, pp.290-1; Paul J.M. Maas & Hiltje Maas, 'Flora da Reserva Ducke, Brasil: Strelitziaceae', *Rodriguésia*, 56 (86): 205, (2005); Robert Chodat & Emile Hassler, 'Plantae Hasslerianae', *Bulletin de L'Herbier Boissier*, vol.3, 2nd series, p.1107; flowering of *Phenakospermum guyanense*, reported https://plus.google.com/+Hawaiigarden/posts, 10 December 2013, and www.hawaiigarden.com, 20 December 2013, accessed 12 January 2014; Laura Ponsonby, *Marianne North at Kew Gardens*, (London, 1990), p.27.
[4] Fawcett, *The banana*, p.260; ed. William Fawcett, *Bulletin of the Botanical Department, Jamaica*, new series, vol.2, (Kingston, 1896), p.47, 71, & 192; Anon., 'Fawcett, William' [online]: http://www.nhm.ac.uk/research-curation/library/archives/catalogue/dserve.exe?dsqServer=placid&dsqIni=Dserve.ini&dsqApp=Archive&dsqDb=Persons&dsqSearch=Code==%27PX965%27&dsqCmd=Show.tcl, accessed 27 February 2017.
[5] J.P. Lotsy, *Vorträge über botanische stammesgeschichte, Ein lehrbuch der pflanzensystematik, Cormophyta Siphonogamia*, p.53, (Jena, 1911); Fawcett, p.260; Hubert Winkler, *Botanisches hilfsbuch für pflanzer, kolonialbeamte, tropenkaufleute und forschungsreisende*, p.222, (Wismar, 1912); Gunn & Codd/Glen & Germishuizen, 'Lotsy, Johannes', p.273; Global Plants, 'Winkler, Hubert', [online]: http://plants.jstor.org/stable/10.5555/al.ap.person.bm000009351, accessed 27 February 2017.
[6] Desmond, *Dictionary*, 'Wallace, Alfred Russel', pp.635-636; Anon., [online]: http://wallacefund.info/, accessed 16 October 2015.
[7] Alfred Russel Wallace, *Palm trees of the Amazon and their uses*, (London, 1853), pp.iii-iv.
[8] P.J.M. Maas, 'Musaceae', in ed. A.R.A. Görts-van-Rijn, *Flora of the Guianas*, series A, Phanerograms, (Koenigstein, 1985), pp.21-2; Gustavo Politis, *Nukak, Ethnoarchaeology of an Amazonian people*, (Walnut Creek, 2009), p.285; Eglée Zent & Stanford Zent, 'Floristic

composition, structure, and diversity of four forest plots in Sierra Maigualida, Venezuelan Guyana', *Biodiversity and conservation*, 13: 2453-84, (2004).
[9] Michael Kessler, Jon-Arvid Grytnes, Stephan R.P. Halloy, Jürgen Kluge, Thorsten Krömer, Blanca León, Manuel J. Macía & Kenneth R. Young, 'Gradients of plant diversity, Local patterns and processes', in eds. S.K. Herzog, R. Martínez, P.M. Jørgenson & H. Tiessen, Climate change effects on the biodiversity of the tropical Andes, (Inter-American Institute of Global Change Research, São José dos Campos, 2011), pp.204-19.
[10] K: *Phenakospermum guianense*, M.W. Chase, DNA Bank no.3905, Kew 1988-4879, cultivated.
[11] Whitehead, p.435.
[12] R.E. Holttum to Mrs Ng Siew Yin, Botanic Gardens, Singapore, 11 July 1990, Kew Archive.
[13] L.J. Gillespie, 'Botanical collecting in Guyana', appendix to Tom Hollowell et al, 'Smithsonian plant collections, Guyana,' pp.99-102.
[14] Paulus Maas to Himansu Baijnath, email, 10 January 2010.
[15] William T. Stearn, 'Maria Sibylla Merian (1647-1717) as a botanical artist', *Taxon*, (August 1982), vol.31, no.3, pp.529-534; Catherine Grimm, 'To see for herself: Maria Sibylla Merian's research journey to Suriname, 1699-1701', [online]: http://sophie.byu.edu/texts/see-herself-maria-sibylla-merians-research-journey-suriname-1699-1701-article-catherine-grimm-, accessed 1 July 2017.
[16] H. Walter Lack, 'Jacquin's "Selectarum stirpium americanarum historia"', *Curtis's Botanical Magazine*, vol.15, (August 1998), pp.194-214.
[17] Donal P. McCracken, *Gardens of empire*, p.103; Anon., 'Joseph Martin', [online]: https://en.wikipedia.org/wiki/Joseph_Martin_(gardener), accessed 18 January 2014.
[18] Angel Martínez Salazar, 'José de Iturriaga: En los límites del Orinoco', *El País*, 7 December 2003; Eduardo Medina Rubio, 'Pehr Löfling, a botanist travelling through Cumaná and Guayana', [online]: http://web.comhem.se/embavene-sweden/ensvensk.htm, accessed 24 May 2017.
[19] Professor Long, G.R. Porter & George Tucker, *America and the West Indies, Geographically Described* (London, 1845), p.483; Pons, pp.335 & 342; Long et al, p.487; Wilfred Blunt, *Linnaeus, The Compleat Naturalist* (London, 2004), pp.190, 192; Antonio González Bueno, *La expedición botánica al virreinato del Perú* (1777-8), (Barcelona, 1988).
[20] Pavord, p.303, plate 118; R.S. Brown, 'Nicolás Bautista Monardes: Su vida y su obra', *Cambridge Journal of Medical History*, vol.7 (2), April 1963, http://www.ncbi.nlm.nih.gov/pmc/articles/PMC1034819/, accessed 22 May 2014; http://www.mcnbiografias.com/app-bio/do/show?key=monardes-nicolas-bautista, accessed 22 May 2014; Anon., 'Alessandro Malaspina', [online]: http://www.imalaspina.com/en/great-figures/article/alessandro-malaspina.html , accessed 22 May 2014.
[21] Anon., 'Huánaco', see www.encyclopediabritannica.com; González Bueno, p.192; Global Plants, 'Dombey, Joseph', [online]: http://plants.jstor.org/stable/10.5555/al.ap.person.bm000002124, accessed 6 October 2015.
[22] González Bueno, p.192.
[23] John Pontin, 'Kingdom of the ants: Jose Celestino Mutis and the dawn of natural history in the New World by Edward O. Wilson and Jose M. Gomez Duran', in *Zoological Journal of the Linnean Society*, 2011, 163, pp.315-317; Daniela Bleichmar, *Visible empire: Botanical expeditions and visual culture in the Hispanic Enlightenment*, (Chicago, 2012), pp.145-147; González Bueno, p.192.
[24] http://www.mundotnc.org/donde-trabajamos/americas/colombia/lugares/riomagdalena.xml, accessed 5 January 2014; Long et al, p.479; Bleichmar, pp.39, 84-90, 96.
[25] Global Plants, 'Bonpland, Aimé', [online]: https://plants.jstor.org/stable/10.5555/al.ap.person.bm000000885, accessed 14 December 2013.
[26] Johann Baptist von Spix and Carl Friedrich von Martius, 'Urania amazonica Mart', *Reise in Brasilien* vol.III (Munich, 1831), p.XX, & vol.IV, *Atlas* Tab, 1, fig. VI.
[27] Carl F. Ph. von Martius, *Die pflanzen und thiere des tropischen America, ein Naturgemälde*, (Munich, 1831), p.12
[28] Lemaire, *L'Illustration Horticole*, vol.7, (1860), 'Ravenala madagascariensis', t.234, 'Phenacospermum guianense', t.239.
[29] Susan Buchanan, *Burchell's travels*, (Cape Town, 2015), chapter 2.
[30] http://www.oum.ox.ac.uk/learning/pdfs/burchell.pdf, accessed 4 January 2014; Gunn & Codd/Glen & Germhuizen, pp.117-118; Buchanan, pp.204-205.
[31] K: *Phenakospermum guyannense*, Burchell no.9691; Lyman B. Smith & Ruth C. Smith, 'Itinerary of William John Burchell in Brazil, 1825-1830', *Phytologia*, vol.14, no.8, pp.492-506; Buchanan, pp.203-204.
[32] Global Plants, 'Richard, Achille', [online]: http://plants.jstor.org/stable/10.5555/al.ap.person.bm000006989; & 'Richard, Louis Claude', [online]: http://plants.jstor.org/stable/10.5555/al.ap.person.bm000006990, both accessed 2 March 2014.
[33] A. Richard, *Nova acta Phys-med. Acad. Caes. Leop.-Carol. Nat. Cur.* 15 (Suppl), pp.21-2, tab.vi.vii.
[34] A. Richard, *De musaceae commentatio botanica*, (Vratislava, 1831), pp.13, 21-2.
[35] Stephan Endlicher, *Prodromus Florae Norfolkicae*, p.35; Stephan Endlicher, *Enchiridion botanicum*, (Vienna, 1841), pp.123-4; New Advent, 'Endlicher, Stephan', [online]: http://www.newadvent.org/cathen/05421a.htm, accessed 27 February 2017; Ernst von Steudel, *Nomenclator botanicus*, vol.II, p.439, (Cassellis, 1874).
[36] HUH, 'Miquel, Friedrich', [online]: http://kiki.huh.harvard.edu/databases/botanist_search.php?botanistid=978, accessed 16 January 2014; F.A.G. Miquel, *Botanische Zeitung*, vol.iii, (1845), pp.345-7; Ludwig Friedrich Splitgerber, 'Description du genre Urania non Schreber nec Richard', *Het Instituut, of Verslagen en mededeelingen*, pp.20, 304-10, (Amsterdam, 1843); Miquel, 'Symbolae ad Floram Surinamensem', *Linneae*. *Ein Journal für die botanik*, vol.18, (1844), pars vi, 'Musaceae Ag Phenkospermum Endl', p.603; Global Plants, 'Splitgerber, Frederik', [online]: http://plants.jstor.org/stable/10.5555/al.ap.person.bm000008053, accessed 19 January 2017.
[37] HUH, 'Von Steudel, Ernst', [online]: http://kiki.huh.harvard.edu/databases/botanist_search.php?botanistid=2231, and http://test.botanicus.org/creator/558, accessed 18 January 2014; Steudel, *Nomenclator botanicus*, ed. Ludovicus Pfeiffer, vol.II, p.1532, (Cassellis, 1874); Richard Schomburgk, *Versuch einer fauna und flora von Britisch-Guiana*, 'Ordo Musaceae, Tribus Uranieae', p.1070, (Leipzig, 1848); Richard Schomburgk, *Reisen in Britisch-Guiana in den jahren 1840-1844*, (Leipzig, 1848), pp.79 & 240; HUH, 'Schomburgk, Moritz Richard', [online]: http://kiki.huh.harvard.edu/databases/botanist_search.php?botanistid=2931; History of Guyana, 'Robert Schomburgk', [online]: http://www.guyana.org/suriname/schomburgk.html, both accessed 17 May 2014.
[38] HUH, 'Splitgerber, Frederik', [online]: http://plants.jstor.org/person/bm000008053, accessed 17 May 2014; Anon., 'Schomburgk, Richard', [online]: http://adb.anu.edu.au/biography/schomburgk-moritz-richard-4543, accessed 17 May 2014.
[39] F.A.G. Miquel, *Stirpes surinamenses selectae*, (Leiden, 7th ed.), pp.212-3, tab.62-3.
[40] Pavord, p.398; Anon., 'de Candolle, Alphonse', [online]: http://www.mcnbiografias.com/app-bio/do/show?key=candolle-alphonse-de, accessed 27 May 2014.
[41] The genus name Ravenala had been proposed in 1777 and *Ravenala madagascariensis* was named in 1782; Kress & Stone, p.291; W. John Kress, 'New taxa of heliconia (Heliconaceae) from Ecuador', *Brittonia*, (1991), vol.43 (4), pp.253-260; William T. Stearn, 'A survey of the tropical genera Oplonia and Psilanthele (Acanthaceae)', in *Bulletin of the British Museum (Natural History) Botany*, vol.4, no.7, 1971, p.292.
[42] K: *Urania amazonica*, Herb. Sag. No.578, Aracouany, 1857.
[43] Donal P. McCracken, *Gardens of empire*, pp.80, 85.
[44] K: *Ravenala guianensis*, Gray Herbarium No.435, W.E. Broadway, vicinity of Cayenne, 6 June 1921; M.S. Hoogmoed, 'On the presence of *Bufo nasicus* Werner in Guiana, with a redescription of the species on the basis of recently collected material', *Zoologische mededelingen*, 51, 1977, p.272.
[45] Desmond, *Dictionary*, 'Thiselton-Dyer, Sir William', p.605.
[46] Anon., 'Hart, John Hinchley', [online]: http://www.nhm.ac.uk/research-curation/library/archives/catalogue/dserve.exe?dsqServer=placid&dsqIni=Dserve.ini&dsqApp=Archive&dsqDb=Persons&dsqSearch=Code=%27PX10687%27&dsqCmd=Show.tcl, accessed 27 February 2017.
[47] Anon., 'List of collectors whose plants are in the herbarium of the Royal Botanic Gardens, Kew, to 31st December, 1899', *Bulletin of miscellaneous information*, nos.169-171, Jan-Mar 1901, p.11; Broadway, 'as far as can be ascertained', supplied 157 specimens.
[48] HUH, 'Freeman, William', [online]: http://kiki.huh.harvard.edu/databases/botanist_search.php?botanistid=1061, accessed 1 March 2014.
[49] Yasmin S. Baksh-Comeau, 'Walter Elias Broadway: Botanist/Naturalist extraordinaire', http://sta.uwi.edu/herbarium/broadway.asp; HUH, 'Broadway, William', [online]: http://kiki.huh.harvard.edu/databases/botanist_search.php?botanistid=19222; http://plants.jstor.org/person/bm000023491, both accessed 9 August 2013; BM(NH), *Ravenala madagascariensis*, W.E. Broadway, Botanic Station, planted, Tobago West Indies, 8 November 1932, also stamped 'Missouri Botanical Gardens 1035221'.
[50] K: *Ravenala guianensis*, Julian Steyermark, 58067, Caño Negro, Venezuela, 28-29 August 1944; L.C. Richard, Achille Richard's father, died in 1821, 10 years before his son first published on the name.
[51] Kenan Heise, 'Botanist Julian A. Steyermark', *Chicago Tribune*, 20 October 1988.
[52] Bruce Hoist quoted in George Yatskievych and Luther J. Raechal, 'In memoriam / Julian Alfred Steyermark, 1909-1988', *Annals of the Missouri Botanic Gardens*, vol.77, winter 1990, p.5; see also Anon, 'Steyermark recollections', vol.76, no.3, fall 1989, pp.627-651.
[53] Heise, *Chicago Tribune*, 20 October 1988, http://articles.chicagotribune.com/1988-10-20/news/8802080881_1_plant-species-flora-missouri-botanical-gardens, accessed 4 August 2013; Yatskievych and Raechal, p.5; John Hays, 'Steyermark, Julian (1909-1988), *Dictionary of Missouri biography*, (University of Missouri Press, pp.720-1); http://plants.jstor.org/person/bm000008146?history=true&, accessed 11 August 2013; http://kiki.huh.harvard.edu/databases/botanist_search.php?botanistid=1531, accessed 11 August 2013.
[54] Lemaire, 'Phenacospermum guianense'.
[55] J. Medley Wood, *List of trees* (1915), p.69; Justus Karl Hasskarl, *Catalogus plantarum in horto botanico bogoriensi cultarum* (Batavia, 1844), pp.53 & 390; F.A.W. Miquel, 'Musaceae', *Flora van Nederlandsch Indië*, part 2, (Leipzig, 1855), pl.587; G. Haberlandt, *Eine Botanische Tropenreise*, (Leipzig, 1893); Jean Masart, 'Un botaniste en Malaisie', *Bulletin de la Société Royale de Botanique de Belgique*, vol.34, (1895), p.156. Desmond, *Dictionary*, 'Ridley, Henry', p.520. BM (NH), *Strelitzia nicolai*, 1909, marked 'H.N. Ridley'

and 'Cultivated in Botanic Gardens, Singapore'.
[56] F.W. Barclay, 'Ravenala madagascariensis', in ed. L.H. Bailey,*The standard encyclopedia of horticulture*, (New York, 1942), pp.2914-5, 1958; BM(NH), *Ravenala madagascariensis*, A.H.G. Alston, Bogor Botanic Gardens, Java, 15 November 1923; 'Ravenala guyanensis', Holttum to Ng Siew Yin, 18 July 1998, Kew Archive; M.R. Henderson to F. Thorns, 10 January 1946, F. Thorns to M.R. Henderson, 1 February 1946, M.R. Henderson to F. Thorns, 24 May 1946, Kew Archive; Gunn & Codd/Glen & Germhuizen, pp.209, 421.

## CHAPTER 8

[1] 'Tributes to Adrian Warren, 1949-2011', http://www.lastrefuge.co.uk/adrian.php, accessed 14 August 2013.
[2] K: *Phenakospermum guinanensis*, D. Philcox, Mato Grosso, Brazil, 4231-4232, n.d.; this also follows the L.C. Richard attribution; http://plants.jstor.org/person/bm000006504, http://kiki.huh.harvard.edu/databases/botanist_search/php?, both accessed 9 August 2013; M. John Whitehead, 'Plant collecting for Kew', in *Kew Guild Newsletter*, 2009, http://www.kewguild.org.uk/media/pdfs/v15s114p372-63.pdf, accessed 9 August 2013.
[3] K: *Phenakospermum guianense*, Lister 96, Santa Rosa de Tenuca, Venezuelan Amazonas, 2 December 1975; Simon Mayo to Himansu Baijnath, 7 November 2013.
[4] K: *Phenakospermum guianense*, Colchester, 2261, Caño Mosquito, Venezuelan Amazonas, 25 February 1976; Anon., 'Marcus Colchester', [online]:
http://www.pewenvironment.org/research-programs/marine-fellow/id/8589942524;
http://www.forestpeoples.org/background/staff-and-board;
http://www.rt10.rspo.org/sp.php?sid=30; Hoogmoed, p.266.
[5] K: *Phenakospermum guyannense*, James L. Zarucchi, 1612, Mitú, Colombia, 20 May 1976; HUH, 'Zarucchi, James', [online]:
http://kiki.huh.harvard.edu/databases/botanist_search.php?botanistid=39496;
http://www.efloras.org/person_profile.aspx?person_id=1071;
http://www.mobot.org/mobot/webmo/mostaff.html#zardini, all accessed 9 August 2013.
[6] HUH, 'Maas, Paulus', http://plants.jstor.org/person/bm000005245;
http://kiki.huh.harvard.edu/databases/botanist_search.php?mode=details&id=3467, & 'Maas-van de Kamer, Hillegonda', http://plants.jstor.org/person/bm000080529;
http://kiki.huh.harvard.edu/databases/botanist_search.php?mode=details&id=81026, both accessed 9 August 2013; Global Plants, 'Boyan, Rufus', http://plants.jstor.org/person/bm000033521, accessed 9 August 2013.
[7] Wade Davis, 'Timothy Charles Plowman, 17 November 1944-7 January 1989', *Economic Botany*, July/September 1989, vol.43, issue 3, pp.416-8; Kenan Heise, 'Timothy Plowman, Expert on coca plant', *Chicago Tribune*, 14 January 1989, http://articles.chicagotribune.com/1989-01-14/news/8902250317_1_cocaine-abuse-field-museum-chewing;
http://kiki.huh.harvard.edu/databases/botanist_search.php?botanistid=11596;
http://plants.jstor.org/person/bm000006585, all accessed 9 August 2013.
[8] K: *Phenakospermum guyannense*, J.J. Strudwick & G.L. Sobel 3616, 26 Jul 1981; Global Plants, 'Strudwick, J.J.', [online]:
http://plants.jstor.org/person/bm000117955?history=true, accessed 9 August 2013.
[9] Daniela Zappi & William Milliken, 'New guidebook for Brazilian biodiversity hotspot': http://www.kew.org/news/kew-blogs/herbarium/brazilian-guidebook.htm, accessed 1 March 2014.
[10] Maurizio Guido Paoletti, Darna L. Dufour, Hugo Cerda, Franz Torres, Laura Pizzoferrato & David Pimentel, 'The importance of leaf- and litter-feeding invertebrates as sources of animal protein for the Amazonian Amerindians', *Royal Society Proceedings: Biological Sciences*, vol.267, no.1459 (Nov 2000), pp.2247-2252.
[11] Paul Hennings, 'Fungi amazonici III, a cl. Ernesto Ule collecti', *Hedwigia*, series 43, no.6, (3 September 1904), p.364; C.G. Hansford, 'Australian fungi: New species and revisions: 1. The Meliolaceae', *Proceedings of the Linnean Society of New South Wales*, vol.78, parts 3-4, nos.367-368, p.81.
[12] Kress & Stone, pp.293-300; Theodore H. Fleming & Vinicio J. Sosa 'Effects of nectarivorous and frugivorous mammals on reproductive success of plants', *Journal of Mammalogy*, 75(4), 845-851, 1994; W. John Kress, 'Bat pollination of an Old World *Heliconia*', *Biotropica*, vol.17, no.4, (Dec. 1985), pp.302-8.
[13] Nancy B. Simmons, Robert S. Voss, Scott A. Mori, 'Bats as pollinators of plants in the lowland forests of Central French Guiana', [online]:
https://www.nybg.org/bsci/french_guiana/bat_poll.html, accessed 9 August 2013.
[14] Cheryl S. Roesel, W. John Kress and Brunella Martire Bowditch, 'Low levels of genetic variation in *Phenakospermum guyannense* (Strelitziaeceae), a widespread bat-pollinated Amazonian herb', *Plant systematics and evolution*, vol.199, no.1-2, Mar 1996, pp.1-15.
[15] Roesel, Kress & Bowditch, 'Low levels of genetic variation in *Phenakospermum guyannense*'; Gustave Politis, *Nukak*, p.251.
[16] Kirsten M. Silvius, 'Spatio-temporal patterns of palm endocarp use by three Amazonian forest mammals: Granivory or 'grubivory'?', *Journal of Tropical Ecology*, vol.18, no.5, (Sept 2002), pp.716, 718-9; Theodore H. Fleming and Vinicio J. Sosa, 'Effects of nectarivorous and frugivorous mammals on reproductive success of plants', Journal of Mammalogy, vol. 75, No. 4 (Nov 1994), pp.845-851.
[17] Andrés Link & Pablo R. Stevenson, 'Fruit dispersal syndromes in animal disseminated plants at Tinigua National Park, Colombia', *Revista chilena de historia natural*, 77, 2004, pp.319-334.
[18] Douglas T. Levey, 'Seed size and fruit-handling techniques of avian frugivores', The *American Naturalist*, vol.129, no.4, April 1987, pp.471-485.
[19] M.M. Grandtner, *Elsevier's dictionary of trees*, vol.1, p.591, no.5283.
[20] 'Occurrence, body mass and biomass of *Syntermes* spp. (Isopter: Termitidae) in Reserva Ducke, Central Amazonia', *Acta Amazonica*, 28 (3): 319-324, 1998; Mario G. Paoletti, Erika Buscardo & Darna L. Dufour, 'Edible invertebrates among Amazonian Indians: A critical review of disappearing knowledge', *Environment, Development and Sustainability*, vol.2, issue 3, pp.195-225. See also Jennifer Gonzalez Covarrubias, 'Diners are catching the insect bug', *The Mercury*, 10 October 2013, p.13.
[21] Marcus Colchester, review: '*Yanomami warfare: A political history* by R. Brian Ferguson', *Journal of the Royal Anthropological Institute*, vol.2, no.3, (Sept. 1996), pp.549-550; Maurice Seiji, Tomioka Nilsson & Philip Martin Fearnside, 'Yanomami mobility and its effects on the forest landscape', *Human Ecology*, vol.39, no.3, (June 2011), p.235; Darna L. Dufour & James L. Zarucchi, '*Monopteryx angustifolia* and *Erisma japura*: their use by indigenous peoples in the northwestern Amazon', *Botanical Museum Leaflets Harvard University*, vol.27, no.3-4 (March-April 1979), p.79.
[22] Mario Paoletti, Erika Buscardo, Dorothy J. Vanderjagt, A. Pastuszyn, L. Pizzoferrato, Y.S. Huang, L.-T. Chuang, R.H. Glew, M. Millson & Hugo Cerda, 'Nutrient content of termites (Syntermes soldiers) consumed by Makiritare Amerindians of the Alto Orinoco of Venezuela', *Ecology of Food & Nutrition*, 42(2), (Mar 2003), pp.177-191; Maurizio Paoletti, Erika Buscardo & Darna Dufour, 'Edible invertebrates among Amazonian Indians: A critical review of disappearing knowledge, *Environment, Development and Sustainability*, vol. 2(3), (Sept 2000), pp.195-225; Juan Alvaro Echeverri & Oscar Enokakuiodo Román-Jitdutjaaño, 'Ash salts and body affects', *Environmental Research Letters*, vol.8, (2013), 015034, pp.2, 6-7 & 10-11; O. Enokakuiodo, S. Roman, A.A. Lopez, B. Weniger & J.A. Echeverri, 'Vegetable salt: An ignored resource', in eds. J. Fleurentin, J.M. Pelt & G. Mazars, *Des sources du savoir aux médicaments du futur: Actes du 4e congrès européen d'ethnopharmacologie*, (Metz, 2002), pp.353-354.
[23] Katherine Milton, C.D. Knight & I. Crowe, 'Comparative aspects of diet in Amazonian forest-dwellers', *Philosophical transactions of the Royal Society of Biological Sciences London B*, 334, 253-63, 1991.
[24] Politis, p.246; Joanna Eede, 'Dr William Milliken: "The Yanomami are great observers of nature"', Survival International [online]:
http://www.survivalinternational.org/articles/3162-yanomami-botanical-knowledge, accessed 12 October 2013.
[25] Politis, p.87; Gustavo G. Politis, 'Moving to produce: Mobility and settlement patterns in Amazonia', *World Archaeology*, vol.27, no.3, Hunter-Gatherer Land Use (Feb. 1996), pp.492-511).
[26] Politis, pp.51 & 53.
[27] Politis, pp.115-119.
[28] Politis, pp.107-113.
[29] Politis, pp.210-213.
[30] Politis, pp.256-257.
[31] Seiji et al, 'Yanomami mobility', pp.240-242, 245 & 249.
[32] Politis, p.147.
[33] Politis, p.116.
[34] Philip Comptom, 'Fruit in the soil of magic: Horticultural practices as socially conditioned techniques in the formation of anthropogenic Amazonia', *Revista de Antropologia Social dos Alunos do PPGAS-OFCar*, v.1, n.2, Jul.-Dez., 2009, pp.20-44.
[35] Jacques Huber, 'La vegetation de la vallée Rio Purus (Amazone)', *Bulletin de l'Herbier Boissier*, vol.6, no.4, (1906), p.266.
[36] Comptom, pp.26 & 35.
[37] Politis, pp.281-285.
[38] Anon., 'Baudin, Nicolas', [online]:
http://adb.anu.edu.au/biography/baudin-nicolas-thomas-1753, accessed 13 January 2014; Bleichmar, p.22; Michael Dettelbach, 'Global physics and aesthetic empire: Humboldt's physical portrait of the tropics', in eds. David Philip Miller & Peter Hanns Reill, *Visions of empire*, (Cambridge, 1996); Anon., 'Bonpland, Aimé', [online]:
http://www.mcnbiografias.com/app-bio/do/show?key=bonpland-aime, accessed 13 October 2013; HUH, 'Kunth, Karl', [online]:
http://kiki.huh.harvard.edu/databases/botanist_search.php?botanistid=308, accessed 21 January 2014; Thomas R. Defler & Jorge I. Hernández-Camacho, 'The true identity and characteristics of *Simia albifrons* Humboldt, 1812: Description of neotype', *Neotropical primates*, 10(2), August 2002, pp.49, 51 & 59.
[39] Sue Boinski, Robert P. Quatrone & Hilary Swartz, 'Substrate and tool use by brown capuchins in Suriname: Ecological contexts and cognitive bases', *American Anthropologist*, 102(4), pp.741-61.
[40] Boinski et al, pp.746, 749 & 753.
[41] Boinski et al, pp.749, 751 & 781-783.
[42] Boinski et al, pp.749 & 751.

## CHAPTER 9

[1] Gunn & Codd/Glen & Germishuizen: 'Wallich, Nathaniel', p.451; 'Roupell, Arabella', pp.367-368; 'Harvey, William', pp.206-208.
[2] Arabella Roupell, *Specimens of the Cape flora by a lady*, (London, 1849); Gunn & Codd/Glen & Germishuizen, 'Holland, Maria', p.217; 'Strelitzia reginae Aiton, from South Africa', BMHOL13259015, Holland collection, Collection of Drawings of South African Plants, BM(NH); R.A. Lubke & Y. van Wijk, 'Terrestrial plants and coastal vegetation', in

ed. Roy Lubke, Fred Gess & Mike Bruton, *A field guide to the Eastern Cape coast*, Grahamstown, 1988.
3 Lighton, pp.36-7; Gunn & Codd/Glen & Germishuizen, pp.102-3. Gunn & Codd date the trip in 1876, but Lighton in 1878.
4 Lighton, p.35.
5 E. Percy Phillips, 'Contribution of South Africa to botany', presidential address to Section C of the South African Association for the Advancement of Science, delivered July 9th, 1930, (South African Association for the Advancement of Science, 1930), p.11; Lighton, pp.35-6; Gunn & Codd/Glen & Germishuizen, p.102-3.
6 Donal P. McCracken, 'Plant hunting and the provenance of South African plants in British botanical magazines, 1787-1910', in ed. Mary N. Harris, *Sights and Insights: Interactive images of Europe and the wider world*, (Pisa, 2007), p.80.
7 Lighton, p.36; P. Barry Tomlinson to W. John Kress, no date (probably early 2002): http://botany.si.edu/cuatrecasas/medal2002.cfm, accessed 17 April 2016.
8 Harry Bolus, 'Sketch of the flora of South Africa', in ed. John Noble, *Official handbook: History, production and resources of the Cape of Good Hope*, (Cape Town, 1886), pp.288-319; Gunn & Codd/Glen & Germishuizen, p.201.
9 Burchell, *Travels in the interior of southern Africa*, vol.1, pp.21-22, (London, 1822); McCracken, 'Plant hunting', pp.75-89.
10 McCracken, 'Plant hunting', p.87.
11 H.H.W. Pearson, 'Presidential address Section C: A national botanic garden', *Report of the South African Association for the Advancement of Science*, vol.7, (1910), pp.37-54.
12 *Kew Bulletin*, (1892), pp.10-14.
13 McCracken, 'Plant hunting', p.80; Botanic Gardens Conservation International: https://www.bgci.org/, accessed 19 March 2016; Christopher Willis & Augustine Morkel, 'National botanic gardens: South Africa's urban conservation refuges', *BGjournal*, vol.5, no.2, (July 2008): https://www.bgci.org/resources/article/0588/, accessed 15 March 2015.
14 Pearson, 'National botanic garden', p.50.
15 Pearson, 'National botanic garden', pp.50 & 54.
16 Anon, 'The national botanic garden of South Africa, p.309, *Bulletin of Miscellaneous Information*, (Royal Botanic Gardens, Kew), vol.8, (1913), pp.309-314; McCracken, *Gardens of empire*, p.44.
17 Eileen M. McCracken & Donal P. McCracken, *The way to Kirstenbosch*, pp.107-8; Huntley, *Kirstenbosch*, p.49; Anon., 'History: Pretoria National Botanical Garden', [online]: https://www.sanbi.org/gardens/pretoria/history/, accessed 14 April 2018.
18 Anon., 'History of the National Herbarium', [online]: https://www.sanbi.org/biodiversity/biosystematics-collections/the-national-herbarium/history-of-the-national-herbarium/, accessed 2 March 2017; Gunn & Codd/Glen & Germishuizen, 'Pole Evans, Illtyd Buller', pp.343-346.
19 I.E. Codd, 'Obituary, Robert Allen Dyer (1900–1987)', *Bothalia*, 18,1 (1988), pp.131-3.
20 R.A. Dyer, 'Plate 995. Strelitzia alba', 'Plate 996. Strelitzia nicolai' & 'Plate 997. Strelitzia caudata', all *Flowering Plants of Africa*, 1946; R.A. Dyer, 'Strelitziaceae' in *The genera of southern African flowering plants*, vol.2: 'Gymnosperms and Monocotyledons', (Pretoria, 1976), pp.982-983; D.J. Killick, 'History of *The Flowering Plants of Africa* and an index to volumes 1 to 49', *Flowering Plants of Africa*, vol.50, (1988-1989).
21 Chris Long, 'Swaziland's flora: siSwati names and uses', [online]: http://www.sntc.org.sz/flora/clffamilies.asp?fid=101, compiled 2005; accessed 21 March 2017; PRE: *Strelitzia caudata*, J. Burtt Davy 2623, PRE0090727, Westphalia, Pietersburg, Soutpansberg, 4 June 1906; A.E. Grewcock 27482, PPRE0090799, Haenertsburg, Pietersburg, 30 September 1930;.
22 Gunn & Codd/Glen & Germishuizen, 'Galpin, Ernest', pp.181-184; C. Plug, 'Galpin, Henry', [online]: http://www.s2a3.org.za/bio/print_action.php?sernum=1018&q_name=./includes/biofinal_query.inc&pf_name=./content/biofinal/biofinal_display.inc, accessed 1 March 2017; PRE: *Strelitzia caudata*, E.E. Galpin 27483, PRE0090798, Zoutpansberg, 27 March 1935.
23 PRE: *Strelitzia caudata*, T.B. Verschuur 27162, PRE0090797, Geluk, Zoutspanberg district, July 1943; Gillian Condy, pers. comm., 15 February 2016; J.H. Keet to R.A. Dyer, R.701, 5 November 1943; PRE: *Strelitzia alba*, J.H. Keet 27476, PRE0090891, Groot River Pass, Knysna, November 1943.
24 Keet to Dyer, 5 November 1943; R.A. Dyer, 'Plate 997. Strelitzia caudata', *Flowering plants*; Elsa Pooley, *The complete guide to the trees of Natal, Zululand & Transkei*, (Durban, 1993), p.60.
25 PRE: *Strelitzia caudata*, R.H. Compton 26955, Millers Falls, Swaziland, 6 April 1957; & J.C. Scheepers 403, Letaba, 31 July 1958.
26 See Moeketsi Letsela & Yvonne Reynolds, 'Strelitzia nicolai Regel & Koern', [online]: http://www.plantzafrica.com/plantqrs/strelitznichol.htm, updated June 2002, accessed 9 August 2016.
27 PRE: *Strelitzia caudata*, D. Heenan 31952, Pigg's Peak, Swaziland, September 1971; Hugh F. Glen, *Sappi What's in a name: The meanings of the botanical names of trees* (Johannesburg, 2004), p.32; *Bothalia*, vol.10, 1969, p.539.
28 R.A. Dyer, 'Strelitzia caudata', *Flowering plants*.
29 L. Crook, Melsetter, Sept 1950, to Kew Herbarium.
30 BM(NH), *Strelitzia caudata*, C.E.O. Swynnerton, no.6051, 12-14 October 1908, Mount Pene, Mozambique; this herbarium also has duplicates of the Crook specimens.
31 Braam van Wyk & Piet van Wyk, *Field guide to the trees of southern Africa*, (Cape Town, 1997), p.56; Pooley, p.90.
32 K: *Strelitzia caudata*, Albert Crook, Melsetter, 1954.
33 Stephen Taylor, *Defiance: The life and choices of Lady Anne Barnard*, (London, 2016),

p.22-23; Gunn & Codd/Glen & Germishuizen, p.91; John Barrow, *Travels into the interior of South Africa*, London, 1806, vol.1, pp.72-3.
34 Barrow, p.73.
35 Taylor, *Defiance*, pp.216-217 & 239; Dorothea Fairbridge, *Lady Anne Barnard at the Cape of Good Hope, 1797-1802*, (Oxford, 1924), pp.50 & 52; Barrow, p.179.
36 K: *Strelitzia parvifolia* var. *juncea*, Ker, W.J. Burchell, S4438.3, South Africa; *Strelitzia juncea*, South Africa; Strelitzia juncea Ait, Drège, Afr. austr.; *Strelitzia juncea forma angustifolia*, Drège, K000308006, Afr. austr.
37 P. Hiepko 'The collections of the botanical museum, Berlin', *Englera*, no.7 (1987), 219-252, 1987; H.F. Link, *Enumeratio plantarum horti regii botanici berolinensis altera*, Berlin, 1821, part 1, p.150, nos. 1381, 1385 & 1386; C.H. Wright, *Flora Capensis*, vol.5, part 3, p.3; also noted by R.A. Dyer in *S. juncea* specimens file, PRE; HUH, 'Maire, René', [online]: http://kiki.huh.harvard.edu/databases/botanist_search.php?botanistid=473, accessed 2 March 2017; René Maire, *Flore de l'Afrique du Nord*, (Paris, 1952).
38 René Maire & Marc Weiller, Flore de l'Afrique du Nord, vol.6, (Paris, 1960 ed.), p.215; R.A. Dyer, 'Plate 1804 Strelitzia juncea', *Flowering plants of Africa*, 1975; H.E. Moore & Peter A. Hyypio, 'Strelitzia', *Baileya*, 1970, vol.17, p.65 & 71; R.A. Dyer, 'Strelitziaceae: The status of *Strelitzia juncea*', *Bothalia*, 1975, vol.11, pp.519-520.
39 H.A. Venter, J.G.C. Small & P.J. Robbertse, 'Notes on the distribution and comparative leaf morphology of the acaulescent species of *Strelitzia* Ait.', *South African Journal of Botany*, vol.41 (1), 1975, pp.1-16; Dyer, 'Status'. p.520.

## CHAPTER 10

1 Carlo Brandelli: 'Six of the best: Design influences', 19 March 2014, http://www.thetimes.co.uk/tto/life/fashion/mensstyle/article4031092.ece, accessed 22 June 2014; Anon., 'Bird of paradise', [online]: http://originallaflowermarket.com/flower/bird-of-paradise/, accessed 9 June 2018.
2 E.H. Gombritch, *The story of art*, (Oxford, Phaidon, 13th ed., 1978), pp.425-7; 442-6; 470-1; 'New York: The Market', *The Florists' Review*, p.68, 9 December 1920.
3 Ocean Institute, 'Gold Diggers Express', Ocean Institute.
4 Anon., 'A history of Watauga County, NC' [North Carolina], p.8, Watauga Archives, http://files.usgwarchives.net/nc/watauga/history/ch1.txt,
5 BM(NH), *Strelitzia reginae*, 'Duplicate from the Mo. Bot. Gard. Herbarium', 1827; Elizabeth McClintock (University of California, Berkeley, Herbarium) to Susyn Andrews, 6 August 1992, Andrews archive; Emily Green, 'A glorious flock', *Los Angeles Times*, 24 March 2005 [online]: http://articles.latimes.com/2005/mar/24/home/hm-paradise24, accessed 9 November 2014; P. Ridgway & J. Works, *Sending flowers to America*, (Los Angeles, 2008), p.6.
6 Ridgway & Works, p.5.
7 Ridgway & Works, p.5.
8 Anon., 'Kate Sessions', [online]: http://www.sandiegohistory.org/archives/biographysubject/sessions/, accessed 10 December 2016.
9 'A landscape of art & culture', www.balboapark.org/info/history, accessed 14 June 2014.
10 John McQuaid, 'The secrets behind your flowers', *Smithsonian Magazine*, February 2011; Jim Wyss, 'A Valentine'S tale of how Colombia bloomed into a USA flower power', *International Herald Tribune*, 13 February 2015.
11 'Store level: Tesco – Bloomin' marvellously' in *Talking Retail*, 24 March 2005; Dave Marshall, 'Steady increase in flower sales', www.flowerexperts.com/flower_growth.asp, accessed 19 April 2014; Danie Jordaan, pp.234-5.
12 Jordaan, pp.234-5.
13 H.M. Butterfield, *Home floriculture in California*, California Agricultural Extension Service Circular 53, (June 1931), p.38.
14 Ridgway & Works, p.184; 'Rare floral beauties seen at exhibition', *California Garden*, March/April 2009, p.35.
15 Anon, 'Bird of Paradise', [online]: http://www.imdb.com/title/tt0022689/, accessed 11 March 2016; Green, 'A glorious flock'.
16 Grace Vawter Bicknell, *The Vawter family in America*, (Indianapolis, 1905), 'Williamson Dunn Vawter', pp.124-125; Marguerita Wuellner, Amanda Kainer & Virginia Harness, 'City landmark assessment and evaluation report: E.J. Vawter House', City of Santa Monica Planning Division, (Sept 2014), p.5.
17 Green, 'A glorious flock', quoting Victoria Padilla, *Southern California gardens*, (Berkeley, 1961).
18 Interviews by Suzanne B. Riess, 'Toichi Domoto: A Japanese-American nurseryman's life in California: floriculture and family, 1883-1992", California Horticulture Oral History Series, pp.173-7, http://cdnc.ucr.edu/cgi-bin/cdnc?a=d&d=SFC19120211.2.50.3&srpos=2&e=-------en--20--1--txt-txIN-domoto-----#; Yutaro Suzuki, *The strelitzia world*, (Chiba, 1982), p.107.
19 Ridgway & Works, p.184; Green, 'Glorious flock'; Anon, 'Friends along the way', *Lasca Leaves*, vol.XXIII, no.1, (March 1973), p.50; Joshua Siskin, 'Everything you need to know about bird of paradise', [online]: http://thesmartergardener.com/everything-you-need-to-know-about-bird-of-paradise/, posted 24 November 2013, accessed 5 February 2017.
20 'Pasadena, Cal', *The Florists' Review*, 11 November 1920.
21 Ridgway & Works, pp.76 & 182.
22 Catherine Ziegler, *Favored flowers: Culture and economy in a global system* (Durham,

USA, 2007), quoted in Ridgway & Works, p.99; Frank Taylor, 'The flowers known as birds', *Saturday Evening Post*, 11 March 1961, quoted in Ridgway & Works, p.160.
[23] Richard Criley, email to Himansu Baijnath, 22 September 2013; Marta Pizano, 'International market trends – Tropical flowers' in *Acta Hort* 683, 2005.
[24] Ridgway & Works, pp.208-209.
[25] Chere Brown, World War II's effect on Japanese nurseries in 'Japanese flower industry remembered', http://thecrackerladyshouse.blogspot.com/2009/10/richmond-ramblings-japanese-flower.html, accessed 15 June 2014.
[26] Brown, 'Japanese flower industry'.
[27] Ridgway & Works, p.210.
[28] Ridgway & Works, p.211.
[29] Richard Criley, e-mail to Himansu Baijnath, 23 September 2013.
[30] Mike Mellano jr, e-mail to Himansu Baijnath, 2 February 2013.
[31] Ridgway & Works, p.184.
[32] USDA, 'Weekly ornamental shipment'.
[33] Ridgway & Works, pp.155 & 185.
[34] Mike Mellano jr, e-mail to Himansu Baijnath, 2 February 2013.
[35] Mellano to Baijnath, 2 February 2013.
[36] Joshua Siskin, 'Birds of paradise', [online]: http://thesmartergardener.com/birds-of-paradise-loved-and-loathed/, posted 11 November 2013, accessed 5 February 2017.
[37] Anon., 'New sex pheromone allows early detection of banana borer', www.floraculture.eu/?p=5235, posted 27 October 2009, accessed 27 June 205.
[38] H.A. van de Venter, J.G.C. Small & A.H. Halevy, 'Some characteristics of leaf and inflorescence production of *Strelitzia reginae* in Port Elizabeth', *Journal of South African Botany*, 46 (3), (1980), p.305.
[39] R.A. Dyer, 'Vegetative multiplication of *Strelitzia reginae* and its allies', *Bothalia*, 10, (1972), 4:575-578.
[40] Van de Venter et al, 'Leaf and inflorescence production of *Strelitzia reginae* in Port Elizabeth', p.305.
[41] K. Jainag, K.V. Jayaprasad, R. Krishna Manohar, Shivanand Hongal & K. Prakash, 'Effect of levels of fertigation on growth and yield of bird-of-paradise [*Strelitzia reginae* Ait.], Asian Journal of horticulture, vol.6, no.1, (June 2011), pp.118-21.
[42] D.J. Simpson, M.R. Baqar & T.H. Lee, 'Ultrastructure and carotenoid composition of chromoplasts of the sepals of *Strelitzia reginae* Aiton during floral development', *Annals of Botany*, 39, 1975, pp.175-83.
[43] H.A. van de Venter & J.G.C. Small, 'Dormancy in seeds of *Strelitzia* Ait', *South African Journal of science*, vol.70, 1974, pp.216-7.
[44] Kiyotake Ishihata, 'Studies on the promoting germination and the growth of seedling in *Strelitzia reginae* Banks', Bulletin of the Faculty of Agriculture, Kagoshima University, (20 March 1976), pp.1-15.
[45] H.A. van de Venter, 'Effect of various treatments on germination of dormant seeds of *Strelitzia reginae* Ait', Journal of South African botany, 44 (2), (1978): 103-110.
[46] M. Kantharaju, B.S. Kulkarni, B.S. Reddy, R.V. Hegde, B.C. Patil & J.D. Adiga, *Karnataka journal of agricultural science*, 21 (2): 2008, pp.324-5.
[47] Franklin W. Martin, 'The systematic exudate of *Strelitzia*', *Phyton*, 27 (1), (1970), pp.47-53.
[48] Eva Kronestedt, B. Walles & I. Alkemar, 'Structural studies of pollen tube growth in the pistil of *Strelitzia reginae*,' *Protoplasma*, 131, (1986), 224-232.
[49] M. Hesse, S. Vogel & H. Halbritter, 'Thread-forming structures in angiosperm anthers: Their diverse role in pollination ecology', *Plant systematics and evolution*, 222 (2000), pp.281-292.
[50] H.A. van de Venter, 'Microsporogenesis and gametogenesis in *Strelitzia reginae* Ait', *Journal of South African Botany*, 42 (1), (1976), 25-31.

## CHAPTER 11

[1] Melanie Eva Boehi, 'Being/becoming the "Cape Town flower sellers": The botanical complex, flower selling and floricultures in Cape Town', MA dissertation, University of the Western Cape, December 2010, p.17: http://etd.uwc.ac.za/xmlui/bitstream/handle/11394/2546/Boehi_MA_2010.pdf?sequence=1, accessed 27 December 2014; Lighton, p.7; Lance van Sittert, 'Making the Cape Floral Kingdom: The discovery and defence of indigenous flora at the Cape ca. 1890-1939', *Landscape Research*, 28:1, 2003, pp.113-129; *Cape Argus*, 2 October 1886, quoted in Lighton, p.7; J.C. Quinton, 'Thorne, Sir William', *Dictionary of South African biography*, vol.4, (Durban, 1981), pp.654-655.
[2] Burchell, p.22 note.
[3] http://www.britannica.com/event/Cape-Frontier-Wars, accessed 10 August 2015.
[4] Quoted in A. Wilmot & John Centlivres Chase, *History of the colony of the Cape of Good Hope*, Cape Town, 1869, p.169; A.L. Muller, 'Coastal shipping and the early development of the southern Cape', *Contree* 18, p.10; Morris Viljoen, 'The Cape fold mountains', in eds. C.R. Anhaeusser, M.J. Viljoen & R.P. Viljoen, *Africa's top geological sites*, (Cape Town, 2016), pp.103-109.
[5] Marion Whitehead, *Passes & poorts*, [Western Cape], Cape Town, 2010, p.65; Tony Murray, *Mega structures and master minds*, Cape Town, 2015, p.66; Graham Ross, *Romance of the Cape mountain passes*, Cape Town, 2004, pp.202-3, 205 & 209; Murray, *Mega structures*, pp. 54 & 72; Malcolm Mitchell, 'A brief history of transport infrastructure in South Africa up to the end of the 20th century: 1', *Civil Engineering*, Jan/Feb 2014, pp.37-8.
[6] Malcolm Mitchell, 'A brief history of transport infrastructure in South Africa up to the end of the 20th century: 2', *Civil Engineering*, March 2014, pp.74-5.
[7] Lizette Rabe, 'Living history – The story of Adderley Street's flower sellers', *SA Tydskrif vir kultuurgeskiedenes*, 24(1): 83-104; Boehi, pp.11-12 & 69-70; in George Bernard Shaw's Pygmalion, Act II, flowergirl Eliza explains she wants to improve her accent to improve her prospects, moving from a street-seller to a flower shop.
[8] Boehi, p.17, referring to House of Assembly, 'Saturday, August 1, 1894', Debates in the House of Assembly, (Cape Town, 1894), 1894, 479; Lighton, p.75; van Sittert, 'Making the Cape floral kingdom', pp.113 & 120; Lance van Sittert, 'From "mere weeds" and "bosjes" to a Cape floral kingdom: The re-imagining of indigenous flora at the Cape, c.1890-1939', *Kronos*, no.28, (November 2002, p.112.
[9] Boehi, pp.17-18; Lighton, p.182-3.
[10] Lighton, pp.67-69, 73 & 76-77.
[11] Lighton, pp.80 & 82.
[12] Rabe, pp.101 & 103.
[13] See Mountain Club of South Africa website: http://cen.mcsa.org.za/, accessed 30 March 2016; Clem Thomas & Greg Thomas, *125 years of the British and Irish Lions*, Johannesburg, 2013, p.61; https://www.westerncape.gov.za/sites/www.westerncape.gov.za/files/documents/2010/7/constitution_of_the_western_cape_act_1_of_1998.pdf, accessed 10 August 2015.
[14] Theunis Jordaan, 'Wild fynbos harvester survey assesses extent of broadcast sowing', *Fynbos Voice*, Bulletin no.6, August 2013, p.1, accessed 15 Jun 2015: http://www.fynbosvoice.co.za/documents/fynbosvoice6_august2013.pdf; see also Michael J. Mangnall & Timothy M. Crowe, 'The effects of agriculture on farmland bird assemblages on the Agulhas Plain, Western Cape, South Africa', *African Journal of Ecology*, vol.41, (2003), pp.266-76.
[15] Agriculture, Nature and Food Quality Department, 'The South African flower industry', Embassy of the Kingdom of the Netherlands, Pretoria, pp.24 & 32; Farmer A interviewed by Himansu Baijnath, 4 September 2013, as part of a survey of selected strelitzia farmers. Most preferred to remain anonymous.
[16] Farmer A, 4 September 2013.
[17] Farmer A, 4 September 2013.
[18] Jaco Duif, pers comm, 22 June 2015.
[19] Johannes Maree & Ben-Erik van Wyk, *Cut flowers of the world*, (Pretoria, 2010), pp.14 & 18; see 'Cooperation in Campania', Floraculture International, posted 18 February 2008, http://www.floraculture.eu/?p=4082, accessed 27 June 2015; 'FloraHolland and Cultra hold their inaugural WinterPlaza show', posted 26 January 2014, http://www.floraculture.eu/?p=18337, accessed 27 June 2015.
[20] Agriculture, Nature and Food Quality Department, 'SA flower industry', pp.16-17 & 22; Jordaan, p.238.
[21] Maryke Middelmann, *Proteas: The birth of a worldwide industry*, (Cape Town, 2012), p.211; Agriculture, Nature and Food Quality Department, 'SA flower industry', p.25.
[22] Maree & van Wyk, p.342.
[23] Maree & van Wyk, p.14; Terry Macalister, 'What caused the 1970s oil price shock?', *Guardian*, 3 March 2011; Daryl Worthington, 'OPEC and the 1979 oil shock', http://www.newhistorian.com/opec-1979-oil-shock/2423/, accessed 29 June 2015.
[24] Agriculture, Nature and Food Quality Department, 'SA flower industry', p.14. For South African gross domestic product, 1995 and 2008, see: http://www.tradingeconomics.com/south-africa/gdp; for relative currency values, see 'SA rand value: 1994–2012', http://businesstech.co.za/news/general/82273/sa-rand-value-1994-2015/; all accessed 27 December 2015. Note World Bank GDP figures are expressed in US dollars.
[25] Maree & van Wyk, pp.16-17; Agriculture, Nature and Food Quality Department, 'SA flower industry', p.23.
[26] Jordaan, p.238. See also http://www.multiflora.co.za/history.php, accessed 14 June 2015.
[27] Farmer A, 4 September 2013; Multiflora history [online].
[28] Anon., 'Where our flowers come from', [online]: www.flowersandplants.org.uk, accessed 13 June 2015; Agriculture, Nature and Food Quality Department, 'SA flower industry', p.31; *Fynbos Bulletin*, no.6
[29] Marshall, 'Steady increase in flower sales'.
[30] 'SA flower exports wilt', http://www.moneyweb.co.za/moneyweb-south-africa/sa-flower-exports-wilt, accessed 1 June 2014; 'Ash hits SA flower, fruit exports', http://www.fin24.com/Business/Ash-hits-SA-flower-fruit-exports-20100419, accessed 1 June 2014; Middelmann, p.210.
[31] Agriculture, Nature and Food Quality Department, 'SA flower industry', p.22.
[32] Agriculture, Nature and Food Quality Department, 'SA flower industry', p.22.
[33] https://weatherspark.com/history/29022/2013/Nelspruit-Mpumalanga-South-Africa, accessed 8 August 2015.
[34] Maree & van Wyk, p.24.
[35] Farmer A interview, 4 September 2013.
[36] Maree & van Wyk, p.31.
[37] Farmer A interview, 4 September 2013; Maree & van Wyk, p.342.
[38] Maree & van Wyk, p.24.
[39] Maree & van Wyk, pp.24-25.
[40] Farmer A interview, 4

41 Maree & van Wyk, p.32.
42 Maree & van Wyk, p.32.
43 Agriculture, Nature and Food Quality Department, 'SA flower industry', p.19; http://www.fedprimerate.com/crude-oil-price-history.htm; Fynbos Marketing Forum, May 2014.
44 Ridgway & Works, p.128.
45 Maree & van Wyk, p.342.
46 Maree & van Wyk, pp.24, 25.
47 Maree & van Wyk, pp.27 & 42-43.
48 Maree & van Wyk, pp.23, 25-26, 30, 34-35 & 342.
49 Maree & van Wyk, p.32; Farmer A interview, 4 September 2013.
50 Maree & van Wyk, p.25.
51 Farmer A interview, 4 September 2013Maree & van Wyk, p.25.
52 Maree & van Wyk, p.34.
53 'Cooperation in Campania'.
54 Jianjun Chen, Dennis B. McConnell, Richard J. Henny, David J. Norman, 'The foliage plant industry', *Horticultural Reviews*, vol.31, p.100; see www.dummenusa.com
55 Agriculture, Nature and Food Quality Department, 'SA flower industry', p.27; Patrick Labaste ed., *The European horticulture market: Opportunities for sub-Saharan African exporters*, World Bank Working Paper No.63, World Bank, (Washington, 2005), p.52.
56 Labaste, p.52.
57 Johannes Maree, pers comm, 22 June 2015; Agriculture, Nature and Food Quality Department, 'SA flower industry', pp.31 & 33.
58 See South African Flower Export Council Market Research, 'Opportunities in Asia and the Middle East markets for South African cut flowers', p.14.
59 Agricultural Products Standards Act 1990 (Act 119 of 1990), Standards and Requirements regarding the Export of Fresh Cut Flowers and Fresh Ornamental Foliage, Part 3.15, p.32.
60 Farmer B interviewed by Himansu Baijnath, 4 October 2013.
61 Farmer C interviewed by Himansu Baijnath, 29 September 2014.
62 Farmer C, 29 September 2014.
63 Farmer C, 29 September 2014.
64 Based on sources canvassed by Himansu Baijnath, 2010–2012.
65 See Van Sittert, 'Making of the Cape Floral Kingdom', p.120.
66 See 'Emerging from poverty', *Floraculture International*, 25 March 2008, http://www.floraculture.eu/?p=4133, accessed 1 June 2014.
67 Alida Alberts, interview with Himansu Baijnath, 3 September 2013.
68 Farmer A interview.
69 Maree & van Wyk, p.32.
70 Farmer A interview.

## CHAPTER 12

[1] Anon., 'The travellers' tree', *The Garden*, 29 August 1874, p.205.
[2] Anon., 'Revue des plantes nouvelles ou intéressantes', *La Belgique Horticole*, vol.9, (1859), p.324-325.
[3] H.W., 'Ravenala madagascariensis', *Annales d'Horticulture et de Botanique*, vol.3, (1860), p.16; Ernest Faivre, 'Jardin-des-plantes et herbiers de Grenoble', *Bulletin de la Société Botanique de France*, vol.7, (1860), p.822; H.R. Goeppert, 'Bericht über den gegenwärtigen Zustand des botanischen Gardens in Breslau', *Archiv der Pharmacie*, [about 1860], 185, series, 1, no.2, p.24; Miguel Colmeiro, *Diccionario de los diversos nombres vulgares de muchas plantas usuales o notables*, (Madrid, 1871), p.23.
[4] Hladik et al, 'L'arbre du voyageur', 2, pp.19 & 26; Julius Bouché, 'Zwei nusbaume der tropen, eds. J. Bouché & R. Herrmann, *Jahrbuch für Gardenfunde und Botanif*, (Bonn, 1885), p.337; http://www.moneysorter.co.uk/calculator_inflation2.html#calculator, accessed 2 May 2014; 'Un arbre de prix', in 'Chronique Horticole', *L'Illustration Horticole*, 15 June 1896; 'Een kostbare plant', *Teysmannia*, part 7, (1897), p.673.
[5] Robert D. Hoyt, *American Exotic Nurseries: Annual illustrated and descriptive catalogue of new, rare and beautiful plants and seeds*, (Seven Oaks, 1894), p.64.
[6] L. Sonthonnax, 'Deux mois aux Antilles Françaises', *L'Echange Revue Linnéenne*, no.171, (March 1899), p.22; Paul C. Standley, 'Flora of Costa Rica', part 1, *Publications of Field Museum of Natural History*, botanical series, vol.18, (October 1937), p.186; New York Botanic Gardens virtual herbarium: P.E. Meleady, Daniel Atha & Javier Cordero 44601-4, bank of Macal river, near Ix Chel Farm, San Ignacio, Cayo district, Belize, 'Cultivated. Growing amongst other ornamentals', [online]: http://sweetgum.nybg.org/science/vh/specimen_details.php?irn=27884; R.O. Williams, 'Descriptive nursery stock list', Bulletin of the Department of Agriculture, Trinidad and Tobago, vol.20, part 1, (1922), p.21; Patrick Leigh Fermor, *The Traveller's Tree* (London, 1950); Ian Fleming, *Live and let die* (London, 1954); Patricia McCracken, observation, 30 November 2012.
[7] Ponsonby, pp.24 & 47; Philateca, 'North Borneo, Traveler's palm', [online]: http://www.philateca.com/stamp/29/2_cents_-_Traveler's_Palm_(Ravenala_madagascariensis), accessed 15 September 2016; William Roxburgh, 'Urania speciosa', Flora indica, vol.2, (Serampore, 1832), p.114; J. Medley Wood, *A guide to the Natal Botanic Gardens*, (Durban 1883), pp.42 and 45; J. Medley Wood, *Guide to the trees and shrubs of the Natal Botanic Gardens, Durban*, (Durban, 1897), p.24; Wood, *List of plants*,(1915), p.69.
[8] 'Royal Botanic Gardens', in *Gardeners' Chronicle*, 11 December 1905, p.340.
[9] Proschavsky, p.273; Review of M.P. de Tchihatchef, Lettres à Michel Chevalier, in *Bulletin de la Société Botanique de France*, (1880), vol.28, 2nd series, pp.146-148; Anon., 'Toronto', *The American Florist*, 29 September 1892, p.192; William Saunders, *U.S. Department of Agriculture catalogue of economic plants*, (Washington, 1891), p.36; Anon., 'Seeds and plants imported during the period from September, 1900, to December, 1903', *U.S. Department of Agriculture Bureau of Plant Industry Bulletin*, (Washington, 1905), no.66, p.211; B.E. Dahlgren, 'Mural painting of Madagascar traveler's tree in hall of plant life', *Field Museum News*, (September 1936), vol.7, no.9, p.1.
[10] Ciska Kay, 'Rebel with a cause' Get It Lowveld: http://lowveld.getitonline.co.za/2013/12/09/rebel-cause/, posted 9 December 2013, accessed 28 June 2014.
[11] J.G. Frazier, *The golden bough*, London, 1894.
[12] Suzuki, preface, p.107, 1982; combined Japanese and English text; photograph captions, acknowledgements, preface and species descriptions are all in English but the bulk of the text, including growing notes, is in Japanese.
[13] Suzuki, pp.2-3, 24, 40, 44, 57-8 & 61.
[14] Suzuki, pp.22, 24 & 29.
[15] Suzuki pp.12-22 & 31.
[16] Suzuki, pp.1 & 28.
[17] Suzuki, pp.35, 38, 63-65 & 69.
[18] Stephen Lacey, 'The very English gardens of the Riviera', *Daily Telegraph* (London): http://www.telegraph.co.uk/gardening/gardenstovisit/11196281/The-very-English-gardens-of-the-Riviera.html, posted 1 November 2014, accessed 28 March 2016; Charles Quest-Ritson, *The English garden abroad*, (London, 1992), p.63.
[19] Quest-Ritson, p.64; Ray Desmond, *Sir Joseph Dalton Hooker: Traveller and plant collector*, (London, 1999), pp.211-215; F. Ballard, 'The Hanbury Garden at La Mortola', *Journal of the Kew Guild*, 1963, vol.8, series 67, pp.198-201; Berger, pp.x-xi.
[20] Alwin Berger, *Hortus Mortolensis*, (London, 1912), p.xi.
[21] Quest-Ritson, pp.65-6; Ballard, p.198; Berger, p.xii; G. Gale, 'Saving the vine from Phylloxera' in eds Merton Sandler & Roger Pinder, *Wine: A scientific exploration*, 2002, pp.70-91; DOI: 10.1201/9780203361382.ch4
[22] Berger, pp.x-xii; Gunn & Codd/Glen & Germishuizen, 'Berger, Alwin', p.96.
[23] Anon., 'Hanbury botanical gardens', UNESCO World Heritage Centre: http://whc.unesco.org/en/tentativelists/336/, published 1 June 2006, accessed 28 March 2016; and see www.grandigiardini.it and www.giardinihanbury.com, accessed 3 April 2016; Berger, p.xiv & note, p.432.
[24] Quest-Ritson, p.66. Fr Hoffmann, 'Friedrich August Flückiger', *American Journal of Pharmacy*, February 1895, pp.65-71; the entire July 1895 issue of *Archiv der Pharmacie* was also devoted to a 'biographical sketch' of Flückiger, the *American Journal of Pharmacy* reported in September 1895 (p.498).
[25] Berger, p.xiii.
[26] Quest-Ritson, pp.68-70; Berger, pp.xvi-xv; Ballard, p.199; Anon, 'Man who gifted RHS its Wisley garden': http://www.getsurrey.co.uk/news/local-news/man-who-gifted-rhs-wisley-4850631, 21 May 2004, accessed 6 April 2016.
[27] Berger, pp.vi-vii.
[28] Berger, pp.v-vi; Quest-Ritson, p.66.
[29] Ballard, pp.199-201; Joan Marble, *Notes from an Italian garden*, (London, 2000), p.315; UNESCO submission, 'Hanbury botanical gardens': www.grandigiardini.it & www.giardinihanbury.com, both accessed 3 April 2016; Pier Giorgio Campodonico, Elena Zappa, Anna Luisa Carboni & Daniela Guglielmi, *Hanbury Botanic Garden*, (Genoa, n.d), p.15; see also www.giardinihanbury.com
[30] Anon., 'Jardin Thuret': http://www6.sophia.inra.fr/jardin_thuret/Historique; http://www.gardenvisit.com/garden/jardin_thuret, both accessed 5 April 2016.
[31] Berger, p.xii.
[32] Anon., 'French government continues its aid program for cut flower growers', Floraculture International, 16 May 2008, http://www.floraculture.eu/?p=4228, accessed 1 June 2014; 'Floral rendezvous in Angers, Floraculture International, 28 April 2008, http://www.floraculture.eu/?p=4189, accessed 1 June 2014.
[33] Quest-Ritson, pp.65-66; 'Production and innovation'.
[34] Anon., 'Production and innovation on the French Riviera', Floraculture International, 29 November 2007, http://www.floraculture.eu/?p=4036, accessed 1 June 2014.
[35] Michael Schuermann, 'In the footsteps of Monet at Bordighera', http://www.huffingtonpost.com/michael-schuermann/just-follow-the-monet_b_5641170.html: published 8 January 2014, accessed 10 April 2016; Anon, 'Empress Eugénie', [online]: https://www.britannica.com/biography/Eugenie, accessed 9 June 2018; Mauro Giorgio Mariotti, 'Ludwig Winter, gardener', [online]: http://www.nuovositoamici.oranjuice.org/index.php?option=com_content&view=article&id=84:ludwig-winter-gardener&catid=21&lang=en&Itemid=300, posted 23 August 2014, accessed 2 March 2017.
[36] Lacey, 'Very English gardens'; UNESCO submission, 'Hanbury botanical gardens'; Berger, p.v.; Anon., 'Alassio's Villa della Pergola opens its gates', FloraCulture International: http://www.floraculture.eu/2014/02/alassios-villa-della-pergola-opens-its-gates-from-29th-march-to-the-end-of-october/, posted 26 February 2014, accessed 1 June 2014.
[37] Anon., 'Apuglia region announces Open Days', FloraCulture International: www.floraculture.eu/?p=16878, posted 25 February 2014, accessed 27 June 2015.
[38] Marco Enfossi quoted in Anon., 'Niche exports settle into Italian strategy', Floraculture International: http://www.floraculture.eu/?p=4011, posted 26 September 2007, accessed 27 June 2015.

39 Anon., 'Cooperation in Campania', www.floraculture.eu/?p=4082, posted 18 February 2008, accessed 27 June 2015; 'Niche exports'.
40 Tamasi Koley, Puja Sharma & Y.C. Gupta, 'Bird of paradise: A specialized cut flower for commercial cultivation', *Floriculture Today*, (February 2015), pp.58-60; McQuaid, 'Secrets behind your flowers'; Wyss, 'How Colombia bloomed into a US flower power'.
41 McQuaid, 'Secrets behind your flowers'.
42 McQuaid, 'Secrets behind your flowers'; Ridgway & Works, 'Trade agreements', *Sending flowers to America*, pp.128-129.
43 Anon., 'Flowers from Colombia', [online]: www.flowerexperts.com/colombian_flowers.asp, accessed 19 April 2014; 'Flower facts'.
44 'Flowers from Colombia'; Anon., 'More business, more international trade', Floraculture International, 30 October 2007 [online]: http://www.floraculture.eu/?p=4020, accessed 1 June 2014; 'Weekly ornamental shipment (movement) report, All commodities', USDA, Tues Apr 08, 6.20pm CDT, www.heartlandcoop.com.
45 Chiew & Chang, p.360; Maree & van Wyk, p.17, 'Flower facts'.
46 Gisela Yll, 'Strelitzia', [online]: https://www.floraqueen.es/blog/strelitzia-la-flor-ave-del-paraiso, accessed 20 May 2017; Anon., 'Flora ornamental', [online]: http://www.jardincanario.org/galeria-de-flora-ornamental, accessed 2 February 2018; Criley to Baijnath, 23 September 2013.
47 Koley et al, pp.58-59.
48 Koley et al, pp.58-59.
49 Koley et al, p.60.
50 Agricultural Trade Office, 'A Snapshot of the Taiwan Market', Taipei, 1995, pp.3-4.
51 Hwang-Jaw Lee, 'The development and expansion strategy for Taiwan's floriculture industry', Taiwan Development Association, 20 November 2014. All exchange equivalents in this section are approximate, calculated using: Lawrence H. Officer, 'Exchange Rates Between the United States Dollar and Forty-one Currencies', Measuring Worth, 2017, [online]: http://www.measuringworth.com/exchangeglobal/
52 'Taiwan market', pp3-4.
53 'Taiwan market', pp.6-7.
54 'Taiwan's floriculture industry', p.3.
55 Taiwan market', p.7; 'Taiwan's floriculture industry', pp.6-8.
56 Lim Heng Jong, Mohd. Ridzuan Mohd Saad & Nor Auni Hamir, 'Cut flower production in Malaysia', Malaysian Agricultural Research and Development Institute, [online]: https://archive.org/stream/agriculturalbull07unse/agriculturalbull07unse_djvu.txt, accessed 19 April 2014, pp.2-3.
57 Eds. Fatimah Mohamed Arshad, Nik Mustapha Raja Adullah, Bisant Kaur, Amin Mahir Abdullah, *50 years of Malaysian agriculture*, Universiti Putra Malaysia Press, Serdang, 2007, p.xi; Eddie Chiew & Fook Chang, 'The Malaysian floriculture industry', p.366, in Arshad, Adullah, Kaur & Abdullah eds., *50 years of Malaysian agriculture*.
58 'Malaysia', pp.xii, 1, 7-8; Chiew & Chang, pp.359 & 366.
59 'Malaysia', p.9; Chiew & Chang, pp.360-361 & 368.
60 http://www.ebay.com/sch/i.html?_from=R40&_trksid=p2050601.m570.l1313.TR0.TRC0.H0.Xstrelitzia.TRS0&_nkw=strelitzia&_sacat=0, accessed 28 June 2015; Himansu Baijnath, pers. obs., 7 June 1999; Januarius Rafael, Antonio V. Lamaris & Denilson Lamaris, 'Social and economic importance of species: Strelitzia', *International Journal of Horticulture and Floriculture*, (April 2014), vol.2 (4), pp.76-78; Jerzy Hetman and Elzbieta Pogroszewska, Strelicja, (Warsaw, 1990). See also Koley et al, p.58.
61 'Flower facts', Sustainable Skills Project: Floristry Topic, accessed 19 April 2014.
62 Lemaire, 'Phenacospermum guianense'.
63 Louis van Houtte, 'Strelitzia nicolai', Flore des Serres, vol.13, (1858), pp.121-123.

## CHAPTER 13

1 Elsa Pooley, pers. comm., 31 March 2016; 'Glorious flock'; Siskin, 'Birds of paradise'.
2 Chen, McConnell, Henny, Norman, p.100.
3 John Rourke, 'Botsoc remembers John Winter', *Veld & Flora*, September 2014, pp.122-3; John van der Linde, 'Tribute to John Winter', International Clivia Forum, accessed 12 March 2016.
4 Rourke, 'John Winter', p.123.
5 John Winter to Himansu Baijnath, email, 12 July 2010.
6 Rourke, 'John Winter', p.123.
7 Van Sittert, 'Making the Cape Floral Kingdom', p.115.
8 McCracken, *Gardens of empire*, pp.43-4.
9 Van Sittert, 'Making the Cape Floral Kingdom', p.122; Lance van Sittert, 'The intimate politics of the Cape Floral Kingdom', *South African Journal of Science*, (Mar-Apr. 2010), vol.106, no.3-4, pp.1-2.
10 Lighton, p.40; Gunn & Codd/Glen & Germishuizen, 'Mathews, Joseph', p.290.
11 Hadfield, *History*, p.361.
12 Lighton, pp. 98 & 182.
13 Lighton, p.103.
14 Boehi, pp.1-2, 5 & 21; Gunn & Codd/Glen & Germishuizen, 'Bolus, Harriet', pp.101-102; Dr Hugh Glen, pers. comm., 29 January 2017.
15 Lighton, p.105.
16 John Rourke & Caroline Voget, 'Botanical Society of South Africa: Highlights of the century', *Veld & Flora*, June 2013, pp.54-9.

17 Susyn Andrews, 'Golden Bird of Paradise', *The Garden* (UK), May 1994, vol.119, no.5, pp.228-30; Jonathan Petre, Untitled, *Daily Telegraph* (London), p.6, 23 December 1992.
18 Gunn & Codd/Glen & Germishuizen, 'Andrews, Susyn', pp.78-9.
19 Andrews, 'Golden bird', p.228; Susyn Andrews to John Winter, 7 May 1992; Andrews to Winter, 21 January 1993, both Andrews archive.
20 John Ashmore to Susyn Andrews, 24 May 1994; Winter to Andrews, 26 May 1992 & 16 May 1994, Andrews archive; John Winter, 'Going for gold', *Veld & Flora*, March 1994, p.22; Kay Montgomery, 'Go for gold', *Weekend Star: Life*, p.4, 30 April 1994.
21 Winter, 'Going for gold'.
22 Winter to Baijnath, 12 July 2010.
23 Winter to Baijnath, 12 July 2010.
24 Winter, 'Going for gold'; Rourke, p.123.
25 Winter to Andrews, 26 May 1992.
26 Gunn & Codd/Glen & Germishuizen, 'Bayer, Bruce', pp.93 & 'Stayner, Frank', 404-405; Winter to Baijnath, 12 July 2010.
27 Gunn & Codd/Glen & Germishuizen, p.93; Bruce Bayer to Susyn Andrews, 4 December 1992, Andrews archive; Winter to Andrews, 21 January 1993.
28 Winter to Andrews, 14 April 1993.
29 Bayer to Andrews, 4 December 1992.
30 Bayer to Andrews, 4 December 1992; Andrews to Winter, 21 January 1993; H.L. Reyneke to Susyn Andrews, 18 May 1993; Susyn Andrews to H.L. Reyneke, 20 April 1993; Reyneke to Andrews, 18 May 1993; Susyn Andrews to E. Dodds, 24 June 1993; all Andrews archive.
31 Winter to Andrews, 14 April 1993.
32 Andrews, 'Golden bird', p.228; Mark Sparrow to Susyn Andrews, 5 January 1993, Andrews archive; Reader's Digest, *Complete guide to gardening in South Africa*, Cape Town, 1971, vol.2, pp.419-20; Kristo Pienaar, *The South African What Flower Is That?*, Cape Town, 1984, p.318.
33 Elizabeth McClintock to Susyn Andrews, 6 August 1992, Andrews archive; Dr Eleanor Bennett to Susyn Andrews, 8 October 1992, Andrews archive; S. Grayer, 'The Royal Horticultural Society's Colour Chart: An everyday tool for use in the herbarium. Its past, present and future', *NatSCA News*, issue 18, p.20; Andrews to Winter, 21 January 1993; Winter to Andrews, 14 April 1993, Andrews archive.
34 Dr Eleanor Bennett to Susyn Andrews, 28 July 1992, Andrews archive; Ashmore to Andrews, 24 May 1994.
35 Brian J. Huntley, *Kirstenbosch: The most beautiful garden in Africa*, Cape Town, 2012, p.99; Gillian Condy interviewed by Patricia McCracken, 13 April 2016.
36 Huntley, *Kirstenbosch*, p.99.
37 http://blogs.nybg.org/plant-talk/2014/02/adult-education/tales-from-south-africas-kirstenbosch/; accessed 4 February 2016;
http://www.plantzafrica.com/plantqrs/strelitregmandelagold.htm; accessed 4 February 2016.
38 Condy interview, 13 April 2016.
39 Condy interview, 13 April 2016.
40 Nicky Coningsby, 'Madiba strikes gold', *Veld & Flora*, December 1996, p.132.
41 Huntley, *Kirstenbosch*, p.99.
42 Phakamani M'Afrika Xaba, '*Strelitzia juncea* Centenary Gold', *Veld & Flora*, pp.76-77.
43 Xaba, p.77; Winter to Baijnath, 12 July 2010; Van der Linde, 'Tribute'; Joy Woodward to Gigi Laidler, email, 12 March 2016; Rourke, p.123.
44 Xaba, p.77.
45 Xaba, pp.76-77.
46 Xaba, p.77.
47 Xaba, p.77.
48 Xaba, pp.76-77.
49 Xaba, p.77.
50 Xaba, p.77.

## CHAPTER 14

1 Ernst van Jaarsveld, 'The Mzimvubu River botanical expedition', *Veld & Flora*, (September 2003), pp.101-5.
2 Van Jaarsveld, 'Mzimvubu', p.102; C.J. Skead, *Historical plant incidence*, p.175; Donal P. McCracken & Patricia A. McCracken, *Natal, The Garden Colony*, (Sandton, 1990), p.21.
3 Bruce Hopwood, pers. comm., 20 March 2016; Global Plants: 'Ernst Jacobus van Jaarsveld': http://plants.jstor.org/stable/10.5555/al.ap.person.bm000010029, accessed 19 February 2016; Ernst van Jaarsveld, 'The Slangkloof expedition', *Veld & Flora*, (December 2005), pp.172-7.
4 Ernst van Jaarsveld, 'Cliff dwelling succulent and bulbous plants', *Botanical Art Association of South Africa (BAASA) Newsletter*, (March 2012), pp.3-4.
5 Van Jaarsveld, 'Kaokoveld', p.154.
6 Van Jaarsveld, 'Mzimvubu', p.101; Ernst van Jaarsveld & Jeanette Loedolff, 'A new subspecies of *Strelitzia reginae*', *The Plantsman*, (September 2007), pp.180-183.
7 Van Jaarsveld, 'Mzimvubu', p.104; Ernst van Jaarsveld interviewed by Patricia McCracken, 9 March 2016.
8 Van Jaarsveld, 'Mzimvubu', pp.101, 103.
9 Ernst van Jaarsveld, 'Strelitzia reginae Banks ex Aiton subsp. mzimvubuensis Van Jaarsv.', http://www.plantzafrica.com/plantqrs/strelitregmzimvu.htm, pp.1-6, accessed 16 March

2016, uploaded March 2008; van Jaarsveld & Loedolff, 'New subspecies', p.181; Michelle Nel, 'The extraordinary floral riches of Pondoland', *Veld & Flora*, (September 2003), pp.96-99.
[10] Van Jaarsveld, interview, 9 March 2016; van Jaarsveld, 'Mzimvubu', pp.101-2; Van Jaarsveld, 'Slangkloof', p.176.
[11] Van Jaarsveld, 'Mzimvubu', p.101.
[12] Van Jaarsveld, 'Mzimvubu', p.103; van Jaarsveld, interview, 9 March 2016.
[13] Van Jaarsveld, interview, 9 March 2016.
[14] Van Jaarsveld, 'Mzimvubu', p.101; van Jaarsveld, interview, 9 March 2016.
[15] Van Jaarsveld, 'Mzimvubu', p.102.
[16] Van Jaarsveld, 'Mzimvubu', pp.102-3.
[17] Van Jaarsveld, 'Mzimvubu', p.103.
[18] Van Jaarsveld, interview, 9 March 2016.
[19] Van Jaarsveld, 'Mzimvubu', pp.104-105.
[20] Van Jaarsveld, 'Mzimvubu', p.105; van Jaarsveld, interview, 9 March 2016.
[21] Van Jaarsveld & Loedolff, 'New subspecies', p.180; van Jaarsveld, 'Strelitzia reginae . . . subsp. mzimvubuensis': Derivation of name and historical aspects; van Jaarsveld, 'Mzimvubu', p.105; van Jaarsveld, Gunn & Codd/Glen & Germishuizen, 'Styles, David', p.410; van Jaarsveld, interview, 9 March 2016.
[22] Van Jaaarsveld, 'Mzimvubu', p.101; van Jaarsveld, interview, 9 March 2016.
[23] Van Jaarsveld, interview, 9 March 2016.
[24] Van Jaarsveld & Loedolff, 'New subspecies', p.180.
[25] Sparrman, *Travels*, vol 1, pp.65-6; Van Jaarsveld, interview, 9 March 2016.
[26] Van Jaarsveld, 'Strelitzia reginae . . . subsp. mzimvubuensis': Derivation of name.
[27] Van Jaarsveld & Loedolff, 'New subspecies', pp.180-181, 183.
[28] Van Jaarsveld & Loedolff, 'New subspecies', p.180
[29] Van Jaarsveld, 'Strelitzia reginae . . . subsp. mzimvubuensis': Propagation.
[30] Van Jaarsveld & Loedolff, 'New subspecies', pp.182-3; Van Jaarsveld, 'Strelitzia reginae . . . subsp. mzimvubuensis': Description.
[31] Van Jaarsveld, interview, 9 March 2016; Van Jaarsveld & Loedolff, 'New subspecies'. p.183.
[32] Van Jaarsveld, interview, 9 March 2016.
[33] Anon., 'William Milliken', [online]:
http://www.kew.org/science/directory/people/Milliken_William.html;
http://www.kew.org/discover/blogs/kew-science/medicinal-knowledge-amazon;
http://www.kew.org/discover/blogs/kew-science/blood-sweat-and-dna-exploring-unknown-amazon
[34] United Nations Development Programme, 'Conservation and sustainable use of biodiversity on the South African Wild Coast', Global Environmental Facility Facility Proposal for PDF B Funding, November 2004, p.4; Nel, 'Floral riches', pp.98-99.
[35] Van Jaarsveld, 'Slangkloof', p.175.
[36] John Manning, *Field guide to fynbos*, (Cape Town, 2018; 2nd ed.), p.21; Glynis V. Cron, Cary Pirone, Madelaine Bartlett, W. John Kress & Chelsea Specht, 'Biogeography' in 'Phylogenetic relationships and evolution in the Strelitziaceae (Zingiberales)', *Systematic Botany*, 2012, vol.37(3), pp.606-19.
[37] Cron et al, pp.615-616; John P. Rourke, '*Clivia mirabilis* (Amaryllidaceae: Haemanthene) a new species from Northern Cape, South Africa', *Bothalia*, 2002, vol.32.1, pp.1-7.
[38] Rourke, '*Clivia mirabilis*', p.1.
[39] H.P. Linder & C.R. Hardy, 'Evolution of the species-rich Cape flora', *Philosophical Transactions of the Royal Society London B*,29 October 2004, vol.359, no.1450, Plant Phylogeny and the origin of major biomes, pp.1623-32, published online 8 October 2004: doi: 10.1098/rstb.2004.1534; accessed 7 February 2016; Manning, pp.21-22; Cron et al, pp.606-19; Rourke, '*Clivia mirabilis*', pp.1-7.
[40] Linder & Hardy, pp.1629-30.
[41] Nel, quoting Braam van Wyk, 'Floral riches', p.97.
[42] Nel, 'Floral riches', p.98.
[43] Van Jaarsveld, interview, 9 March 2016; Pamela Ann Vass, 'Plant diversity and spatial discontinuities of the Albany Centre in the south-eastern Cape, South Africa', PhD thesis: School of Oriental and African Studies, University of London, December 2004, http://www.ambiotek.ccm/theses/pam_vass_thesis_final.pdf, accessed 16 March 2016.
[44] UNDP, 'Conservation and sustainable use', p.1.
[45] UNDP, 'Conservation and sustainable use', pp.4 & 8.
[46] UNDP, 'Conservation and sustainable use', pp United Nations Development Programme, 'Conservation and sustainable use', p.4.
[47] Nel, 'Floral riches', p.98.
[48] Van Jaarsveld, 'Mzimvubu', pp.103, 105.

# CHAPTER 15

[1] Anon., 'The role of botanic gardens', https://www.bgci.org/plant-conservation/botanic-gardens/, accessed 19 March 2016; Royal Botanic Gardens, Kew, *State of the world's plants 2017*, (London, 2017,) p.10.
[2] Kew, *State of world's plants 2017*, p.3.
[3] Janet M. Chernela, '*Guyana, Fragile frontier: Loggers, miners, and forest peoples* by Marcus Colchester', review, *Hispanic American Historical Review*, vol.79, no.1 (Feb. 1999), pp.170-171; Stephen G. Perz, 'The changing social contexts of deforestation in the Brazilian Amazon', *Social Science Quarterly*, March 2002, vol.83, no.1, pp.35-52.
[4] Anon., 'The economy booms, the trees vanish', *The Economist*, 19 May 2005.
[5] Lee Hannah, Makihiko Ikegami, David G. Hole, Changwan Seo, Stuart H.M. Butchart, A. Townsend Peterson & Patrick R. Roehrdanz, 'Global climate change adaptation priorities for biodiversity and food security', *Plos One*, August 2013, vol.8, issue 8, e72490; Kew, *State of world's plants 2017*, p.37; Claire Kremen, Isaia Raymond, Kate Lance & Andrew Weiss, 'Monitoring natural resource on the Masaola peninsula, Madagascar: A tool for managing integrated conservation and development projects', *Conservation Biology*, vol.12, issue 3, pp.63-82.
[6] Kew, 'Country focus – Status of knowledge of Madagascan plants', *State of the world's plants*, p.371; Kremen et al, p.63.
[7] Kremen et al, p.64.
[8] Hans Ter Steege, Rens W. Vaessen, Dairon Cárdenas-López, Daniel Sabatier, Alexandre Antonelli, Sylvia Mota de Oliveira, Nigel C.A. Pitman, Peter Meller Jørgensen & Rafael P. Salomão, 'The discovery of the Amazonian tree flora with an updated checklist of all known tree taxa', www.nature.com/scientificreports, doi:10.1038/srep29549, published 13 July 2016, accessed 26 February 2017.
[9] Perz, 'Changing social contexts of deforestation', pp.35-6; William F. Laurance & G. Bruce Williamson, 'Positive feedbacks among forest fragmentation, drought, and climate change in the Amazon', *Conservation Biology*, December 2001, vol.15, no.6, pp.1529-1535.
[10] Jessica Aldred, 'Purring monkey and vegetarian piranha among 400 new Amazon species', www.guardian.com, posted 23 October 2013, accessed 18 March 2016; 'The economy booms', *Economist*; Perz, 'Changing social contexts of deforestation', pp.43 & 47; Hans ter Steege, Nigel C.A. Pitman, Timothy J. Killeen, William F. Laurance, Carlos A. Peres, Juan Ernesto Guevara et al, 'Estimating the global conservation status of more than 15,000 Amazonian tree species', *Science Advances*, November 2015, 1:e1500936; Richard E. Zeebe, Andy Ridgwell & James C. Zachos, 'Anthropogenic carbon emission rate unprecedented during the past 66 million years', *Nature Geoscience*, vol.9, (2016), pp.325-329; World Wildlife Fund: http://wwf.panda.org/what_we_do/where_we_work/amazon/about_the_amazon/why_amazon_important/, accessed 18 March 2016.
[11] William Milliken, Daniela Zappi, Denise Sasaki, Mike Hopkins & R. Toby Pennington, 'Amazon vegetation: How much don't we know and how much does it matter?', *Kew Bulletin*, 2010, vol.65, no.4, pp.691-709.
[12] Milliken et al, 'Amazon vegetation', p.692; Perz, 'Changing social contexts of deforestation', p.36; Royal Botanic Gardens, Kew, *State of the world's plants 2016*, (London, 2016), p.34; Laurance & Williamson, 'Positive feedbacks', p.1532.
[13] Yu-shan Wu, 'Local values key to wildlife survival', *Natal Mercury*, 20 September 2016; Aldred, 'Purring monkey'.
[14] Aldred, 'Purring monkey'.
[15] Daniel Fleming, head of programmes for Brazil and Amazonia, World Wide Fund For Nature UK, quoted in Aldred, 'Purring monkey'.
[16] William Milliken, 'Blood, sweat and DNA', Kew blogs: http://www.kew.org/discover/blogs/kew-science/blood-sweat-and-dna-exploring-unknown-amazon, published 4 February 2016, accessed 16 April 2016.
[17] Milliken, 'Blood, sweat and DNA': 'Exploring the border' & 'Botanical challenges'.
[18] Milliken, 'Blood, sweat and DNA': 'Technical solutions'; William Milliken, 'Kew Gardens: The unknown Amazon', https://www.youtube.com/watch?v=2u_xunMOVYE, accessed 18 March 2016.
[19] G.T. Prance, 'Floristic inventory of the tropics: Where do we stand?', *Annals of the Missouri Botanic Gardens*, 1977, vol.64, pp.659-84; Milliken, 'Unknown Amazon'.
[20] Kew, *State of world's plants 2016*, pp.57-48; 'December 16th, 1886', *Proceedings of the Linnean Society*, Session 1886-87; Joseph F. Rock, 'Preliminary list of plants growing in Mrs Mary E. Foster's grounds, Nuuanu Avenue, Honolulu', *Hawaiian Forester and Agriculturist*, (vol.13, no.4, (April 1916), p.116; L. Sonthonnax, 'Deux mois aux Antilles Françaises', *L'Echange Revue Linnéenne*, no.171, (March 1899), p.22; Paul C. Standley, 'Flora of Costa Rica', part 1, *Publications of Field Museum of Natural History*, botanical series, vol.18, (October 1937), p.186; Paul C. Standley & Julian Steyermark, 'Flora of Guatemala', *Fieldiana: Botany*, vol. 24, part 3, (April, 1952), p.191; *Official guide to the Botanic Gardens, Dominica*, (Dominica, 1922), p.38; Hubert Winkler, 'Ravenala madagascariensis', *Botanisches Hilfsbuch für Pflanzer, Kolonialbeamte, Tropenkaufleute und Forschungreisende*, (Wismar, 1912), p.222. M. Calley, R.W. Braithwaite & P.G. Ladd, 'Reproductive biology of *Ravenala madagascariensis* Gmel. as an alien species', *Biotropica*, 25 (1), pp.61-72, 1993; Vikash Tatayah to Himansu Baijnath, email, 2 January 2012; 'Ravenala madagascariensis', Pacific Island Ecosystems at Risk: http://www.hear.org/pier/species/ravenala_madagascariensis.htm, updated 11 February 2013, accessed 15 May 2016; Anon., 'Ravenala madagascariensis', Global compendium of weeds: http://www.hear.org/gcw/species/ravenala_madagascariensis/, updated 9 October 2007, accessed 15 May 2016; Dennis Hansen to Aliens listserve, quoted in 'Ravenala madagascarensis', PIER.
[21] Adapted from [online]: www.hear.org/pier/species/ravenala_madagascariensis.htm, accessed 7 September 2007.
[22] See *Uebersicht der Aemter-Vertheilung und wissenshchaflichen Thätatigkeit des Naturwissenschaftlichen Vereins zu Hamburg-Altona*, (January 1869), p.15; Thomas J. Carlson, Barry Mamadou Foula, Julie A. Chinnock, Steven R. King, Gandeka Abdourahmaue, Bah Mamadou Sannoussy, Amadou Bah, Sekou Ahmed Cisse, Mohamed 54 Camara & Rowena K. Richter, 'Case study on medicinal plant research in

Guinea: Prior informed consent, focused benefit sharing, and compliance with the Convention on Biological Diversity', *Economic Botany*, 55(4), 2001, pp.478-91; Bob Allkin, Kristina Patmore, Nicholas Black, Anthony Booker, Catia Canteiro, Elizabeth Dauncey, Sarah Edwards, Felix Forest, Peter Giovannini, Melanie-Jayne Howes, Alex Hudson, Jason Irving, Christine Leon, William Milliken, Eimear Nic Lughadha, Uwe Schippmann & Monique Simmonds, 'Useful plants – Medicines', *State of the world's plants 2017*, (Kew, 2017), p.22; World Health Organization, 'Malaria': http://www.who.int/malaria/areas/treatment/overview/en/, accessed 21 March 2016.
[23] David R. Thomas, Claire A. Penney, Amrita Majumder & Amanda M. Walmsley, 'Evolution of plant-made pharmaceuticals', *International Journal of Molecular Sciences*, 2011, vol.12, pp.3322-3336.
[24] John C. Tilburt & Ted Kaptchuk, 'Herbal medicine research and global health: An ethical analysis', *Bulletin of the World Health Organization*, August 2008, vol.86, no.8, (August 2008), pp.577-656; Kew, *State of world's plants 2017*, pp.22 & 24; Kate Wilkinson, 'Do 80% of S. Africans regularly consult traditional healers?': https://africacheck.org/reports/do-80-of-south-africans-regularly-consult-traditional-healers-the-claim-is-false/, posted 31 July 2013, accessed 20 April 2016.
[25] David Ewing Duncan, 'A shaman's cure: Can a west African plant slow the scourge of diabetes?' *Life*, Special issue: 'Medical miracles', fall 1998, pp.26-28.
[26] Sharmin et al, 'Evaluation of antimicrobial activities', p.170.
[27] Helen Coster, 'One pharma entrepreneur's never ending quest', www.forbes.com, posted 12 March 2010, accessed 21 March 2016.
[28] Sanjeev Neelakantan, 'Counting on Crofelemer', *Future Woman*, posted 19 December 2012, accessed 22 March 2016.
[29] Damaris Colhoun, 'The 25-year battle for a folk remedy from the rainforest to gain FDA approval': http://www.atlasobscura.com/articles/the-25-year-battle-for-a-folk-remedy-from-the-rainforest-to-gain-fda-approval, posted 9 November 2015, accessed 22 March 2016.
[30] S. Sakthi Priyadarsini, R. Vadiva & N. Jayshree, 'Pharmacognostical standardization of leaves of *Ravenala madagascariensis* Sonn.', *Research Journal of Pharmacognosy and Phytochemistry*, July - August, 2010, vol.2 (4), pp.288-292.
[31] Betsy Levy, 'Shaman Pharmaceuticals researches African Traveler's Palm for potential diabetes drug', www.herbclip.com, posted 15 August 1999, accessed 8 August 2001.
[32] Carlson et al, 'Medicinal plant research in Guinea', p.480, table 3.
[33] Carlson et al, 'Medicinal plant research in Guinea', pp.478 & 480.
[34] Duncan, 'A shaman's cure', pp.26-29.
[35] Dr Ronald Khan, quoted in Duncan, 'A shaman's cure', p.29.
[36] Carlson et al, 'Medicinal plant research in Guinea', p.478; Levy, 'Shaman Pharmaceuticals researches'; Duncan, 'A shaman's cure, pp.29, 36.
[37] S. Sakthi Priyadarsini, R. Vadivu & N. Jayshree, '*In vitro* and In vivo antidiabetic activity of the leaves of *Ravenala madagascariensis* Sonn., on alloxan induced diabetic rats', *Journal of pharmaceutical science and technology*, vol.2 (9), 2010, pp.312-317 & referencing *The wealth of India: A dictionary of Indian raw materials and industrial products: Raw materials*, National Institute of Science Communication and Information Resources (2003), vol.8, p.391; S. Sakthi Priyadarsini, R. Vadivu & N. Jayshree, 'Hypolipidaemic and renoprotective study on the ethanolic & aqueous extracts of leaves of *Ravenala madagascariensis* Sonn., on alloxan induced diabetic rats', *Journal of pharmaceutical science and technology*, vol.2 (1), 2010, pp.44-50; Maliheh Najari Beidokhti, Eva S. Lobbens, Philippe Rasovaivo, Dan Staerk & Anna K. Jager. 'Inhibitory potential of 40 medicinal plant extracts from Madagascar against enzymes linked to type 2 diabetes', poster published by Department of Drug Design & Pharmacology, Faculty of Health & Medical Sciences, University of Copenhagen; Tasnuva Sharmin, Sharmin Reza Chowdhury, Md. Yeunus Mian, Masbahul Hoque, Md. Sumsujjaman & Faijun Nahar, 'Evaluation of antimicrobial activities of some Bangladeshi medicinal plants', *World journal of pharmaceutical sciences*, 2014, 2(2): pp.170-5.
[38] Katy Moran, 'Moving on: Less description, more prescription for plants', *Journal of the Environment Liaison Center International*, January 1998, vol.21, no.4, featured on the World Bank website 'Health : *Indigenous knowledge*, equitable benefits', in Indigenous Knowledge (IK) Notes, Washington, no.15/23428, pp.1-4: http://documents.worldbank.org/curated/en/1999/12/1671225/health-indigenous-knowledge-equitable-benefits, published December 1999, accessed 22 March 2016.
[39] Moran, 'Moving on', p.1.
[40] Conte interviewed by Helen Coster, 'One Pharma Entrepreneur's Never Ending Quest', www.forbes.com, posted 30 December 2012, accessed 21 March 2016; article 10e of the Convention on Biological Diversity, 1992, quoted by Carlson et al, 'Medicinal plant research in Guinea', pp.480-481.
[41] Joan Martínez Alier, 'International biopiracy versus the value of local knowledge', *Capitalism, Nature, Socialism*, vol.11, no.2 (2000), pp.59-66; http://www.etcgroup.org/users/pat-mooney, accessed 9 June 2016; Moran, 'Moving on', p.2.
[42] Carlson et al, 'Medicinal plant research in Guinea', p.486, table 14.
[43] Carlson et al, 'Medicinal plant research in Guinea', p.486.
[44] Anon., 'What is a botanical drug?', US Food and Drug Administration: http://www.fda.gov/AboutFDA/CentersOffices/OfficeofMedicalProductsandTobacco/CDER/ucm090983.htm, last updated 27 November 2015, accessed 22 March 2016.
[45] 'Botanical drug'.
[46] Colhoun, 'The 25-year battle for a folk remedy'.
[47] Colhoun, '25-year battle'.

[48] 'Lisa Conte rides animal health toward a $70m IPO for Jaguar', *San Francisco Business Times*, 28 August 2014: http://www.bizjournals.com/sanfrancisco/blog/biotech/2014/08/jaguar-animal-health-jagx-lisa-conte.html, accessed 21 March 2016; Karen Weintraub, 'Efficacy of anti-ageing pill may take a lifetime to prove', *Business Day*, 5 March 2015.
[49] Cary Pirone, David Lee, Leung Kim & John Kress, 'Aril structure and pigments in the Strelitziaceae', abstract 564, oral paper, Botanical Society of America conference, 2006: http://www.2006.botanyconference.org, accessed 23 January 2009.
[50] Jeffrey A. Crowther & Graeme R. Hanson, 'Chemistry of the aril of *Ravenala madagascariensis*', abstract SP60, *Proceedings of the Australian Society for Biochemistry and Molecular Biology*, vol.23, 1991.
[51] Crowther & Hanson, 'Chemistry of aril'; Pirone et al, 'Aril structure and pigments'.
[52] Harold H. Strain, *Chloroplast pigments and chromatographic analysis*, (Pennsylvania, 1958); Pirone et al, 'Aril structure and pigments'.
[53] Pirone et al, 'Aril structure and pigments'.
[54] Cary Pirone, Jodie V. Johnson, J. Martin E. Quirke, Horacio A. Priestap & David Lee, 'The animal pigment bilirubin identified in *Strelitzia reginae*, the bird-of-paradise flower', in *HortScience September 2010 vol. 45 no. 9 1411-1415*; Anon, 'Previously known as animal-only pigment, bilirubin now confirmed in Bird of Paradise flower', https://www.sciencedaily.com/releases/2010/09/100908160356.htm, posted 3 September 2010, accessed 1 January 2016.
[55] Cary Pirone, Jodie V. Johnson, J. Martin E. Quirke, Horacio A. Priestap & David Lee, 'Bilirubin present in diverse angiosperms', *AoB Plants*, 2010: plq020, doi:10.1093/aobpla/plq02, published 28 October 2010, accessed 1 January 2016.
[56] Cary Pirone, J. Martin E. Quirke, Horacio A. Priestap & David W. Lee, 'The animal pigment bilirubin discovered in plants', *Journal of the American Chemistry Society*, 4 March 2009, 131(8), p. 2830; Depika Dwarka, Himansu Baijnath, Veenesha Thaver & Mickey Naidu, 'New insights into the presence of bilirubin in a plant species *Strelitzia nicolai* (Strelitziaceae)', *African Journal of Traditional, Complementary and Alternative Medicines*, 14(2), (2017), pp.253-262; F. Howard & J. Yudkin, 'Effects of dietary change upon the amylase and trypsin activities of the rat pancreas', *British Journal of Nutrition*, 1963, vol.17, pp.28-37.
[57] Susan Milius, 'Animals jaundice pigment found in plants': https://www.sciencenews.org/article/animals%E2%80%99-jaundice-pigment-found-plants, posted 20 February 2009, accessed 18 October 2011; Cary Pirone, David Lee, Martin Quirke, Horacio Priestap, 'A mammalian pigment in the plant kingdom', abstract, http://2009.botanyconference.org/, accessed 21 January 2016.
[58] Pirone et al, 'A mammalian pigment' abstract.
[59] Dwarka et al, 'New insights into bilirubin', pp.1-2 & 4-5; L.E. Otterbein & A.M. Choi, 'Heme oxygenase: Colors of defense against cellular stress', *Journal of Lung, Cell and Molecular Physiology*, 2000, 279 (6), pp.29-37.
[60] Alan R. Battersby, 'Tetrapyrroles: The pigments of life', *Natural Product Reports*, December 2000, vol.17, no.6, pp.507-26: http://www.ncbi.nlm.nih.gov/pubmed/11152419, accessed 19 March 2016;; Pirone et al, 'Bilirubin in diverse angiosperms', pp.1-2; R. Tanaka & A. Tanaka, 'Tetrapyrrole biosynthesis in higher plants', *Annual Review of Plant Biology*, vol.58, 2007, pp.321-46; Michael R. Moore, 'an historical introduction to porphyrin and chlorophyll synthesis', in eds. Martin J. Warren & Alison G. Smith, T*etrapyrroles: Birth, life and death*, 2009, chapter 1; Dwarka et al, p.15.
[61] Milius, 'Animals' jaundice pigments'; Pirone et al, 'Bilirubin in diverse angiosperms', pp.1-2.
[62] Pirone et al, 'Bilirubin in diverse angiosperms', p.1; Milius, 'Animals jaundice pigment'; Dean DellaPenna interviewed by Susan Milius, quoted in 'Animals' jaundice pigment found'; Pirone et al, 'Animal pigment bilirubin discovered in plants'; Anon, 'Bilirubin now confirmed in Bird of Paradise flower'.
[63] Pirone et al, 'Bilirubin in diverse angiosperms', p.2, table 1.
[64] Pirone et al, 'Bilirubin in diverse angiosperms', p.6.
[65] Subtitle adapted from Battersby, 'Tetrapyrroles: The pigments of life'; Dwarka et al, p.6; Jed W. Fahey, Katherine K. Stephenson, Albena T. Dinkova-Kostova, Patricia A. Egner, Thomas W. Kensler & Paul Talalay, 'Chlorophyll, chlorophyllin and related tetrapyrroles are significant inducers of mammalian phase 2 cytoprotective genes', *Carcinogenesis*, 2005, vol.27, no.7, pp.1247-55.
[66] Dwarka et al, pp.6-8 & 10; S.I. Yamagishi, S. Maeda, T. Matsui & S. Ueda, 'Role of advanced glycation end products (AGEs) and oxidative stress in vascular complications of diabetes', *Biochim Biophys Acta*, 2011, pp.20-27; Bhavisha Bakrania, Eugene F. du Toit, Karl-Heinz Wagner, John P. Headrick & Andrew C. Bulmer, 'Pre- or post-ischemic bilirubin ditaurate treatment reduces oxidative tissue damage and improves cardiac function', *International Journal of Cardiology*, 2016, 202: 27 DOI: 10.1016/j.ij-card.2015.08.192.
[67] Jakob Vinther, Ian Bull & Dianne Edwards, 'Colouring the past': http://www.bristol.ac.uk/media-library/sites/earthsciences/documents/Colouring-the-past-Preservation-potential-and-palaeobiological-implications-of-carotenoid-and-porphyrin-pigments.pdf, accessed 23 March 2016; Pirone et al, 'Bilirubin in diverse angiosperms', p.6.
[68] Philip Ludwig, '*Strelitzia nicolai* seed aril extract degrades bilirubin to increase skin brilliance', HPC Today, Body care, grooming, protection and hygiene, vol.8 (2), pp.8-11.
[69] Ludwig, 'Aril extract', pp.9 & 11.
[70] Ludwig, 'Aril extract degrades bilirubin', p.11

Clinton Friedman, *Strelitzia nicolai*, 120 film photograph, 2005

# SELECT BIBLIOGRAPHY

**MANUSCRIPT SOURCES**

BRITISH MUSEUM (NATURAL HISTORY): ARCHIVE & HERBARIUM
- Correspondence of the Honourable Sir Joseph Banks Bart, vol.X, 1796-1797
- Holland Collection of Drawings of South African Plants
- Franz Bauer Collection: *Strelitzia depicta* drawings
- Strelitziaceae specimens (BMNH Herbarium)

ARCHIVE OF ROYAL BOTANIC GARDENS, KEW
- Kew record books
- Strelitziaceae specimen (K Herbarium, including carpological collection)

ARCHIVE OF SANBI NATIONAL HERBARIUM, PRETORIA
- Artists' files
- Holdings of Strelitziaceae illustrations
- Strelitziaceae specimens (PRE Herbarium)

**PERIODICALS**

*La Belgique Horticole*
*Blumengarten*
*Botanical Register*
*Bulletin of Miscellaneous Information*
*Curtis's Botanical Magazine*
*The Garden*
*Gardeners' Chronicle*
*Flore des Serres et des Jardins de l'Europe*
*Kew Bulletin*
*L'Illustration Horticole*
*Journal of Botany, British and Foreign*
*Journal of South African Botany*
*Cape Argus*
*Chicago Tribune*
*Business Day*
*Daily Telegraph*
*International Herald Tribune*

**ONLINE SOURCES**

http://www.academie-francaise.fr/
http://www.archivesdefrance.culture.gouv.fr/
http://www.archive.org
http://www.bergianska.se/english/collections/ (Bergius Herbarium)
http://www.biodiversitylibrary.org (Biodiversity Heritage Library)
http://www.bibliomonde.com/
http://www.bgci.org/worldwide/ (Botanic Gardens Conservation International)
http://www.bmlisieux.com/normandie/
http://www.botanischestaatssammlung.de/herbarium/plants/ (Munich Herbarium online)
http://www.bu.edu/missiology/missionary-biography/
http://cths.fr/ (Comité des Travaux Historiques et Scientifiques)
http://www.dacb.org/ (Dictionary of African Christian Biography)
http://data.bnf.fr/ (Bibliothèque National, France)
http://www.encyclopedia.com/
http://www.econlib.org/library/Enc/ (Concise Encyclopedia of Economics)
http://evolution.berkeley.edu/evolibrary/article/0_0_0/history_08 (Berkeley University: Understanding Evolution)
https://global.britannica.com/ (Encyclopaedia Britannica)
http://www.gbif.org/ (Global Biodiversity Information Facility)
http://gutenberg.net
http://www.hear.org/pier/ (Pacific Island Ecosystems at Risk)
https://www.histoire-image.org/
http://www.history.com/
http://www.historyofparliamentonline.org/
https://www.hortusleiden.nl/en/ (Leiden Botanic Gardens)
http://kiki.huh.harvard.edu/databases/botanist_search (Harvard University Herbarium online)
http://www.iep.utm.edu/ (Internet Encyclopedia of Philosophy)
http://www.imalaspina.com/en/ (Malaspina Online)
http://www.iucnredlist.org/
http://www.larousse.fr/encyclopedie/

http://www.linnaeus.uu.se/online/ (Linné Online)
http://www.marinersmuseum.org/
http://www.maritimeheritage.org/ports/madagascar.html
http://www.nationsencyclopedia.com/Africa/Madagascar-HISTORY.html
http://www.nhm.ac.uk/our-science/departments-and-staff/library-and-archives/collections/cook-voyages-collection/endeavour-botanical-illustrations/about.dsml (HMS Endeavour illustrations)
http://www.newworldencyclopedia.org/
www.pierre-poivre.fr
http://www.theplantlist.org/ (World Checklist)
http://plants.jstor.org/compilation/ (Global Plants online)
http://pza.sanbi.org (Plantz Africa)
http://libweb5.princeton.edu/visual_materials/maps/websites/pacific/bougainville/bougainville.html, (Princeton University, Bougainville Online)
http://www.s2a3.org.za/bio/ (SA Association for the Advancement of Science)
www.sanbi.org
http://broughttolife.sciencemuseum.org.uk/broughttolife/people/ (Science Museum, London)
http://www.slsa.sa.gov.au/encounter/flindersvoyage.htm (Flinders's voyage)
http://www.societe-belge-de-malacologie.be/
http://southseas.nla.gov.au/biogs/P000328b.htm, (South Seas Companion)
http://www.tropicos.org/
http://apps.kew.org/wcsp/ (World Checklist of Selected Plant Families online)

## BOTANICAL & HORTICULTURAL HISTORY & ART
### BOOKS

Ed. Arnold, Marion, *South African botanical art*, (Cape Town, 2001).
Banks, Joseph, *Delineations of exotic plants*, (London, 1796).
Berger, Alwin, *Hortus Mortolensis*, (London, 1912).
Biffen, Rowland H., *The auricula*, (Cambridge, 2014).
Bleichmar, Daniela, *Visible empire: Botanical expeditions and visual culture in the Hispanic Enlightenment*, (Chicago, 2012).
Blunt, Wilfrid, & William T. Stearn, *The art of botanical illustration*, (London, 1950).
Blunt, Wilfrid, *In for a penny*, (London, 1978).
Campbell-Culver, Maggie, *The origin of plants*, (London, 2012).
Campodonico, Pier Giorgio, Elena Zappa, Anna Luisa Carboni & Daniela Guglielmi, *Hanbury Botanic Garden*, (Genoa, n.d).
Coats, Alice M., *The book of flowers*, (London, 1984 ed.).
De Nave, F., & D. Imhof, *Botany in the Low Countries*, (Antwerp, 1993).
Desmond, Ray, *Dictionary of British and Irish botanists and horticulturists*, (London, 1977).
*A celebration of flowers*, (London, 1987).
*Kew*, (London, 1995).
*Sir Joseph Dalton Hooker: Traveller and plant collector*, (London, 1999).
Duncan, Graham, Barbara Jeppe & Leigh Voigt, *The Amaryllidaceae of southern Africa*, (Pretoria, 2016).
Forbes, V.S., *Pioneer travellers in South Africa*, (Cape Town, 1965).
González Bueno, Antonio, *La expedición botánica al vicerreinato del Perú (1777-8)*, (Barcelona, 1988).
Gorer, Richard, *The growth of gardens*, (London, 1978).
Grove, Richard H., *Green imperialism* (Delhi, 1995).
Green, Lawrence G., *Grow lovely, growing old*, (Cape Town. 1952 ed.).
Mary Gunn & Leslie Codd, eds. Hugh Glen & Gerrit Germishuizen, *Botanical exploration of southern Africa*, (Pretoria, 2010, 2nd edition) as *Strelitzia* 26.
Hayden, Ruth, *Mrs Delany: Her life and her flowers*, (London, 1980; 2005 edition).
Hadfield, Miles, *A history of British gardening*, (London, 1969).
& Robert Harling & Leonie Highton, *British gardeners: A biographical dictionary*, (London, 1980).
Jerzy Hetman & Elzbieta Pogroszewska, *Strelicja*, (Warsaw, 1990).
Hoyt, Robert D., *American Exotic Nurseries: Annual illustrated and descriptive catalogue of new, rare and beautiful plants and seeds*, (Seven Oaks, 1894).
Huigen, Siegfried, *Knowledge and colonialism: Eighteenth-century travellers in South Africa*, (Leiden, 2009).
Huntley, Brian J., *Kirstenbosch: The most beautiful garden in Africa*, (Cape Town, 2012).
Karsten, Mia, *The old Company's garden at the Cape & its superintendents*, (Cape Town, 1951).
Kirkby, Mandy, *The language of flowers: A miscellany*, (London, 2001).
Laws, Bill, *Fifty plants that changed the course of history*, (London, 2012 ed.).
Lighton, Conrad, *Cape floral kingdom*, (Cape Town, 1960).
Mallary, Peter & Frances, *A Redouté Treasury*, (New York, 1985).
Maree, Johannes, & Ben-Erik van Wyk, *Cut flowers of the world*, (Pretoria, 2010).
McCracken, Donal P., *A new history of Durban Botanic Gardens*, (Durban, 1996).
*Gardens of empire*, (Leicester, 1997), pp.100-3.
& P.A. McCracken, *Natal, The Garden Colony*, (Johannesburg, 1990).
McCracken, Eileen M., & Donal P. McCracken, *The way to Kirstenbosch*, (Cape Town, 1988), Annals of Kirstenbosch Botanic Gardens, vol.18.
Musgrave, Toby, Chris Gardner & Will Musgrave, *The plant collectors*, (London, 1998).
Padilla, Victoria, *Southern California gardens*, (Berkeley, 1961).
Pavord, Anna *The naming of names*, (London, 2005).
Quest-Ritson, Charles, *The English garden abroad*, (London, 1992).
Schiebinger, Londa, & Claudia Swan, *Colonial botany: Science, commerce and politics in the early modern world*, (Philadelphia, 2007).
Sherwood, Shirley, & Martyn Rix, *Treasures of botanical art* (London, 2008).

Smith, Lyman B., & Ruth C. Smith, 'Itinerary of William John Burchell in Brazil, 1825-1830', *Phytologia*, vol.14, no.8, pp.492-506.
Stearn, William T., *The Natural History Museum at South Kensington*, (London, 1981).
Stearn, William T., *Botanical Latin*, (Portland, 2004 ed.).
Suzuki, Yutaro, *The strelitzia world*, (Chiba, 1982).
Ziegler, Catherine, *Favored flowers: Culture and economy in a global system* (Durham, USA, 2007).

## ARTICLES & CHAPTERS

Agricultural Trade Office, 'A Snapshot of the Taiwan Market', (Taipei, 1995).
Agriculture, Nature and Food Quality Department, 'The South African flower industry', Embassy of the Kingdom of the Netherlands, (Pretoria, 2010).
Andrews, Susyn, 'Golden Bird of Paradise', *The Garden* (UK), May 1994, vol.119, no.5, pp.228-30.
Anon., 'List of collectors whose plants are in the herbarium of the Royal Botanic Gardens, Kew, to 31st December, 1899', *Bulletin of miscellaneous information*, nos.169-171, Jan-Mar 1901.
Ballard, F., 'The Hanbury Garden at La Mortola', *Journal of the Kew Guild*, 1963, vol.8, series 67, pp.198-201.
Barnard, T.T., 'The story of *Nerine undulata* (L.) Herb.', *Nerine Society Bulletin*, 3, 1968.
Bayer, A.W., 'Discovering the Natal flora', *Natalia*, vol.4, (1974), p.42-48.
Bretschneider, Emil, 'Dr Fischer and the Imperial Botanical Gardens, St Petersburg', *History of European botanical discoveries in China* (London, 1898).
Britten, James, 'John Frederick Miller and his "Icones"', *Journal of Botany, British and Foreign*, 1913, vol.51, p.256-257.
    'The herbarium of John Lightfoot', *Journal of Botany, British & Foreign*, vol.53, (1915), pp.269-271.
    '"John Frederick Miller and his Icones"', *Journal of Botany*, (1919), p.353.
Boddy-Evans, Marian, 'Camera lucida: An optical illusion for artists', [online]: http://painting.about.com/od/oldmastertechniques/ss/camera_lucida.htm, accessed 16 September 2016.
Buchanan, Handasyde, 'On collecting old flower books', in ed. Miles Hadfield, *The gardener's album*, (London & Wisbech, 1954).
Buckley, Julia, 'A royal flower – Bauer's Strelitzias', [online]: http://www.kew.org/discover/blogs/library-art-and-archives/royal-flower-%E2%80%93-bauer%E2%80%99s-strelitzias, posted 19 November 2014, accessed 4 May 2016.
Carter, H.B., 'Sir Joseph Banks and the plant collection from Kew sent to the Empress Catherine II of Russia 1795', *Bulletin of the British Museum (Natural History)*, historical series, vol.4, no.5, London, 1974, pp.281-385.
Chen, Jianjun, Dennis B. McConnell, Richard J. Henny, David J. Norman, 'The foliage plant industry', *Horticultural Reviews*, vol.31, (2003), pp.45-110.
Chiew, Eddie, & Fook Chang, 'The Malaysian floriculture industry', in eds. Arshad, Fatimah Mohamed, Nik Mustapha Raja Adullah, Bisant Kaur, Amin Mahir Abdullah, 50 years of Malaysian agriculture, (Serdang, 2007).
Coningsby, Nicky, 'Madiba strikes gold', *Veld & Flora*, December 1996, p.132.
Cross, Anthony, 'Russian gardens, British gardeners', *Garden History*, vol.19, no.1, (spring, 1991), pp.12-20.
Dettelbach, Michael, 'Global physics and aesthetic empire: Humboldt's physical portrait of the tropics', in eds. David Philip Miller & Peter Hanns Reill, *Visions of empire*, (Cambridge, 996).
Edwards, Phyllis I., 'Sloane Manuscript 5286: An important source for Decade 9, 'Herbarium Capense' of Petiver's Gazophylacium Naturae and Artis', *Journal of South African Botany*, vol.xxxiv, part iv, (July 1968), pp.243-253.
Grayer, Susan, 'The Royal Horticultural Society's Colour Chart: An everyday tool for use in the herbarium. Its past, present and future', *NatSCA News*, issue 18, p.19-26.
Halliwell, Brian, 'Some Australian plants in cultivation in England by 1800', [online]: http://anpsa.org.au/APOL29/mar03-4.html, posted March 2003, accessed 23 August 2016.
Hamilton, H.F., *A handbook of tropical gardening and planting*, (Colombo, 1914, 2nd edition), pp.312-314.
Hayden, Peter, 'The Russian Stowe', *Garden History*, vol.19, no.1, (spring, 1991), pp.21-27.
Hemsley, W. Botting, 'Margaret Meen', *Gardeners' Chronicle*, 17 February 1894, pp.197-198.
Jong, Lim Heng, Mohd. Ridzuan Mohd Saad & Nor Auni Hamir, 'Cut flower production in Malaysia', *Malaysian Agricultural Research and Development Institute*, Agricultural Bulletin 7, n.d.
Jordaan, Theunis, 'Wild fynbos harvester survey assesses extent of broadcast sowing', *Fynbos Voice*, Bulletin no.6, August 2013, p.1.
Karsten, Mia C., 'Francis Masson, A gardener-botanist who collected at the Cape: Masson's journeys at the Cape', *Journal of South African Botany*, vol.XXV, pp.169-170.
Koley, Tamasi, Puja Sharma & Y.C. Gupta, 'Bird of paradise: A specialized cut flower for commercial cultivation', *Floriculture Today*, (February 2015), pp.58-60.
Labaste, Patrick ed., *The European horticulture market: Opportunities for sub-Saharan African exporters*, World Bank Working Paper No.63, World Bank, (Washington, 2005).
Lack, H. Walter, & Victoria Ibáñez, 'Recording colour in late eighteenth century botanical drawings', *Botanical Magazine*, vol.14, no.2, May 1997, pp.87-100.
    'Jacquin's "Selectarum stirpium americanarum historia"', *Curtis's Botanical Magazine*, vol.15, (August 1998), pp.194-214.
Lee, Hwang-Jaw Lee, 'The development and expansion strategy for Taiwan's floriculture industry', Taiwan Development Association, (November 2014).
Mabberley, David, 'A note on some adulatory botanical plates distributed by Sir Joseph Banks', *Kew Bulletin*, vol.66, 2011, pp.475-477.
MacOwan, Peter, 'President's annual address, July 28, 1886: Personalia of botanical collectors', *Transactions of the South African Philosophical Society*, vol.IV, (1888), pp.xxx-liii.
Maddison, R.E.W., & Raymond E. Maddison, 'Spring Grove, the country house of Sir Joseph Banks, Bart, PRS', *Notes and Records of the Royal Society of London*, vol.11, no.1, (Jan. 1954), p.93.
Madriñan, Santiago, 'The illustrious Nikolaus Jacquin', *Botanica Collected*, March 2014, p.30.
Prain, David, *Bengal plants*, (Calcutta, 1903).
McCracken, Donal P., 'Plant hunting and the provenance of South African plants in British botanical magazines, 1787-1910', in ed. Mary N. Harris, *Sights and Insights: Interactive images of Europe and the wider world*, (Pisa, 2007), p.80.
    'The old Durban forest', in ed. Mark Mattson, *The Durban forest*, (Durban, 2015), pp.45-89.
Musselman, Elizabeth Green, 'Plant knowledge at the Cape: A study in African and European collaboration', *International Journal of African Historical Studies*, vol.36, no.2, (2003), pp.367-392.
Nel, Michelle, 'The extraordinary floral riches of Pondoland', *Veld & Flora*, September 2003, pp.96-99.
Nelson, E.C., 'William Ramsay McNab's herbarium in the National Botanic Gardens, Glasnevin', Glasna, new ser., 1 (1990), pp.1-7.
Rabe, Lizette, 'Living history – The story of Adderley Street's flower sellers', *SA Tydskrif vir kultuurgeskiedenes*, 24(1), (2010), pp.83-104.
Rafael, Januarius, Antonio V. Lamaris & Denilson Lamaris, 'Social and economic importance of species: Strelitzia', *International Journal of Horticulture and Floriculture*, vol.2 (4), (April 2014), pp.76-78.
Rourke, John P., 'Thunberg's botanical achievements at the Cape and their implications', in ed. Forbes, Thunberg, *Travels*
    'Beauty in truth', in ed. Marion Arnold, *South African botanical art*.
    'Botsoc remembers John Winter', Veld & Flora, September 2014, pp.122-3.
    '*Clivia mirabilis* (Amaryllidaceae: Haemanthene) a new species from Northern Cape, South Africa', *Bothalia*, 2002, vol.32.1, pp.1-7.
    & Caroline Voget, 'Botanical Society of South Africa: Highlights of the century', *Veld & Flora*, June 2013, pp.54-9.
Saltmarsh, Anna C., 'Francis Masson: Collecting plants for king and country', *Curtis's Botanical Magazine*, vol.20, no.4, (November 2003), pp.225-244.
Shuker, Karl, Shuker Nature [blog], 'Sonnerat's non-existent penguins (and kookaburra) of New Guinea', [online]: http://karlshuker.blogspot.com/2013/05/sonnerats-non-existent-penguins-and-kookaburra.html, posted 1 May 2013, accessed 23 December 2013.
South African Flower Export Council Market Research, 'Opportunities in Asia and the Middle East markets for South African cut flowers', (Pretoria, 2013).

Sprague, T.A., 'Bibliographical notes: Cl. John Sebastian Miller's 'Icones Novae', *Journal of Botany, British and Foreign*, vol.74, 1936, pp.208-209.
Stafleu, F.A., 'Dates of botanical publications 1788-1792', *Taxon*, 12 (2), 1963, p.50.
Stearn, William T., 'Maria Sibylla Merian (1647-1717) as a botanical artist', *Taxon*, (August 1982), vol.31, no.3, pp.529-534.
Svedelius, Nils, 'Carl Peter Thunberg (1743-1828)', *Isis*, vol.35, no.2 (spring, 1944), pp.128-134.
Taylor, George W., 'Poor men's treasures', in ed. Miles Hadfield, *Album*, pp.142-152.
United Nations Development Programme, 'Conservation and sustainable use of biodiversity on the South African Wild Coast', Global Environmental Facility Facility Proposal for PDF B Funding, (November 2004).
Van Jaarsveld, Ernst, 'The Slangkloof expedition', *Veld & Flora*, December 2005, pp.172-7.
'Cliff dwelling succulent and bulbous plants', *Botanical Art Association of South Africa (BAASA) Newsletter*, March 2012, pp.3-4.
& Jeanette Loedolff, 'A new subspecies of *Strelitzia reginae*', *The Plantsman*, September 2007, pp.180-183.
'The Mzimvubu River botanical expedition', *Veld & Flora*, (Sept. 2003), pp.101-5.
Van Sittert, Lance, 'From "mere weeds" and "bosjes" to a Cape floral kingdom: The re-imagining of indigenous flora at the Cape, c. 1890-1939', Kronos, no.28, (November 2002), pp.102-126.
'Making the Cape Floral Kingdom: The discovery and defence of indigenous flora at the Cape ca. 1890-1939', Landscape Research, 28:1, (2003), pp.113-129.
Winter, John, 'Going for gold', *Veld & Flora*, March 1994, p.22.
Xaba, Phakamani M'Afrika, '*Strelitzia juncea* Centenary Gold', *Veld & Flora*, (June 2013), pp.76-77.

## BIOGRAPHY

Dictionary of South African biography, vol.4, (Durban, 1981)
Blunt, Wilfrid, *Linnaeus, The Compleat Naturalist* (London, 2004).
Buchanan, Susan, *Burchell's travels*, (Cape Town, 2015).
Carter, Harold, *Sir Joseph Banks*, (London, 1988).
De Montessus, F.B., *Martyrologie et biographie de Commerson*, (Châlon-sur-Saone, 1889).
Fairbridge, Dorothea, *Lady Anne Barnard at the Cape of Good Hope, 1797-1802*, (Oxford, 1924).
Fara, Patricia, *Erasmus Darwin*, (Oxford, 2012).
Filliozat, Manonmani, 'D'Après de Mannevillette, Captain and hydrographer to the French East India Company (1707-1780)', *Indian Journal of History of Science*, vol.29, no.2, (1994), pp.329-342.
Hedley, Olwen, *Queen Charlotte*, (London, 1975).
Monnier, Jeannine, Anne Lavondès, Jean-Claude Jolinon & Pierre Elouard, *Philibert Commerson, Le découvreur de Bougainvillier*, (Chatillon-sur-Chalaronne, 1993).
ed. Smith, Lady [Pleasance], *Memoir and correspondence of the late Sir James Edward Smith, M.D.*, (London, 1832).
Storrar, Clare D., *The four faces of Fourcade*, (Cape Town, 1990),
Taylor, Stephen, *Defiance: The life and choices of Lady Anne Barnard*, (London, 2016).
Wood, Dennis, 'Jacques-Henri Bernardin de Saint Pierre', *New Oxford companion to French literature*, (Oxford, 1995).

## PRINTED PRIMARY SOURCES

Adanson, Michel, *Familles naturelles des plantes*, (Paris, 1763).
Afzelius, Karl, *Fauna och Flora*, (Uppsala, 1914).
Aiton, William, *Hortus kewensis*, (London, 1789).
Ed. Aiton, William Townsend, *Hortus kewensis*, (London, 1811, 2nd edition).
Alexander, J.E., *An expedition of discovery into the interior of Africa, through the hitherto undiscovered countries of the great Namaquas, Boschmans and Hill Damaras*, (London, 1840).
Anon., *Seeds and plants imported during the period from July 1 to September 30, 1911, Inventory no.28; Nos.31371 to 31938*, (Washington, 1912).
Baines, Thomas, *Greenhouse and stove plants*, (London, 1894).
Baker, J.G., 'Flora of Madagascar', in Samuel Pasfield Oliver, *Madagascar*, (London, 1886).
Banks, Joseph, *The Endeavour journal of Sir Joseph Banks* [online]: http://gutenberg.net.au
Barrow, John, *Travels in the interior of southern Africa*, (London, 1806).
Bean, W.J., *Royal Botanic Gardens, Kew*, (London, 1908).
Bergius, Peter, *Descriptiones plantarum ex Capite Bonae Spei*, (Stockholm, 1767).
Bolus, Harry, 'Sketch of the flora of South Africa', in ed. John Noble, *Official handbook: History, production and resources of the Cape of Good Hope*, (Cape Town, 1886).
Burchell, William J., *Travels in the interior of South Africa*, (London, 1822).
Bureau of Plant Industry, 'Plant material introduced by the Division of Foreign Plant Introduction, Bureau of Plant Industry, April 1 to June 30, 1932', *United States Department of Agriculture Inventory No.111*, (May 1934), p.28.
Campbell, Belle McPherson, *Madagascar*, Missionary Annals (Chicago, 1889).
Ed. Chambers, Neil, *Letters of Sir Joseph Banks: A selection, 1768-1820*, (London, 2000).
Chapman, Olive Murray, 'Primitive tribes in Madagascar', *Geographical Journal*, vol.96, no.1, (June, 1940), pp.14-25.
Eds. Craig, Adrian, & Chris Hummel, Johan August Wahlberg, *Travel journals and some letters, 1836-1856*, Van Riebeeck Society series 2, no.23, (Cape Town, 1994 for 1992).
Eds. Crampton, Eds. Hazel, Jeff Peires & Carl Vernon, *Into the hitherto unknown*, (Cape Town, 2013), Van Riebeeck 2nd series, no.44, (Cape Town, 2013).
ed. Dawson, Warren R., *The Banks Letters*, (London, 1958).
Darwin, Charles, *The effects of cross and self fertilisation in the vegetable kingdom*, (London, 1876)
De Courset, F.L.M. du Mont, *Le botaniste cultivateur*, vol.VII (Supplement), (Paris, 1814, 2nd ed.).
De Flacourt, Le Sieur, *Histoire de la Grande Isle Madagascar*, (Paris, 1658 ed.).
Delegorgue, Adulphe, *Travels in southern Africa*, translated Fleur Webb, (Pietermaritzburg, 1990).
Delpino, F., 'Rapporti tra insetti e nettari extranuziali nelle piante', *Bollettino della Società Entomologica italiana*, vol.6, (1874), pp.234-239.
Delpino, F., 'Funzione mirmecofila nel regno vegetale', *Memorie della Reale Accademia delle Scienza di Bologna*, vol.7, (1886), pp.215-323.
Ed. Dorrel, Ruth, 'Biographies of Deceased United Brethren Ministers', based on Adam Byron Condo, *History of the Indiana Conference of the Church of The United Brethren in Christ*, (1926).
Dyer, W.T. Thiselton-, 'Historical account of Kew to 1841', *Bulletin of Miscellaneous Information*, no.60, (December 1891), pp.279-327.
Ellis, William, *History of Madagascar*, (London, [1838]).
Endlicher, Stephan, Prodromus Florae Norfolkicae, (Vienna, 1833).
Endlicher, Stephan, Enchiridion botanicum, (Vienna, 1841).
Fawcett, William, *The banana: Its cultivation, distribution and commercial uses*, (London, 1921).

Ed. Forbes, V.S., Anders Sparrman, *A voyage to the Cape of Good Hope*, Van Riebeeck Society, 2nd series, no.6, (Cape Town, 1975).
    Carl Thunberg, *Travels at the Cape of Good Hope, 1772-5*, Van Riebeeck Society, 2nd series, no.17, (Cape Town, 1986).
Fourcade, H.G., *Checklist of the flowering plants of the divisions of George, Knysna, Humansdorp, and Uniondale*, Memoirs of the Botanical Society of South Africa no.20, (Pretoria, 1941).
Haberlandt, G., *Eine Botanische Tropenreise*, (Leipzig, 1893).
Hasskarl, Justus Karl, *Catalogus plantarum in horto botanico bogoriensi cultarum* (Batavia, 1844).
Huber, Jacques, 'La vegetation de la vallée Rio Purus (Amazone)', *Bulletin de l'Herbier Boissier*, vol.6, no.4, (1906).
Jacquin, Nikolaus, *Plantarum rariorum horti caesarei schoenbrunnensis*, (Vienna, 1797).
Lair, Pierre-Aimé, 'Description des jardins de Courset, aux environs de Boulogne-sur-mer', *Mémoires de la Société Royal d'Agriculture et de Commerce de Caen*, vol.4, (1836).
Lichtenstein, Hinrich, *Travels in southern Africa*, translated by Anne Plumptre (London, 1812-1815; 1930 reprint).
Eds. Lindley, John, & Thomas Moore, *Treasury of botany*, (London, 1866).
Link, H.F., *Enumeratio plantarum horti regii botanici berolinensis altera*, (Berlin, 1821)/
Linné, Car. O., *Mantissa plantarum*, (Holm, 1771).
Linné [Linnaeus], Carolo a, *Supplementum plantarum*, (Braunschweig, 1781).
Long, (Professor), G.R. Porter & George Tucker, *America and the West Indies, Geographically Described* (London, 1845), p.483.
Lotsy, J.P., *Vorträge über botanische stammesgeschichte, Ein lehrbuch der pflanzensystematik, Cormophyta Siphonogamia*, (Jena, 1911).
Loudon, John Claudius, *Arboretum et fruticetum Britannicum*, (London, 1838).
Maire, René, & Marc Weiller, *Flore de l'Afrique du Nord*, vol.6, (Paris, 1960 ed.).
Ed. Mann, R.J., Henry Brooks, *Natal: A history and description of the colony*, (London, 1876).
Masson, Francis, 'An account of three journeys from the Cape Town into the Southern Parts of Africa undertaken for the Discovery of New Plants, towards the Improvement of the
    Royal Botanic Gardens at Kew. By Mr Francis Masson. One of His Majesty's Gardeners. Addressed to Sir John Pringle, Bart. P.R.S.' *Philosophical Transactions of the Royal Society of
    London*, volume 66 (1776), pp.268-317.
Ed. Mossop, E.E., *Journals of the expeditions of the Honourable Ensign Olof Bergh (1682 and 1683) and the Ensign Isaq Schriver (1689)*, (Cape Town, 1931), Van Rieebeck Society, series 1, no.12.
Eds. John C. Manning & Nicola C. Anthony, Cuthbert J. Skead, *Historical plant incidence in South Africa*, (Pretoria, 2009), as *Strelitzia* 24.
Miquel, F.A.W., *Botanische Zeitung*, vol.iii, (1845).
    *Stirpes surinamenses selectae*, (Leiden, 1850; 7th ed.).
    'Musaceae', *Flora van Nederlandsch Indië*, part 2, (Leipzig, 1855).
Mittermeier, Russell E., 'Primate diversity and the tropical forest', in ed. E.O. Wilson, *Biodiversity*, (Washington, 1988), pp.145-154.
Ed. Nash, M.D., *The last voyage of the Guardian, Lieutenant Riou, Commander, 1789-1791*, Van Riebeeck Society, series 2, no.20, (Cape Town, 1990).
Ed. George Nicholson, *Illustrated dictionary of gardening*, (London, 1887).
Paxton, Joseph, 'Select list of stove perennials', *Paxton's Botanical Magazine*, (1836), vol.2, p.207.
Pearson, H.H.W., 'Presidential address Section C: A national botanic garden', *Report of the South African Association for the Advancement of Science*, vol.7, (1910), pp.37-54.
Petersen, Otto Georgius, 'Musaceae, Zingiberaceae, Cannaceae, Marantaceae' in eds. A. Engler & K. Prantl, *Die natürlichen plantzenfamilien*, (Leipzig, 1889), vol.II, section 6, pp.1-43.
Philips, E. Percy, 'Contribution of South Africa to botany', presidential address to Section C of the South African Association for the Advancement of Science, delivered July 9th, 1930,
    (South African Association for the Advancement of Science, 1930).
Politis, Gustavo, *Nukak, Ethnoarchaeology of an Amazonian people*, (Walnut Creek, 2009).
Pulteney, Richard, with William George Maton, *A general view of the writings of Linnaeus*, (London, 1805 ed.).
Raueschel, Ernst, *Nomenclator botanicus omnes plantas*, (Liepzig, 1797, ed.3).
Regel, E., *Catalogus plantarum quae in horto Aksakoviano coluntur*, (St Petersburg, 1860).
Richard, Achille, 'Urania ravenala', *Nova Acta Physico-Medica Academiae Caesareae Leopoldino-Carolinae Naturae Curiosorum Exhibentia Ephemerides sive Observationes Historias et
    Experimenta*, vol.15, (1831), p.15-22.
Richard, Achille, *De musaceae commentatio botanica*, (Vratislava, 1831).
Ridgway, P., & J. Works, *Sending flowers to America*, (Los Angeles, 2008).
Roupell, Arabella, *Specimens of the Cape flora by a lady*, (London, 1849).
Roxburgh, William, *Flora indica* (Serampore, 1832).
Salisbury, Richard A., *Prodromus stirpium in horto ad Chapel Allerton vigentium*, (London, 1796).
Saunders, William, *U.S. Department of Agriculture catalogue of economic plants*, (Washington, 1891).
Schomburgk, Richard, *Versuch einer fauna und flora von Britisch-Guiana*, (Leipzig, 1848).
    *Reisen in Britisch-Guiana in den jahren 1840-1844*, (Leipzig, 1848).
Schreber, Johann, *Genera plantarum*, (Frankfurt, 1789).
Schumann, K., 'Musaceae', in A. Engler, *Das pflanzenreich*, IV.45, (1900), pp.28-33.
Scopoli, Giovanni, *Introductio ad historiam naturalem, sistens genera lapidum, plantarum et animalium hactenus detecta, caracteribus essentialibus donate, in tribus divisa, subinde ad
    leges naturae* (Prague, 1777).
Scott-Elliot, G.F., 'Ornithophilous flowers in South Africa', *Annals of Botany*, vol.IV, no.XIV, (May 1890), pp.265-280.
    'Note on the fertilisation of Musa, Strelitzia reginae, and Ravenala madagascariensis', *Annals of Botany*, vol.IV, no.XIV, (May 1890), pp.260-261.
Sonnerat, Pierre, *Voyage aux Indes orientales & à la Chine*, (Paris, 1782).
Splitgerber, Ludwig Friedrich, 'Description du genre Urania non Schreber nec Richard', in *Het Instituut, of Verslagen en mededeelingen*, (Amsterdam, 1843).
Standley, Paul C., 'Flora of Costa Rica', part 1, *Publications of Field Museum of Natural History*, botanical series, vol.18, (October 1937), p.186.
    & Julian Steyermark, 'Flora of Guatemala', *Fieldiana: Botany*, vol. 24, part 3, (April, 1952).
Thunberg, Carl, *Resa uti Europa, Africa, Asia, forrättad åren 1770-1779*, (Uppsala, 1788).
Thunberg, Charles, *Travels in Europe, Africa, and Asia, made between the years 1770 and 1779*, (London, 1795, 2nd edition).
Thunberg, Carl P., *Prodromus plantarum capensis* (Uppsala, 1794).
Thunberg, Carl Peter, & Eric Carl Travenfeldt, *Nova genera plantarum*, part 7, (Uppsala, 1792).
Thunberg, Carl Peter & Nicolaus Gustavus Bodin, *Genera nova plantarum*, part 9, (Uppsala, 1798).
Von Martius, Carl F. Ph., *Die pflanzen und thiere des tropischen America, ein Naturgemälde*, (Munich, 1831).
Von Spix, Johann Baptist, & Carl Friedrich von Martius, 'Urania amazonica Mart', Reise in Brasilien vols.III & IV, *Atlas*, (Munich, 1831).
Von Steudel, Ernst, *Nomenclator botanicus*, (Stuttgart, 1874).
Wallace, Alfred Russel, *Palm trees of the Amazon and their uses*, (London, 1853).

Wallace, Alfred Russel, *Island life*, (London, 1892, 2nd ed.).
Werth, Emil, 'Blütenbiologische Fragmente aus Ostafrika. Ostafrikanische Nectarinienblumen und ihre Kreuzungsvermittler. Ein Beitrag zur Erkenniniss der Wechselbeziehungen zwischen Blumen- und Vogelwelt', *Voranisches Centralblatt*, January 1900, LXXXIV, pp.188-192.
ed. Willdenow, Carl Ludwig, *Carl Linnaeus Species plantarum*, (Berlin, 1798-1799).
Williams, R.O., 'Descriptive nursery stock list', Bulletin of the Department of Agriculture, Trinidad and Tobago, vol.20, part 1, (1922).
Winkler, Hubert, *Botanisches hilfsbuch für pflanzer, kolonialbeamte, tropenkaufleute und forschungsreisende*, (Wismar, 1912).
Wood, J. Medley, *A guide to the Natal Botanic Gardens containing plan of the gardens, byelaws etc., & catalogue of plants*, (Durban, 1883).
Wood, J. Medley, *List of trees, shrubs and a selection of herbaceous plants, growing in the Durban Municipal Botanic Gardens*, (Durban, 1897 & 1915 eds.)
ed. Woolsey, Sarah Chauncey, *The autobiography and correspondence of Mrs. Delany*, (Boston, 1879).
Wright, C.H., *Flora Capensis*, vol.5, (Cape Town, 1913).

## MODERN SCIENTIFIC STUDIES

Aluri, Raju J.S., & C. Subba Reddi, 'Explosive pollen release and pollination in flowering plants', *Proceedings of the National Academy of Sciences, India*, (1995), B61, no.4, pp.323-332.
Baagøe, Hans J., 'Observations on the biology of the banana bat, *Pipistrellus nanus*', *Proceedings of the Fourth International Bat Conference*, eds. R.J. Olembo, J.B. Castelino & F.A. Mutere, (Nairobi, 1978), pp.275-282.
Baccarini, P., 'Intorno ad un singolare accumulo d'acqua nel sistema lacunare delle guaine foliari di una *Musa ensete*', *Bulletino della Società Botanica Italiana*, 1904, pp.276-281.
Bakrania, Bhavisha, Eugene F. du Toit, Karl-Heinz Wagner, John P. Headrick & Andrew C. Bulmer, 'Pre- or post-ischemic bilirubin ditaurate treatment reduces oxidative tissue damage and improves cardiac function', *International Journal of Cardiology*, 202 (2016), p.27.
Eugene F. du Toit, Karl-Heinz Wagner, John P. Headrick & Andrew C. Bulmer, 'Pre- or post-ischemic bilirubin ditaurate treatment reduces oxidative tissue damage and improves cardiac function', *International Journal of Cardiology*, 2016, 202: 27 DOI: 10.1016/j.ij-card.2015.08.192.
Battersby, Alan R., 'Tetrapyrroles: The pigments of life', *Natural Product Reports*, vol.17, no.6, (Dec. 2000), pp.507-26.
Beaumont, Angela J., H. Baijnath & T.J. Edwards, 'Ravenala madagascariensis', *Flowering Plants of Africa*, (2005), vol.59, pp.44-50.
Beidokhti, Maliheh Najari, Eva S. Lobbens, Philippe Rasovaivo, Dan Staerk & Anna K. Jager. 'Inhibitory potential of 40 medicinal plant extracts from Madagascar against enzymes linked to type 2 diabetes', poster published by Department of Drug Design & Pharmacology, Faculty of Health & Medical Sciences, University of Copenhagen.
Birkinshaw, Christopher R., & Ian C. Colquhoun, 'Pollination of *Ravenala madagascariensis* and *Parkia madagascariensis* by *Eulemur macaco* in Madagascar' *Folia Primatologica*, vol. 69, issue 5, (Oct 1998), p.252-259.
Blanc, Patrick, Annette Hladik & Claude Marcel Hladik, 'L'arbre du voyageur dans toute sa diversité', 1er partie, *Hommes & Plantes*, pp.38-47.
& A. Hladik, N. Rabendrinaina, J.S. Robert & C.-M. Hladik, 'Strelitziaceae: The variants of *Ravenala* in natural and anthropogenic habitats', in eds. S.M. Goodman & J. Benstead, *The natural history of Madagascar*, University of Chicago Press, (London, 2003).
Boinski, Sue, Robert P. Quatrone & Hilary Swartz, 'Substrate and tool use by brown capuchins in Suriname: Ecological contexts and cognitive bases', *American Anthropologist*, 102(4), (Dec. 2000), pp.741-61.
Boyd, Austin, '*Musopsis* N. Gen.: A banana-like leaf genus from the early tertiary of eastern North Greenland', *American Journal of Botany*, vol.79, issue 12, (1992), pp.1359-1367.
Calley, M., R.W. Braithwaite & P.G. Ladd, 'Reproductive biology of *Ravenala madagascariensis* Gmel. As an alien species', *Biotropica*, 25 (1), (1993), pp.61-72.
Carlson, Thomas J., Barry Mamadou Foula, Julie A. Chinnock, Steven R. King, Gandeka Abdourahmaue, Bah Mamadou Sannoussy, Amadou Bah, Sekou Ahmed Cisse, Mohamed Camara & Rowena K. Richter, 'Case study on medicinal plant research in Guinea: Prior informed consent, focused benefit sharing, and compliance with the Convention on Biological Diversity', *Economic Botany*, 55(4), (2001), pp.478-91.
Chen, Stephanie T., & Selena Y. Smith, 'Phytolith variability in Zingiberales: A tool for the reconstruction of past tropical vegetation', *Palaeogeography, Palaeoclimatology, Palaeoecology*, vol.370, (2013), pp.1-12.
Comptom, Philip, 'Fruit in the soil of magic: Horticultural practices as socially conditioned techniques in the formation of anthropogenic Amazonia', *Revista de Antropologia Social dos Alunos do PPGAS-OFCar*, v.1, n.2, (Jul.-Dez. 2009), pp.20-44.
Congo Biodiversity Initiative, 'Botanical gardens of Eala-Mbandaka', [online]: http://www.congobiodiv.org/en/infrastructure/botanical-gardens/eala-mbandaka, accessed 28 December 2013.
Coombs, G., & C.I. Peter, 'Do floral traits of *Strelitzia reginae* limit nectar theft by sunbirds?', *South African Journal of Botany*, vol.75, (2009), pp.751-756.
Cowgill, R., S.B. Davis & D. Harebottle, 'Firefinches and sunbirds on the move', *Bird numbers*, vol.12, no.1, (July 2003), p.39.
Cron, Glynis V., Cary Pirone, Madelaine Bartlett, W. John Kress & Chelsea Specht, 'Biogeography' in 'Phylogenetic relationships and evolution in the Strelitziaceae (Zingiberales)', *Systematic Botany*, vol.37(3), (2012), pp.606-19.
Cronk, Quentin, Cronk & Isidro Ojeda, 'Bird-pollinated flowers in an evolutionary and molecular context', *Journal of Experimental Botany*, vol.59, no.4, (2008), pp.715-719.
Crowther, Jeffrey A., & Graeme R. Hanson, 'Chemistry of the aril of *Ravenala madagascariensis*', abstract SP60, *Proceedings of the Australian Society for Biochemistry and Molecular Biology*, vol.23, (1991).
Cumberlidge, Neil, Danté B. Fenolio, Mark E. Walvoord & Jim Stout, 'Tree-climbing crabs (Potamonautidae and Sesarmidae) from phytotelmic microhabitats in rainforest canopy in Madagascar', *Journal of Crustacean Biology*, vol.25, issue 2, (2005), pp.302-308.
Czapek, Friedrich, 'Die Verhältnisse im reifen Samen', section 2, chapter 25, subsection 1, pp.372-373, *Biochemie der pflanzen* (Jena, 1925).
Dahlgren, R., H.T. Clifford & P.F. Yeo, 'Strelitziaceae' in *The families of Monocotyledons*, (Berlin, 1985).
Decary, R., 'Les formations littorals dans la region de Mananara', *Bulletin de l'Académie Malgache*, new series, vol.6, (1922-1923), pp.63-66.
Decrock, E., 'Sur l'assise silicifère du tegument seminal des Ravenala', Proceedings of 22 May 1911 meeting, Académie des Sciences, Paris, in *Comptes Rendus Hebdomadaires des Séances de l'Académie des Science*, vol.152, Jan-Jun 1911, pp.1406-1407.
Defler, Thomas R., & Jorge I. Hernández-Camacho, 'The true identity and characteristics of *Simia albifrons* Humboldt, 1812: Description of neotype', *Neotropical primates*, 10(2), August 2002, pp.49-63.
De Wit, Maarten, Margaret Jeffrey, Hugh Bergh & Louis Nicolaysen (University of Witwatersrand), 'Gondwana reconstruction and dispersion', http://www.searchanddiscovery.com/documents/97019/index.htm, accessed 21 December 2013.
DuChemin, J.B., O. Ramiljaona Ravoahangimalala & G. Le Goff, 'Culicidae, mosquitoes, Moka Gasy', in eds. Goodman & Benstead, *Natural history of Madagascar*, pp.714-715.
Dufour, Darna L. & James L. Zarucchi, '*Monopteryx angustifolia* and *Erisma japura*: Their use by indigenous peoples in the northwestern Amazon', *Botanical Museum Leaflets Harvard University*, vol.27, no.3-4 (March-April 1979), pp.69-91.
Dwarka, Depika, Himansu Baijnath, Veenesha Thaver & Mickey Naidu, 'New insights into the presence of bilirubin in a plant species *Strelitzia nicolai* (Strelitziaceae)', *African Journal of Traditional, Complementary and Alternative Medicines*, 14(2), (2017), pp.253-262.
Dyer, R.A., 'Plate 995. Strelitzia alba', 'Plate 996. Strelitzia nicolai' & 'Plate 997. Strelitzia caudata', all *Flowering Plants of Africa*, (1946).
'Vegetative multiplication of *Strelitzia reginae* and its allies', *Bothalia*, 10, (1972), 4:575-578.
'Strelitziaceae: The status of *Strelitzia juncea*', *Bothalia*, vol.11, (1975), pp.519-520.

'Plate 1804 Strelitzia juncea', *Flowering plants of Africa*, (1975)

*The genera of southern African flowering plants*, vol.2: 'Gymnosperms and Monocotyledons', (Pretoria, 1976).

Echeverri, Juan Alvaro, & Oscar Enokakuiodo Román-Jitdutjaaño, 'Ash salts and bodily affects', *Environmental Research* Letters, vol.8, (2013), pp.2-11.

Eger, J.L., & L. Mitchell, 'Systematic accounts: Bats', in eds. Goodman & Benstead, *Natural history of Madagascar*, pp.1291-1292.

Ekstrom, J.M., 'Psittaciformes: *Coracopsis* spp., Parrots', in eds. Goodman & Benstead, *Natural history of Madagascar*, pp.1098-1099.

Enokakuiodo, O., S. Roman, A.A. Lopez, B. Weniger & J.A. Echeverri, 'Vegetable salt: An ignored resource', in eds. J. Fleurentin, J.M. Pelt & G. Mazars, *Des sources du savoir aux médicaments du futur : Actes du 4e congrès européen d'ethnopharmacologie*, (Metz, 2002).

Fahey, Jed W., Katherine K. Stephenson, Albena T. Dinkova-Kostova, Patricia A. Egner, Thomas W. Kensler & Paul Talalay, 'Chlorophyll, chlorophyllin and related tetrapyrroles are significant inducers of mammalian phase 2 cytoprotective genes', *Carcinogenesis*, 2005, vol.27, no.7, pp.1247-55.

Feeley-Harnik, Gillian, '*Ravenala madagascariensis* Sonnerat: The historical ecology of a "Flagship Species" in Madagascar', *Ethnohistory*, 48:1-2 (winter-spring 2001), pp.31-86.

Fleming, Theodore H., & Vinicio J. Sosa, 'Effects of nectarivorous and frugivorous mammals on reproductive success of plants', *Journal of Mammalogy*, vol. 75, No. 4 (Nov 1994), pp.845-851.

Fölling, Markus, Christoph Knogge & Wolfgang Böhme, 'Geckos are milking honeydew-producing planthoppers in Madagascar', in *Journal of Natural History*, 2001, 35, pp.279-284.

Foord, S.H., R.J. van Aarde & S.M. Ferreira, 'Seed dispersal by vervet monkeys in rehabilitating coastal dune forests at Richards Bay', *South African Journal of Wildlife Research*, vol.24, no.3, (1994), pp.56-59.

Frost, S.K., & P.G.H. Frost, 'Sunbird pollination of *Strelitzia nicolai*', *Oecologica*, vol.49, no.3, (1981), pp.379-384.

Givnish, Thomas J., Timothy M. Evans, J.C. Pires & K.J. Sytsma, 'Polyphyly and convergent morphological evolution in Commelinales and Commelinidae', *Molecular Phylogenetics and Evolution*, vol.12, no.23, (August 1999), pp.360-85.

 & J. Chris Pires, Sean W. Graham, Marc A. McPherson, Linda M. Prince, Thomas B. Patterson, Hardeep S. Rai, Eric H. Roalson, Timothy M. Evans, William J. Hahn Kendra C. Millam, Alan W. Meerow, Mia Molvray, Paul J. Kores, Heath E. O'Brien, Jocelyn C. Hall, W. John Kress & Kenneth J. Systma, 'Repeated evolution of net venation and fleshy fruits among monocots in shaded habitats confirms *a priori* predictions: Evidence from an ndhF phylogeny', *Proceedings of the Royal Society*, (2005, 272), pp.1481-99.

Goodman, Steven M. & Beverley A. Lewis, 'Description of the Réserve Naturelle Intégrale d'Andringitra, Madagascar', *A floral and faunal inventory of the eastern slopes of the Réserve Naturelle Intégrale d'Andringitra, Madagascar, Fieldiana* (Zoology), (1996), new series no.85.

Gunnell, Gregg F., 'Biogeography and the legacy of Alfred Russel Wallace', in *Geologica Belgica*, (2013), 16/4: 211-216.

Galley, C., & H.P. Linder, 'Geographical affinities of the Cape flora, South Africa', in *Journal of Biogeography*, vol.33, no.2, (Feb. 2006), pp.236-250.

Hannah, Lee, Makihiko Ikegami, David G. Hole, Changwan Seo, Stuart H.M. Butchart, A. Townsend Peterson & Patrick R. Roehrdanz, 'Global climate change adaptation priorities for biodiversity and food security', *Plos One*, vol.8, issue 8, (Aug. 2013), pp.10.

Helme, Nick A. Helme & Pierre Jules Rakotomalaza, 'An overview of the botanical communities of the Réserve Naturelle Intégrale d'Andohahela', *Fieldiana* (Zoology), (June 1999), new series no.94, pp.11-23.

Hesse, M., S.Vogel & H. Halbritter, 'Thread-forming structures in angiosperm anthers: Their diverse role in pollination ecology', Plant systematics and evolution, 222 (2000), pp.281-292.

Hladik, Annette, Patrick Blanc, Nicolas Dumetz, Vololoniaina Jeannoda, Nelson Rabenandrinina & Marcel Hladik, 'Données sur la repartition géographique du genre *Ravenala* et sur son rôle dans la dynamique forestière à Madagascar', *Mémoires de la Société de Biogéographie de Paris*, (2000), pp.93-104.

Hladik, Claude Marcel, Patrick Blanc & Annette Hladik, 'L'arbre du voyageur: Reproduction, usages et diffusion horticole', 2eme partie, in *Hommes & Plantes*, no.41, (Apr.-Jun. 2002), pp.18-27.

Hollowell, Tom, Lynn J. Gillespie, V.A. Funk & Carol L. Kelloff, 'Smithsonian Plant Collections, Guyana: 1989-1991, Lynn J. Gillespie', *Contributions from the United States National Herbarium*, vol.44, (Washington, 2003).

Hoogmoed, M.S., 'On the presence of *Bufo nasicus* Werner in Guiana, with a redescription of the species on the basis of recently collected material', *Zoologische mededelingen*, 51, (1977), p.272.

Howard, F., & J. Yudkin, 'Effects of dietary change upon the amylase and trypsin activities of the rat pancreas', *British Journal of Nutrition*, vol.17, (1963), pp.28-37.

Hutcheon, J.M., 'Frugivory by Malagasy bats', in eds. Goodman & Benstead, *Natural history of Madagascar*, p.1205-1207.

Hutchinson, J., 'The flora of Madagascar', *Nature*, vol.145, (23 March 1940), pp.448-451.

'Zingiberales' in *The families of flowering plants*, (1959), vol.II: Monocotyledons, pp.581-90.

Ishihata, Kiyotake, 'Studies on the promoting germination and the growth of seedling in Strelitzia reginae Banks', Bulletin of the Faculty of Agriculture, Kagoshima University, (20 March 1976,) pp.1-15.

Jaing, K., K.V. Jayaprasad, R. Krishna Manohar, Shivanand Hongal & K. Prakash, 'Effect of levels of fertigation on growth and yield of bird-of-paradise [Strelitzia reginae Ait.], Asian journal of horticulture, vol.6, no.1, (June 2011), pp.118-21.

Kantharaju, M., B.S. Kulkarni, B.S. Reddy, R.V. Hegde, B.C. Patil & J.D. Adiga, Karnataka journal of agricultural science, 21 (2), (2008), pp.324-5.

Kessler, Michael, Jon-Arvid Grytnes, Stephan R.P. Halloy, Jürgen Kluge, Thorsten Krömer, Blanca León, Manuel J. Macía & Kenneth R. Young, 'Gradients of plant diversity, Local patterns and processes', pp.204-19.

Knapp, Sandra, Gerardo Lamas, Eimear Nic Lughadha & Gianfranco Novarino, 'Stability or stasis in the names of organisms: The evolving codes of nomenclature', *Philosophical Transactions of the Royal Society London* B, vol.359, (2004), pp.611-622.

Kremen, Claire, Isaia Raymond, Kate Lance & Andrew Weiss, 'Monitoring natural resource use on the Masaola peninsula, Madagascar: A tool for managing integrated conservation and development projects', in eds. K. Saterson, R. Margoluis & N. Salafsky, *Managing conservation impact*, (Washington, 1999).

Kress, W. John, 'The phylogeny and classification of the Zingiberales', *Annals of the Missouri Botanic Gardens*, (1990), pp.678-721.

'Bat pollination of an Old World *Heliconia*', *Biotropica*, vol.17, no.4, (Dec. 1985), pp.302-308.

'New taxa of heliconia (Heliconaceae) from Ecuador', *Brittonia*, (1991), vol.43 (4), pp.253-260.

 & Donald E. Stone, 'Morphology and floral biology of *Phenakospermum* (Strelitziaceae), an arborescent herb of the neotropics', *Biotropica*, 25(3), (1993), pp.290-300.

 & G.E. Schatz, M. Adrianifahanana & H.S. Morland, 'Pollination of *Ravenala madagascariensis* by lemurs in Madagascar: Evidence for an archaic coevolutionary system?', in *American Journal of Botany*, 81, (1994), pp.542-551.

'Phylogeny of the Zingiberaneae: Morphology and molecules', in eds. P.J. Rudall, P.J. Cribb, D.F. Cutler & C.J. Humphries, *Monocotyledons: Systematics and evolution*, (London, 1995).

 & Linda M. Prince, William J. Hahn & Elizabeth A. Zimmer, 'Unraveling the evolutionary radiation of the families of Zingiberales using morphological and molecular evidence', *Systematic Biology*, vol.50, no.6, (Nov-Dec 2001), pp.926-944.

Kronestedt, Eva, B. Walles & I. Alkemar, 'Structural studies of pollen tube growth in the pistil of Strelitzia reginae', *Protoplasma*, 131, (1986), 224-232.

Lane, Irwin E., 'Genera and generic relationships in Musaceae', *Mitteilungen der Botanischen Staatssammlung München*, vol.13, (1955), pp.114-31.

Laurance, William F., & G. Bruce Williamson, 'Positive feedbacks among forest fragmentation, drought, and climate change in the Amazon', *Conservation Biology*, (December 2001), vol.15, no.6, pp.1529-1535.

Levey, Douglas T., 'Seed size and fruit-handling techniques of avian frugivores', *The American Naturalist*, vol.129, no.4, (April 1987), pp.471-485.

Linder, H.P., & C.R. Hardy, 'Evolution of the species-rich Cape flora', *Philosophical Transactions of the Royal Society London B*, vol.359, no.1450, Plant Phylogeny and the origin of major biomes, (October 2004), pp.1623-32.

Link, Andrés, & Pablo R. Stevenson, 'Fruit dispersal syndromes in animal disseminated plants at Tinigua National Park, Colombia', *Revista chilena de historia natural*, 77, (2004), pp.319-334.

Eds. Lubke, Roy, Fred Gess & Mike Bruton, *A field guide to the Eastern Cape coast*, (Grahamstown, 1988).

Ludwig, Philip, '*Strelitzia nicolai* seed aril extract degrades bilirubin to increase skin brilliance', *HPC Today, Body care, grooming, protection and hygiene*, vol.8 (2), (2013), pp.8-11.

P.J.M. Maas, 'Musaceae', in ed. A.R.A. Görts-van-Rijn, *Flora of the Guianas*, series A, Phanerograms, (Koenigstein, 1985), pp.21-2.

& Hiltje Maas, 'Flora da Reserva Ducke, Brasil: Strelitziaceae', *Rodriguésia*, vol.56, no.86, (2005), pp.125-130.

MacKinnon, J.L., C.E. Hawkins & P.A. Racey, 'Pteropodidae, Fruit Bats, *Fanihy*, *Angavo*', in eds. Goodman & Benstead, *Natural history of Madagascar*, p.1302.

Mancuso, Stefano, 'Federico Delpino and the foundation of plant biology', *Plant Signaling & Behavior*, vol.5, no.9, (September, 2010), pp.1067-1071.

Manning, John, *Field guide to fynbos*, (Cape Town, 2018; 2nd ed.).

& Peter Goldblatt, 'Chromosome number in *Phenakospermum* and *Strelitzia* and the basic chromosome number in Strelitziaceae (Zingiberales)', *Annals of Missouri Botanic Gardens*, , vol.76, (1989), pp.932-3.

Martin, Franklin W., 'The systematic exudate of Strelitzia', *Phyton*, 27 (1), (1970), pp.47-53.

Martínez Alier, Joan, 'International biopiracy versus the value of local knowledge', *Capitalism, Nature, Socialism*, vol.11, no.2 (2000), pp.59-66.

Martius, Christopher, 'Occurrence, body mass and biomass of *Syntermes* spp. (Isopter: Termitidae) in Reserva Ducke, Central Amazonia', *Acta Amazonica*, 28 (3), (1998), pp.319-324.

Messmer, Nathalie, Pierre Jules Rakotomalaza & Laurent Gautier, 'Structure and floristic composition of the vegetation of the Parc National de Marojejy, Madagascar', *Fieldiana* (Zoology), (August 2010), new series no.97, pp.41-104.

Meve, Ulrich, & Sigrid Liede, 'Floristic exchange between mainland Africa and Madagascar', in *Journal of Biogeography*, vol.29, no.27, (Jul.2002), pp.865-873.

Milliken, William, Daniela Zappi, Denise Sasaki, Mike Hopkins & R. Toby Pennington, 'Amazon vegetation: How much don't we know and how much does it matter?', *Kew Bulletin*, (2010), vol.65, no.4, pp.691-709.

Milton, Katherine, C.D. Knight & I. Crowe, 'Comparative aspects of diet in Amazonian forest-dwellers', *Philosophical transactions of the Royal Society of Biological Sciences London B*, 334, (1991), pp.253-263.

Moore, H.E. & Peter A. Hypio, 'Strelitzia', *Baileya*, 1970, vol.17, p.65 & 71.

Moore, Michael R., 'An historical introduction to porphyrin and chlorophyll synthesis', in eds. Martin J. Warren & Alison G. Smith, *Tetrapyrroles: Birth, life and death*, 2009, ch.1.

Nakai, Takenosin, 'Essential results obtained from my observation on tropical plants in Java, Galang Island of Rio Archipelago, and on Japanese plants in the surroundings of Beppu Hotspring, province of Bungo, Kyusyu (XII), An observation on wild Nepenthes in Galang Island', *Bulletin of Tokyo Science Museum*, no.22, (1948), pp.39-43.

Otterbein, L.E., & A.M. Choi, 'Heme oxygenase: Colors of defense against cellular stress', *Journal of Lung, Cell and Molecular Physiology*, 2000, 279 (6), pp.29-37.

Palmer, Eve, *A field guide to the trees of southern Africa*, (London, 1977).

Paoletti, Maurizio Guido, Darna L. Dufour, Hugo Cerda, Franz Torres, Laura Pizzoferrato & David Pimentel, 'The importance of leaf- and litter-feeding invertebrates as sources of animal protein for the Amazonian Amerindians', *Royal Society Proceedings: Biological Sciences*, vol.267, no.1459 (Nov 2000), pp.2247-2252.

& Erika Buscardo & Darna L. Dufour, 'Edible invertebrates among Amazonian Indians: A critical review of disappearing knowledge', *Environment, Development and Sustainability*, vol.2, no.3, (Sept. 2000), pp.195-225.

& Erika Buscardo, Dorothy J. Vanderjagt, A. Pastuszyn, L. Pizzoferrato, Y.S. Huang, L.-T. Chuang, R.H. Glew, M. Millson & Hugo Cerda, 'Nutrient content of termites (Syntermes soldiers) consumed by Makiritare Amerindians of the Alto Orinoco of Venezuela', *Ecology of Food & Nutrition*, 42(2), (Mar 2003), pp.177-191.

Perz, Stephen G., 'The changing social contexts of deforestation in the Brazilian Amazon', *Social Science Quarterly*, vol.83, (March 2002), no.1, pp.35-52.

Piaulian R., & P. Viette, 'An introduction to terrestrial and freshwater invertebrates', in eds. Goodman & Benstead, *Natural history of Madagascar*, ch.8.

Pirone, Cary Jodie V. Johnson, J. Martin E. Quirke, Horacio A. Priestap & David Lee, 'The animal pigment bilirubin identified in *Strelitzia reginae*, the bird-of-paradise flower', in *HortScience*, vol. 45 no. 9, (Sept. 2010), pp.1411-1415.

& Jodie V. Johnson, J. Martin E. Quirke, Horacio A. Priestap & David Lee, 'Bilirubin present in diverse angiosperms', *AoB Plants*, 2010: plq020, doi:10.1093/aobpla/plq02, published 28 October 2010, accessed 1 January 2016.

& J. Martin E. Quirke, Horacio A. Priestap & David W. Lee, 'The animal pigment bilirubin discovered in plants', *Journal of the American Chemistry Society*, 131(8), (4 March 2009), p. 2830.

Politis, Gustavo G., 'Moving to produce: Mobility and settlement patterns in Amazonia', *World Archaeology*, vol.27, no.3, Hunter-Gatherer Land Use (Feb. 1996), pp.492-511.

Pooley, Elsa, *The complete guide to the trees of Natal, Zululand & Transkei*, (Durban, 1993).

Prance, G.T., 'Floristic inventory of the tropics: Where do we stand?', *Annals of the Missouri Botanic Gardens*, vol.64, (1977), pp.659-84.

Priyadarsini, S. Sakthi, R. Vadiva & N. Jayshree, 'Pharmacognostical standardization of leaves of *Ravenala madagascariensis* Sonn.', *Research Journal of Pharmacognosy and Phytochemistry*, vol.2 (4), (Jul-Aug 2010), pp.288-292.

& R. Vadivu & N. Jayshree, 'In vitro and *In vivo* antidiabetic activity of the leaves of *Ravenala madagascariensis* Sonn., on alloxan induced diabetic rats', *Journal of Pharmaceutical Science and Technology*, vol.2 (9), (2010), pp.312-317.

& R. Vadivu & N. Jayshree, 'Hypolipidaemic and renoprotective study on the ethanolic & aqueous extracts of leaves of *Ravenala madagascariensis* Sonn., on alloxan induced diabetic rats', *Journal of Pharmaceutical Science and Technology*, vol.2 (1), (2010), pp.44-50.

Rakotomalaza, Pierre Jules, 'An overview of the botanical communities of the Réserve Naturelle Intégrale d'Andohahela', *Fieldiana* (Zoology), new series no.94, (June 1999), pp.12-13.

Randrianarivelojosia, Milijoana, 'History of malaria treatment in Madagascar', [online]: http://www.mmv.org/sites/default/files/uploads/docs/artemisi nin/2010_Madagascar/History_of_Malaria_Treatment.pdf, accessed 25 December 2013.

Raselimanana, A.P., & D. Rakotomalala, 'Systematic accounts: Skinks', in eds. Goodman & Benstead, *Natural history of Madagascar*, pp.989-992.

Raxworthy, Christopher J., Colleen M. Ingram, Nirhy Rabibisoa & Richard G. Pearson, 'Applications of ecological niche modelling for species delimitation: A review and empirical evaluation using day geckos (Phelsuma) from Madagascar', in *Systematic Biology*, 56:6, December 2007, 907-923.

Roesel, Cheryl S., W. John Kress & Brunella Martire Bowditch, 'Low levels of genetic variation in *Phenakospermum guyannense* (Strelitziaceae), a widespread bat-pollinated Amazonian herb', *Plant systematics and evolution*, vol.199, no.1-2, (Mar. 1996), pp.1-15.

Rowan, M.K., 'Bird pollination of *Strelitzia*', *Ostrich*, vol.45, (1974), p.40.

Royal Botanic Gardens, Kew, *State of the world's plants 2016*, (London, 2016).

Royal Botanic Gardens, Kew, *State of the world's plants 2017*, (London, 2017).

Schliemann, H., & S.M. Goodman, '*Myzopoda aurita*, Old World sucker-footed bat' in eds. Goodman & Benstead, *Natural history of Madagascar*, pp.1303-1306.

Seiji, Maurice, Tomioka Nilsson & Philip Martin Fearnside, 'Yanomami mobility and its effects on the forest landscape', *Human Ecology*, vol.39, no.3, (June 2011), pp.235-256.

Sharmin, Tasnuva, Reza Chowdhury, Md. Yeunus Mian, Masbahul Hoque, Md. Sumsujjaman & Faijun Nahar, 'Evaluation of antimicrobial activities of some Bangladesh medicinal plants', *World journal of pharmaceutical sciences*, 2(2), (2014), pp.170-5.

Silvius, Kirsten M., 'Spatio-temporal patterns of palm endocarp use by three Amazonian forest mammals: Granivory or 'grubivory'?', *Journal of Tropical Ecology*, vol.18, no.5, (Sept 2002), pp.716-719.

Simmons, Nancy B., Robert S. Voss, Scott A. Mori, 'Bats as pollinators of plants in the lowland forests of Central French Guiana', [online]: https://www.nybg.org/bsci/french_guiana/bat_poll.html, accessed 9 August 2013.

Simpson, D.J., M.R. Baqar & T.H. Lee, 'Ultrastructure and carotenoid composition of chromoplasts of the sepals of Strelitzia reginae Aiton during floral development', *Annals of botany*, 39, 1975, pp.175-83.

C.J. Skead, 'Sunbirds and strelitzias', *Bokmakierie*, vol. 14, no. 23, (1963), p. 70.
Stearn, William T., 'A survey of the tropical genera Oplonia and Psilanthele (Acanthaceae)', in *Bulletin of the British Museum (Natural History) Botany*, vol. 4, no. 7, 1971, pp. 259-323.
Tanaka, R., & A. Tanaka, 'Tetrapyrrole biosynthesis in higher plants', *Annual Review of Plant Biology*, vol. 58, (2007), pp. 321-346.
Ter Steege, Hans, Rens W. Vaessen, Dairon Cárdenas-López, Daniel Sabatier, Alexandre Antonelli, Sylvia Mota de Oliveira, Nigel C.A. Pitman, Peter Meller Jørgensen & Rafael P. Salomão, 'The discovery of the Amazonian tree flora with an updated checklist of all known tree taxa', www.nature.com/scientificreports, doi:10.1038/srep29549, published 13 July 2016, accessed 26 February 2017.
 & Nigel C.A. Pitman, Timothy J. Killeen, William F. Laurance, Carlos A. Peres, Juan Ernesto Guevara et al, 'Estimating the global conservation status of more than 15,000 Amazonian tree species', *Science Advances*, (November 2015), 1:e1500936;
Thomas, David R., Claire A. Penney, Amrita Majumder & Amanda M. Walmsley, 'Evolution of plant-made pharmaceuticals', *International Journal of Molecular Sciences*, vol. 12, (2011), pp. 3322-3336.
Tilburt, John C. & Ted Kaptchuk, 'Herbal medicine research and global health: An ethical analysis', *Bulletin of the World Health Organization*, August 2008, vol. 86, no. 8, (August 2008), pp. 577-656
Tomlinson, P. Barry, 'An anatomical approach to the classification of Musaceae', *Journal of the Linnean Society*, Botany, vol. 55, (1959), pp. 779-809.
 'Strelitziaceae', *Anatomy of the monocotyledons*, (London, 1969), pp. 325-333.
Van de Venter, H.A., & J.G.C. Small, 'Dormancy in seeds of *Strelitzia* Ait', *South African Journal of science*, vol. 70, (1974), pp. 216-7.
 & J.G.C. Small & P.J. Robbertse, 'Notes on the distribution and comparative leaf morphology of the acaulescent species of *Strelitzia* Ait.', *South African Journal of Botany*, vol. 41 (1), (1975), pp. 1-16.
 'Microsporogenesis and gametogenesis in Strelitzia reginae Ait', *Journal of South African Botany*, 42 (1), (1976), ppp. 25-31.
 'Effect of various treatments on germination of dormant seeds of Strelitzia reginae Ait', *Journal of South African Botany*, 44 (2), (1978): 103-110.
 & J.G.C. Small & A.H. Halevy, 'Some characteristics of leaf and inflorescence production of Strelitzia reginae in Port Elizabeth', Journal of South African Botany, 46 (3), (1980), pp. 305-311.
Van der Merwe, Mac, & Rebecca L. Stimemann, 'Group composition and social events of the banana bat, *Neoromicia nanus*, in Mpumalanga, South Africa', *South African Journal of Wildlife Research*, vol. 39, no. 1, (April 2009), pp. 48-56.
Van Oye, Paul, 'Recherches sur la biologie de Ravenala madagascariensis Sonner.', in *Revue de la Zoologie Africaine*, XI, 2, Suppl. Bot., (15 July 1923), pp. 18-34.
Waser, Nickolas M., Jeff Ollerton & Andreas Erhardt, 'Typology in pollination biology: Lessons from an historical critique', *Journal of Pollination Ecology*, vol. 3, no. 1, (2011), pp. 1-7.
Willis, Christopher, & Augustine Morkel, 'National botanic gardens: South Africa's urban conservation refuges', *BGjournal*, vol. 5, no. 2, (July 2008): https://www.bgci.org/resources/article/0588/, accessed 15 March 2015.
Yamagishi, S.I., S. Maeda, T. Matsui & S. Ueda, 'Role of advanced glycation end products (AGEs) and oxidative stress in vascular complications of diabetes', *Biochim Biophys Acta*, (2011), pp. 20-27.
Zachos, James C., 'Anthropogenic carbon release rate unprecedented during the past 66 million years', *Nature Geoscience*, vol. 9, (2016), pp. 325-329.
Zent, Eglée & Stanford Zent, 'Floristic composition, structure, and diversity of four forest plots in Sierra Maigualida, Venezuelan Guyana', *Biodiversity and conservation*, no. 13, (2004), pp. 2453-2484.

## GENERAL BACKGROUND

Agricultural Products Standards Act 1990 (Act 119 of 1990 [South Africa]), Standards and Requirements regarding the Export of Fresh Cut Flowers and Fresh Ornamental Foliage.
Bailey, L.H., *The standard encyclopedia of horticulture*, (1958).
Brown, Mervyn, *A history of Madagascar*, (Cambridge, 1995).
Campbell, Gwyn, 'Imperial rivalry in the western Indian Ocean and schemes in colonised Madagascar, 1769-1826', *Revue des mascareignes*, no. 1, (1999), pp. 75-97.
Central Intelligence Agency, 'Africa – Madagascar', *The world factbook*, [online]: https://www.cia.gov/library/publications/the-world-factbook/geos/ma.html, updated 15 August 2016; accessed 11 November 2017.
Davenport, T.R.H., *South Africa: A modern history*, (Cape Town, 1988 edition).
Duminy, Andrew, *Mapping South Africa*, (Johannesburg, 2011).
Glen, Hugh F., *Sappi What's in a name: The meanings of the botanical names of trees* (Johannesburg, 2004).
Howells, Robin, 'Bernardin de Saint-Pierre's founding work: The *Voyage à l'île de France*', *Modern Language Review*, vol. 107, issue 3, (July 2012), p. 756;
Jones, Steve, *No need for geniuses*, (London, 2016).
Library of Congress Country Studies, 'Madagascar: Precolonial era, prior to 1894', [online]: www.loc.gov, posted August 1994, accessed 11 November 2013.
Lionel Kochan, *The making of modern Russia*, (London, 1962).
Larson, Pier Martin, 'Colonies lost: God, hunger, and conflict in Anosy (Madagascar) to 1674', *Comparative Studies of South Asia, Africa and the Middle East*, vol. 27, no. 2, (2007), pp. 345-366.
Mitchell, Malcolm, 'A brief history of transport infrastructure in South Africa up to the end of the 20th century: 1', *Civil Engineering*, Jan/Feb 2014, pp. 37-38.
 'A brief history of transport infrastructure in South Africa up to the end of the 20th century: 2', *Civil Engineering*, (March 2014), pp. 74-75.
Muller, A.L., 'Coastal shipping and the early development of the southern Cape', *Contree* 18, pp. 5-9.
Murray, Tony, *Mega structures and master minds*, (Cape Town, 2015).
Padmore, George, 'Madagascar fights for freedom', *Left*, no. 132, October 1947, transcribed by Christian Hogsbjerg for the Marxists' Internet Archive 2007, [online]: https://www.marxists.org/archive/padmore/1947/madagascar.htm, accessed 13 November 2013.
Ed. Pama, C., *Genealogies of old Cape families*, (Cape Town, 1966).
Pienaar, Kristo, *The South African What Flower Is That?*, (Cape Town, 1984).
Ralaimihoatra, Edouard, *Histoire de Madagascar*, (Paris, 1958).
Reader's Digest, *Complete guide to gardening in South Africa*, (Cape Town, 1971).
Réau, Louis, *L'Europe française au siècle des lumières*, (Paris, 1938).
Rookmaker, L.C., *The zoological exploration of southern Africa*, (Rotterdam, 1989).
Ross, Graham, *Romance of the Cape mountain passes*, (Cape Town, 2004).
Thompson, Leonard, *A history of South Africa*, (Sandton, 1990).
Van Ass, Jo, Johann du Preez, Leslie Brown & Nico Smit, *The story of life & the environment: An African perspective* (Cape Town, 2012).
Van Wyk, Braam, & Piet van Wyk, *Field guide to the trees of southern Africa*, (Cape Town, 1997).
Viljoen, Morris, 'The Cape fold mountains', in eds. C.R. Anhaeusser, M.J. Viljoen & R.P. Viljoen, *Africa's top geological sites*, (Cape Town, 2016).
Whitehead, Marion, *Passes & poorts*, [Western Cape], (Cape Town, 2010).
Wilmot, A., & John Centlivres Chase, *History of the colony of the Cape of Good Hope*, (Cape Town, 1869).
Yule, Henry, *The book of Ser Marco Polo*, (London, 1871).
Zakharov, Viktor N., Gelina Harlaftis & Olga Katsiardi-Hering, *Merchant colonies in the early modern period*, (Cambridge, 2012).

Clinton Friedman, *Strelitzia juncea,* digital photograph, 2018

# INDEX

Adanson, Michel 90-92, 95, 97, 100
Adderley Street Flower Market 192-93
Aiton snr, William 10, 13, 15, 26, 33, 35, 63, 66, 71, 79, 174
Aiton jnr, William Townsend 15, 19, 26-27, 33, 49-50, 76
Aksakov, Nikolai 53
Alexander, Sir James 76
Allamand, Jean Nicholas 26
Andrews, Henry 14, 40
Andrews, Susyn 227-229, 232
Andrianahifotsi, (King) 121
Angas, George French 44-45, 190-191
Arnold Arboretum 215
Ashmore, John 228, 232
Auge, Johann Andreas 21, 25, 30-35
Augusta, (Princess) 11, 18, 33, 36
Bäck, Abraham 128
Baines, Thomas 76
Baker, John Gilbert 121, 142
Balfour, Dr Isaac 101
Banks, Sir Joseph 10-19, 23, 26-27, 32-33, 46-53, 56, 63, 66-67, 70-71, 79, 97, 100, 173, 192
Baret, Jeanne 93-94
Barnard, Lady Anne 173
Barrère, Pierre 127
Barrow, Sir John 32, 35, 37, 173, 238
Bartholinus, Thomas 18
Bashee (Mbashe) River 247
Bass, Eileen 196
Batten, Auriol 20, 27
Bauer, Ferdinand 67, 69, 74
Bauer, Franz 28, 31, 37, 67-72, 81, 250
Bayer, Bruce 230
Beaumont, Angela 8, 41, 50-51, 111, 113, 166, 183, 222, 231
Beenke, Hendrik 25
Benedict, John 82
Bennett, Dr Eleanor 232
Bentham, George 63
Bentham Herbarium 38
Berger, Alwin 212
Bergius Herbarium 21, 37, 40, 76
Bergius, Peter 22, 32-35
Betia, (Princess) 101
Beutler, August 25
Bladh, Pehr 23
Bligh, (Captain) William 24, 49
Boddy, Robert 186
Bolus, Harry 160, 162-63, 168, 193, 214, 227
Bombay Natural History Society 113
Bonpland, Aimé 129
Bösenberg, De Wet 234
*Botanical Cabinet* 75, 163
*Botanical Register* 35, 163
*Botanists' Repository* 14
Botha's Farm, Jacobus 32
Bourne, (Sir) Alfred Gibbs & (Lady) 120
Boyd, Austin 82
Breslau Botanic Gardens 38, 208
Brewer, William 179

Breyne, Jacob 18
Briggs, Donald F. 182
British East India Company 160
Broadway, Walter 136, 138
Brooks, Henry 60
Brougham, Henry 212
Brown, D.S. 209
Brunnthaler, (Dr) 214
Bruton, Gerry 195
Buchan, Alexander 16
Buffeljagts River 30
Buitenzorg Botanic Gardens 134, 139
Bujumbura 209
Buller, A.C. 197
*Bulletin of Miscellaneous Information* 79
Burchell, William 17, 19, 30, 131-32, 135, 162-64, 192, 225
Burger, Alwin 42
Burges, Edith 42, 62
Burges, (Sir) James Bland 47-48
Burman, Johannes 21, 23, 30, 32, 36
Burman, Nicolaas 23, 26
Burtt Davy, Joseph 61, 165, 167, 214
Bute, (Lord) 11, 18
Calcutta Botanic Gardens 94
Caledon 193
California Flower Growers Association 182
California Horticultural Society 182
Campbell, Belle McPherson 104-5
*Cape Argus* 192
Cape Town Dutch EIC Garden/Botanic Gardens 21, 23-25, 30-35, 56, 163
Carlos III 128
Carlson, (Dr) Tom 258-60
Carmichael, J.R. 120
Carter, Humphrey B. 46
Cassidy, G.E. 71
Catherine the Great, (Empress) 46-56
Cavanilles, Antonio José 129
'Centenary Gold' ch.13
Chapman, Olive Murray 104-5
Charlotte of Mecklenburg-Strelitz, (Queen) 10-14, 33, 36, 46, 53-54, 66, 69-72
Chelsea Flower Show 14, 215
Cheever, David 180, 216
Chelsea Physic Garden 12, 18, 76, 228
Childs, Ozro W. 179
Clifford, H.T. 25
Clifford, George 14, 26
Clusius, Carolus 16, 197
Cochrane, Sue 217, 229
*Codex Comptoniana* 26
Colchester, Marcus 142
Colvill's, Mesrs 14
Commelin, Jan 18, 23
Commerson, Philibert 92, 94-97, 108, 113
Compton, (Bishop) Henry 26
Compton, (Prof.) Robert H. 225
Condy, Gillian 233-34
Conte, Lisa 257-60
Cook, (Captain) James 16, 18, 24, 97

Cooper, Thomas 17
Copland, Samuel 117
Correia, M.F. 172
Cracraft, Joel 43
Croft-Cook, Rupert 108
Crook, Albert & Lucy 170-72
Crowther, Jeffrey A.
*Curtis's Botanical Magazine* 14, 22, 33, 37-39, 53-55, 61, 67, 76, 91, 163
Curtis, William 14, 67, 72
Cuvier, Georges 95
Czapek, Friedrich 91
D'Alembert, Jean Le Rond 95
Da Orta, Garcia 87
Darwin, Charles 56, 72
Darwin, Erasmus 72
Davis, Wade 144
De Balboa, Nuñez 180
De Bougainville, (Comte) Louis-Antoine 92
De Candolle, Alphonse 43
De Candolle, Augustin Pyramus 43
De Chalain, Pearl 181
De Choiseul, (Duc) Etienne-François 92
De Clugny, (Baron) Marc Antoine 95
De Cossigny, Joseph-François Charpentier 93-94
Decrock, E. 91
De Flacourt, Etienne 88-89, 98, 104, 120
De Grainville, Lt-Col. René 101
De Hartekamp 26
D'Houdetot, (Baron) 76
De Iturriaga José 128
De Jossigny, Paul Sauguin 94, 97
De Jussieu, Antoine Laurent, 14, 100
De la Bâthie, Henri Perrier 117
De la Billaderie, Charles-Claude Flahaut 98
De la Caille, (Abbé) Nicolas 23, 25
De Lamarck, Jean Baptiste 43
Delany, (Mrs) Mary 66
Delegorgue, Adulphe 53-59
Dellapenna, Dean 264
Delpino, Federico 56
De Mannevillette, J-B d'Après 25
De Saint-Pierre, Jacques-Henri Bernardin 93
*Descriptiones plantarum ex Bonae Spei* (Thunberg) 35
Desroches, (Governor) Julien 96
De Sessé y Lacasta, Martin 178
De Smidt, Gail 210
Dietrich, D. 37
*Discovery*, HMS 48
Dodds, Eric 230
Dombey, Joseph 128
Domoto family 182
Drake, Sir Francis 16
Drège, Carl Friedrich 38, 58, 173
Drège, Johann Frantz 17, 38, 58
Dryander, Jonas 12-13
Du Courset, (Baron) F. du Mont 38
Du Jussieu, Antoine & Bernard 91
Dumas, Daniel 92
Durban (Port Natal) 4, 58, 60, 119, 164, 209, 238
Dürer, Albrecht 74

Dutch East India Company (VOC) 15-26, 32, 36, 88, 126, 164, 193
Dyer, (Dr) Robert 58-61, 165-174, 188
Dyer, (Sir) William Thiselton- 15, 42, 136
Ecca Pass 211
Edwards, Sydenham 67
Ehret, Georg 69
Ekeberg, Carl Gustaf 23
Eala 112, 209
Elizabeth, (Princess) 66, 69
Ellis, (Rev.) William 89, 96, 108, 110
Elsinore 52
*Endeavour*, HMS 16, 18, 21, 97
Endlicher, Stephan 133
Eugene, (Duke) Charles 50
*Exoticorum libri decem* 16
*Exotic plants … Kew* (Meen) 67
Fairchild, David 39
*Familles naturelles des plantes* (Adanson) 91
Fauré, A.A. 30-31
Fawcett, William 76, 79, 112, 124
Federation of Dutch Flower Auctions 202
Feilden, G. St Clair 38
Fereira, A. 142
Fermor, Patrick Leigh 209
Filet, (Caporal) Jean-Onésime 101, 121
Fischer, (Dr) Friedrich 56
Fischer, (Dr) Sebastian 56
Fish River (Great) 17, 25, 56, 173
Fitch, Walter Hood 39, 67, 71
Fleming, Ian 209
*Flora brasiliensis* (Martius) 133
*Flora capensis* (Thunberg) 36, 38
*Flora capensis* (Harvey) 168
*Flore des Serres et des Jardins de l'Europe* 39, 41, 53, 77
*Flowering Plants of Africa* 42, 61, 113, 165, 167, 225
Flückiger, Fredrich 213
Forero, Enrique 142
Forster jnr, Edward 48
Fort Dauphin 88-89, 94, 104
Fourcade, (Dr) Henri 42, 168
Fowler, Henry 136
Fraser, John 53
Frederickson, August Daniel 112
French East India Company 25, 88
Friedman, Clinton 283, 293
Fusée-Aublet, Jean-Baptiste 127
Fyodorovna, (Grand Duchess) Maria 51, 53
Galpin, Ernest 167-68
Gandolfo, Nick 185
*Gardeners' Chronicle* 76, 208-9
Genoa Botanic Gardens 213
Gascoyne, Bamber 12-14, 17
Gawler, John Ker 174
George 32, 42
George III, (King) 10-11, 46-47, 50-53, 66, 71
Gerrard, William 96
Gide, André 112
Gill, Dr William 17
Gillett, Margaret 17
*Glorieus* 25

Gmelin, Johannes 27, 100
Goldblatt, Peter 82
Goodenough, (Bishop) Samuel: Herbarium 17
Göppert, (Prof.) Heinrich 38
Goukamma River 32
Gould, Helen 209
Gower, William 24
Gowie, W. & C. 230
Graeme, (Col.) David 11
Grahamstown Botanic Gardens 78
Grenville, Lord William 46
Grewcock, A.E. 167-168
Grober, (Dr) H. 211
Groen, Cornelius J. 184
Grubb, Michael 22
*Guardian*, HMS 34
Gustav III, (King) 36
Guthrie, Francis 162-163
Hanbury family/Villa 212-15
Harkerville 41
Harold Porter Botanic Gardens 234
Harrower, Adam 239-240
Hart, John 136
Hartog, Jan 23
Harvey, William 160, 168
Hawaii Tropical Botanic Gardens 124
Heatley, Margaret 1722
Heenan, Denis 169
Herbert, George 162
Hermann, Paul 18, 22, 26
Holland, Maria 160-61
Holttum, Richard & Ursula 71
Hooker, (Sir) Joseph 54, 60-63, 162-163, 213
Hooker, (Sir) William 17, 38-39, 56, 110, 135, 160, 173
Hop, (Capt.) Hendrik 32
*Hortus cliffortianus* 26, 69
*Hortus kewensis* 10, 12-15, 19, 21, 33, 35-36, 40, 50, 74, 77, 174
*Hortus Medicus* 18
Hottentots Holland Mts 22, 193
Huber, Charles 179, 214
Hubner, Hermanus 25
Hull, Ebraime 234
Humansdorp 32, 42
Huntley, (Prof.) Brian 232
Hutchinson, John 17, 63
*Icones novae* (Miller) 13-14
*Icones plantarum* 33
Immelman, D.F. 20, 33
Iqbal, Farhat, 59
Janssens, J.W. 30-31
Japan 33-36
Jiménez, José Antonio Pavón 128
Johnes, Thomas 14
Jones, Mary 203
Jordaan, Danie 180
Judd, Albert F. 120
Jumelle, Henri 117
Karoo Botanic Gardens 165, 229, 230
Keet, J.H. 42, 168-69
Keiskamma River 17

Keit, Wilhelm 164
Kensit, William 162
Kew, Royal Botanic Gardens 10-15, 19, 35, 43, 46-50, 54, 60, 66-67, 70-72, 77, 79, 90, 120, 135-36, 160, 162-65, 173, 208, 228, 244
*Kew Guild* 142
King, Steven 257-58, 260
Kirstenbosch Botanic Gardens 6, 82, 139, 165, 168, 225, 227-29, 235, 241, 244
'Kirstenbosch Gold' ch.13
Kluge, Leon 210, 215
Knysna 41-42
König, Johann 23
Körnicke, Friedrich 53
Krauss, Christian 58
Kress, W. John 260
Krige, Japie 227
Kronstadt 52
Kunth, Carl 130
Kwelegha River 211
*La Boudeuse* 92
*Lady's Magazine of Gardening* 76
Lagarde, Alex 77
*La Garonne* 96
Lair, Pierre-Aimé 38
Lalande, Jérôme 94-95
Lamarck, Jean-Baptiste 95
Lambert, Aylmer Bourke 128
Lang, Herbert 41
Larson, Mark 187
Lee (nurseryman), James 34
Leiden Botanic Gardens 25-26, 30
Leiden 18, 23
Lemaire, Charles 39, 41, 90, 107, 221
Lemoinier, M. 77, 79, 81, 228
Leonardi, C.H. 20
Leopoldina, Maria 130
*Les liliacées* (Redouté) 69
Lichtenstein, (Dr) Hinrich 30-32
Liège 208
Lightfoot, (Rev.) John 66
Link, Heinrich 27, 173
Linnean Society 10, 48, 95, 167
Linnaeus jnr, Carl 13, 36-37, 40, 74, 91, 129
Linnaeus snr, Carolus 12-13, 16, 21, 23, 27, 40, 43, 63, 128
Loedolff, Jeanette 236, 246
Löfling, Pehr (The Great Vulture) 127-28
Loher, August 120
*London Journal of Botany* 77
Long, F.R. 227, 230
Los Angeles Flower Market 185
Lotsy, Johannes 124
Loudon, Jane 76
Luton Hoo Botanic Garden 11
Lyons, Israel 16
Maas, Paulus & Hillegonda 143
McClintock, Elizabeth 232
McKen, Mark 164, 238
McMurdo, (General) William 215
McNab, William 26
MacOwen, Peter 32, 57, 163, 165, 168, 214, 225
Madagascar 16, ch.5, 251

Madrid 128, 178
Maire, René 174
Malan, (Mrs) D.J. 194
Malaysia 220
Malmaison, Chateau de 70, 130
Mandela, (President) Nelson 232-34
'Mandela's Gold' 6, 195, 205, 222, ch.13
Manning, John 81
*Mantissa plantarum* 13
Maria, (Grand Duchess) 51, 53
Marseille Botanic Gardens 117
Masson, Francis 10, 12, 15-16, 19, 23-25, 30, 33-35, 38, 40, 49, 56, 74, 76, 173
Marloth, Dr Rudolf 57, 194, 214
Martin, Joseph 127
Mathews, Joseph 225-226
*Maund's Botanical Garden* 163
Maurice, Rev. Dr 72
Mauritius 94-98, 101, 208, 255
Maximilian of Leuchtenberg, (Duke) 56
Mbutse, Mluleki 234
Mellano family 185-187, 212
Meen, Margaret 66-67, 74
Mendonça, Frederico de Ascenção 172
Merian, Maria Sibylla 127
Meyberg, Manfred 182
Miellez, M. 77
Miller, Philip 12-14, 74
Milliken, William 144, 244, 254
Milne, (Dr) Lindsey 211
Missouri Botanic Gardens 82, 136, 138-139, 143, 179
Mittermeier, Russell A. 108
Miquel, Friedrich 133-135, 139
Montpellier 212
Mooney, Pat 259
Moore, Harold 174
Moran, Katy 259
Morren, Charles 27, 76-81
Moss, Charles 17
Murray, Johan Anders 14, 66
Mutis y Bosio, José Celestino 128
Mzimvubu River ch.14
Mzukwa, Thandile 234
Nahoon River 76
Nakai, Takenosin 63
Natural History Museum (London) 23, 32-33, 36, 38, 58, 94, 100, 139, 170
Nelson, David 24
Nelson, William 23
Netherlands 10, 14, 18, 26, 143
Nikolai, (Grand Duke) 54
Noe, George 49-50, 52
North, Marianne 209
*Nova genera plantarum* (pamphlets) 36
Odell, J.R. 211
Oldenburg, Frans 32
Oldenland, Heinrich 18, 23, 26, 38, 96
Oliver, Samuel Pasfield 104-107
Orinoco River 128
O.R. Tambo Airport 195
Ortega, Casimiro Gómez 128, 178
Osbeck, Pehr 23

Osborne, William B. 179
Oxford Botanic Gardens 67
Palmengarten (Frankfurt) 79
Palmer, Eve 58
Pamplemousses Botanic Gardens 208
Parc National de Marojejy 121
Parkinson, Sydney 16
Paul, (Grand Duke) 51-53
Pavlovsk Estate 46, 49, 51-52
Paxton, Joseph 76
Pedley brothers 182
Peradeniya Botanic Gardens 94, 110, 112
Pearson, H.H.W. 164, 194, 225, 227
Petersen, Otto 63
Petershof (Petrodvorets) 46
Peterson, Dickie 230, 234
Petiver, James 18, 23
Philcox, David 142
Philip, Ethel & Elsie 136
Philips, John 41
Pienaar, Cecilia 226
Pienaar, Kristo 230
Piesang River 30, 32
Pirone, (Dr) Cary L. 260, 263
Planchon, (Prof.) Jules-Emile 77, 213
Plant, Robert 238
Plowman, Tim 143-44
Plettenberg Bay 17, 20, 30, 32, 41
Pogroszewska, (Dr) Elzbieta 221
Poissonnier, Pierre-Isaac 92
Poivre, Pierre 92-97, 208
Pole Evans, (Dr) I.B. 165, 168, 225
Pondoland 25, 244, 246-247
Port Alfred 17
Potter, Thomas 74
Prain, David 94
Prance, (Sir) Ghillean 142
Pretorius, Fransiolemtru 158-59
*Prodromus plantarum capensis* (Thunberg) 36, 38, 53
Price, Charles T. 88
Pringle, Sir John 19
Quesnay, François 92
Rabary, (Pastor) 110
Radama I, (King) 104
Raeüschel, Ernst 100
Raikes, Thomas 49
Rainero, John 185
Ramos, José 142
Ranavalona II, (Queen) 121
Ratsimilao, (King) 101
Raubenheimer, Margaret 194
Razumovski, Count Alexei 56
Redouté, Pierre-Joseph 5, 67, 69, 70-71
Rees, Abraham 10
*Reliance*, HMS 48
Repton, Humphry 26
Retzius, Anders 23
Réunion 96
Revesby Abbey 16
Reyneke, H.L. 230

Richard, Achille 27, 100, 133
Richard, J.J. 15, 63
Richard, Louis-Claude 133
Ridley, Henry 139
Rijksherbarium (Leiden) 124
Riou, (Capt.) Edward 33-34
Rivieras 212
Robinson, William 226
Robberg 30
Rockingham, (Marquis) Charles 12-13
Rockingham, (Marchioness) Mary 13-14
Rodrigues 101
Roth, Johanny 187
Roupell, Arabella 160
Rourke, (Dr) John 229, 246
Rousseau, Jean-Jacques 92-93
Roxburgh, William 94
Royal Geographical Society 32, 105, 144, 173
Royal Horticultural Society 74
Royal Navy 18
Royal Society 15-16, 18, 100-101
Russian Company 48-49
Russian Horticultural Society 234
Rycroft, (Prof.) Brian 164
Salisbury, Richard 14
Sand, George 214
Schomburgk, Richard 134
Schönbrunn Imperial Botanic Gardens 67, 93
Schönland, (Prof.) Selmar 165, 167, 214
Schultes, Richard Evans 143
Schumann, K. 63, 262
Schweickerdt, Herold Georg 120
Scopoli, Giovanni Antonio 90
Scott-Elliot, George 56-58, 76
Selznick, David 181
Sessions, Kate 180
Seven Years War 93
Shaw, Dr 72
Sharp, Helen 185
Sibree, (Rev.) James 105
Sibthorp, (Prof.) John 67
Siebert, August 78-79
Siebrecht's House of Flowers 181
Singapore Botanic Gardens 139
Skeels, Homer Collar 38-40
Smith, Sir James Edward 10, 14, 23, 66
Smith, Matilda 55, 61
Smuts, General Jan 17
Sobodie, Marie Elizabeth 101
Solander, (Dr) Daniel 18, 32-33
Sonnerat, Pierre 23, 92, 94, 96-98, 100, 106, 108, 133, 248
SA Flower Growers Assn 196
SA National Biodiversity (prev. Botanical) Institute 5, 41-42, 61, 233
*Southampton* 74
Sowerby, James 13, 67
Sparrman, Anders 20, 192, 238, 242
Splitgerber, Frederik 133-34
Spöring, Herman 16
St Anne's Kew 71
St Petersburg 41, 46-56, 61

Standley, Paul C. 254
Stayner, Frank 230
Stearn, William T. 135
Steyermark, Julian A. 138-39, 142
Stole, Hieremias 18
Strain, Harold 262
Strelitzia:
    Bilirubin 263-265
    Birds 57-59, 76, 116, 145-48, 253-257, 264
    Class/species 80
    Conservation ch15
    Floriculture ch.10-12
    Frogs 115, 142, 149, 253
    Indigenous peoples 59, 90, 100, 118, 147, 150-151, 154, 251
    Insects 60, 114, 253-54
    Mammals 60-61, 98, 115-18, 145-149, 151, 155-57, 253
    Online sales 221
    Pharmaceutical use ch.15
    Reptiles and amphibians 115, 155
    Seed yields 204
    Stem sales/Flower sales ch.12
*Strelitzia depicta* (Bauer) 4, 70, 250
Struben, Edith 225-26
Stuttaford, Samson R. 192
Styles, David 241
Surinam 13, 124-26, 154
Suzuki, Yutaro 81, 210-212, 228
Swartz, Olof Peter 37, 40
Swedish Academy of Science 21, 23, 37
Swedish East India Company 22
Swedish Museum of Natural History 37
Swellengrebel, Hendrik 30
Swiel, Basia Hitcock 20, 83, 175
Swisher, Lowell 181, 212
Taiwan 218
*Temple of Flora* (Thornton) 72
The Hague Botanic Gardens 26
Thomas, Vicki 234
Thompson, Adrian 143
Thorne, William 192
Thorns, Frank 139
Thunberg, Carl Pehr 17-25, 30-40, 43, 53-54, 74, 96, 162
Thuret, Gustave 214
Tidmarsh, Edwin 78
Tomlinson, P. Barry 81
Torre, A.R. 172
Townsend, William J. 110
Townshend, Sally 245
Tretchikoff, Vladimir 5
Tulbagh 194
Tulbagh, Governor Rijk 23, 30, 32
Travenfeldt, Eric Carl 36-40, 43
*Treasury of Botany* (Lindley) 76
Tsarkoe Selo Palace 46, 49, 51
Uppsala 14, 21, 36
Urton, Noel 174
Usteri, Paulus 100
Van Cassel, F. 76
Vander Bruggen, Waandert J. 184, 187
Van der Stel, Willem Adriaan 23

Van der Venter, H.A. 174, 235
Van Houtte, Louis 39, 208, 221
Van Jaarsveld, Ernst ch.14
Van Oye, Dr Paul 110, 112-14, 209
Van Plettenberg, Governor 96
Van Riebeeck, Jan 16
Van Royen, David 25
Van Wyk, Braam 171
Van Wyk, Piet 171
Vawter, Edwin James 182
*Venus*, HMS 49, 51
Verdonck, Abbé 76
Verschaffelt, Alexandre Ambroise 77
Verschuur, T.B. 168-69
Vidor, King 181
Villett, C.J. 173
Visser, Gerrit 242
Vlok, Ellen 226
Von Humboldt, Alexander 129-130, 156
Von Jacquin, Baron Joseph Franz 67, 74
Von Jacquin, Nikolaus Joseph 67, 84-85, 127
Von Martius, Carl 13C-34
Von Regel, Eduard 53-56, 58, 60-61
Von Schreber, Johann 100
Von Spix, Johann Baptist 130
Von Strudel, Ernst 133
Von Waldheim, Aliksandr Fischer 53
Von Willdenow, Carl 37, 100
Voss, Andreas 79
Wahlberg, Johann 59
Walker, Col. George 120
Wallace, Alfred Russel 86, 125
Wallich, Nathaniel 160
Warren, Adrian N. 142
Warren, Col. 179
Watson, William 78
Watson-Wentworth, Charles 12
Watteau, Jean-Antoine 92
Webb, Philip Barker 128
Weiller, Marc 174
Wells, Edgar 180, 216
Westmacott, (Sir) Richard 71
White, Thomas Tew 101
Whitworth, (Sir) Charles 46, 49-52
Wikström, Johan Emanuel 37
Willemet, Pierre Rémi 100
Wilson, Julie 219
Wilson, (Mrs) 67
Winkler, Hubert 63, 124, 254
Winter, John 228-30, 233-234
Winter, Ludwig 214
Wisley, RHS Garden 213
Woestyne, Auguste Vande 77
Wood, (Dr) John Medley 4, 60, 139, 164, 168, 214
Worchester Veld Reserve 230
World War I 194, 214
World War II 182, 185, 214
Wüttemberg 50-53
Xaba, M'Afrika Phakamani 234-235, 239, 241
Yoshida, Mas 187

Zarucchi, L. 143
Zola, Emile 209
Zwide, Godfrey 238

## PLANTS
*Amaryllis disticha* 20
Baobab 117
Canna 101
*Clivia miniata* 234
*Cussonia puniculata* 32
*Cyrthanthus montanus* 41
*Dalembertia uranoscopa* 95, 97, 100
*Dias cotinifolia* 25
*Disa uniflora* 194
*Encephalartos caffer* 20
*Encephalartos longifolius* 20
Erythrina 31
*Heliconia alba* 13, 31, 33, 36-37
*Heliconia albiflora* 32, 40
*Heliconia aurantica* 76
*Heliconia bihai* 13, 15, 27, 33, 35, 40, 63
*Heliconia caffra* 13
*Heliconia ravenala* 100
*Heliconia strelitzia* 27
*Ischnosiphon gracilis* 154
Musa (banana) 30, 56, 100, 170, 264
*Nerine undulate* 25
*Phenakospermum guyannense* ch.7&8, 91, 250, 253, 257, 264
*Protea neriifolia* 16
*Ravenala madagascariensis* (traveller's tree) ch.5&6, 82, 208-209, 248, 250, 255-256, 259, 262
*Smilax cumanensis* 154
*Spartium junceum* 10
*Strelitzia alba* ch.2, 54, 76, 79, 82, 168, 171-72
*Strelitzia angustiflolia* 174
*Strelitzia augusta* 14, 32, 37-41, 53-54, 59, 70, 79, 209, 213, 262
*Strelitzia caudata* ch.9, 38, 43, 82, 211, 247
*Strelitzia farinosa* 27, 60, 209
*Strelitzia glauca* 27
*Strelitzia humilis* 27, 53
*Strelitzia juncea* 43, 60, 76, 82, 192, 195, 205, 209, 211, 213, 217, 221, 234-235
*Strelitzia latifolia* 174
*Strelitzia lutea* 74, 76
*Strelitzia nicolai* ch.3, 25, 32, 38, 41, 43, 46, 82, 139, 158, 161, 167-168, 195, 204-5, 209, 211, 217, 221, 262, 264
*Strelitzia ovata* 27, 53, 76, 174
*Strelitzia parvifolia juncea* 76, 79, 174, 213
*Strelitzia reginae* ch.1, 30, 33, 35, 43, 46, 53, 60, 66-70, 72, 74-82, 91, 139, 160, 172, 174, 178-79, 185, 187, 189, 195-96, 205, 209, 211, 213, 217, 219, 221, 226-230, 238, 242, 245-247, 261-262, 264
*Strelitzia reginae* var. *citrina* 78
*Strelitzia lutea* 230
*Strelitzia reginae* var. *lemoinierii* 77-88
*Strelitzia reginae* var. *mzimvubuensis* 43, 81, ch.14
*Strelitzia reginae* var. *reginae* 43, 76, 243-44
*Strelitzia regis* 35
*Strelitzia rutilans* 27
*Strelitzia x kewensis* 79
Tulip 10
*Urania ravenala* 100

CONTRIBUTING TO DURBAN BOTANIC GARDENS'
TRADITION OF BOTANICAL PUBLISHING EXCELLENCE
SINCE THE 19TH CENTURY

PUBLISHED BY

**Durban Botanic Gardens Trust**

LEAD SPONSOR

**Stanley Smith (UK) Horticultural Trust**

CORPORATE CONTRIBUTORS

Angiyati Flowers
Fotomax
Rockjumper Tours

PREMIER SPONSOR

Pat Long

SILVER SPONSORS

Peter Adams
Mark Liptrot
Evelyn McGuire
Norman Pammenter